Connexin Cell Communication Channels

Roles in the Immune System and Immunopathology

Connexin Cell Communication Channels

Roles in the Immune System and Immunopathology

Edited by
Ernesto Oviedo-Orta
Brenda R. Kwak
William Howard Evans

CRC Press
Taylor & Francis Group
Boca Raton   London   New York

CRC Press is an imprint of the
Taylor & Francis Group, an **informa** business

CRC Press
Taylor & Francis Group
6000 Broken Sound Parkway NW, Suite 300
Boca Raton, FL 33487-2742

First issued in paperback 2017

ISBN-13: 978-1-4398-6257-5 (hbk)
ISBN-13: 978-1-138-19950-7 (pbk)

Library of Congress Cataloging-in-Publication Data

Connexin cell communication channels : roles in the immune system and
    immunopathology / editors, Ernesto Oviedo-Orta, Brenda R. Kwak, William Howard
    Evans.
       p. ; cm.
    Includes bibliographical references and index.
    ISBN 978-1-4398-6257-5 (hardcover : alk. paper)
    I. Oviedo-Orta, Ernesto. II. Kwak, Brenda R. III. Evans, W. Howard.

    [DNLM: 1. Connexins--immunology. 2. Gap Junctions--immunology. 3.
    Inflammation--physiopathology. QU 55.7]
    571.9'6364--dc23   2012050715

Visit the Taylor & Francis Web site at
http://www.taylorandfrancis.com

and the CRC Press Web site at
http://www.crcpress.com

Contents

SECTION 1   Immunology and Cell Biology

SECTION 2   Inflammation and Inflammatory Diseases

SECTION 3   Connexin-Based Therapeutical Approaches in Inflammatory Diseases

Preface

Cells collaborate and communicate, working as a team. This is especially evident in our tissues and organs. The key proteins acting as membrane channels that enable direct communication or cross talk between attached cells are connexins (Cxs), a family of around 22 closely homologous proteins in vertebrates that oligomerize generating a hexameric arrangement in the membrane with a central pore. Aligned head to head, groups of these Cx hemichannels form a gap junction—a well-established structure that provides a conduit connecting directly cell interiors.

An important and surprising finding in the cell communication/signalling field has been widespread detection of functional Cx channels in cells comprising the immune system. Morphologically mature gap junctions are rarely recognized in the immune system except in lymphoid tissues, and intensive recent research has pointed to the functional roles of Cx hemichannels residing in the cell's plasma membrane. Normally closed, these hemichannels when they become leaky present cells with a severe metabolic disability as they lose control of the dynamics between the cytoplasm and the surrounding milieu. Environmental insults such as hypoxia, oxidative, osmotic, and mechanical stresses can provide challenges to the gating integrity of hemichannels as they become permeable for example, allowing, the leakage of ATP and glutamate. In the immune system especially the loss of purinergic molecules poses signalling problems and can modify the interaction of not only immune cells but also with endothelial and other cells depending on location.

The identification of Cx hemichannels, especially those built up of Cx43, as key cell communication regulators and their easy accessibility to direct modulation by "chemical blockage" has opened up the field to an ever-increasing translational applications and these are described in specific chapters. A wide range of "mimetic peptide" reagents are finding potentially important applications in diseases induced by hypoxia in the brain and cardiovascular systems. Also, the use of antisense RNA reagents provides an alternative and complementary way of addressing the consequences of Cx channel dysfunction in several diseases as illustrated in many chapters and especially in relation to diabetes. The development of peptide mimetics that confine their targeting to hemichannels and avoid deleterious action on gap junction functions (e.g., as in addressing cardiac arrhythmia as opposed to hypoxic damage) are on the horizon and will provide an added level of selectivity in targeting hemichannels. These can aid in dissecting the physiological and pathological roles of Cx channels especially in the immune system itself and in lymphoid cells and organs. Finally, the implication of pannexin (Panx) channels similar younger but unrelated cousins of the Cx hemichannels are considered by various authors.

Editors

Ernesto Oviedo-Orta received his MD from the Havana Medical School in Cuba where he also specialized in clinical immunology. During his postgraduate specialist training, he engaged in vaccine research related to diverse infectious agents and diseases, including measles and hepatitis C. Dr. Oviedo-Orta carried out his PhD research work in the laboratory of Professor W. Howard Evans at the Department of Medical Biochemistry, University of Wales College of Medicine (Cardiff, UK) focused on the role of gap junction intercellular communication in the immune system. He then moved to work as a postdoctoral fellow and, subsequently, as British Heart Foundation junior research fellow at the Bristol Heart Institute (University of Bristol, UK). His work at Bristol allowed him to gain expertise in the development of vaccines to prevent and treat noncommunicable diseases such as atherosclerosis. He then worked as lecturer in immunology at the University of Surrey, where he led a team of scientists dedicated to investigate the role of the immune system in inflammation linked to atherosclerosis, with a special interest in the discovery of new immunomodulatory molecules to combat cardiovascular disease. During this appointment Dr. Oviedo-Orta continues working on gap junction intercellular communication but with special attention to their role in immune regulation. Dr. Oviedo-Orta now works as a clinical sciences expert at Novartis Vaccines & Diagnostics, Siena, Italy, where he is a medical and scientific advisor supporting new and ongoing projects in research and providing early clinical/scientific input, and contributing to clinical development plans.

Brenda R. Kwak studied at the University of Amsterdam, the Netherlands, where she also obtained her PhD. Dr. Kwak carried out her PhD research work in the laboratory of Professor Habo J. Jongsma at the Department of Physiology focusing on the regulation of cardiac gap junction channels using a dual patch–clamp approach. She then continued her work for several years as a postdoctoral fellow at the University of Utrecht, the Netherlands. After moving to the University of Geneva, she axed her gap junction work more toward the vascular system, first in the endothelium only at the Department of Morphology, and worked later on the regulation of major histocompatibility complex class II (MHC-II) in the context of atherosclerosis, a chronic immuno-inflammatory disease of large- and medium-sized arteries at the Department of Internal Medicine—Cardiology. On receiving a professorship from the Swiss National Science Foundation in 2003, she established an independent research group in the same department to dissect the role of Cxs in atherosclerosis. Since 2009, she is affiliated as associate professor to both the Department of Pathology and Immunology and to the Department of Internal Medicine—Cardiology at the University of Geneva and focuses more broadly on the role of Cxs in cardiovascular physiology and pathology, work that still includes atherosclerotic disease but also restenosis, ischemia–reperfusion injury of the heart, platelet activation and thrombosis as well as the lymphatic system.

William Howard Evans, PhD, is professor of medical biochemistry in the Institute of Infection and Immunity at Cardiff University Medical School, United Kingdom. His interest in intercellular communication via gap junctions spans from 1972 and his group has studied many facets of the structure and function of these Cx-based channels in a variety of normal and pathological situations. His recent studies focus on addressing their roles of Cx channels in translational medicine. This has involved the design and development of small Cx mimetic peptides that block specifically the operation of these channels and especially their use in addressing cardiac pathology.

Contributors

David L. Becker
Department of Cell and
    Developmental Biology
University College London
London, United Kingdom

Marie Billaud
Robert M. Berne Cardiovascular
    Research Center
University of Virginia School of
    Medicine
Charlottesville, Virginia

Laurent Burnier
Depatment of Pathology and
    Immunology
and
Department of Internal
    Medicine—Cardiology
University of Geneva
Geneva, Switzerland

Marc Chanson
Department of Pediatrics
Geneva University Hospitals
and
University of Geneva
Geneva, Switzerland

Jeremy E. Cook
Department of Cell and
    Developmental Biology
University College London
London, United Kingdom

Peter Cormie
Department of Cell and
    Developmental Biology
University College London
London, United Kingdom

Helen V. Danesh-Meyer
Department of Ophthalmology
University of Auckland
Auckland, New Zealand

Steven Donnelly
Department of Biological and
    Biomedical Sciences
Glasgow Caledonian University
Glasgow, Scotland, United Kingdom

William Howard Evans
Institute of Infection and Immunity
Cardiff University Medical School
Cardiff, Wales, United Kingdom

Nicolas Froger
Centre Interdisciplinaire de Recherche
    en Biologie (CIRB)
and
MEMOLIFE Laboratory of
    Excellence and Paris Science
    Lettre
Collège de France
Paris, France

Christian Giaume
Centre Interdisciplinaire de Recherche
    en Biologie (CIRB)
and
MEMOLIFE Laboratory of Excellence
    and Paris Science Lettre
Collège de France
Paris, France

Aaron M. Glass
State University of New York Upstate
    Medical University
Syracuse, New York

Robert G. Gourdie
VirginiaTech Research Institute and
    School of Medicine
Roanoke, Virginia

Colin R. Green
Department of Ophthalmology
University of Auckland
Auckland, New Zealand

Katherine R. Heberlein
Robert M. Berne Cardiovascular
    Research Center
and
Department of Molecular Physiology
    and Biological Physics
University of Virginia School of
    Medicine
Charlottesville, Virginia

Malcolm B. Hodgins
Department of Biological and
    Biomedical Sciences
Glasgow Caledonian University
Glasgow, Scotland, United Kingdom

Brant E. Isakson
Robert M. Berne Cardiovascular
    Research Center
and
Department of Molecular Physiology
    and Biological Physics
University of Virginia School of
    Medicine
Charlottesville, Virginia

Scott R. Johnstone
Robert M. Berne Cardiovascular
    Research Center
University of Virginia School of
    Medicine
Charlottesville, Virginia

Gergo Kiszner
1st Department of Pathology and
    Experimental Cancer Research
Semmelweis University
Budapest, Hungary

Michael Koval
Departments of Medicine and Cell
    Biology
Emory University School of
    Medicine
Atlanta, Georgia

Tibor Krenacs
1st Department of Pathology and
    Experimental Cancer Research
Semmelweis University
Budapest, Hungary

Brenda R. Kwak
Department of Pathology and
    Immunology
and
Department of Internal
    Medicine—Cardiology
University of Geneva
Geneva, Switzerland

Darcy Lidington
Department of Physiology
and
Toronto Centre for Microvascular
    Medicine
University of Toronto
and
The Li Ka Shing Knowledge
    Institute at St. Michael's
    Hospital
Toronto, Ontario, Canada

Patricia E. Martin
Department of Biological and
    Biomedical Sciences
Glasgow Caledonian University
Glasgow, Scotland, United kingdom

Ariadna Mendoza-Naranjo
Department of Cell and Developmental
    Biology
University College London
London, United Kingdom

Jacques Neefjes
Division of Cell Biology
The Netherlands Cancer Institute
Amsterdam, the Netherlands

Thien D. Nguyen
State University of New York Upstate
    Medical University
Syracuse, New York

Louise F. B. Nicholson
Department of Anatomy
University of Auckland
Auckland, New Zealand

Simon J. O'Carroll
Department of Anatomy
University of Auckland
Auckland, New Zealand

Michael P. O'Quinn
Department of Regenerative Medicine
    and Cell Biology
Medical University of South Carolina
Charleston, South Carolina

Emily L. Ongstad
Department of Bioengineering
Clemson University
Clemson, South Carolina

Juan Andrés Orellana
Departamento de Fisiología
Pontificia Universidad Católica de
    Chile
Santiago, Chile

Ernesto Oviedo-Orta
Vaccines Research
Novartis Vaccines & Diagnostics
Siena, Italy

Joseph A. Palatinus
Department of Regenerative
    Medicine and Cell Biology
Medical University of
    South Carolina
Charleston, South Carolina

Baoxu Pang
Division of Cell Biology
The Netherlands Cancer Institute
Amsterdam, the Netherlands

Anthony R. J. Phillips
School of Biological Sciences
University of Auckland
Auckland, New Zealand

Mauricio Retamal
Laboratorio de Fisiología
Clínica Alemana-Universidad del
    Desarrollo
Santiago, Chile

J. Matthew Rhett
Department of Regenerative
    Medicine and Cell Biology
Medical University of
    South Carolina
Charleston, South Carolina

Martin Rosendaal
1st Department of Pathology and
    Experimental Cancer Research
Semmelweis University
Budapest, Hungary

Juan Carlos Sáez
Departamento de Fisiología
Pontificia Universidad Católica de
    Chile
Santiago, Chile
and
Instituto Milenio, Centro
    Interdisciplinario de
    Neurociencias de Valparaíso
Valaparaiso, Chile

Pablo J. Sáez
Departamento de Fisiología
Pontificia Universidad Católica de
    Chile
Santiago, Chile

Flavio A. Salazar-Onfray
Institute of Biomedical Sciences
University of Chile
Santiago, Chile

Antonio E. Serrano
Department of Cell and Developmental
  Biology
University College London
London, United Kingdom

Kenji F. Shoji
Departamento de Fisiología
Pontificia Universidad Católica de
  Chile
Santiago, Chile

Adam C. Straub
Robert M. Berne Cardiovascular
  Research Center
University of Virginia School of
  Medicine
Charlottesville, Virginia

Steven M. Taffet
State University of New York Upstate
  Medical University
Syracuse, New York

Christopher Thrasivoulou
Department of Cell and Developmental
  Biology
University College London
London, United kingdom

Karel Tyml
Lawson Health Research Institute
and
Department of Medical Biophysics
and
Department of Physiology and
  Pharmacology
University of Western Ontario
London, Ontario, Canada

Henri C. Van der Heyde
La Jolla Infectious Disease Institute
San Diego, California

Maurice A. M. van Steensel
Department of Dermatology
and
GROW School for Oncology and
  Developmental Biology
Maastricht University
  Medical Center
Maastricht, the Netherlands

Catherine S. Wright
Department of Biological and
  Biomedical Sciences
Glasgow Caledonian University
Glasgow, Scotland, United kingdom

Jie Zhang
Department of Ophthalmology
University of Auckland
Auckland, New Zealand

Ivett Zsakovics
1st Department of Pathology and
  Experimental Cancer Research
Semmelweis University
Budapest, Hungary

Prequel: Gap Junctions, Hemichannels, and Cell-to-Cell Signalling

William Howard Evans and Ernesto Oviedo-Orta

CONTENTS

I.1 INTRODUCTION

Gap junctions are strongly adhesive junctions of porous construction that enable cells to communicate/signal directly with each other. Gap junctions consist of closely packed arrays of hexameric protein channels that provide direct inter-cytoplasmic continuity in cells (Figure I.1). The term "gap" is a historic misnomer and derives from observations that lanthanum salts revealed a narrow 2 nm extracellular space between attached cells now known to be bridged by the component channels; at the time, these studies identified important morphological and functional differences between gap and tight junctions. In reality, gap junctions underpin cell–cell cooperation that is essential for orchestrating metabolic, electrical, and mechanical cell behavior in tissues and organs. A large number of channel units are compressed into a plaque in the membrane, but this does not necessarily mean that all channels are open and functional. Thus, small gap junction plaques size found in components of immune and other systems need not compromise the extent, and the importance of intercellular communication for electrophysiological evidence shows that as few as 20% of the channel structures in a large gap junction are active, with these being mainly those newly recruited to the edges of junctional plaques (Gaietta et al., 2002). Gap junction channels are weakly selective and allow small molecules and ions (generally up to 1200 Da) to pass from cell to cell while restricting the passage of larger molecules that might compromise cellular individuality.

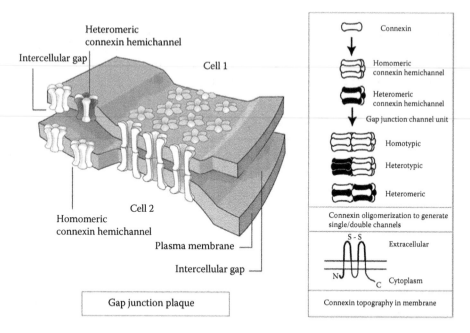

FIGURE I.1 Diagrammatic structure of a gap junction linking two cells and the composite hemichannels. Cxs oligomerize to form homomeric or heteromeric hemichannels. These dock to form the basic gap junctional unit that may be homomeric, heterotypic, or heteromeric depending on the class of Cxs made by the cooperating cells and their arrangements in the channel. Gap junction plaques vary in size depending on the number of functional units.

Connexin (Cx) hemichannels (connexons) are biogenetic precursors of gap junctions and, unsurprisingly, the two categories of channels display similar permeability properties (see Table I.1). Cx hemichannels dock head to head with closely aligned partner hemichannels in neighboring cells as they become attached to the rims of preexisting gap junction plaques. Hemichannels were once regarded as nonfunctional closed channels in transit prior to their incorporation after docking into gap junctions. It was strongly argued that were hemichannels to assume an open configuration they could seriously compromise cell permeability and viability and lead ultimately to apoptosis. However, it is now well established that they display an array of distinctive functions, with channel gating sensitive to external/internal metabolic, oxidative, osmotic, and mechanical stresses (Evans et al., 2012). Indeed, it is now becoming clear that hemichannel operation is controlled by an array of intracellular protein kinase cascades and intracellular calcium-dependent processes that are fundamental to cell signalling (Figure I.2).

I.2 CONNEXINS, PANNEXINS, AND INNEXINS

These are biochemically different classes of channel forming membrane-embedded proteins. Cxs oligomerize to form hexameric channels with the subunits surrounding

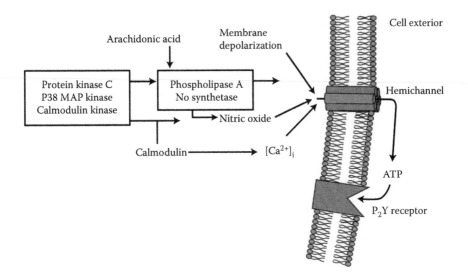

FIGURE I.2 Schematic showing a phalange of signalling components converging at a Cx hemichannel in the plasma membrane. A major regulator is calcium but other factors controlling hemichannels include phosphorylation especially by protein kinase C of the cytoplasmically located carboxyl tail of Cx43 as well as nitrosylation and membrane polarization. (Modified from de Vuyst, E. et al., 2009. *Cell Calcium*, 46, 176–87.)

a central pore. In vertebrates, over 20 Cxs have been found and they are expressed in all nucleated cells except sperm and adult skeletal muscle (Sohl and Willecke, 2004). Cxs display tetraspan membrane topography with cytoplasmic amino and carboxyl tails and a single intracellular loop (Figure I.1). The two disulfide-linked extracellular loops are highly conserved in all Cxs and are crucial for the docking of the hemichannel pairs that form the basic gap junction double-channel unit. The main determinant accounting for the size range of Cxs (23–58 kDa) is the length of the cytoplasmically orientated carboxyl tail where phosphorylation sites are located; in Cx43, there can be up to 11 sites, mainly on serine and threonine residues (Solan and Lampe, 2009; Jeyaraman et al., 2011; Marquez-Rosado et al., 2011). Most Cxs are phosphoproteins excepting Cx26 that has a short cytoplasmic tail. Phosphorylation is likely to modulate Cx channel gating, for example, phosphorylation of Cx43 by protein kinase C is implicated in cardiac ischemia (Saurin et al., 2002; Heyman et al., 2009). Hemichannel operation is also directly regulated by calcium and membrane depolarization (Saez et al., 2005) (Figure I.2). The cytoplasmic carboxyl tail of Cx43 interacts with the protein's intracellular loop, a process proposed to explain how the channel opens and closes in gap junctions as well as in Cx hemichannels (Delmar et al., 2004; Ponsaerts et al., 2010). The cytoplasmic tail also interacts with several cytoskeletal and scaffolding proteins such as tubulin, ZO1, and catenin (Giepmans, 2004).

Cx hemichannels allow cells to monitor the external environment and to respond to unfavorable contingencies (Scemes et al., 2009). The release of ATP across these channels provides a mechanism for para-cellular purinergic signalling (Figure I.3).

TABLE I.1
Properties of Gap Junction, Cx, or Panx Channels

Gap Junctions	Cx Channels	Panx Channels
Dodecameric paired hexamers	Hexameric structures. Six extracellular loop cysteines	Octameric structures. Four extracellular loop cysteines
Composed of over 20 Cxs; wide tissue distribution	Composed of 22/23 Cxs; wide tissue distribution	Panx1, Panx2, Panx3. Especially abundant in neural tissues
Nonglycosylated; phosphorylated	Nonglycosylated; phosphorylated	Panx1 is glycosylated and palmitoylated Panx2 is palmitoylated
Connect directly cytoplasms of attached cells	Connect cytoplasm with extracellular milieu; closure is protective	Mainly closed; opening allows cells to connect with external environment and signal by a paracellular mechanism
Pharmacological manipulation difficult	Blocked quickly by, e.g., Cx mimetic peptides	Inhibited by probenecid, carbenoxalone
Permeable to cAMP, calcium, IP$_3$	Permeable to ATP, glutamate, glutathione, calcium	Permeable to ATP; Open Panx channels may generate "death pore"
Closed internally by high calcium	Opened by low external calcium. Calmodulin and calcium regulate internally	Relatively unaffected by external calcium. Calmodulin independent
	Unitary conductance 15–300 ps	Unitary conductance 475–550 ps

Roles for Cx hemichannels in cell adhesion have been proposed (Bao et al., 2011; Cotrina et al., 2008; Elias et al., 2007; Wong et al., 2006), and they may also act as sensors and regulators of cell cycling (Vinken et al., 2011).

Most cells express more than one Cx, and a large number of permutations of various Cx components that make up hemichannels and gap junctions can provide a basis for channel selectivity, a possibility supported by the varying electrical characteristics and dye coupling properties of channels constructed of different Cxs (Bedner et al., 2011; Heyman et al., 2009).

Pannexins (Panxs) are not so well studied as Cxs, and the proteins do not share amino acid sequence homology (D'hondt et al., 2011; Shestopalov and Panchin, 2008). Three Panxs, Panx1, Panx2, and Panx3, are found in vertebrates (Table I.1). Panx1, in contrast to all Cxs, is glycosylated on the second extracellular loop possibly explaining why Panxs function as channels in the plasma membrane and are disinclined to form gap junctions (Sosinsky et al., 2011). Little information is currently available on whether Panxs are phosphorylated. Cx and Panx channels are highly expressed in neurons and astrocytes where they are associated with various physio-pathological roles (Thompson et al., 2008). Panx channels also participate in purinergic intercellular signalling in the immune system, especially in T-cell receptor regulation (Bopp et al., 2007). Cx and Panx channels release ATP that acts as a paracellular messenger. Release of ATP and its metabolites that bind to P2Y and P2Xn receptors (Figure I.3) provides these channels with intercellular signalling potential and a basis for playing key roles in inflammatory events that underpin a range of pathologies, also involving cytokines and other pro-inflammatory molecules (Scemes et al., 2009) (see also

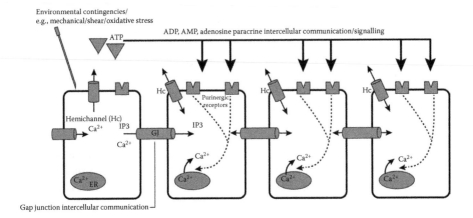

Environmental contingencies/
e.g., mechanical/shear/oxidative stress

ATP

ADP, AMP, adenosine paracrine intercellular communication/signalling

Hc

Purinergic
receptors

Hc

Hc

Hemichannel (Hc)
Ca²⁺   IP3

IP3

GJ

Ca²⁺

Ca²⁺

Ca²⁺

Ca²⁺

Ca²⁺   ER

Ca²⁺

Ca²⁺

Ca²⁺

Gap junction intercellular communication

FIGURE I.3 Propagation of Ca waves across linked cells occurs directly via gap junctions and/or indirectly via a paracellular route. An increase in Ca in a stimulated cell spreads to neighboring cells directly via gap junctions and involves the spread of metabolic mediators such as IP3. Para-cellular communication mechanisms involves the release of ATP across hemichannels which then acts on purineric receptors in neighboring cells leading ultimately to intercellular propagation of the Ca signalling response. (Modified from Evans, W. H. and Leybaert, L. 2007. *Cell Communication and Adhesion*, 14, 265–73.)

Chapters 8, 10, 11, and 15). Cx and Panx channels remaining in open configuration can lead to apoptosis (Saez et al., 2010). Cleavage of Panx1 channels by caspases causes channel opening and appears to be a key step linking their operation to cytoplasmic activation (Chekeni et al., 2010). Panx1 forms hexameric oligomers and Panx2 is reported to be an octameric channel (Sosinsky et al., 2011).

Innexins are found in invertebrates. They demonstrate some amino acid sequence homology with Panxs but little homology with Cxs. All three classes of proteins show tetraspan topography in the membrane (Fushiki et al., 2010).

I.3 CLINICAL IMPORTANCE AND PROSPECTS FOR THERAPEUTIC MANIPULATION OF CX AND PANX CHANNELS

Gap junctions have been largely out of bounds to direct pharmacological manipulation. However, undocked Cx and Panx channels in nonjunctional areas of the plasma membrane are more accessible. Short peptides mimicking sequences in the two extracellular loops of Cxs named Gap 26 and 27 (Evans and Leybaert, 2007) selectively, rapidly, and reversibly block Cx hemichannels; gap junctions are also inhibited by the peptides at later times following application of the peptides depending on such factors as cell geometry and the balance between hemichannel and gap junction communication (Bodendiek and Raman, 2010; Evans and Leybaert, 2007; Herve and Dhein, 2010). A number of Cx channel functions have been studied using these mimetic peptides and include the acceleration of spinal cord repair (see Chapter 14), brain and wound healing (Chapter 15), as well as retinal (Danesh-Meyer et al., 2012) and cardiac ischemia (Hawat et al., 2010), lung inflammation

(Sarieddine et al., 2009), and adhesion of inflammatory cells (Wong et al., 2006). The mimetics increase human dermal fibroblast migration implying a role for Cx hemichannels in cell movement and contact (Wright et al., 2012). These channels also underpin neural transmission linked to fear and memory (Bissiere et al., 2011) and epileptiform activity in the hippocampus (Thompson et al., 2008). The mimetic peptides used in these broad investigations may act by inducing "virtual" docking, inhibiting gap junction formation and/or by blocking molecular and ionic traffic through Cx hemichannels. By blocking the release of ATP and glutamate from cells, the peptides provide a useful strategy to reduce cell death especially in neurons and may thus become useful in addressing many pathologies such as Alzheimer's disease, diabetes, and ischemia. Peptides designed from the intracellular loop of Cxs also disrupt Cx hemichannels and limit ATP permeation with minimal effects on blockage of gap junctions and these mimetic peptides show therapeutic potential for they also reduce damage caused by cardiac hypoxia (De Vuyst et al., 2009). Short peptides corresponding to the cytoplasmic tail of Cx43 influence cardiac remodelling and arrhythmia (O'Quinn et al., 2011). See also Chapter 13. Cx mimetic peptides have been used to unravel complex metabolic and mechanical relationships between smooth muscle and endothelium and in diabetes. A complementary and alternative approach for disrupting gap junctional communication uses antisense mRNA probes to Cxs (Chapters 14 and 15).

I.4   STRUCTURE, ASSEMBLY, AND BREAK DOWN OF GAP JUNCTION CHANNELS

Progress in obtaining high-resolution models of gap junctions has been slow. Models of gap junctions constructed of Cx43 (Gaietta et al., 2002, Unger et al., 1999), Cx26 (Maeda et al., 2009), and Cx32 (Pantano et al., 2008) have helped elucidate how Cxs are arranged in the membrane to form a pore extending across essentially a double membrane in gap junctions. The crystal structure of Cx43 has provided new inputs into the molecular determinants of ion permeation characteristics such as conductance, current–voltage relationships, and charge selectivity of the channel (Harris, 2001).

Cxs are rapidly degraded with half-lives of 1–5 h (except for Cx46 in the lens that has an extended life time). Short protein life times allow cells to regulate more precisely expression levels of the channels. In contrast, Panxs have longer half-lives. Cxs are co-translationally inserted by ribosomes into the endoplasmic reticulum, and they oligomerize into closed hemichannels as they traverse along the secretory pathway to the plasma membrane (Martin and Evans, 2004). Some Cxs (e.g., Cx26) may also utilize a more ancient posttranslational mechanism with the protein directly inserted into the plasma membrane (Ahmad and Evans, 2002). Degradation of gap junctions commences by removal into cells of nonfunctioning units from the central areas of gap junction plaques generating annular structures that then disperse and disappear (Laird, 2010). This autophagic process is a key regulatory act in ensuring rapid turnover of gap junctions (Lichtenstein et al., 2011). There is also proteosomal break-down of Cxs.

I.5   CX MODIFICATIONS IN DISEASE

Mutations in Cx genes are linked to human disease (Dobrowolski and Willecke, 2009) (Table I.2). Over 200 mutations in Cx32 are linked to the peripheral demyelinating neuropathy Charcot–Marie–Tooth X-linked disease. These mutations impact on channels in the paranodal loops and Schmidt–Landermann incisures of myelinated Schwann cells. The mutations (missense, frameshift, deletion, and nonsense) lead to a failure of Cx32 to oligomerize into functional hemichannels and/or to traffic to the plasma membrane. Mutations in Cx26 and 30 are associated with autosomal recessive nonsyndromal hearing abnormalities. Of around 50 mutations found in Cx26, many are related to the functionality of the cochlea featuring potassium circulation from the interstitial space and involving cochlear supporting cells. Mutations in Cx26, Cx30, and Cx31 are also associated with skin disorders (see Chapters 12 and 15). Cx47 mutations are associated with Pelizaeus–Mezbacher disease (Ostergaard et al., 2011) and also primary lymphedema (Ferrell et al., 2010). Cx43 is widely expressed and is the default Cx identified in many tissues. Mutations in Cx43 give rise to oculodentodigital dysplasia, a syndromic condition affecting a range of cells in the body. Despite its wide distribution, fewer mutations are associated with Cx43, but it is more prone to dysregulated expression.

TABLE I.2
Properties of Major Mouse/Human Cxs

Cx	Chromosome Mouse	mRNA (kb)	Cells/Tissues Where Present	Disease Involvement
Cx26	14	2.4	Mammary gland, liver, skin, cochlea	Deafness, palmoplantar hyperkeradosis
Cx30	14	20.23	Skin, cochlea	Deafness, ectodermal dysplasia, hair loss
Cx31	4	1.9, 2.3	Skin, cochlea	Deafness, erythrokeratoderma variabilis
Cx32	X	1.6	Liver, oligodendrocytes	Peripheral neuropathy; Charcot–Marie–Tooth X-linked disease
Cx36	2	2.9	Neurons, retina	Visual defects
Cx37	4	1.7	Endothelium, ovaries	Bleeding
Cx40	3	3.5	Heart, endothelium	Atrial arrhythmia, hypertension
Cx43	10	3	Over 37 tissues and organs	Atherosclerosis, cardiac arrhythmia, and oculodentodigital dysplasia
Cx45	11	2.2	Heart, brain thalamus, bladder	Atrial fibrillation
Cx46	14	2.8	Lens	Cataract
Cx47	—	—	Lymph nodes, myelin	Lymphedema, Pelizaeus–Mezbacher
Cx50	3	8.5	Lens	Cataract

I.6   CXs IN THE VASCULATURE AND IMMUNE SYSTEMS

Endothelial cells of the vasculature express Cx37, Cx40, and Cx43 and smooth muscle cells express Cx43 and Cx45 (Johnstone et al., 2009). In the heart, there is variable expression of Cx40, Cx43, and Cx45 in different anatomical regions of the organ (Jansen et al., 2010). Knockout of cardiac Cx expression has unveiled the complex interplay of the various Cx channels regulating propagation of vasomotor activity, the pathology of hypertension and vasomotor function in renal hemodynamics (Brisset et al., 2009). Even platelets express Cx37 suggesting a role in the pathogenesis of thrombosis (Angelillo-Scherrer et al., 2011).

The widening roles of Cxs and their partner proteins, Panxs, in the immune system and inflammatory responses are increasingly appreciated and prompted this monograph. Cx43 is emerging as a key player in the immune system, for example, in T-cell activation (Oviedo-Orta et al., 2010), a process that is associated with the immunological synapse (Mendoza-Naranjo et al., 2011; Woehrle et al., 2010). Further roles for Cxs in the immune system (Chapters 1 to 6) and in inflammation (Chapters 7–12) are described and discussed in this book . Inflammatory pathologies appear to be related mainly to changes in expression levels of the channels and therefore the intensity of intercellular communication occurring within specific time frames and operating within and across physiological niches.

I.7   EFFECTS OF TOXINS, CYTOKINES, VIRUSES, AND NANOPARTICLES

Pathogenic bacteria produce toxins that influence Cx/Panx-dependent cell signalling (Ceelen et al., 2011). A plethora of cellular mechanisms are influenced directly or indirectly by metals, many released from environmental and anthropogenic sources (Vinken et al., 2010). Calcium enables important signalling functions inside cells as its concentration varies between different membrane-bound compartments leading to important effects on the operation of membrane channels and receptors, not least Cx hemichannels (de Vuyst et al., 2009). Metal toxicity is frequently carcinogenic (Naus and Laird, 2010) with Cxs implicated in chemically induced glioma and breast cancer possibly via indirect effects on cell migration and cell cycle events (Kameritsch et al., 2011). Cytokines are potent modulators in inflammation and cell proliferation; interleukins and TNF-α have been shown to influence gap junctional communication (Lemenager et al., 2011).

Hemichannels and gap junctions are penetrated by human immunodeficiency virus (HIV-1) that acts on astrocytes and leads to disruption of the blood–brain barrier (Eugenin et al., 2011). Also, in HIV-1 infection, ATP is released from cells via Panx channels and acts on purinergic receptors leading to activation of a protein kinase and resulting in membrane depolarization (Seror et al., 2011).

Nanoparticles released from implanted metal joints (e.g., cobalt-chromium orthopedic joint replacements) damage DNA and chromosomes especially in cell bilayers but not in monolayers. These nanoparticles disrupt Cx/Panx ATP-dependent cell signalling/communication that are key processes involved in metabolically integrating various cell bilayers in body organs (Bhabra et al., 2009; Sood et al., 2011).

Thus, in a range of tissues and organs including those of the immune system emphasized here, the involvement of Cx and Panx hemichannels and gap junctions in endogenous and environmentally induced physio-pathological situations is widely evident. Information detailed in this prequel should reinforce the key roles of these membrane channels that fulfil ever-widening activities in addition to their well-established fundamental intercellular communication/signalling functions.

REFERENCES

Ahmad, S. and Evans, W. H. 2002. Post-translational integration and oligomerization of connexin 26 in plasma membranes and evidence of formation of membrane pores: Implications for the assembly of gap junctions. *Biochemical Journal,* 365, 693–9.

Angelillo-Scherrer, A., Fontana, P., Burnier, L. et al. 2011. Connexin 37 limits thrombus propensity by downregulating platelet reactivity. *Epub,* 124, 930–9.

Bao, B., Jiang, J., Yanase, T. et al. 2011. Connexon-mediated cell adhesion drives microtissue self-assembly. *FASEB Journal,* 25, 255–64.

Bedner, P., Steinhäuser, C., Theis, M. 2011. Functional redundancy and compensation among members of gap junction protein families? *Biochimica et Biophysica Acta,* 1818(8), 1971–84.

Bhabra, G., Sood, A., Fisher, B. et al. 2009. Nanoparticles can cause DNA damage across a cellular barrier. *Nature Nanotechnology,* 4, 876–83.

Bissiere, S., Zelikowsky, M., Ponnusamy, R. et al. 2011. Electrical synapses control hippocampal contributions to fear learning and memory. *Science,* 331, 87–91.

Bodendiek, S. B. and Raman, G. 2010. Connexin modulators and their potential targets under the magnifying glass. *Current Medicinal Chemistry,* 17, 4191–230.

Bopp, T., Becker, C., Klein, M. et al. 2007. Cyclic adenosine monophosphate is a key component of regulatory T cell-mediated suppression. *The Journal of Experimental Medicine,* 204, 1303–10.

Brisset, A. C., Isakson, B. E., and Kwak, B. R. 2009. Connexins in vascular physiology and pathology. *Antioxidants and Redox Signaling,* 11, 267–82.

Ceelen, L., Haesebrouck, F., Vanhaecke, T. et al. 2011. Modulation of connexin signaling by bacterial pathogens and their toxins. *Cellular and Molecular Life Sciences: CMLS,* 68, 3047–64.

Chekeni, F. B., Elliott, M. R., Sandilos, J. K. et al. 2010. Pannexin 1 channels mediate "find-me" signal release and membrane permeability during apoptosis. *Nature,* 467, 863–7.

Cotrina, M. L., Lin, J. H. C., and Nedergaard, M. 2008. Adhesive properties of connexin hemichannels. *Glia,* 56, 1791–8.

Danesh-Meyer, H. V., Kerr, N. M., Zhang, J., Eady, E. K., O'Carroll, S. J., Nicholson, L. F., Johnson, C. S., and Green, C. R. 2012. Connexin43 mimetic peptide reduces vascular leak and retinal ganglion cell death following retinal ischaemia. *Brain,* 135(Pt 2), 506–20.

D'hondt, C., Ponsaerts, R., de Smedt, H. et al. 2011. Pannexin channels in ATP release and beyond: An unexpected rendezvous at the endoplasmic reticulum. *Cellular Signalling,* 23, 305–16.

de Vuyst, E., Wang, N., Decrock, E. et al. 2009. Ca^{2+} regulation of connexin 43 hemichannels in C6 glioma and glial cells. *Cell Calcium,* 46, 176–87.

Delmar, M., Coombs, W., Sorgen, P. et al. 2004. Structural bases for the chemical regulation of Connexin43 channels. *Cardiovascular Research,* 62, 268–75.

Dobrowolski, R. and Willecke, K. 2009. Connexin-caused genetic diseases and corresponding mouse models. *Antioxidants & Redox Signaling,* 11, 283–95.

Elias, L. A., Wang, D. D., and Kriegstein, A. R. 2007. Gap junction adhesion is necessary for radial migration in the neocortex. *Nature,* 448, 901–7.

Eugenin, E. A., Clements, J. E., Zink, M. C. et al. 2011. Human immunodeficiency virus infection of human astrocytes disrupts blood–brain barrier integrity by a gap junction-dependent mechanism. *Journal of Neuroscience: The Official Journal of the Society for Neuroscience*, 31, 9456–65.

Evans, W. H., Bultynck, G., and Leybaert, L. 2012. Manipulating connexin communication channels: Use of peptidomimetics and the translational outputs. *J Membr Biol*, 245(8), 437–49.

Ferrell, R. E., Baty, C. J., Kimak, M. A. et al. 2010. GJC2 missense mutations cause human lymphedema. *American Journal of Human Genetics*, 86, 943–8.

Fushiki, D., Hamada, Y., Yoshimura, R. et al. 2010. Phylogenetic and bioinformatic analysis of gap junction-related proteins, innexins, pannexins and connexins. *Biomedical Research-Tokyo*, 31, 133–42.

Gaietta, G., Deerinck, T. J., Adams, S. R. et al. 2002. Multicolor and electron microscopic imaging of connexin trafficking. *Science*, 296, 503–7.

Giepmans, B. N. 2004. Gap junctions and connexin-interacting proteins. *Cardiovascular Research*, 62, 233–45.

Harris, A. L. 2001. Emerging issues of connexin channels: Biophysics fills the gap. *Quarterly Reviews of Biophysics*, 34, 325–472.

Hawat, G., Benderdour, M., Rousseau, G. et al. 2010. Connexin 43 mimetic peptide Gap26 confers protection to intact heart against myocardial ischemia injury. *Pflugers Archiv: European Journal of Physiology*, 460, 583–92.

Herve, J. C. and Dhein, S. 2010. Peptides targeting gap junctional structures. *Current Pharmaceutical Design*, 16, 3056–70.

Heyman, N. S., Kurjiaka, D. T., Ek Vitorin, J. F. et al. 2009. Regulation of gap junctional charge selectivity in cells coexpressing connexin 40 and connexin 43. *American Journal of Physiology. Heart and Circulatory Physiology*, 297, H450–9.

Jansen, J. A., van Veen, T. A., de Bakker, J. M. et al. 2010. Cardiac connexins and impulse propagation. *Journal of Molecular and Cellular Cardiology*, 48, 76–82.

Jeyaraman, M. M., Srisakuldee, W., Nickel, B. E., and Kardami, E. 2012. Connexin43 phosphorylation and cytoprotection in the heart. *Biochimica et Biophysica Acta*, 1818(8), 2009–13.

Johnstone, S., Isakson, B., and Locke, D. 2009. Biological and biophysical properties of vascular connexin channels. *International Review of Cell and Molecular Biology*, 278, 69–118.

Kameritsch, P., Pogoda, K., and Pohl, U. 2012. Channel-independent influence of connexin 43 on cell migration. *Biochimica et Biophysica Acta*, 1818(8), 1993–2001.

Laird, D. W. 2010. The gap junction proteome and its relationship to disease. *Trends in Cell Biology*, 20, 92–101.

Lemenager, T., Richter, A., Reinhard, I. et al. 2011. Impaired decision making in opiate addiction correlates with anxiety and self-directedness but not substance use parameters. *Journal of Addiction Medicine*, 5, 203–13.

Lichtenstein, A., Minogue, P. J., Beyer, E. C. et al. 2011. Autophagy: A pathway that contributes to connexin degradation. *Journal of Cell Science*, 124, 910–20.

Maeda, S., Nakagawa, S., Suga, M. et al. 2009. Structure of the connexin 26 gap junction channel at 3.5 A resolution. *Nature*, 458, 597–602.

Márquez-Rosado, L., Solan, J. L., Dunn, C. A., Norris, R. P., and Lampe, P. D. 2012. Connexin43 phosphorylation in brain, cardiac, endothelial and epithelial tissues. *Biochimica et Biophysica Acta*, 1818(8), 1985–92.

Martin, P. E. and Evans, W. H. 2004. Incorporation of connexins into plasma membranes and gap junctions. *Cardiovascular Research*, 62, 378–87.

Mendoza-Naranjo, A., Bouma, G., Pereda, C. et al. 2011. Functional gap junctions accumulate at the immunological synapse and contribute to T cell activation. *Journal of Immunology*, 187, 3121–32.

Naus, C. C. and Laird, D. W. 2010. Implications and challenges of connexin connections to cancer. *Nature Reviews. Cancer,* 10, 435–41.

O'Quinn, M. P., Palatinus, J. A., Harris, B. S., Hewett, K. W., and Gourdie, R. G. 2011. A peptide mimetic of the connexin43 carboxyl terminus reduces gap junction remodeling and induced arrhythmia following ventricular injury. *Circ Res,* 108(6), 704–15.

Ostergaard, P., Simpson, M. A., Brice, G. et al. 2011. Rapid identification of mutations in Gjc2 in primary lymphoedema using whole exome sequencing combined with linkage analysis with delineation of the phenotype. *Journal of Medical Genetics,* 48, 251–5.

Oviedo-Orta, E., Perreau, M., Evans, W. H. et al. 2010. Control of the proliferation of activated CD4+ T cells by connexins. *Journal of Leukocyte Biology,* 88, 79–86.

Pantano, S., Zonta, F. and Mammano, F. 2008. A fully atomistic model of the Cx32 connexon. *PloS One,* 3, e2614.

Ponsaerts, R., de Vuyst, E., Retamal, M. et al. 2010. Intramolecular loop/tail interactions are essential for connexin 43-hemichannel activity. *The FASEB Journal: Official Publication of the Federation of American Societies for Experimental Biology,* 24, 4378–95.

Saez, J. C., Retamal, M. A., Basilio, D. et al. 2005. Connexin-based gap junction hemichannels: Gating mechanisms. *Biochimica et Biophysica Acta,* 1711, 215–24.

Saez, J. C., Schalper, K. A., Retamal, M. A. et al. 2010. Cell membrane permeabilization via connexin hemichannels in living and dying cells. *Experimental Cell Research,* 316, 2377–89.

Sarieddine, M. Z., Scheckenbach, K. E., Foglia, B. et al. 2009. Connexin43 modulates neutrophil recruitment to the lung. *Journal of Cellular and Molecular Medicine,* 13, 4560–70.

Saurin, A. T., Pennington, D. J., Raat, N. J. et al. 2002. Targeted disruption of the protein kinase C epsilon gene abolishes the infarct size reduction that follows ischaemic preconditioning of isolated buffer-perfused mouse hearts. *Cardiovascular Research,* 55, 672–80.

Scemes, E., Spray, D. C., and Meda, P. 2009. Connexins, pannexins, innexins: Novel roles of "hemi-channels". *Pflugers Archiv: European Journal of Physiology,* 457, 1207–26.

Seror, C., Melki, M. T., Subra, F. et al. 2011. Extracellular ATP acts on P2Y2 purinergic receptors to facilitate HIV-1 infection. *Journal of Experimental Medicine,* 208, 1823–34.

Shestopalov, V. I. and Panchin, Y. 2008. Pannexins and gap junction protein diversity. *Cellular and Molecular Life Sciences: CMLS,* 65, 376–94.

Sohl, G. and Willecke, K. 2004. Gap junctions and the connexin protein family. *Cardiovascular Research,* 62, 228–32.

Solan, J. L. and Lampe, P. D. 2009. Connexin43 phosphorylation: Structural changes and biological effects. *The Biochemical Journal,* 419, 261–72.

Sood, A., Salih, S., Roh, D. et al. 2011. Signalling of DNA damage and cytokines across cell barriers exposed to nanoparticles depends on barrier thickness. *Nature Nanotechnology,* 6, 824–33.

Sosinsky, G. E., Boassa, D., Dermietzel, R. et al. 2011. Pannexin channels are not gap junction hemichannels. *Channels,* 5, 193–7.

Thompson, R. J., Jackson, M. F., Olah, M. E. et al. 2008. Activation of pannexin-1 hemichannels augments aberrant bursting in the hippocampus. *Science,* 322, 1555–9.

Unger, V. M., Kumar, N. M., Gilula, N. B. et al. 1999. Three-dimensional structure of a recombinant gap junction membrane channel. *Science,* 283, 1176–80.

Vinken, M., Ceelen, L., Vanhaecke, T., and Rogiers, V. 2010. Inhibition of gap junctional intercellular communication by toxic metals. *Chemical Research in Toxicology,* 23(12), 1862–7.

Vinken, M., Decrock, E., de Vuyst, E. et al. 2011. Connexins: Sensors and regulators of cell cycling. *Biochimica et Biophysica Acta,* 1815, 13–25.

Woehrle, T., Yip, L., Elkhal, A. et al. 2010. Pannexin-1 hemichannel-mediated ATP release together with P2X1 and P2X4 receptors regulate T-cell activation at the immune synapse. *Blood,* 116, 3475–84.

Wong, C. W., Christen, T., Roth, I. et al. 2006. Connexin37 protects against atherosclerosis by regulating monocyte adhesion. *Nature Medicine,* 12, 950–4.

Wright, C. S., Pollok, S., Flint, D. J., Brauder, J. M., and Martin P.E.M. 2012. The connexin mimetic peptide Gap27 increases human dermal fibroblast migration in hyperglycaemic and hyperinsulinaemic conditions *in vitro. Journal of Cellular Physiology,* 227, 77–87.

Section 1

Immunology and Cell Biology

1 Communication in the Immune System by Connexin Channels

Ernesto Oviedo-Orta and William Howard Evans

CONTENTS

1.1 INTRODUCTION

In recent years, we have seen a burst of evidence supporting the role of gap junction intercellular channels in immune functions, and their roles as fundamental structures underpinning immune regulation have become increasingly evident. Advances in connexin biology proceeded in parallel with discoveries in the field of immunology that have revolutionized our understanding of the subject. The discovery and expansion of knowledge related to T cells with suppression activity (now known as regulatory T cells), the realization that antigen cross-presentation is a fundamental mechanism to prime and maintain immune responses, and the discovery of Th17 cells that have been shown to play a key role in bridging innate and adaptive immune responses are just a few examples.

Immune responses and immunovigilance are established not only to protect us from diseases caused by the invasion of microorganisms but also to maintain internal homeostasis that keep at bay autoimmune processes with the potential to lead to life-threatening pathological conditions. From the differentiation of pluripotent stem cells located in primary lymphoid organs to the production of well-differentiated effector T and B lymphocytes, immune responses rely on a complex network of interactions ranging from direct cell-to-cell contacts to responses triggered by soluble factors such as cytokines or growth factors secreted in a paracrine or endocrine way. Both arms of the immune system, namely innate and adaptive, are required for the successful outcome of any of these responses and also to reestablish the lost equilibrium disturbed by the insult in the first place. Both arms are equipped with "sensor" molecules on the plasma membrane of leukocytes that regulate in a positive (activation) or a negative (suppression) way their multiple internal metabolic responses. Most importantly, some of these receptors permit the identification of the primary insult (i.e., the T or B cell receptors) allowing its specific and efficient elimination. Cross-talk between lymphocytes and their surrounding cellular and molecular environment would not be possible without these membrane receptors.

This chapter goes beyond the traditional structural description of gap junction channels and connexins in cells and organs of the immune system, and places them in the functional environment in which they develop as active contributors to metabolic and cellular responses and analyzes their importance in immunity. We describe some of the most important historical discoveries and provide a prospective analysis of the role of these channels in immunoregulation and highlight important questions that will require answers in the next decade.

1.2   EARLY OBSERVATIONS POINTING TO ROLES FOR CONNEXINS AND GAP JUNCTIONS IN THE IMMUNE SYSTEM

The observation that lymphocytes can form aggregates after stimulation with phytohemagglutinin (PHA) and other similar polyclonal stimulators was the trigger for Hülser and Peters (1972) to study whether direct electrical communication occurred between these cells. These studies were carried out at a time when evidence of the existence of gap junction channels was already emerging in tissues such as liver and heart and when their discovery was driven by the use of a combined approach based on electrophysiological and morphological studies, especially electron microscopy. Hülser and Peters (1972) also used bovine lymphocytes immobilized in agar and stimulated with PHA to measure variations in the electrical current compared to control nonstimulated cells. They observed a high ohmic resistance in the membrane of stimulated cells that had established close contact and related this to the possibility that direct intercellular transfer of metabolites was occurring between cells. These findings were corroborated by Oliveira-Castro and coworkers in 1973 (Oliveira-Castro et al., 1973) using peripheral blood-derived human lymphocytes. In addition to the lower membrane potentials and high ohmic resistances observed, these authors showed that the effect was maintained for long period of time (up to 48 h) and related these observations with mechanisms involved in metabolic cooperation and cell division in these cells (Oliveira-Castro et al., 1973).

Following these observations, other investigators identified structures that could account for the permeability characteristics noted using the same or similar sources of cells. Zones where intercellular contacts were seen as early as 1 h after PHA stimulation in pseudopod-like processes. These membrane areas were separated by a gap of 30 Å, and the thickness of the junctional complex was approximately 200 Å, corresponding with the earliest gap junction structures (Gaziri et al., 1975). Similar structures were also found by electron microscopy in macrophages in culture oriented in a linear chain (Levy et al., 1976).

The first ultrathin sections and freeze-fracture electron microscopy were carried out using rabbit agglutinated lymphocytes obtained from peripheral blood and the spleen. After 4 h in culture, particles were occasionally observed as aggregates of small clusters between areas of membrane contacts (Kapsenberg and Leene, 1979). Gap junction-like structures were also detected *in vitro* by freeze-fracture electron microscopy in monocyte-macrophage colony-forming cells derived from canine bone marrow (Porvaznik and MacVittie, 1979). In this case, endotoxin-stimulated dog serum was used as a stimulator, which induced the appearance of approximately 9.3 nm hexagonal particles with a center-to-center spacing of 10.4 nm on the membrane P-face fractures. On the E-faces, highly ordered arrays of pits with 8.7 nm center-to-center spacing were noted (see also Chapter 2) (Porvaznik and MacVittie, 1979). These observations suggested that the synthesis of gap junctions was occurring in macrophages and it was induced by soluble factors in inflammatory serum, known today to be related to secreted cytokines and growth factors.

1.3   DISCOVERY OF GAP JUNCTIONS AND CONNEXINS IN PRIMARY AND SECONDARY LYMPHOID ORGANS

Following Porvaznik's observations (Porvaznik and MacVittie, 1979) described above, Watanabe (1985) showed for the first time, using electron microscopy, that gap junctions were present between bone marrow stromal cells. These studies were followed by others that showed the presence of gap junctions using electron microscopy and intercellular dye transfer using murine bone marrow-derived cells (Umezawa and Hata, 1992; Watanabe, 1985). A series of elegant experiments by Rosendaal and Krenacs (Krenacs and Rosendaal, 1995), and described in Chapter 2, demonstrated the role that gap junctions and connexins played in communication between stromal cells and also between stromal and hematopoietic cells. These were some of the earliest functional demonstrations showing that hematopoietic stem cell growth was likely to be underpinned by gap junctions and used a wide range of techniques, including electron and fluorescence microscopy and dye transfer (Krenacs and Rosendaal, 1995, 1998a,b; Krenacs et al., 1997).

Initially, it was believed that Cx43 was the only connexin expressed by the hematopoietic system. However, further studies reported the expression of other connexin proteins, and encoding mRNAs for Cx26, Cx30.3, Cx31, Cx31.1, Cx32, Cx37, Cx40, Cx45, and Cx50 in murine bone marrow primary cells (Cancelas et al., 2000; Hurtado et al., 2004; Paraguassu-Braga et al., 2003).

Gap junction proteins are also expressed in the thymus, a primary lymphoid organ responsible for the maturation of T lymphocytes. Ultrastructural data first showed the existence of gap junctions in the thymus between thymocytes and nurse epithelial cells (Carolan and Pitts, 1986). The presence of direct cell-to-cell communication mediated by gap junctions was also demonstrated in human and murine thymic epithelial cells by means of *in situ* and *in vivo* immunohistochemical labeling as well as *in vitro* dye transfer and double whole-cell patch clamping experiments (Alves et al.,

1994, 1995, 1996). These experiments highlighted for the first time the presence and likely importance of gap junction channels and metabolic endocrine coupling of thymic cells that underpinned the organ's development. They also pointed toward the potential selectivity of the expression and function of these structures as poor intercellular coupling was found between epithelial and thymocyte pairs. Moreover, no intercellular communication was demonstrated between cultured phagocytic cells and thymic cells in *in vitro* cultures (Alves et al., 1995). Most recently, it has been shown using a mouse model in which Cx43 was deleted in the immune system, while maintaining normal cardiac function after birth (because Cx43 KO mice dye around birth time), that there were no differences in thymocyte development or in the ability of lymphocytes to transmigrate to peripheral lymphoid organs (Nguyen and Taffet, 2009). This study contrasted with previous and some recent data showing that gap junction-mediated communication in the thymic epithelium plays a role in exerting a regulatory role in cell coupling (Nihei et al., 2010; Savino et al., 1999), an effect that seems to be phylogenetically conserved in the thymus as it was demonstrated in both mouse and human thymic epithelial cell preparations (Nihei et al., 2010).

The above studies provided a starting point to carry out further research into how gap junction channels and connexins impinged on other components of the immune system, including secondary lymphoid organs. These comprise a network of encapsulated tissues distributed throughout the body, which play a fundamental role in the initiation, maintenance, and maturation of immune responses. Secondary lymphoid organs such as lymph nodes and spleen contain stromal supporting cells, antigen-presenting cells (APCs) such as follicular dendritic cells, and macrophages and immature lymphocytes distributed in well-defined areas. Lymph nodes "filter" antigens derived from the external environment, which gain access to these organs through mucosal tissues, while blood-born antigens are dealt with by the spleen. In these organs, immature B and T cells encounter antigens and acquire the necessary properties to fight them in specific ways. A similar process occurs in what are called tertiary tissues, which are mainly confined to the mucosal surfaces and which are made of nonencapsulated aggregates of cells with a similar makeup. The latter are also involved in tolerance and homeostasis as they are located in places where a large number of diverse, but harmless, antigens enter the body constantly.

In 1995, the demonstration of gap junctions in secondary lymphoid organs was provided by Krenacs and Rosendaal (Krenacs and Rosendaal, 1995). These authors investigated the expression of Cx26, Cx32, and Cx43 in normal, activated, and diseased human lymphoid tissue using single and double immunolabeling and confocal laser scanning microscopy (Krenacs and Rosendaal, 1995). Cx43 was expressed in follicular dendritic cells, around lymphoendothelial cells, and in the vascular endothelium, including high endothelial venules. Cx26 and Cx32 were not detected but Cx26 was found expressed in tonsil epithelium. Most Cx43 was expressed in the sinus lining cells of lymph nodes involved in malignancies and in follicular dendritic cells in the light zone of germinal centers where maturating, but still proliferating, lymphocytes are localized (Krenacs and Rosendaal, 1995).

These findings were extended using a combination of techniques to study the role of direct intercellular communication in the formation of germinal centers. Experiments using antigenic challenge studied the formation of secondary follicles containing a meshwork of Cx43 and noted that the amount of gap junctions made of this protein was higher in the light zone of germinal centers. At the same time, gap junctions were found between follicular dendritic cells and B lymphocytes paving the way for further studies aiming to elucidate their role in antibody production (Krenacs and Rosendaal, 1995; Oviedo-Orta et al., 2001).

1.4 PERIPHERAL BLOOD LEUKOCYTES: ARE THERE CONNEXINS?

Why should floating, circulating, cells need to express connexins if they physically do not interact with each other? Or do they? This fundamental question delayed for years the acceptance by the immunologists of the expression of connexins by peripheral blood lymphocytes. The evidence accumulated pointed to the fact that only when cells are present in a tridimensional environment that facilitates tight proximity among them, and provided that other conditions such as exposure to appropriate stimuli are met, they are capable of expressing connexins and communicating directly via gap junction channels. It has now become clear that the expression of connexins is highly regulated in circulating leukocytes. Pioneering work by Saez's group provided preliminary evidence concerning the expression and modulation of Cx43 in leukocytes isolated from peripheral blood and stimulated with lipopolysaccharide (LPS). They measured the subcellular distribution of Cx43 in hamster leukocytes before and after activation with endotoxin in *in vitro* and *in vivo*. Cx43 was detected in peritoneal-derived macrophages 5–7 days after LPS-induced inflammation and also after ischemia–reperfusion using a hamster cheek pouch model (Jara et al., 1995). These studies provided evidence suggesting a role of gap junction channel formation between transmigrating leukocytes and endothelial cells. Using electron microscopy, they demonstrated the formation of small appositions between these two types of cells and also between activated leukocytes (Jara et al., 1995).

The expression of connexins and the establishment of gap junction channels between defined subpopulations of human peripheral blood-derived T and B lymphocytes and also natural killer (NK) cells soon followed (Oviedo-Orta et al., 2000). This work provided evidence that Cx43 was the major connexin expressed at mRNA and protein levels in lymphocytes and that Cx40 expression by these cells was confined to secondary lymphoid organs such tonsils, which may relate to inflammation. This work also showed for the first time that both homologous and heterotypic cell populations can establish a regulated bidirectional gap junction-mediated communication, which can be blocked by the use of connexin mimetic peptides, highly specific gap junction channel blockers (Desplantez et al., 2012; Evans et al., 2006; Oviedo-Orta et al., 2000). Most importantly, blockage of gap junction intercellular communication between lymphocytes by the mimetic peptide channel blockers reduced the synthesis of IgM, IgG, and IgA, and completely blocked the synthesis of the mRNA of important lymphocyte cytokines such as interleukin-10 (IL-10) and significantly reduced the synthesis of interleukin-2 (IL-2) and interferon gamma (IFNγ) (Oviedo-Orta et al., 2001).

These observations paved the way to formulate experimental strategies aiming to advance knowledge on the contribution of connexin channels in immunoregulation. New experimental tools have been deviced or adapted to study gap junction intercellular communication and connexin expression in the immune system and in the context of related pathologies.

An element of great importance to this discussion is the finding that bidirectional intercellular communication through gap junction channels can potentially underpin lymphocyte development and effector responses. Intercellular channel rectification exists during heterologous lymphocyte interactions, suggesting a role for these intercellular connexin channels in the polarization of immune responses (Oviedo-Orta et al., 2000). Driven by these results and by the demonstration that gap junctions are involved in antigen cross-presentation (Neijssen et al., 2005), the differential expression of Cx43 in mouse-derived CD4[+] Th0, Th1, and Th2 lymphocyte subpopulations was investigated and demonstrated that gap junction channel communication was operational in primary macrophages *in vitro*. While all lymphocyte subsets expressed Cx43, CD4[+] Th1 cells showed the highest levels (see also Chapter 5). Likewise, dye transfer between lymphocytes and macrophages occurred in all groups with a higher degree of communication observed with Th1 lymphocytes. Under inflammatory conditions induced by coculturing cells in the presence of LPS, dye transfer between cells was inhibited and was similar between macrophages and CD4[+] Th1 and Th2 lymphocytes (Bermudez-Fajardo et al., 2007). These results suggest that gap junction-mediated communication can be modulated by lymphocytes in the absence of specific antigenic stimulation and provided further support for connexin channel involvement in immune regulation. However, these basic observations merit further experimental evaluation. The demonstration that lymphocyte gap junctional communication significantly contributes to subset polarization or the determination of the fate of specific immune responses will add a new physiological and therapeutic dimension to their role. In this regard, further studies are required to address and extend many of these fundamental questions and thus position gap junction channels and connexins among the key molecular components of the immune system.

1.5   LYMPHOCYTE GAP JUNCTIONAL COMMUNICATION AND ANTIGEN-MEDIATED RESPONSES

The immune system has evolved a specialized system capable of recognizing, assessing, and dealing with self- and foreign antigens in a specific way. Such specializations rely on molecular structures and mechanisms present in the principal cellular components of the system, namely T and B cells and APCs. Cx43 is the main protein in gap junctions in the immune system and has been recently shown to form part of the immunological synapse (Mendoza-Naranjo et al., 2011). It appears increasingly likely that Cx43 is involved in intracellular and intercellular signaling leading to lymphocyte activation (Machtaler et al., 2011; Oviedo-Orta et al., 2010). Cx43 is also likely to play a crucial role in phagocytosis (Anand et al., 2008) and antigen cross-presentation (Neijssen et al., 2005) and to be involved in antigen-specific effector T cell responses, including cytotoxic T cell (CTL) killing of tumor cells (Benlalam et al., 2009). Despite extensive evidence, practical barriers such as the inability to adhere lymphocytes to tissue culture supports or the short life span of primary cells have limited the study of the role of gap junction channels during lymphocyte development. Animal models, especially the use of tissue targeted knockout mice, have helped to shed light on the potential role of Cx43 in central (thymic) and peripheral immune responses in physiological and inflammatory conditions (Nguyen and Taffet, 2009).

Of special note is the demonstration of Cx43 channels within the immunological synapse (Mendoza-Naranjo et al., 2011) and its involvement in CD4[+] T cell proliferation and sustained clonal expansion after CD3 and CD28 engagement (Elgueta et al., 2009; Oviedo-Orta et al., 2010), suggesting for the first time that in addition to peptide antigen presentation and recognition, intercellular signaling involving metabolic exchange through gap junction channels is a requirement to achieve cross-cellular activation. This connexin-mediated communication maximizes the use of energy resources required by cells. Perturbation of gap junction channels by blocking agents can alter the expression of costimulatory molecules (Elgueta et al., 2009) but had a limited effect on cytokine production by T cells (Oviedo-Orta et al., 2010). Such observations also raise fundamental questions about the contribution of these intercellular channels in specific stages of cell activation.

Finally, the analysis of how these connexin channels are regulated or contribute to the regulation of immune responses cannot obviate the contribution of the antigen. Perhaps the next generation of studies will focus on this fundamental element to pinpoint functional and even structural differences in the context of immune responses and will open up new avenues for the application of this vast knowledge to specific pathological conditions. Recently, Benlalam et al. (2009) have shown that gap junctions made of Cx43 could be part of the principal components involved in antitumor responses upon bacterial infection by facilitating the transfer of processed antigenic peptides from tumor cells to DCs and enhancing their elimination by CTLs. This study reinforces the possibility that gap junctions also contribute to enhance antigen-specific antitumor killing by bacteria or bacteria-derived components (Fuchs and Bachran, 2009; Terman et al., 2006).

1.6 PANNEXIN CHANNELS: IMPLICATIONS FOR THE IMMUNE SYSTEM

Pannexins (Panx) are, in many ways, considered as proteins with some functions that are similar to those carried by connexins. Although their amino acid sequences are entirely different, both proteins have a similar topology in the membrane, crossing it four times and with cytoplasmic N and C termini, two extracellular loops, and a single intracellular loop. Both proteins oligomerize intracellularly to generate channels in the membrane that are poorly selective. There are three Panx proteins currently identified, whereas there are up to 22 Cx proteins with Cx43 being the most widely distributed and as described above is present in cells of the immune system. For example, Panx1 and Panx3 are glycosylated, a posttranslational modification required for trafficking to the plasma membrane, whereas Cx are not glycosylated but are subject to phosphorylation and nitrosylation. Cx hemichannels in the plasma membrane dock with partner proteins in neighboring cells to form gap junction intercellular channels whereas Panx do not and are thus referred to as merely channels (see Penuela et al. (2012) and this volume's Prequel).

Panx are present at high levels in the central nervous system. Recently, several reports have appeared indicating their presence in cells of the immune system. Here, they are implicated in purinergic signaling involving the release of ATP through their channels. For example, ATP release by Panx channels activates T cells and results in Ca influx (Schenk et al., 2008). Similarly, Panx1 is a channel involved in ATP release in apoptosis, although it is not involved in the activation of the inflammasome (Chekeni et al., 2010; Qu et al., 2011). ATP release through pannexin channels act on P2Y2 receptors that facilitate HIV infection (Seror et al., 2011).

Panx1 has been implicated in the Toll-like receptor-independent inflammasome (Kanneganti et al., 2007). There is much debate concerning the relative roles of Panx1 channels and Cx hemichannels in the release of ATP by cells and further work is required to resolve these issues. Panx1 channels are found at higher levels intracellularly than Cx hemichannels that engage in the biogenesis of gap junctions. Moreover, Panx channels and Cx hemichannels also react in different manners to changes in Ca^{2+} levels, suggesting different roles in Ca^{2+} signaling. The area of Panx functions is rapidly advancing and their role in the activation of immune cells is sure to progress.

1.7 WHERE DO WE GO FROM HERE?

Since connexins and gap junction channels are present and are functional in cells of the immune system and contribute to the initiation and maintenance of immune responses, future research is likely to focus on providing answers to the following questions:

Do lymphocyte gap junctions equip the immune system with an extra level of selectivity? It is known that the existence of homotypic versus heterotypic or homomeric versus heteromeric connexin channels (see this volume's Prequel) may provide one element of selectivity in these channels. However, even in the presence of homotypic/homomeric interactions, the establishment of gap junction channels between heterotypic cells can also contribute to another level of regulation. This analysis leads us to formulate another important question.

What molecular interactions and pathways underpin gap junction channel regulation in different subsets of immune cells? This question is starting to receive some answers. Machtaler et al. (2011) have recently evaluated the regulatory role of Cx43 in the development of responses by B cells. These authors found that Cx43 is required to restore and sustain B cell receptor (BCR)-mediated Rap1 activation, a GTPase that participates in BCR activation and cell adhesion. More specifically, the C-terminal tail of Cx43 was responsible for the interaction with LFA-1 and CXCL12, which mediate spreading and adhesion of these cells to endothelial cells (Machtaler et al., 2011). Other studies show that specific T cell subsets respond specifically to macrophages previously activated with atherogenic stimuli generating differential intracellular signaling events classically linked with T cell receptor (TCR) engagement and lymphocyte activation (Rueban Jacob Anicattu Issac, Alexandra Bermudez-Fajardo, Emmanuel Dupont, Christopher Fry, and Ernesto Oviedo-Orta, unpublished data). Exploration of these molecular components of intracellular regulation will potentially open up therapeutic avenues and shed light on this yet unexplored field.

Since connexins form part of the immunological synapse, how do connexins interact with membrane receptors and signaling proteins and how do connexins influence synapse regulation? The evidence so far points to a closer link with the adhesion components of the synapse, and therefore, further elucidation of the macromolecular structure of the overall synapse is needed.

Is there a partial or absolute contribution of connexins to the activation, differentiation, and maturation of cells during the development of immune responses? Most of the knowledge produced so far represents data of snapshots taken from *in vitro* or *ex vivo* experiments. The use of *in vivo* models and/or, for example, three-dimensional cultures may also provide us with valuable information on the temporal changes occurring during activation, and maturation of immune responses.

What roles do hemichannels play in the development of the immune response? Evidence is already in hand for their role in CD4 T cell activation (Oviedo-Orta et al., 2010). Hemichannels are constructed of connexins or pannexins (see this volume's Prequel) and important differences occur between direct communication across gap junctions and purinergic signaling involving ATP release.

Finally, studies designed to *assess the role of connexins and gap junctions in immune regulation, namely regulatory T cell responses*, are crucial to know if they play a role in maintaining homeostasis and in pathological inflammatory or tumoral immune responses and many of these are considered in other chapters.

REFERENCES

Alves, L. A., A. C. Campos de Carvalho, E. O. Cirne Lima, C. M. Rocha e Souza, M. Dardenne, D. C. Spray, and W. Savino. 1995. Functional gap junctions in thymic epithelial cells are formed by connexin 43. *Eur J Immunol* 25: 431–437.

Alves, L. A., A. C. de Carvalho, L. Parreira-Martins, M. Dardenne, and W. Savino. 1994. Intrathymic gap junction-mediated communication. *Adv Exp Med Biol* 355: 155–158.

Alves, L. A., R. Coutinho-Silva, P. M. Persechini, D. C. Spray, W. Savino, and A. C. Campos de Carvalho. 1996. Are there functional gap junctions or junctional hemichannels in macrophages? *Blood* 88: 328–334.

Anand, R. J., S. Dai, S. C. Gribar, W. Richardson, J. W. Kohler, R. A. Hoffman, M. F. Branca et al. 2008. A role for connexin43 in macrophage phagocytosis and host survival after bacterial peritoneal infection. *J Immunol* 181: 8534–8543.

Benlalam, H., A. Jalil, M. Hasmim, B. Pang, R. Tamouza, M. Mitterrand, Y. Godet et al. 2009. Gap junction communication between autologous endothelial and tumor cells induce cross-recognition and elimination by specific CTL. *J Immunol* 182: 2654–2664.

Bermudez-Fajardo, A., M. Yliharsila, W. H. Evans, A. C. Newby, and E. Oviedo-Orta. 2007. CD4+ T lymphocyte subsets express connexin 43 and establish gap junction channel communication with macrophages in vitro. *J Leukoc Biol* 82: 608–612.

Cancelas, J. A., W. L. Koevoet, A. E. de Koning, A. E. Mayen, E. J. Rombouts, and R. E. Ploemacher. 2000. Connexin-43 gap junctions are involved in multiconnexin-expressing stromal support of hemopoietic progenitors and stem cells. *Blood* 96: 498–505.

Carolan, E. J., and J. D. Pitts. 1986. Some murine thymic lymphocytes can form gap junctions. *Immunol Lett* 13: 255–260.

Chekeni, F. B., M. R. Elliott, J. K. Sandilos, S. F. Walk, J. M. Kinchen, E. R. Lazarowski, A. J. Armstrong et al. 2010. Pannexin 1 channels mediate "find-me" signal release and membrane permeability during apoptosis. *Nature* 467: 863–867.

Desplantez, T., V. Verma, L. Leybaert, W. H. Evans, and R. Weingart. 2012. Gap26, a connexin mimetic peptide, inhibits currents carried by connexin43 hemichannels and gap junction channels. *Pharmacol Res* 65: 546–552.

Elgueta, R., J. A. Tobar, K. F. Shoji, J. De Calisto, A. M. Kalergis, M. R. Bono, M. Rosemblatt, and J. C. Saez. 2009. Gap junctions at the dendritic cell-T cell interface are key elements for antigen-dependent T cell activation. *J Immunol* 183: 277–284.

Evans, W. H., E. De Vuyst, and L. Leybaert. 2006. The gap junction cellular internet: Connexin hemichannels enter the signalling limelight. *Biochem J* 397: 1–14.

Fuchs, H., and C. Bachran. 2009. Targeted tumor therapies at a glance. *Curr Drug Targets* 10: 89–93.

Gaziri, I. F., G. M. Oliveira-Castro, R. D. Machado, and M. A. Barcinski. 1975. Structure and permeability of junctions in phytohemagglutinin stimulated human lymphocytes. *Experientia* 31: 172–174.

Hurtado, S. P., A. Balduino, E. C. Bodi, M. C. El-Cheikh, A. C. Campos de Carvalho, and R. Borojevic. 2004. Connexin expression and gap-junction-mediated cell interactions in an *in vitro* model of haemopoietic stroma. *Cell Tissue Res* 316: 65–76.

Hülser, D. F., and J. H. Peters. 1972. Contact cooperation in stimulated lymphocytes. II. Electrophysiological investigations on intercellular communication. *Exp Cell Res* 74: 319–326.

Jara, P. I., M. P. Boric, and J. C. Saez. 1995. Leukocytes express connexin 43 after activation with lipopolysaccharide and appear to form gap junctions with endothelial cells after ischemia-reperfusion. *Proc Natl Acad Sci USA* 92: 7011–7015.

Kanneganti, T. D., M. Lamkanfi, Y. G. Kim, G. Chen, J. H. Park, L. Franchi, P. Vandenabeele, and G. Nunez. 2007. Pannexin-1-mediated recognition of bacterial molecules activates the cryopyrin inflammasome independent of Toll-like receptor signaling. *Immunity* 26: 433–443.

Kapsenberg, M. L., and W. Leene. 1979. Formation of B type gap junctions between PHA-stimulated rabbit lymphocytes. *Exp Cell Res* 120: 211–222.

Krenacs, T., and M. Rosendaal. 1995. Immunohistological detection of gap junctions in human lymphoid tissue: Connexin43 in follicular dendritic and lymphoendothelial cells. *J Histochem Cytochem* 43: 1125–1137.

Krenacs, T., and M. Rosendaal. 1998a. Gap-junction communication pathways in germinal center reactions. *Dev Immunol* 6: 111–118.

Krenacs, T., and M. Rosendaal. 1998b. Connexin43 gap junctions in normal, regenerating, and cultured mouse bone marrow and in human leukemias: Their possible involvement in blood formation. *Am J Pathol* 152: 993–1004.

Krenacs, T., M. van Dartel, E. Lindhout, and M. Rosendaal. 1997. Direct cell/cell communication in the lymphoid germinal center: Connexin43 gap junctions functionally couple follicular dendritic cells to each other and to B lymphocytes. *Eur J Immunol* 27: 1489–1497.

Levy, J. A., R. M. Weiss, E. R. Dirksen, and M. R. Rosen. 1976. Possible communication between murine macrophages oriented in linear chains in tissue culture. *Exp Cell Res* 103: 375–385.

Machtaler, S., M. Dang-Lawson, K. Choi, C. Jang, C. C. Naus, and L. Matsuuchi. 2011. The gap junction protein Cx43 regulates B-lymphocyte spreading and adhesion. *J Cell Sci* 124: 2611–2621.

Mendoza-Naranjo, A., G. Bouma, C. Pereda, M. Ramirez, K. F. Webb, A. Tittarelli, M. N. Lopez et al. 2011. Functional gap junctions accumulate at the immunological synapse and contribute to T cell activation. *J Immunol* 187: 3121–3132.

Neijssen, J., C. Herberts, J. W. Drijfhout, E. Reits, L. Janssen, and J. Neefjes. 2005. Cross-presentation by intercellular peptide transfer through gap junctions. *Nature* 434: 83–88.

Nguyen, T. D., and S. M. Taffet. 2009. A model system to study Connexin 43 in the immune system. *Mol Immunol* 46: 2938–2946.

Nihei, O. K., P. C. Fonseca, N. M. Rubim, A. G. Bonavita, J. S. Lyra, S. Neves-dos-Santos, A. C. de Carvalho, D. C. Spray, W. Savino, and L. A. Alves. 2010. Modulatory effects of cAMP and PKC activation on gap junctional intercellular communication among thymic epithelial cells. *BMC Cell Biol* 11: 3.

Oliveira-Castro, G. M., M. A. Barcinski, and S. Cukierman. 1973. Intercellular communication in stimulated human lymphocytes. *J Immunol* 111: 1616–1619.

Oviedo-Orta, E., M. Perreau, W. H. Evans, and I. Potolicchio. 2010. Control of the proliferation of activated CD4+ T cells by connexins. *J Leukoc Biol* 88: 79–86.

Oviedo-Orta, E., P. Gasque, and W. H. Evans. 2001. Immunoglobulin and cytokine expression in mixed lymphocyte cultures is reduced by disruption of gap junction intercellular communication. *Faseb J* 15: 768–774.

Oviedo-Orta, E., T. Hoy, and W. H. Evans. 2000. Intercellular communication in the immune system: Differential expression of connexin40 and 43, and perturbation of gap junction channel functions in peripheral blood and tonsil human lymphocyte subpopulations. *Immunology* 99: 578–590.

Paraguassu-Braga, F. H., R. Borojevic, L. F. Bouzas, M. A. Barcinski, and A. Bonomo. 2003. Bone marrow stroma inhibits proliferation and apoptosis in leukemic cells through gap junction-mediated cell communication. *Cell Death Differ* 10: 1101–1108.

Penuela, S., R. Gehi, and D. W. Laird. 2012. The biochemistry and function of pannexin channels. *Biochim Biophys Acta*, doi:10.1016/j.bbamem.2012.01.017.

Porvaznik, M., and T. J. MacVittie. 1979. Detection of gap junctions between the progeny of a canine macrophage colony-forming cell in vitro. *J Cell Biol* 82: 555–564.

Qu, Y., S. Misaghi, K. Newton, L. L. Gilmour, S. Louie, J. E. Cupp, G. R. Dubyak, D. Hackos, and V. M. Dixit. 2011. Pannexin-1 is required for ATP release during apoptosis but not for inflammasome activation. *J Immunol* 186: 6553–6561.

Savino, W., E. Arzt, and M. Dardenne. 1999. Immunoneuroendocrine connectivity: The paradigm of the thymus-hypothalamus/pituitary axis. *Neuroimmunomodulation* 6: 126–136.

Schenk, U., A. M. Westendorf, E. Radaelli, A. Casati, M. Ferro, M. Fumagalli, C. Verderio, J. Buer, E. Scanziani, and F. Grassi. 2008. Purinergic control of T cell activation by ATP released through pannexin-1 hemichannels. *Sci Signal* 1: 1–12.

Seror, C., M. T. Melki, F. Subra, S. Q. Raza, M. Bras, H. Saidi, R. Nardacci et al. 2011. Extracellular ATP acts on P2Y2 purinergic receptors to facilitate HIV-1 infection. *J Exp Med* 208: 1823–1834.

Terman, D. S., G. Bohach, F. Vandenesch, J. Etienne, G. Lina, and S. A. Sahn. 2006. Staphylococcal superantigens of the enterotoxin gene cluster (egc) for treatment of stage IIIb non-small cell lung cancer with pleural effusion. *Clin Chest Med* 27: 321–334.

Umezawa, A., and J. Hata. 1992. Expression of gap-junctional protein (connexin 43 or alpha 1 gap junction) is down-regulated at the transcriptional level during adipocyte differentiation of H-1/A marrow stromal cells. *Cell Struct Funct* 17: 177–184.

Watanabe, Y. 1985. Fine structure of bone marrow stroma. *Nihon Ketsueki Gakkai Zasshi* 48: 1688–1700.

2 Gap Junctions and Connexins in the Hematopoietic-Immune System
Structural Considerations

Tibor Krenacs, Ivett Zsakovics, Gergo Kiszner, and Martin Rosendaal

CONTENTS

2.1 INTRODUCTION

Maturation, functional activation, and migration of cells of the hematopoietic-immune system are based on interactions between the mobile immune cells and a dynamic framework formed by stromal, endothelial, and dendritic cells (DCs) in the central and peripheral immune tissues (Delves and Roitt, 2000). Their interactions involving soluble growth factors and adhesion molecules act locally either on adjacent cells or on those at short distance, which do not explain alone the high degree of structural and functional order, observed in hematopoiesis and the immune response in these tissues. Gap junction direct cell–cell communication channels can couple large compartments of the scaffolding cells into networks for coordinating their functions, while also mediating mutual interactions with hematopoietic/immune cells (Rosendaal et al., 1991; Krenacs et al., 1997). Functioning gap junctions (and partly hemichannels too) have been established and their related connexins (Cxs) detected at the mRNA, protein, and ultrastructural levels in the hematopoietic-immune system, including the bone marrow, thymus, spleen, lymph nodes, and mucosa-associated lymphoid tissues (MALT) (Krenacs and Rosendaal, 1995; Rosendaal, 1995; Oviedo-Orta and Evans, 2004; Neijssen et al., 2007). By now, all kinds of hematopoietic/lymphoid cells of the innate and adaptive immune system are shown to form Cx channels and functioning gap junctions, except mature red blood cells. Furthermore, connexin expression and function are modulated during development and regeneration, or by pathological conditions of the hematopoietic-immune system (Saez et al., 2003). For instance, they can be upregulated during posttherapy regeneration and downregulated in malignant tumors. These results and Cx gene perturbation experiments underline the significance of Cx channels in the regulation of hematopoiesis and immune functions.

Immune and accessory cells involved in gap junction interactions are localized to selective compartments in central and peripheral lymphoid tissues where dedicated microenvironment supports the finest host defense response. This chapter summarizes the distribution of connexins in the bone marrow, thymus, and peripheral lymphoid organs and briefly highlights the related potential of homo- and heterocellular gap junction interactions in immune functions. Pathological dysregulation of connexin expression in malignant neoplasms of the bone marrow (leukemias) and of the lymphoid germinal center, that is, follicular lymphomas and follicular dendritic cell (FDC) sarcomas, are also briefly addressed.

2.2 RELEVANCE OF STRUCTURAL CORRELATIONS

Both major lines of approaches, that is, structural or functional techniques have got their advantages and limitations in proving relatively rare gap junction structures or coupling events in highly motile immune cells. It is clear by now that functional activation-related upregulation of gap junctions has resulted in the breakthrough in their convincing detection (Rosendaal, 1995; Krenacs et al., 1997; Krenacs and Rosendaal, 1998a; Ovideo-Orta and Evans, 2004; Neijssen et al., 2007; Scheckenbach et al., 2011). Nevertheless, functional results in simplified culture models where cells are broken away from their microenvironment ignore tissue complexity. However, equally limited conclusions can be drawn from structural studies, which do not inform on the mechanisms of action of connexins. As a result, careful analysis and correlation of results gained with both approaches must be considered for reaching the most realistic conclusions for *in vivo* situations, where redundant regulatory mechanisms may further complicate clear views (Laird, 2006).

Their submicroscopic size and relative rarity caused major difficulties in detecting gap junctions in lymphoid tissues and cells (Rosendaal et al., 1994). Ultrastructural methods, including transmission and freeze fracture electron microscopy, were used initially. Techniques of molecular morphology rendering molecular information to traditional histology then prevailed, including immunohistochemistry for *in situ* protein detection and *in situ* hybridization for localizing gene transcripts. However, the limited resolution of light microscopy prevented the detection of submicroscopic signals, particularly where they were rare. The major breakthrough came from the exploitation of immunofluorescence microscopy, particularly when combining with confocal laser scanning. This approach allowed the precise recognition of small fluorescing Cx signals against the dark background throughout the whole sample thickness by multilayer scanning (Z-stacking) and also to reveal the relations of 2–4 target proteins in the same sample by using double-multiple labeling (Krenacs et al., 2010b). However, fluorescent samples may fade quickly, which limits the potential of static field-by-field scanning for large samples or for later reevaluation. Digital microscopy creates permanent archives of microscopic slides that can be analyzed with computer software at any time later (Krenacs et al., 2010a). The Pannoramic Scan instrument (3DHITECH Ltd., Budapest) we use in multichannel, multilayer fluorescent mode allows Cx plaques along with additional phenotypic or functional proteins to be reliably detected within whole slides for convenient examination without time limitations, including automated image/signal analysis. This approach, which is demonstrated in this chapter, allows the systemic screening of connexins in a wide range of normal and diseased tissues. Immunomorphological correlations of connexins and gap junctions in the hematopoietic-immune system offer essential clues for the critical translation of *in vitro* and *in vivo* functional and molecular-genetic results into wider contexts on their roles in tissues and organs.

2.3 CONNEXINS AND GAP JUNCTIONS IN BONE MARROW

Immune cells of either the innate or the adaptive immune system stem from the bone marrow, where pluripotent precursors are produced, get committed into lineage-restricted leukocyte subtypes and mature for effector functions either there or in the thymus (Delves and Roitt, 2000). Blood cell formation and maturation take place within the microenvironment of the intersinusoidal spaces made up by a three-dimensional meshwork of irradiation-resistant stromal cells and endothelial cells (Metcalf, 1989). This stromal scaffolding responds to soluble growth factors, sets up adhesion interactions with migrating blood cells, and mediates direct cell–cell communication through gap junctions (Rosendaal et al., 1991). Accumulating data of genetic and functional perturbations of gap junction coupling and connexin expression suggest the importance of gap junction communication in the regulation of hematopoiesis (Montecino-Rodriguez and Dorshkind, 2001). Further support for this came from the highly upregulated connexin expression in conditions related to enhanced hematopoiesis in the neonatal marrow and post-chemo- or radiotherapy regeneration (Rosendaal et al., 1994; Krenacs and Rosendaal, 1998a).

2.3.1 Structural Evidence of Bone Marrow Gap Junctions

Although electrophysiological signs of gap junctions were noted 50 years ago (Hulser and Peters, 1971, 1972) and ultrastructural hallmarks were also reported soon later in homotypic hematolymphoid cell cultures (Levy et al., 1976; Porvaznik and MacVittie, 1979; Kapsenberg and Leene, 1979; Neumark and Huynh, 1989), the confirmation of gap junctions in the bone marrow tissue was ambiguous for a long time with the available ultrastructural techniques (see Rosendaal et al., 1994). The submicroscopic size of gap junctions, their relative rarity in resting marrow, and the special orientation of cell membranes required for ultrastructural identification may explain these difficulties. In addition, the pentalaminar ultramicroscopic structures resembling gap junctions in mice and chick marrow when treated with tannic or gallic acid could not be validated for some time (Campbell, 1980, 1986; Watanabe, 1985). Nevertheless, ultrastructural confirmation came from bone marrow of upregulated gap junctions, that is, in genetically anemic mice where they were found between reticular cells and between reticular and adventitial cells (Yamazaki, 1988), and in regenerating marrow after cytotoxic treatment showing all possible homo- and heterocellular interactions of stromal and hematopoietic cells (Rosendaal et al., 1994).

The functional evidence of direct cell–cell communication through gap junctions was established between bone marrow stromal cells (Umezawa et al., 1990) and between them and hematopoietic cells in mice long-term cultures using Lucifer yellow dye transfer (Rosendaal et al., 1991; Krenacs and Rosendaal, 1998a; Rosendaal and Krenacs, 2000). Electrophysiological coupling between stromal cells was also confirmed (Dorshkind et al., 1993). Cx43 was then identified as the principal Cx isotype in marrow stromal cells and their adjacent osteoblasts and adipocytes (Figure 2.1a and b) (Umezawa and Hata, 1992; Dorshkind et al., 1993; Rosendaal et al., 1994). It also turned out that osteoblasts also form a functional unit through Cx43 coupling with their other progeny bone-forming osteocytes (Civitelli, 2008). These findings can raise the possibility that the blood-forming and bone-forming cell networks are functionally linked through gap junctions. Cx43 has still remained the major Cx protein in the bone marrow despite one group detecting nine different connexins at the mRNA level in S-17 murine bone marrow stromal cells (Hurtado et al., 2004). However, in mouse marrow and stromal cell lines, only Cx45 and Cx31 mRNA could be convincingly found besides that of Cx43 (Cancelas et al., 2000).

2.3.2 Gap Junctions in Normal, Regenerative, and Leukemic Hematopoiesis

The amount of Cx43 gap junctions in resting bone marrow and the degree of communication in its fresh long-term cultures was low (Rosendaal et al., 1994; Rosendaal, 1995; Krenacs and Rosendaal, 1998a; Montecino-Rodriguez and Dorshkind, 2001). However, coupling in this *ex vivo* bone marrow increased substantially 2 h after harvesting in culture, which suggested that connexins and their gap junction channels are regulated by soluble factors (Rosendaal, 1995). Indeed, the amount of bone-marrow-related gap junctions and direct cell–cell communication could

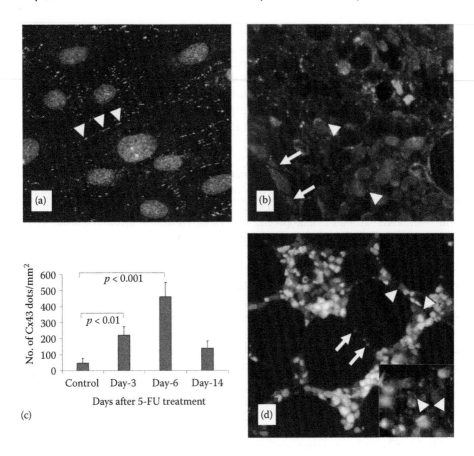

FIGURE 2.1 Cx43 gap junctions in the cell membranes (arrowheads) of mouse FBMD1 bone marrow stromal cells (a), and in osteoblasts (arrows), and between hematopoietic cells (arrowheads) in newborn mouse marrow (b). Cx43 is upregulated significantly ($p < 0.001$–0.01) in mouse marrow regenerating after cytoablative treatment with 5-FU (c). In resting human adult, bone marrow Cx43 is consistently seen on adipocyte membranes (arrows) but only few in between hematopietic cells (d, inset and arrowheads). Confocal laser scanning (a and b) and digital microscopy (d).

be modulated by a number of physiological, pathological, and pharmacological stimuli. For instance, the differentiation-related retinol treatment upregulated while hydrocortisone and IL-1 down-regulated gap junction expression and coupling in murine bone marrow stromal cells (Dorshkind et al., 1993; Hurtado et al., 2004). Also, significantly elevated numbers of gap junctions were noted in genetically anemic Sl/Sld mice compared to their wild-type littermates (Yamazaki, 1988). Human bone marrow stromal cells of acute lymphoid or myeloid leukemic marrow (ALL or AML) showed downregulated Cx43 and gap junction coupling to be normalized in patients' samples showing complete remission upon successful chemotherapy (Liu et al., 2010).

The testing of Cx43 expression in neonatal, adult, and regenerating mice marrow and in normal and leukemic human bone marrow also showed that gap junctions may not play an important role in resting adult hematopoiesis (Rosendaal et al., 1994; Krenacs and Rosendaal, 1998a). However, based on their high upregulation, gap junctions should be important where there is substantial need for progenitor cell proliferation, that is, in developmentally active neonatal marrow in epiphyseal marrow of developing bones and in regeneration following cytoablation with 5-fluorouracil (5-FU) (Figure 2.1b and c). Forced stem cell division resulted in the elevation of gap junctions by two orders of magnitude in bone marrow stroma and osteoblastic cells at very early stage of regeneration before a recognizable number of blood cells formed compared to that of resting untreated marrow (Rosendaal et al., 1994). Gap junctions were also elevated in a few leukemic bone marrows with increased stromal/hematopoietic cell ratio, including myelodysplastic syndrome, and acute lymphoblastic leukemia (ALL) and acute myeloblastic leukemia (AML) (Krenacs and Rosendaal, 1998a).

The considerable irradiation resistance of gap junctions and their cell coupling in marrow stromal cells may contribute to the postirradiation recovery of regenerating bone marrow (Umezawa et al., 1990; Yamazaki and Allen, 1991). In line with this, the preservation of Cx43 plaques on adipocyte membranes in human and mouse bone marrow (Figure 2.1d) (Krenacs and Rosendaal, 1998a) may also assist in regaining hematopoietic phenotype and functions following trauma with massive blood loss, although dye coupling vanished during irradiation-provoked stromal-adipocytic cell transition *in vitro* (Umezawa et al., 1987).

Besides its finding in bone marrow stromal cells, Cx43 gap junction protein was also identified on CD34+ hematopoietic progenitors (Rosendaal et al., 1994) and in Mac1+ myeloid but not in lymphoid (B220+) progenitors (Hurtado et al., 2004). Long-term bone marrow culture experiments on pharmacological blocking of Cx43 gap junctions in stromal cells raised the idea that gap junction communication may regulate the proliferation and maintenance of hematopoietic stem cells (HSC) (Rosendaal et al., 1997; Ploemacher et al., 2000). This was in line with the decreased support of late-appearing clones, which originate from less differentiated precursors, in cobblestone forming area assay (CAFC week 3–5) by genetically Cx43-deficient, thus communication-deprived, stromal cells (Cancelas et al., 2000). Similarly, genetic upregulation of Cx43 gap junctions elevated the undifferentiated myeloid cell fraction, while downregulation of Cx43 supported myeloid cell differentiation in S-17 murine stromal cells-based long-term cultures (Bodi et al., 2004).

2.3.3 HEMATOPOIETIC DEFECTS UPON CONNEXIN GENE ABLATION

Genetically connexin43-deficient homozygous (Cx43$^{-/-}$) and heterozygous (Cx43$^{+/-}$) mice also highlighted the importance of gap junction Cx-s in the generation of hematolymphoid cells. Since Cx43$^{-/-}$ mice die perinatally, hematopoietic organs in the embryo such as liver, spleen, and thymus were studied. The results showed reduced numbers of granulocyte-macrophage colony-forming units (GM-CFUs) and erythroid bursts unit (BFU-Es) in fetal liver from 2-week-old Cx43$^{-/-}$ mice embryos compared to age-matched wild-type controls (Cancelas et al., 2000). Targeted ablation of Cx43 gene also resulted in lower numbers of CD4+ and T cell receptor-expressing thymocytes and surface IgM+ B lymphocytes both in Cx43$^{-/-}$ and in Cx43$^{+/-}$ newborn mice compared to their Cx43$^{+/+}$ littermates (Montecino-Rodriguez et al., 2000). Marrow stromal cell coupling was also impaired. Heterozygous (Cx43$^{+/-}$) stromal cells showed intermediate levels of dye coupling between those of the homozygous knockout (low) and wild-type (sizable) animals (Reaume et al., 1995). However, adult heterozygous animals did not develop hematopoietic defects, which also support the primary importance of gap junction communication in early hematopoiesis and regeneration. In line with this, Cx43$^{+/-}$ mice produced only 20% of blood cells 9 days after 5-FU treatment of that of wild-type animals (Montecino-Rodriguez and Dorshkind, 2000).

Though Cx32 could not been detected in the bone marrow by most groups, Cx32 knockout mice showed impaired hematopoiesis (Hirabayashi et al., 2007). C-Kit positive lineage negative murine HSC were shown to express Cx32 mRNA and protein, although the latter result was not convincing. Nevertheless, Cx32 KO mice exhibited delayed regeneration following chemical cell abrasion with 5-FU and elevated risk for leukemic transformation when treated with the genotoxic methyl nitrosourea (MNU). These findings suggest that Cx32 plays a role in protecting HSCs from chemical abrasion and from leukemogenesis, while supporting regeneration. From the other end, normal bone marrow stroma could inhibit the proliferation of leukemic cells, which, however, increases the resistance of tumor cells to cytotoxic treatment (Paraguassu-Braga et al., 2003).

In conclusion, gap junctions are possibly involved in the early phase of blood formation when forced generation of hematopoietic progenitors is at high demand, for example, in neonatal marrow, in developing epiphyseal marrow, and in regenerating marrow. However, wild-type bone marrow stromal cells can also use alternative regulatory pathways to support hematopoiesis. This was reflected by the elevated blood cell formation from Cx43 heterozygous progenitors compared to wild-type progenitors (Rosendaal and Stone, 2003), and the similar blood-forming capacity of Cx43 wild-type and knockout fetal liver hematopoietic cells on wild-type stroma (Rosendaal and Jopling, 2003).

2.4   GAP JUNCTIONS IN THYMUS

Thymus is the primary lymphoid organ where CD4+/CD8+ double-positive precursor T lymphocytes mature into antigen-specific CD4+ or CD8+ single-positive T cells. Besides T lymphocyte populations, the thymus microenvironment is made up of DCs of bone marrow monocytic origin, mesenchymal fibroblasts, and sinus endothelial cells, and a meshwork of thymic epithelial cells (TECs) (Boyd et al., 1993). Negative selection of T cells reactive for self-antigens and the commitment toward CD4+ or CD8+ mature but naïve T cell fate is based on crucial interactions between thymocytes and TECs. TECs (possibly stromal and endothelial cells too) express Cx43 (Figure 2.2) but neither Cx26 nor Cx32 was detected in the thymus (Alves et al., 1995). TEC showed both homocellular communication and occasional heterocellular coupling with thymocytes. Later, Cx43 and

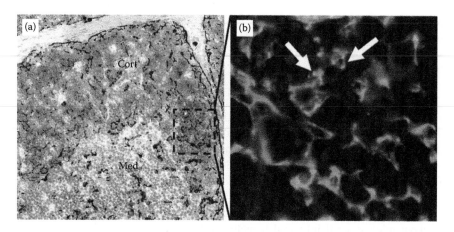

FIGURE 2.2 **(See color insert.)** Cytokeratin-positive TEC processes (a—purple; b—green) between thymocytes in the thymus cortex (Cort) and less in the medulla (Med) (a) are associated with Cx43 gap junctions (b—red; colocalization with green results in yellow color—arrows). Immunoperoxidase and immunofluorescence digital microscopy.

Cx30.3 transcripts and proteins were confirmed in thymocytes but no homocellular coupling between them was observed (Fonseca et al., 2004). Similar to that in bone marrow, thymic recovery in Cx43$^{+/-}$ heterozygous mice after cytoablative treatment with 5-FU was only 30% of that of the Cx43 wild-type animals, which clearly shows the involvement of gap junctions in thymic functions (Montecino-Rodriguez and Dorshkind, 2001).

Hormone secretion and gap junctions in TEC were shown to be mutually regulated. The first proof of this was the reversible inhibition of thymulin production in TECs with the gap junction blocker heptanol (Alves et al., 1995). Then, thymulin secretagogues such as IL-α and IL-1β, progesterone, estrogen, testosterone, adrenocorticotropic hormone, and growth hormone were described to inhibit TEC coupling as opposed to cell proliferation (thymidine incorporation), which was increased for these agents (Head et al., 1997; Alves et al., 2000). On the contrary, locally produced neuropeptides that can influence vascular resistance through smooth muscle such as vasopressin (that may elevate vascular resistance) or vasoactive intestinal polypeptide (VIP; that is known to cause coronary vasodilatation) could upregulate TEC coupling, possibly through elevated c-AMP levels (Alves et al., 2000). TEC functions could be inversely modulated through activating c-AMP, for example, by epinephrine for upregulation of Cx43 protein, mRNA, and cell coupling, or through PKC (protein kinase-C) activation by phorbol ester for downregulating the same features (Nihei et al., 2010). During age-related thymic atrophy, thymocyte output of the thymus is markedly decreased due to the decline of IL-7 by TECs but without significant change in the number of gap junctions (Andrew and Aspinall, 2002). Gap junctions in the thymus, therefore, are likely to be involved in the regulation of TEC functions and the concomitant T cell maturation/selection, which may also be influenced directly through occasional TEC–thymocyte coupling.

2.5   GAP JUNCTIONS SUPPORT INNATE AND ADAPTIVE CELLULAR IMMUNITY ALL OVER THE BODY

Cells of the innate immunity, including neutrophil leukocytes, monocytes/macrophages, and NK cells, can act in any injured or infected tissue and may use gap junctions, which is reviewed both in Chapters 8, 11, and 12 of this book and earlier by Brenda Kwak's group (Chanson et al., 2005; Scheckenbach et al., 2011). For instance, human monocytes and macrophages elevated their Cx43 expression, dye coupling, and transmigration through blood–brain barrier upon functional activation with the combination of lipopolysaccharide (LPS) with IFN-γ or TNF-α with IFN-γ (Eugenin et al., 2003). In atherosclerosis, which represents a local progressive inflammatory disease, Cx43 can be upregulated in activated monocytes/macrophages and Cx37 in atherosclerotic plaques, while Cx43 can also be associated with activated neutrophil leukocytes (Polacek et al., 1993; Morel et al., 2009).

While the effector and memory cells of the humoral immune response are produced in lymphoid organs, cytotoxic T cell activation with viral and tumor antigens can happen anywhere in the body through MHCI molecules by any nucleated cell. The cellular arm of the adaptive immune system mediated by cytotoxic T cells also utilizes gap junction communication as it is reviewed in this book and earlier by Neijssen and coworkers (2007). Since gap junctions can be established between most cell types, including immune cells, cross-presentation of peptide antigens containing 8–10 amino acids has been demonstrated through direct cell coupling (Neijssen et al., 2005). This can secure the elimination of virally infected cells with their adjacent neighbors by cytotoxic T cells.

The benefits of Cx43 gap junction induction in immune cells have been recently related to important tumor-controlling functions. Elevated Cx43 expression and coupling upon priming DCs with melanoma cell lysate in combination with TNF-α showed the facilitation of the cross-presentation of melanoma antigen, which boosted antitumor T cell response (Mendoza-Naranjo et al., 2007). Also, bacterial (*Salmonella*) infection was shown to induce Cx43 expression and cell coupling that are usually lost in murine melanoma cells (Saccheri et al., 2010). Preprocessed tumor antigens were then efficiently cross-presented between DC through gap junctions to activate cytotoxic T cells and control both the infected and adjacent uninfected melanoma cells. These approaches can be exploited in cancer vaccination, for example, in melanoma immunotherapy.

The expression of Cx43 mRNA and protein was recently demonstrated in CD4+ lymphocyte subpopulations (Th0, Th1, and Th2), which could couple to primary macrophages *in vitro*, favoring Th1 cell–macrophage interactions (Bermudez-Fajardo et al., 2007). Also, FoxP3/CD4+ regulatory T cells (T-reg) were also shown that can mediate immune tolerance through Cx43 gap junctions either in conventional CD4+ T helper cells or in antigen-presenting DCs (Bopp et al., 2007). Intimate interactions and gap junction coupling were demonstrated in these cell relations with two-photon laser scanning microscopy showing the intercellular transfer of c-AMP, which suppressed the major Th1 cytokine IL-2 in CD4+ T cells or down-regulated costimulatory molecules and promoted IL-10 (supporting Th2 response) production in DCs. This T-reg control of DC activation through gap junctions was used to inhibit cytotoxic T cell response and edema in experimental contact hypersensitivity (Ring et al., 2010). These data show that the cellular immune response can also be regulated at diverse levels through gap junction communication.

2.6   CONNEXINS AND GAP JUNCTIONS IN PERIPHERAL IMMUNE TISSUES

Peripheral lymphoid organs, that is, the spleen, tonsils, lymph nodes, and MALT, are the catalytic sites of the humoral immune response, which is strictly organized both structurally and functionally (Delves and Roitt, 2000). Antigen-specific B and T lymphocytes representing the two interrelated arms of the adaptive immunity reside here at well-defined regions (Figure 2.3a). Lymphoid tissues act as special filters that trap external antigens that recirculating immune cells continuously survey for neutralization and elimination (Katakai et al., 2004a). Lymphoid tissues offer the finest microenvironment formed by meshworks of fibroblastic reticular cells (FRC), vascular- and lymph-endothelia, and FDC as well as antigen-presenting DC for the most efficient activation and expansion of lymphocytes (Fossum and Ford, 1985). Germinal centers of secondary lymphoid follicles play essential roles in controlling B cell proliferation, selection, and isotype switching, resulting in high-affinity effector and memory B cells (Gatto and Brink, 2010). Random interactions though adhesion molecules and soluble factors (Delves and Roitt, 2000) do not explain alone the high structural order and regulation throughout the humoral immune response even when considering the well-arranged conduits of FRC and collagen matrix (Katakai et al., 2004b). Gap junctions can form in all kinds of accessory and immune cells and through homocellular and heterocellular interactions, they are involved in the mediation and coordination of immune functions (Krenacs and Rosendaal, 1998b; Ovideo-Orta and Evans, 2004; Neijssen et al., 2007).

A series of *in situ* evidence on lymphoid gap junctions has been collected in tissue sections using electron microscopy, immunofluorescence, and *in situ* hybridization. However, T cell-dependent humoral immune response requires the germinal center microenvironment, which is equally difficult to study *in vivo* and to model *in vitro*. Therefore, functional *in vitro* data on gap junctions mainly relate to mitogenic (phytohemagglutinin (PHA)) or LPS-activated B cells or mixed T–B cell populations that largely represent T cell-independent response without the major germinal center functions related to affinity selection/maturation and isotype switching (Gatto and Brink, 2010).

2.6.1   STRUCTURAL EVIDENCE OF GAP JUNCTIONS IN HUMORAL IMMUNE RESPONSE

The progressive upregulation of Cx43 gap junctions in regional lymph nodes of mice upon antigen challenge provided *in vivo* evidence on the potential contribution of gap junctions to the humoral immune response (Krenacs et al., 1997). Repeated immunization with lysozyme resulted in fully formed secondary follicles around FDC meshworks densely decorated with Cx43. Gap junctions were also found all over human secondary lymphoid tissues. In all these sites, high density of Cx43 was identified around FDC and B cells, and in lymphoendothelial cells, less protein in high endothelial venules (HEV) and FRC as they were phenotyped with double labeling

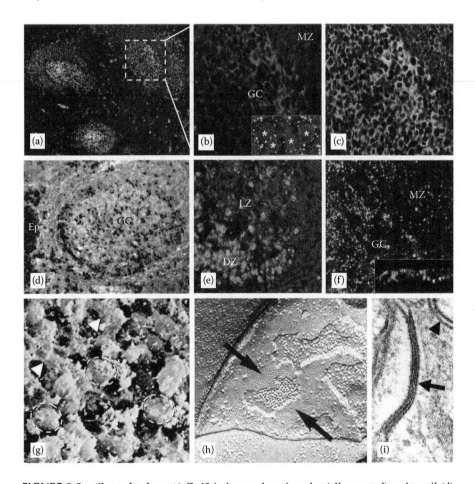

FIGURE 2.3 (**See color insert.**) Cx43 in human lymph nodes (all except d) and tonsil (d). (a) Secondary lymphoid follicles localize most CD20-positive B cells (green) and Cx43 (red-yellow), while the perifollicular T cell region is only sparsely labeled for these antigens. (b) High-power view shows strong and particulate Cx43 labeling around follicular B cells (red) (green) on B cell membranes (inset; green, asterisks) and in the germinal center (GC) and few in the mantle zone (MZ). (c) Cx43 (red) partly colocalizes with CD20 (green) in B cell membranes (yellow). (d) Cx43 transcripts are also found in germinal center cells (GC) and in epithelial (Ep) cells in human tonsil. (e) In the germinal center, the dark zone (DZ) full of proliferating B cells (green) carry much less Cx43 (red) than the light zone (LZ) where B cells mature by interacting with FDC. (f) Colocalization of Cx43 (red) with the FDC-related desmoplakin (green) in the germinal center; inset shows a cross section. (g) Scanning electron microscopy of B lymphocytes (dashed circles) enveloped by FDC (arrowheads) in the germinal center. (h) Freeze fracture hallmarks ordered particles and pits (between arrows) in the FDC cell membrane. (i) Pentalaminar ultramicroscopic structure (middle) consistent with a gap junction and the normal membrane distance (arrowhead) between FDC processes. Multifluorescent digital microscopy (a–c and e–f) and confocal laser scanning (b inset and f).

(Figure 2.3a–c) (Krenacs and Rosendaal, 1995). The intensity and density of Cx43 signal in sinus lining cells adjacent to Hodgkin's lymphoma and Langerhans' cell histiocytosis was proportional to the tumor cell infiltration, suggesting an activation-related role for gap junctions in endothelial lymphoid cell interactions, which was proved later (Ovideo-Orta et al., 2002). Transcripts of the Cx43 gene were revealed in similar tissue associations as the protein (Figure 2.3d). In the germinal center, the distribution of Cx43 protein correlated well with that of FDC by decorating mainly the light zone and much less the highly proliferating dark zone (Figure 2.3e). Accordingly, most Cx43 colocalized with desmosomal proteins (Figure 2.3f) (desmo-plakin and desmoglein) and complement receptors CD21 and CD35, which are known to serve homophile adhesion and antigen binding, respectively, on FDC (Krenacs and Rosendaal, 1995). Freeze fracture and ultrastructural hallmarks, plaques of ordered particles and close membrane appositions, respectively, were also confirmed in and between FDC processes with electron microscopy (Figure 2.3g–i). T helper cells, which are important in the maturation and maintenance of secondary lymphoid fol-licles and in the Ig isotype switching in B cells (de Vinuesa et al., 2000), can also form Cx43 gap junctions (Bermudez-Fajardo et al., 2007; Ovideo-Orta et al., 2010), so as FRCs (fibroblastic reticular cells) and DCs (dendritic cells) (Neijssen et al., 2007). Similar distribution of gap junctions was observed in other lymphoid organs. In human spleen, a high amount of Cx43 was seen in B cell follicles, sinus lining lym-phoendothelial cells, stromal cells, and adipocytes, and some were associated with T cells in the periarteriolar lymphoid sheet (PALS) (Figure 2.4a–f). Recently, Cx45, but not Cx43, was detected in mice lymph node DCs; however, muscle injury could induce Cx43 expression in them in regional lymph nodes (Corvalan et al., 2007).

2.6.2 FUNCTIONAL EVIDENCE OF GAP JUNCTIONS IN HUMORAL IMMUNE RESPONSE

Studies in *ex vivo* germinal centers, set up from low-density cell fractions of human tonsil, including activated B and T cells and FDC, showed that Cx43 gap junctions appeared very early to be progressively upregulated in time (Krenacs et al., 1997). In freshly isolated round cells, Cx43 was detected primarily in IgG+ and less in IgM+ and traces in IgD+ B cells (Krenacs and Rosendaal, 1998b). From 2h on CD35+ FDC processes and membranes gradually enveloping CD19+, B cells were deco-rated with Cx43 plaques at B cell contact areas by forming cell clusters resembling germinal centers. Dye microinjection technique revealed that FDC processes within these clusters were functionally coupled with each other and also with CD19+ B cells (Krenacs et al., 1997). Therefore, gap junction coupling was confirmed in FDC–FDC and FDC–B cell interactions.

Later, a series of functional studies proved the humoral immune response-related roles for gap junctions. Cx43 mRNA and protein were detected both in circulating and in tonsil-derived human T and B lymphocytes as well as in circulating NK cells, while Cx40 was found in isolated tonsil lymphocytes (Ovideo-Orta et al., 2000). Stimulation with PHA or LPS increased Cx40 and Cx43 expression, respectively. Specific gap junction inhibitors connexin mimetic peptide and 18α-glycyrrhetinic acid reduced

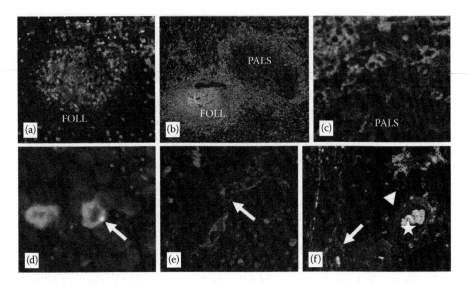

FIGURE 2.4   (See color insert.) (a) In human spleen, most Cx43 (red) is also found in the germinal centers of lymphoid follicles (FOLL) along few T cells (CD3, green). (b) The follicle is full of CD20-positive B cells (green) and Cx43 (red-yellow) while adjacent periarteriolar lymphoid sheet (PALS) include only few B cells and Cx43. (c) Higher-power view of the PALS seen on b. (d) Cx43 protein (red-yellow) in the cell membrane of a CD3-positive (green) T cell (arrow). (e) Dense Cx43 labeling in sinus lining lymphoendothelial cells (arrow). (f) Stromal cells (arrow), adipocytes (arrowhead), and vascular endothelia (asterisk) in the splenic capsule are also strongly Cx43 positive. Multifluorescence digital microscopy.

the secretion of immunoglobulins IgM, IgG, and IgA, both in purified B cells and in mixed lymphocyte cultures (Ovideo-Orta et al., 2001). Inhibition of IL-2 and IL-10 transcripts, which are crucial cytokines of Th1 and Th2 response initiation, was also observed when blocking gap junctions. These findings suggested that gap junction coupling may contribute to direct metabolic cooperation between immune cells.

Bone marrow-derived DCs that catalyze B cell response through priming T helper cells for B cell activation were also shown to require gap junction coupling for their efficient activation (Matsue et al., 2006). LPS plus IFN-γ or TNF-γ plus IFN-γ induced Cx43 mRNA expression, dye-coupling, and upregulation of costimulatory molecules CD80, CD86, and MHCII, which could be inhibited with Cx43 mimetic peptide. In addition, the importance of Cx hemichannel activity was also demonstrated in the expansion of $CD4^+/CD25^-$ T helper cells upon antigen activation (Ovideo-Orta et al., 2010).

Lymphocytes were also shown to communicate through gap junctions with endothelial cells during transendothelial migration both upon entering through HEVs or when leaving the lymphoid organs through medullary sinus endothelial cells (Ovideo-Orta et al., 2002). Furthermore, gap junction expression and coupling could be selectively modulated through histamine receptors (H1 or H2) in human high endothelial cells (HUTEC), which are involved in lymphocyte homing to immune tissues (Figueroa et al., 2004). Activation of H1 or H2 receptors resulting in elevated

Ca^{2+} or c-AMP levels induced the inhibition or support of coupling and membrane bond Cx43 levels, respectively.

2.7 GAP JUNCTIONS IN FOLLICULAR LYMPHOMA AND FDC SARCOMA

Follicular lymphoma is an indolent but incurable B cell lymphoma, which responds to chemotherapy but relapses frequently and can transform to aggressive, diffuse, large B cell lymphoma (Horning and Rosenberg, 1984). Instructive and responsive tumor microenvironment, including FDC, affects follicular lymphoma behavior by direct interactions and growth factors/cytokines (Glas et al., 2007; de Jong et al., 2009). Disrupted FDC pattern revealed with C3d complement receptor (CD21) staining showed correlation with the rapidly transforming disease, while another FDC biomarker, the low-affinity IgE receptor (FcεRII or CD23), did not prove such correlation. This may be explained by the functional heterogeneity and modulation of FDC by the interrelated lymphoma (Glas et al., 2007).

Preliminary data on three cases of follicular hyperplasia and 10 cases of follicular lymphoma showed downregulation of Cx43 gap junctions along the increasing tumor grade by being at the lowest level in grade III lymphoma samples (Figure 2.5a–d). The impairment involved less membrane bond Cx43 along FDC processes

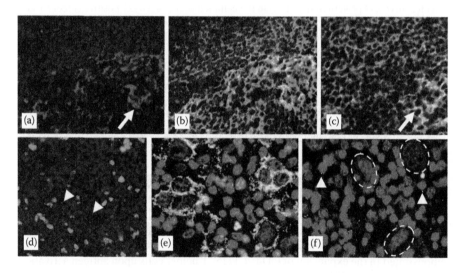

FIGURE 2.5 (**See color insert.**) Cx43 protein in malignant lymphoid disease involving FDC. (a) High density of intracytoplasmic Cx43 (red) in fragmented FDC (arrow) in a grade I follicular lymphoma. (b) Colocalization (yellow) of Cx43 (red) with CD20 (green) in a merged image of (a). (c) Further fragmentation and intracytoplasmic delocalization of Cx43 (red) in FDC in a grade II follicular lymphoma, double stained for B cells (CD20, green). (d) Loss of most Cx43 (red, arrowheads) in a grade III follicular lymphoma of elevated proliferation (Ki67, green). (e–f) CD35-positive (e, green) pleiomorphic FDC sarcoma cells (dashed circle) are fully negative for Cx43 (f, green). Arrowheads show few Cx43 on reactive lymphoid cells. Multifluorescent digital (a–d) and confocal laser scanning (e–f) microscopy.

and cytoplasmic delocalization at still high concentrations of Cx43 in abortive, multinucleated FDC, which then lost most of its processes and gap junctions. The downregulated or missing gap junctions can possibly contribute to FDC destruction and the progression of the follicular lymphoma.

FDC sarcoma (FDCS) results from the malignant transformation of FDC itself, which can recur or form metastases in 20–25% of the cases (Soriano et al., 2007). It can be identified with the traditional FDC markers CD21, CD23, and CD35. In the two low-grade cervical cases, few intracytoplasmic Cx43 particles were detected but Cx43 was completely missing from the high-grade nodal FDCS of poor prognosis, while adjacent processing of reactive stromal cells demonstrated cell membrane positivity as an endogenous control (Figure 2.5e and f). Therefore, similar to other tissues, gap junctions are also impaired in malignant cell transformation involving or affecting FDC in lymphoid tissues.

2.8 STRUCTURAL CORRELATIONS OF GAP JUNCTION FUNCTIONS IN LYMPHOID TISSUES

Cx43 and functional cell coupling has been detected in most resident and mobile cells in the lymphoreticular tissues as it is summarized in Figure 2.6 (Krenacs and Rosendaal, 1995; Oviedo-Orta and Evans, 2004; Neijssen et al., 2007). Therefore, gap junctions may contribute to the coordination of functions within bone marrow and lymphoid tissue compartments through the relevant scaffolding formed by bone marrow stromal cells, FRC, FDC, lymphatic and vascular endothelia, DCs, and TECs (Alves et al., 1995; Rosendaal et al., 1997; Krenacs and Rosendaal, 1998a,b). These accessory cells can also communicate through gap junctions with HSCs in the bone marrow (Rosendaal et al., 1994), recirculating memory B and T lymphocytes in the lymph nodes, and with cells of the innate immunity such as monocytes/macrophages, leukocytes, and NK cells at the periphery (Oviedo-Orta and Evans, 2004; Neijssen et al., 2007; Scheckenbach et al., 2011).

Host defense-related activation is the most common requirement for Cx and gap junction detectability in immune cells due to their upregulated expression and coupling under these conditions. The short half-life of Cx43 between 2 and 4 h fits to the rapid and transient cell interactions and coupling following immunological activation (Evans et al., 2006). Since all immune-related cells utilize Cx43, the potential permutations between cell types to establish gap junctions with each other possibly outnumber those that have been confirmed so far. Also, as multiple rather than single Cx expression is regular in most cell types, more Cx isotypes may be expected in immune tissues.

As immune cells enter or leave lymphoid tissues, they can interact and couple with endothelial cells (Ovideo-Orta et al., 2002). In the thymus, functionally coupled TECs regulate thymocyte maturation (Alves et al., 1995). Both inside and outside peripheral lymphoid tissue gap junctions mediate viral and tumor antigen cross-presentation between nonimmune cells and DCs (Neijssen et al., 2005) and immune tolerance by T-reg cells to DCs (Bopp et al., 2007). Also, blocking of gap junctions reduces immunoglobulin secretion in B cells or IL-2 and IL-10 levels in T cells (Ovideo-Orta et al., 2001). Gap junction coupling is also required between DCs for

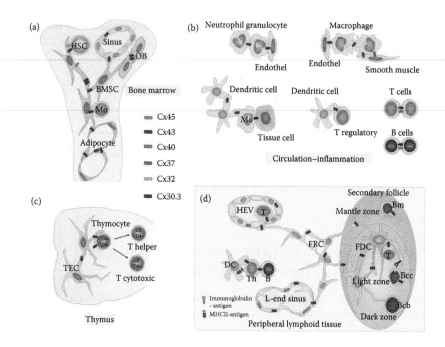

FIGURE 2.6 **(See color insert.)** Summary drawing on published connexin and gap junction interactions in the hemato-lymphoid tissues, and in circulating and local inflammation-related cells as reviewed in this chapter and by Krenacs and Rosendaal (1998b, 2000), Ovideo-Orta et al. (2004), Neijssen et al. (2007), and Scheckenbach et al. (2010). The following direct cell–cell interaction can be potentially mediated through Cx43 gap junctions. (a) In the bone marrow: BMSC (bone marrow stromal cells) with BMSC; with osteoblasts (OB), with sinus endothelial cells, with adipocytes, with HSC (hematopoietic stem cells), and with monocytes (Mo). Cx32, protein, and mRNA (in HSC) and Cx45 mRNA (in BMSC) were also detected. (b) In the spleen: thymus endothelial cells (TEC) with TEC and with thymocytes. Cx30.3 mRNA was also found in thymocytes. (c) In circulating and local immune response-associated cells: granulocyte with granulocyte and with endothelium; macrophage with macrophage, with endothelium and with smooth muscle (Cx37 was also detected in these contexts); antigen-presenting dendritic cell (DC) with monocyte, with tissue cells, and with regulatory T cells. Circulating T cells and B cells (NK cells and mast cells) may also form homocellular and heterocellular coupling. (d) In the peripheral lymphoid organs, DCs with T helper cells and the latter with B cells; Homocellular coupling between lymphoendothelial (L-end) and vascular endothelia (HEV); endothelia with lymphocytes; fibroblastic reticular cells (FRC) with FRC, with endothelial cells, and with follicular dendritic cells (FDC); FDC with FDC, with B cells (centrocytes—Bcc, memory cells—Bm, and less with centroblasts—Bcb) and with follicular T cells (see also Figures 2.1 through 2.5).

their efficient activation and T helper cell priming that catalyze humoral immune response (Matsue et al., 2006). Cx hemichannels may also be important in T helper cell expansion (Ovideo-Orta et al., 2010).

Cx43 gap junctions are highly upregulated in the germinal centers upon immune challenge on FDC meshwork, which forms a functional syncytium and can also couple germinal center B cells (Krenacs et al., 1997). FDC embrace and may also form common Cx channels with follicular T helper cells. However, FDC lose Cx43 gap junction along elevating tumor grades, which is accompanied with more aggressive behavior in follicular lymphoma and FDC sarcoma. Although it is still speculative, these findings raise several functional roles for gap junctions to be further elucidated in the germinal center: (1) FDC may assist as a communication network for propagating local signals to distant lymphocytes of either T or B cell phenotype and FDC throughout germinal centers that can modulate the adherence, migration, and differentiation of activated B cells; (2) direct FDC–B cell coupling in the light zone may transmit rescue signals of unknown nature for positive B cell selection (Lindhout et al., 1995); and (3) gap junction coupling can also be a pathway to control FDC growth and the size of lymphoid follicles that control is lost in follicular lymphoma. High levels of c-AMP in reactive light zone (Knox et al., 1993), which is known to upregulate Cx43 and can induce both cell differentiation (Mehta et al., 1996) and apoptosis (Insel et al., 2012), may support this idea.

Further potential gap junction-mediated interactions in lymphoid organs involve the following: (1) FDC–DC coupling since recent data show DCs to continuously sample FDC-presented antigens, which in turn may induce immune tolerance without T-reg involvement (McCloskey et al., 2011); (2) B cell can also sample FDC-presented antigen in a similar way to DCs, but may not be able to generate the epitope-sized peptides for gap junction cross-presentation; and (3) in the light zone of germinal center B cells, T cells and macrophages contact FDC and can potentially establish Cx43 gap junctions in any combinations.

Further complexity of the immune functions of Cx-s relates to unpaired Cx hemichannels and channel-unrelated interactions. Cx hemichannels can communicate with the extracellular milieu by releasing active metabolites, including ATP, glutamate, NAD^+, and prostaglandin E2, which can activate membrane receptors through autocrine or paracrine interactions (Evans et al., 2006). Cx hemichannel activity has already been confirmed in T helper cells (Ovideo-Orta et al., 2010). A wide range of intracytoplasmic molecular interactions between Cx-s and signaling protein kinases and phosphatases, including those involved in tumor development and promotion or with cell adhesion and cytoskeleton assembly-related proteins, have been revealed but their immune functions still need to be elucidated (Laird, 2006).

Connexins and gap junctions have already been shown to be involved in many aspects of innate and adaptive immune functions either inside or outside lymphoid tissues. Further relevant results are still expected to come and will assist in prevailing over the resistance against gap junctions by the wide immunological community.

ACKNOWLEDGMENTS

The authors are indebted to Edith Parsch and to the development staff of 3DHISTECH Ltd. for their excellent assistance in fluorescent digital microscopy. These works have been supported by OTKA-K62758 and ETT-105 in Hungary.

REFERENCES

Alves, L. A., Campos de Carvalho, A. C., Lima, E. O. C. et al. 1995. Functional gap junctions in thymic epithelial cells are formed by connexin 43. *Eur. J. Immunol.* 25:431–7.

Alves, L. A., Nihei, O. K., Fonseca, P. C., Carvalho, A. C., and Savino, W. 2000. Gap junction modulation by extracellular signaling molecules: The thymus model. *Braz. J. Med. Biol. Res.* 33:457–65.

Andrew, D., and Aspinall, R. 2002. Age-associated thymic atrophy is linked to a decline in IL-7 production. *Exp. Gerontol.* 37:455–63.

Bermudez-Fajardo, A., Yliharsila, M., Evans, W. H., Newby, A. C., and Oviedo-Orta, E. 2007. CD4+ T lymphocyte subsets express connexin 43 and establish gap junction channel communication with macrophages in vitro. *J. Leukoc. Biol.* 82:608–12.

Bodi, E., Hurtado, S. P., Carvalho, M. A., Borojevic, R., and Carvalho, A. C. 2004. Gap junctions in hematopoietic stroma control proliferation and differentiation of blood cell precursors. *An. Acad. Bras. Cienc.* 76:743–56.

Bopp, T., Becker, C., and Klein, M. 2007. Cyclic adenosine monophosphate is a key component of regulatory T cell-mediated suppression. *J. Exp. Med.* 204:1303–10.

Boyd, R. L., Tucek, C. L., and Godfrey, D. I. 1993. The thymic microenvironment. *Immunol. Today.* 14:445–59.

Campbell, F. R. 1980. Gap junctions between cells of bone marrow: An ultrastructural study using tannic acid. *Anat. Rec.* 196:101–7.

Campbell, F. R. 1986. Ultrastructural studies of intercellular contacts (junctions) in bone marrow. A review. *Scan. Electron. Microsc.* (Pt 2):621–9.

Cancelas, J. A., Koevoet, W. L., Koning, A. E., de Mayen, A. E., Rombouts, E. J., and Ploemacher, R. E. 2000. Connexin-43 gap junctions are involved in multiconnexin-expressing stromal support of hemopoietic progenitors and stem cells. *Blood* 96:498–505.

Chanson, M., Derouette, J. P., Roth, I., Foglia, B., Scerri, I., Dudez, T., and Kwak, B. R. 2005. Gap junctional communication in tissue inflammation and repair. *Biochim. Biophys. Acta.* 1711:197–207.

Civitelli, R. 2008. Cell-cell communication in the osteoblast/osteocyte lineage. *Arch. Biochem. Biophys.* 473:188–92.

Corvalan, L. A., Araya, R., Branes, M. C. et al. 2007. Injury of skeletal muscle and specific cytokines induce the expression of gap junction channels in mouse dendritic cells. *J. Cell Physiol.* 211:649–60.

de Jong, D., Koster, A., Hagenbeek, A. et al. 2009. Impact of the tumor microenvironment on prognosis in follicular lymphoma is dependent on specific treatment protocols. *Haematologica* 94:70–7.

Delves, P. J., and Roitt, I. M. 2000. The immune system. First of two parts. *N. Engl. J. Med.* 343:37–49.

de Vinuesa, C. G., Cook, M. C., Ball, J. et al. 2000. Germinal centers without T cells. *J. Exp. Med.* 191:485–94.

Dorshkind, K., Green, L., Godwin, A., and Fletcher, W. H. 1993. Connexin-43-type gap junctions mediate communication between bone marrow stromal cells. *Blood* 82:38–45.

Eugenin, E. A., Branes, M. C., Berman, J. W., and Saez, J. C. 2003. TNF-alpha plus IFN-gamma induce connexin43 expression and formation of gap junctions between human monocytes/macrophages that enhance physiological responses. *J. Immunol.* 170: 1320–8.

Evans, W. H., De Vuyst, E., and Leybaert, L. 2006. The gap junction cellular internet: Connexin hemichannels enter the signalling limelight. *Biochem. J.* 397:1–14.

Figueroa, X. F., Alvina, K., and Martínez, A. D. 2004. Histamine reduces gap junctional communication of human tonsil high endothelial cells in culture. *Microvasc. Res.* 68:247–57.

Fonseca, P. C., Nihei, O. K., and Urban-Maldonado, M. 2004. Characterization of connexin 30.3 and 43 in thymocytes. *Immunol. Lett.* 94:65–75.

Fossum, S., and Ford, W. L. 1985. The organization of cell populations within lymph nodes: Their origin, life history and functional relationships. *Histopathology* 9:469–99.

Gatto, D., and Brink, R. 2010. The germinal center reaction. *J. Allergy Clin. Immunol.* 126:898–907.

Glas, A. M., Knoops, L., and Delahaye, L. 2007. Gene-expression and immunohistochemical study of specific T-cell subsets and accessory cell types in the transformation and prognosis of follicular lymphoma. *J. Clin. Oncol.* 25:390–8.

Head, G. M., Mentlein, R., Kranz, A., Downing, J. E., and Kendall, M. D. 1997. Modulation of dye-coupling and proliferation in cultured rat thymic epithelium by factors involved in thymulin secretion. *J. Anat.* 191:355–65.

Hirabayashi, Y., Yoon, B. I., and Tsuboi, I. 2007. Protective role of connexin 32 in steady-state hematopoiesis, regeneration state, and leukemogenesis. *Exp. Biol. Med. (Maywood).* 232:700–12.

Horning, S. J., and Rosenberg, S. A. 1984. The natural history of initially untreated low-grade non-Hodgkin's lymphomas. *N. Engl. J. Med.* 311:1471–5.

Hulser, D. F., and Peters, J. H. 1971. Intercellular communication in phytohemagglutinin-induced lymphocyte agglutinates. *Eur. J. Immunol.* 1:494–5.

Hulser, D. F., and Peters, J. H. 1972. Contact cooperation in stimulated lymphocytes. II. Electrophysiological investigations on intercellular communication. *Exp. Cell Res.* 74:319–26.

Hurtado, S. P., Balduino, A., Bodi, E. C., El-Cheikh, M. C., Campos de Carvalho, A. C., and Borojevic, R. 2004. Connexin expression and gap-junction-mediated cell interactions in an *in vitro* model of haemopoietic stroma. *Cell Tissue Res.* 316:65–76.

Insel, P. A., Zhang, L., Murray, F., Yokouchi, H., and Zambon, A. C. 2012. Cyclic AMP is both a pro-apoptotic and anti-apoptotic second messenger. *Acta Physiol. (Oxf.).* 204:277–87.

Kapsenberg, M. L., and Leene, W. 1979. Formation of B type gap junctions between PHA-stimulated rabbit lymphocytes. *Exp. Cell Res.* 120:211–22.

Katakai, T., Hara, T., Lee, J. H., Gonda, H., Sugai, M., and Shimizu, A. 2004b. A novel reticular stromal structure in lymph node cortex: An immuno-platform for interactions among dendritic cells, T cells and B cells. *Int. Immunol.* 6:1133–42.

Katakai, T., Hara, T., Sugai, M., Gonda, H., and Shimizu, A. 2004a. Lymph node fibroblastic reticular cells construct the stromal reticulum via contact with lymphocytes. *J. Exp. Med.* 200:783–95.

Knox, K. A., Johnson, G. D., and Gordon, J. 1993. Distribution of cAMP in secondary follicles and its expression in B cell apoptosis and CD40-mediated survival. *Int. Immunol.* 5:1085–91.

Krenacs, T., Krenács, L., and Raffeld, M. 2010a. Multiple antigen immunostaining procedures. In *Immunocytochemical Methods and Protocols.* Ed. C. Oliver and M. C. Jamur, 3rd edition, Totowa, NJ: Humana Press Inc. *Methods. Mol. Biol.* 588:281–300.

Krenacs, T., and Rosendaal, M. 1995. Immunohistological detection of gap junctions in human lymphoid tissue: Connexin43 in follicular dendritic and lymphoendothelial cells. *J. Histochem. Cytochem.* 43:1125–37.

Krenacs, T., and Rosendaal, M. 1998a. Connexin43 gap junctions in normal, regenerating and cultured mouse bone marrow and in human leukaemias: Their possible involvement in blood formation. *Am. J. Pathol.* 152:993–1004.

Krenacs, T., and Rosendaal, M. 1998b. Gap junctional communication pathways in germinal centre reactions. *Dev. Immunol.* 6:111–8.

Krenacs, T., Rosendaal, M., van Dartel, M., and Lindhout, E. 1997. Direct cell-cell communication in the germinal centre: Connexin43 gap junctions functionally couple follicular dendritic cells and follicular dendritic cells with B lymphocytes. *Eur. J. Immunol.* 27:1489–97.

Krenacs, T., Zsakovics, I., Micsik, T. et al. 2010b. Digital microscopy—The upcoming revolution in histopathology teaching, diagnostics, research and quality assurance. In *Microscopy: Science, Technology, Applications and Education*. Volume 2. Ed. A. Méndez-Vilas and J. Diaz, 965–77. Bajadoz, Spain: Formatex Research Center.

Laird, D. W. 2006. Life cycle of connexins in health and disease. *Biochem. J.* 394:527–43.

Levy, J. A., Weiss, R. M., Dirksen, E. R., and Rosen, M. R. 1976. Possible communication between murine macrophages oriented in linear chains in tissue culture. *Exp. Cell Res.* 103:375–85.

Lindhout, E., Lakeman, A., and de Groot, C. 1995. Follicular dendritic cells inhibit apoptosis in human B lymphocytes by a rapid and irreversible blockade of preexisting endonuclease. *J. Exp. Med.* 181:1985–95.

Liu, Y., Zhang, X., Li, Z. J., and Chen, X. H. 2010. Up-regulation of Cx43 expression and GJIC function in acute leukemia bone marrow stromal cells post-chemotherapy. *Leuk. Res.* 34:631–40.

Matsue, H., Yao, J., Matsue, K. et al. 2006. Gap junction-mediated intercellular communication between dendritic cells (DCs) is required for effective activation of DCs. *J. Immunol.* 176:181–90.

McCloskey, M. L., Curotto de Lafaille, M. A., Carroll, M. C., and Erlebacher, A. 2011. Acquisition and presentation of follicular dendritic cell-bound antigen by lymph node-resident dendritic cells. *J. Exp. Med.* 208:135–48.

Mehta, P. P., Lokeshwar, B. L., Schiller, P. C. et al. 1996. Gap-junctional communication in normal and neoplastic prostate epithelial cells and its regulation by cAMP. *Mol. Carcinog.* 15:18–32.

Mendoza-Naranjo, A., Saéz, P. J., Johansson, C. C. et al. 2007. Functional gap junctions facilitate melanoma antigen transfer and cross-presentation between human dendritic cells. *J. Immunol.* 178:6949–57.

Metcalf, D. 1989. The molecular control of cell division, differentiation commitment and maturation in haemopoietic cells. *Nature* 339:27–30.

Montecino-Rodriguez, E., and Dorshkind, K. 2001. Regulation of hematopoiesis by gap junction-mediated intercellular communication. *J. Leukoc. Biol.* 70:341–7.

Montecino-Rodriguez, E., Leathers, H., and Dorshkind, K. 2000. Expression of connexin 43 (Cx43) is critical for normal hematopoiesis. *Blood* 96:917–24.

Morel, S., Burnier, L., and Kwak, B. R. 2009. Connexins participate in the initiation and progression of atherosclerosis. *Semin. Immunopathol.* 31:49–61.

Neumark, T., and Huynh, D. C. 1989. Gap junctions between human T-colony cells. *Acta Morphol. Hungarica* 37:147–53.

Nihei, O. K., Fonseca, P. C., Rubin, N. M. et al. 2010. Modulatory effects of cAMP and PKC activation on gap junctional intercellular communication among thymic epithelial cells. *BMC Cell Biol.* 11:3.

Neijssen, J., Herberts, C., Drijfhout, J. W., Reits, E., Janssen, L., and Neefjes, J. 2005. Cross-presentation by intercellular peptide transfer through gap junctions. *Nature* 434:83–8.

Neijssen, J., Pang, B., and Neefjes, J. 2007. Gap junction-mediated intercellular communication in the immune system. *Prog. Biophys. Mol. Biol.* 94:207–18.

Oviedo-Orta, E., and Evans, H. W. 2004. Gap junctions and connexin-mediated communication in the immune system. *Biochim. Biophys. Acta.* 1662:102–12.

Oviedo-Orta, E., Gasque, P., and Evans, W. H. 2001. Immunoglobulin and cytokine expression in mixed lymphocyte cultures is reduced by disruption of gap junction intercellular communication. *FASEB J.* 15:768–74.

Oviedo-Orta, E., Errington, R. J., and Evans, H. W. 2002. Gap junction intercellular communication during lymphocyte transendothelial migration. *Cell Biol. Int.* 26:253–63.

Oviedo-Orta, E., Hoy, T., and Evans, W. H. 2000. Intercellular communication in the immune system: Differential expression of connexin40 and 43, and perturbation of gap junction channel functions in peripheral blood and tonsil human lymphocyte subpopulations. *Immunology* 99:578–90.

Oviedo-Orta, E., Perreau, M., Evans, W. H., and Potolicchio, I. 2010. Control of the proliferation of activated CD4+ T cells by connexins. *J. Leukoc. Biol.* 88:79–86.

Paraguassu-Braga, F. H., Borojevic, R., Bouzas, L. F., Barcinski, M. A., and Bonomo, A. 2003. Bone marrow stroma inhibits proliferation and apoptosis in leukemic cells through gap junction-mediated cell communication. *Cell Death Differ.* 10:1101–8.

Ploemacher, R. E., Mayen, A. E., De Koning, A. E., Krenacs, T., and Rosendaal, M. 2000. Hematopoiesis: Gap junction intercellular communication is likely to be involved in regulation of stroma-dependent proliferation of hemopoietic stem cells. *Hematology.* 5:133–47.

Polacek, D., Lal, R., Volin, M. V., and Davies, P. F. 1993. Gap junctional communication between vascular cells: Induction of connexin43 messenger RNA in macrophage foam cells at atherosclerotic lesions. *Am. J. Pathol.* 142:593–606.

Porvaznik, M., and MacVittie, T. J. 1979. Detection of gap junctions between the progeny of a canine macrophage colony-forming cell in vitro. *J. Cell Biol.* 82:555–64.

Reaume, A. G., de Sousa, P. A., Kulkarn, S. et al. 1995. Cardiac malformation in neonatal mice lacking connexin43. *Science* 267:1831–4.

Ring, S., Karakhanova, S., Johnson, T., Enk, A. H., and Mahnke, K. 2010. Gap junctions between regulatory T cells and dendritic cells prevent sensitization of CD8(+) T cells. *J. Allergy Clin. Immunol.* 125:237–46.

Rosendaal, M. 1995. Gap junctions in blood forming tissues. *Microscopy Res. Technol.* 31:396–7.

Rosendaal, M., Gregan, A., and Green, C. R. 1991. Direct cell-cell communication in the blood-forming system. *Tissue Cell* 23:457–70.

Rosendaal, M., Green, C. R., Rahman, A., and Morgan, D. 1994. Up-regulation of the connexin43+ gap junction network in haemopoietic tissue before the growth of stem cells. *J. Cell Sci.* 107:29–37.

Rosendaal, M., and Jopling, C. 2003. Hematopoietic capacity of connexin43 wild-type and knock-out fetal liver cells not different on wild-type stroma. *Blood* 101:2996–8.

Rosendaal, M., and Krenacs, T. 2000. Regulatory pathways in blood-forming tissue with particular reference to gap junctional communication. *Pathol. Oncol. Res.* 6:243–9.

Rosendaal, M., Mayen, A., de Koning, A., Dunina-Barkovskaya, T., Krenacs, T., and Ploemacher, R. 1997. Does transmembrane communication through gap junctions enable stem cells to overcome stromal inhibition? *Leukemia.* 11:1281–9.

Rosendaal, M., and Stone, M. 2003. Enhancement of repopulation haemopoiesis by heterozygous connexin 43 stem cells seeded on wild-type connexin 43 stroma. *Clin. Sci. (Lond.).* 105:561–8.

Saez, J. C., Berthoud, V. M., Branes, M. C., Martinez, A. D., and Beyer, E. C. 2003. Plasma membrane channels formed by connexins: Their regulation and functions. *Physiol. Rev.* 83:1359–400.

Saccheri, F., Pozzi, C., Avogadri, F. et al. 2010. Bacteria-induced gap junctions in tumors favor antigen cross-presentation and antitumor immunity. *Sci. Transl. Med.* 2(44):44–57.

Scheckenbach, K. E., Crespin, S., Kwak, B. R., and Chanson, M. 2011. Connexin channel-dependent signaling pathways in inflammation. *J. Vasc. Res.* 48:91–103.

Soriano, A. O., Thompson, M. A., Admirand, J. H. et al. 2007. Follicular dendritic cell sarcoma: A report of 14 cases and a review of the literature. *Am. J. Hematol.* 82:725–8.

Umezawa, A., Harigaya, K., Abe, H., and Watanabe, Y. 1990. Gap-junctional communication of bone marrow stromal cells is resistant to irradiation in vitro. *Exp. Hematol.* 18:1002–7.

Umezawa, A., Harigaya, K., and Watanabe, Y. 1987. Bone marrow stromal cells lose their gap junctional communication *in vitro* during the differentiation to adipocytes. *Hematol. Rev.* 1:277–83.

Umezawa, A., and Hata, J. 1992. Expression of gap-junctional protein (connexin 43 or a1 gap junction) is down-regulated at the transcriptional level during adipocyte differentiation of H-1/A marrow stromal cells. *Cell Struct. Function.* 17:177–84.

Yamazaki, K. 1988. Sl/Sld mice have an increased number of gap junctions in their bone marrow stromal cells. *Blood Cells* 13:421–31.

Yamazaki, K., and Allen, T. D. 1991. Ultrastructural and morphometric alterations bone marrow stromal tissue after 7Gy irradiation. *Blood Cells* 17:527–49.

Watanabe, Y. 1985. Fine structure of bone marrow stroma. *Acta Haematol. Jpn.* 48:1688–700.

3 Approaches for Studying the Role(s) of Gap Junctions in the Immune System

Aaron M. Glass, Thien D. Nguyen, and Steven M. Taffet

CONTENTS

Although gap junctions were first described in the immune system in the 1970s (Dos Reis and De Oliveira-Castro, 1978; Gaziri et al., 1975; Oliveira-Castro et al., 1973), our knowledge of how connexins might participate in immune function remains incomplete. Of all the connexin family members, connexin 43 (Cx43) is the most widely expressed isoform in the mammalian body, including the immune system. Cx43 has been reported to be expressed in lymphocytes (Oviedo-Orta et al., 2000), neutrophils (Diaz et al., 1995; Zahler et al., 2003), natural killer cells (Oviedo-Orta et al., 2000), dendritic cells (Matsue et al., 2006; Neijssen et al., 2005), monocytes/ macrophages (Anand et al., 2008; Bermudez-Fajardo et al., 2007; Chadjichristos et al., 2010; Eugenin et al., 2003; Jara et al., 1995; Neijssen et al., 2005; Polacek et al., 1993), and mast cells (Vliagoftis et al., 1999). Expression of connexins among immune cells is not limited to mice, as studies have identified family members in rats, hamsters, canines, horses, and humans, suggesting that expression of connexins in the immune system is an evolutionarily conserved phenomenon. Based on this widespread expression among numerous immune cell types, and across a spectrum of mammalian species, it is hypothesized that Cx43 serves some purpose in immune function.

Some of the immune functions in which gap junctions and Cx43 have been suggested to act include antigen cross-presentation (Mendoza-Naranjo et al., 2007; Neijssen et al., 2005), regulatory T cell inhibition (Bopp et al., 2007), inflammation

(Martin et al., 1998), phagocytosis (Anand et al., 2008), wound healing (Chanson et al., 2005), dendritic cell maturation (Matsue et al., 2006), B cell antibody production (Oviedo-Orta et al., 2001), and immune responses to pathogens (Anand et al., 2008; Oloris et al., 2007; Sarieddine et al., 2009). Additionally, Cx43 may play an active role in processes that are responsible for the development of the immune system, such as hematopoiesis (Oviedo-Orta and Howard Evans, 2004; Presley et al., 2005; Rosendaal and Stone, 2003; Montecino-Rodriguez and Dorshkind, 2001; Montecino-Rodriguez et al., 2000; Bodi et al., 2004; Cancelas et al., 2000). This becomes evident during events that require increased levels of hematopoiesis, such as neonatal development and hematopoietic recovery after chemotherapy.

Early studies on connexins in immunity relied heavily on chemical inhibitors to close connexin channels. These agents, which included long-chain alcohols and glycyrrhetinic acid (or derivatives thereof), were effective at blocking channels but had the disadvantage of acting promiscuously on multiple connexin isoforms. The pharmacology of gap junctions has been reviewed extensively by Srinivas et al. (2004). Unfortunately, there are few specific drugs for the modulation of gap junction function. Long-chain alcohols (heptanol and octanol) have long been used as uncoupling agents. Their mechanisms of action are unknown, although alteration of membrane fluidity is thought to be the primary target. General anesthetics have also been shown to uncouple gap junctions (Burt and Spray, 1989). Again, the mechanism is unknown and this effect is not specific to gap junction proteins. Some agents such as butanedione monoxime are thought to modify phosphorylation of gap junctions, leading to uncoupling (Herve and Sarrouilhe, 2002). Flufenamic acid has been shown to be an effective inhibitor of gap junctions but the mechanism is thought to be indirect and not specific to any connexin protein (Srinivas and Spray, 2003). Perhaps among the more specific inhibitors are the glycyrrhetinic acids (Davidson et al., 1986, Goldberg et al., 1996), which may also alter phosphorylation (Guan et al., 1996). The most significant shortcomings of the glycyrrhetinic acids are their lack of specificity for different connexin proteins and the lack of a specific mechanism of action. Interestingly, quinine blocks a subgroup of gap junction channels (Srinivas et al., 2001) but does not block gap junctions formed by the connexins found in the heart.

Specific inhibition of gap junction formation has been demonstrated with the use of extracellular loop peptides (Boitano and Evans, 2000; Evans and Boitano, 2001; Kwak and Jongsma, 1999). It is thought that these peptides inhibit gap junctions by preventing connexin docking in the extracellular gap. These effects are slow and the interaction requires high concentrations of peptide (Dahl et al., 1994). The specificity of these peptides has also been called into question: Wang et al. (2007) demonstrated that these peptides targeted connexins less precisely than had been previously suggested.

In particular, the discovery of pannexin channels in immune cells has complicated the interpretation of inhibitor-based studies. Pannexins have been identified in many immune cells (Marina-Garcia et al., 2008, Pelegrin and Surprenant, 2006, 2007, Pelegrin et al., 2008, Woehrle et al., 2010a,b), and pannexin channels are sensitive to many of the inhibitors of connexin channels. Pannexins are susceptible to inhibition by carbenoxolone (Bruzzone et al., 2005, Locovei et al., 2006) and are actually more sensitive to mefloquine than are channels formed by connexins (Iglesias et al., 2008). There is also inhibition of pannexin channels by some of the gap junction "specific" peptides (Wang et al., 2007). Thus, inhibition of junctions by a variety of chemical agents is open to alternative interpretation.

3.1 MOLECULAR MANIPULATION OF CX43 LEVELS IN IMMUNE CELLS

As an alternative to the use of pharmacological inhibitors, the role(s) of connexins in immunity can be studied by altering connexins at the level of gene expression. This can be achieved either through the introduction of specific wild-type or mutant connexin isoforms into target cells, or by silencing the expression of endogenous connexins. Owing to their anionic and hydrophilic nature, the DNA and RNA substrates required to achieve gene-level manipulation are unable to enter cells by diffusion alone. They must, therefore, be introduced into target cells, generally by one of the two distinct strategies: transfection or transduction. Transfection refers to the use of transient disruption of the plasma membrane (and in some cases, the nuclear envelope) to achieve insertion of genetic material into cells, whereas transduction is gene delivery using a viral vector. Each of these two strategies for the introduction of genetic material has inherent advantages and disadvantages that must be weighed before deciding which to employ.

Owing to the lower cost of reagents, greater biosafety, and existence of semioptimized commercially available kits, transfection is often the first choice for genetic manipulation of mammalian cells. Transfection can be carried out via biochemical methods or physical methods. Common chemical methods include transfection mediated by diethylaminoethyl (DEAE)-dextran, calcium phosphate, and cationic lipid reagents. Chemical methods feature a positively charged transfection reagent, which is first combined with the genetic material to be introduced. By virtue of the net negative charge of DNA and RNA, an electrostatic attraction drives the formation of genetic material–transfection reagent complexes. Fortuitously, the positive charge of the complexes facilitates their attachment to the target cell surface, triggering their uptake by endocytosis. Physical methods rely on the disruption of the cell membrane to achieve introduction of genetic material. The only method in widespread use is electroporation. Electroporation involves the delivery of electrical pulses that increase transmembrane potential, ultimately causing the formation of hydrophilic pores that allow the passage of DNA or RNA. If successful, the pores reseal following penetration of the genetic material, leaving the integrity and viability of most target cells intact. A variation of the electroporation technique called *Nucleofection* (Lonza AG) is a proprietary technology that claims to deliver genetic material directly into the target cell nucleus.

Unfortunately, an obstacle to using genetic manipulation in immune studies is presented by the fact that most immune cells and derived cell lines are refractory to transfection. Macrophages exemplify the intransigent nature of immune cells to transfection and, as mentioned before, encompass a central focus of our studies of Cx43 in immune function. Therefore, in this section, we will review the existing literature, with a focus on murine macrophages.

In the late 1990s, a handful of studies were undertaken to determine the most effective methods for transfection of macrophages and macrophage-like cell lines. Mack et al. (1998) compared calcium phosphate-, cationic liposome-, electroporation-, and DEAE-dextran-mediated transfection in cultured human macrophages harvested from peripheral blood. Their findings identified DEAE-dextran to be the most effective method. In contrast, using a similar luciferase reporter gene construct in the mouse macrophage-like cell line RAW264.7, Thompson et al. (1999) found DEAE-dextran to be among the most unreliable methods, instead obtaining the best results from electroporation. Using the RAW264.7 and NR 383 cell lines, Dokka et al. (2000a) found lipofection to be effective, but did not test electroporation. Their work suggests that the most successful transfections of these cell types can be carried out using the Lipofectamine (Invitrogen) system in combination with the arginine-rich DNA-binding protein, protamine. In our own studies, and in the work of others, we found that electroporation has been highly effective in transfecting the RAW264.7 macrophage-like cell line (Tomaras et al., 1999; O'Donnell and Taffet; 2002), but was not successful with primary macrophages or other macrophage cell lines (J774 and P388.D1). The variability in these results may reflect interspecies differences and differences in the cell lines tested. No reports have described the true efficiency of transfection and the relative toxicity of the method. Only Dokka et al. (2000a,b) provided a quantification of the toxicity of the different transfection methods by lactate dehydrogenase release. One point that these authors agree upon is the difficulty of introducing exogenous DNA into macrophages.

Several theories exist to explain the difficulty introducing genetic material into macrophages by transfection. One major cause of low transfection efficiency into macrophages is thought to be the presence of contaminating endotoxin, especially lipopolysaccharide (LPS) in DNA preparations (Dokka et al., 2000a). Macrophages are extremely sensitive to even low levels of this contaminant and respond with the generation of toxic reactive oxygen species, especially hydroxyl radicals, which can lead to cell death.

It has been appreciated for some time that treatment of cultured peritoneal mouse macrophages with LPS results in cell death by apoptosis (Albina et al., 1993). Experiments utilizing bone marrow-derived macrophages from Type I TNF-α receptor knockout (KO) mice suggest that this effect is mediated by TNF-α in an autocrine manner (Xaus et al., 2000). If this is true, LPS treatment or endotoxin contamination results in the secretion of TNF-α from macrophages, leading to subsequent TNF receptor signaling and culminating in the induction of apoptosis.

This may account for the success of employing lipofection in macrophages and macrophage-like cells as compared to cells transfected using other techniques (Dokka et al., 2000a). Lipofection of peritoneal macrophages and a related microglial cell line has been shown to uncouple LPS stimulation from downstream effectors (mitogen-activated protein (MAP) kinases) (Leon-Ponte et al., 2005). This would effectively blunt the macrophage response to endotoxin contamination: reducing cell death and increasing transfection efficiency. As much a cautionary tale as one of success, however, this study implies that polycationic lipids likely have many unintended effects on macrophage biology. These effects should be considered before embarking on functional studies of macrophages transfected with connexin isoforms utilizing these techniques.

Macrophages are not the sole population of immune cells for which transfection remains a technical challenge; the same is true for lymphocytes. The difficulty is compounded in lymphocytes, as it seems that, while lipofection may improve transfection efficiency into macrophages, it does little to ameliorate the production of TNF-α and consequent apoptosis in primary lymphocytes. Several studies have detailed the difficulties in transfecting T-lymphocytes with exogenous cDNA. Ebert and Fink found that cultured human lymphocytes, when lipofected, released high levels of TNF-α and underwent apoptosis (Ebert et al., 1997). Since lymphocytes lack TLR4 and its coreceptor, CD14, and are unable to respond to contaminating LPS, a separate mechanism of apoptosis induction may be necessary to explain these results (Mattern et al., 1998). A simpler explanation may be that the Ebert and Fink study was performed on cultured ficoll-separated mononuclear cells, not pure lymphocytes, meaning that bystander transfection of monocytes could have occurred. These transfection-activated monocytes may have been responsible for the high levels of TNF-α observed.

Another factor that has been proposed to limit the efficiency of transfection of macrophages and other immune cells is methylation-mediated gene silencing (Escher et al., 2005). Escher et al. found that, following transfection with a number of common reagents, the CMV promoter used to drive reporter gene expression became methylated, preventing expression of the reporter gene. Application of the DNA methylation inhibitor 5-azacytidine enhanced reporter gene expression efficiency to nearly 100% of RAW264.7 cells with minimal impact on cell viability. Unfortunately, epigenetic modifying compounds, such as the 5-azacytidine used in this study, undoubtedly produce many off-target effects, confounding the effects of expression or silencing of the gene of interest.

Alternate transfection techniques, such as nucleofection (Lonza), where plasmid DNA is electroporated directly into the nuclei of target cells, have been attempted with varied success. As mentioned above, we and others at our institution have utilized the Nucleofector on both primary macrophages and macrophage lines (Glass and Taffet, unpublished data), generally with disappointing results. A unique variation of this technology was pioneered by Van De Parre et al. (2005) who found that the introduction of plasmid DNA by Nucleofection resulted in apoptosis of J774A.1 macrophage-like cells. In contrast, introduction of mRNA by nucleofection resulted in expression of the reporter gene (CMV-eGFP) in 60–75% of cells without a concomitant reduction in viability. Despite the expectation that (relative to DNA) mRNA is short-lived within cells, expression of an eGFP reporter gene electroporated into monocyte-derived dendritic cells remained detectable (around 30%) after five days (Van Tendeloo et al., 2001). Even considering these initial reports of success, electroporation of mRNA has yet to gain widespread popularity as an alternative to lipofection.

RAW264.7 cells are, in our experience, among the easiest macrophage lines to transfect (Foster et al., 1997; O'Donnell and Taffet, 2002; Taffet et al., 1990; Tomaras et al., 1999). These cells may be ideal for Cx43 studies, as they do not appear to express detectable levels of Cx43 (Glass and Taffet, unpublished results) and can be successfully transfected using a wide variety of techniques. These cells are similar to macrophages but have a number of shortcomings that limit their usefulness.

Although the literature is peppered with a few reports of successful, high-efficiency transfection of immune cells, these remain the exception, rather than the rule. Virtually all of these reports involve the use of cell lines, further intimating at the technical difficulties of transfecting primary immune cells.

Since Cx43 is expressed in macrophages from several sources as well as other immune cells, silencing of endogenous connexin is an ideal approach to teasing out the functional significance of its presence. In contrast to transfection of exogenous genes into macrophages and related lines, silencing of endogenous gene expression has been met with greater success, as is evidenced by several papers that have reported accomplishing satisfactory knockdown of target gene expression. Investigators have used polycationic lipid reagents for delivery of siRNA to RAW264.7 cells (Anand et al., 2008; Amarzguioui et al., 2006) and even bone marrow-derived macrophages (Akinc et al., 2008).

Other groups claim greater success with nucleofection and electroporation for delivery of siRNA. Zhang et al. have reported an almost complete knockdown of target gene expression in bone marrow-derived macrophages, as indexed by both mRNA levels and protein activity (Zhang et al., 2009). One report (Anand et al., 2008) describes the silencing of Cx43 in the RAW264.7 macrophage-like cell line. Interestingly, we could not find any other reports of connexin gene silencing in macrophages or lymphocytes. Ultimately, at this point, several technical hurdles must be overcome before transfection of exogenous genes or silencing constructs into immune cells becomes trivial. The vast majority of studies that have featured these techniques have employed cell lines as opposed to primary macrophages and lymphocytes. A more desirable alternative for the study of macrophages may be mouse or human primary macrophages derived in culture.

The major alternative to transfection for gene delivery or silencing in mammalian cells is viral transduction. Common strategies encompassed by transduction are the use of adenoviral, and retro/lentiviral manipulation. Both these techniques have been attempted on immune cells with varying degrees of success.

Adenoviruses are double-stranded DNA viruses that are minor causes of upper respiratory tract infections in the human population. Viruses of this family have been used for several decades as vectors for the delivery of exogenous genes to cells due to the fact that they offer several advantages over other viral vectors, including (1) a DNA genome, making manipulation practical; (2) an ability to infect many cellular targets with high efficiency, including nondividing cells; and (3) a large capacity for gene inserts. After entry into target cells, genes delivered by adenoviral vectors do not incorporate into a cell's chromosomes, eliminating the possibility of random insertion disrupting the genome of the target cell. This is unfortunate in the case of dividing cell lines since the viral genes become diluted in successive generations. Several studies have employed adenoviral systems for the delivery of Cx43 to cells, including primary rat myoblasts (Reinecke et al., 2004) and macrophages (Anand et al., 2008).

Unfortunately, there are complications to the use of adenoviral vectors for gene delivery into immune cells. Infection of primary immune cells with adenovirus results in the production of inflammatory cytokines and type I interferon (Sakurai et al., 2008, Zsengeller et al., 2000). It has also been reported that adenoviral infection can induce TNF-α production and maturation in dendritic cells, evidence that the effects of this gene delivery strategy may exceed the effects of the delivered gene itself (Philpott et al., 2004). These issues are compounded by the fact that some types of macrophages, such as alveolar, do not express high levels of the Coxsackie-Adenovirus receptor, the main protein required for cellular entry by the virus (Kaner et al., 1999).

Perhaps more common in the manipulation of immune cells is the use of ret-roviruses or lentiviruses, RNA viruses capable of reverse transcribing their RNA genomes into cDNA, which is then integrated into the target cell's genome. Retroviral and lentiviral transduction systems are ideal for the delivery of exogenous genes, as well as silencing of endogenous genes. Lentiviruses offer the advantage of the ability to transduce dividing and nondividing cells alike, while retroviruses are only capable of integration into cells undergoing division.

Retroviruses have been used in several studies to express, or silence, Cx43. As examples, retroviruses have been used to deliver Cx43 to a myoblast cell line (Reinecke et al., 2004), into skeletal muscle myoblasts (Tolmachov et al., 2006), and to deliver a GFP-tagged connexin construct into a human embryonic stem cell line (Moore et al., 2008). Silencing of Cx43 using retro- or lentiviral transduction has been demonstrated in osteoblastic and osteocytic cell lines (Plotkin et al., 2008), and multiple breast cancer cell lines (Shao et al., 2005).

Multiple studies have used retroviral or lentiviral vectors for the genetic manipulation of immune cells. Various retroviruses have been successfully used to transduce bone marrow-derived murine macrophages (Pan et al., 2008; Zeng et al., 2006), human monocyte-derived macrophages (Leyva et al., 2011), bone marrow-derived dendritic cells (Stewart et al., 2003), and human cytotoxic T cells (Zhou et al., 2003).

Biosafety is a concern when using viruses for genetic manipulation since most viral vectors are capable of infecting humans as well as target cultured cells or animals. As a feature to enhance safety, most commercially available adenoviral and lentiviral vectors are engineered to be replication incompetent. Generally, this is achieved by separating genes that are necessary for replication (E1/E3 or gag/pol for adenovirus and lentivirus, respectively) from remaining viral genes. To produce infectious virus, the missing genes are added in *trans*, by using a pretransformed packaging cell line (in the case of adenovirus) or by cotransfecting packaging cells with replication genes simultaneously with viral genes (for lentivirus). Under these conditions, virus replication occurs within the packaging cell line but the viral progeny cannot undergo replication after infecting target cells. Despite this safety feature, working with infectious virus still demands moderate technical care and precautionary biosafety measures, details that must be considered before embarking on connexin research using any viral transduction strategy.

3.2 TRANSGENIC AND KO MICE

The Cx43 KO mouse has been in existence since 1995 but has been of limited use in studying the immune response due to the neonatal lethality of Cx43 deletion (Reaume et al., 1995). Much has been gained by studying Cx43 heterozygotic mice (Anand et al., 2008; Kwak et al., 2003; Montecino-Rodriguez et al., 2000; Montecino-Rodriguez and Dorshkind, 2001; Rosendaal and Stone, 2003). Three approaches will be described to circumvent the issue of neonatal lethality. These are the use of fetal hematopoietic stem cells (HSC) from a KO mouse to repopulate an irradiated host; the direct culturing of fetal HSC to produce macrophages or dendritic cells; and the use of conditional or tissue-specific knockout animals.

Radiation chimeras are a commonly used model to study the role of specific genes in the immune system, where deletion of the gene of interest is lethal and does not allow the transgenic animal to reach adulthood (Duran-Struuck and Dysko, 2009; Iwasaki, 2006; van Os et al., 2001). Normally, bone marrow cells are used to reconstitute lethally irradiated mice. However, since both bone marrow and fetal liver contain abundant pluripotent stem cells, irradiated mice can also be reconstituted with cells derived from fetal liver (Eckardt and McLaughlin, 2008). In cases such as Cx43, where the genetic deletion results in neonatal death, fetal liver cells are preferable to bone marrow cells (Navarro and Touraine, 1989; Nguyen and Taffet, 2009).

In models where wild-type bone marrow or fetal liver cells are transplanted into wild-type hosts, immune cells such as monocytes, dendritic cells, and neutrophils can be found in the spleen by day 7 after implantation of the hematopoietic cells (Ogawa et al., 1996). By day 21, peripheral lymphoid cell reconstitution may be normal, which includes the presence of the NK, T, and B cells. However, these cells may not be fully matured and functional in terms of possessing full effector capabilities (Ojielo et al., 2003). During this recovery period, irradiated mice are immunocompromised and susceptible to opportunistic infections, necessitating a specific pathogen-free environment and a regimen of prophylactic antibiotics.

Ideally, engrafted donor cells should have a marker for easy identification, making it possible to assess the success of implantation and the degree of chimerization. In the case of the model we developed (Nguyen and Taffet, 2009), we have used variations in the CD45 isoforms expressed on the leukocytes of common mouse strains to monitor reconstitution. C57BL/6 mice normally express the CD45 isoform CD45.2 but genetically modified mice express the SJL-associated isoform, CD45.1 (Kiel et al., 2005; Teshima et al., 2002) on a C57BL/6 background. By employing donor and recipient animals with different CD45 isoforms, we were able to monitor the degree of chimerization by assessing CD45.1 and CD45.2 positive cells by flow cytometry.

An alternative donor cell marking system uses CD90.1/CD90.2 (Stohlman et al., 2008). Although hematopoietic stem cells have been shown to express CD90, this marker can be effectively used to track T cell expansion because of its limited cellular distribution (Haeryfar and Hoskin, 2004). Additionally, other proteins that are not normally expressed in murine cells have been used as markers for chimerization. For example, fluorescent markers such as GFP (Janssen et al., 2010) have been used to track the expansion of donor hematopoietic stem cells, in this case requiring genetically modified mice or transduction of the GFP transgene into donor stem cells.

We have successfully used the radiation chimera model (Figure 3.1) to study the repopulation of immune organs in irradiated animals with cells derived from the fetal livers of Cx43 KO mice (Nguyen and Taffet, 2009). These studies demonstrated that immune organs can be efficiently repopulated with donor-derived cells, resulting in host animals that have a normal constitution of immune cells. Using this model, we sought to test the findings of other models of Cx43 in immunity.

It has been suggested that gap junction communication has a role in thymocyte development (Fonseca et al., 2004; Montecino-Rodriguez et al., 2000). Interestingly,

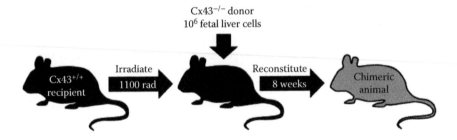

FIGURE 3.1   Radiation chimera model. Fetal liver cells from Cx43$^{-/-}$ mice were used to reconstitute irradiated mice.

it has been reported that Cx43 heterozygote embryos and neonates exhibit increased thymic cellularity when compared to littermates with normal Cx43 expression (Cx43$^{+/+}$) (Montecino-Rodriguez et al., 2000). This includes an increased frequency of double-positive (CD4+/CD8+) thymocytes, but a decreased frequency of CD4+ thymocytes. However, these differences become less apparent over time, with the thymic composition achieving similarity between heterozygotes and wild-type littermates by 4 weeks after birth.

Similarly, in experiments conducted involving reconstituting irradiated mice with either Cx43$^{+/+}$ or Cx43$^{+/-}$ bone marrow, Montecino-Rodriquez et al. (2000) found that reconstituting Cx43$^{+/+}$ hosts resulted in normal thymic cellularity, whether the donor cells were Cx43$^{+/+}$ or Cx43$^{+/-}$. In contrast, there was reduced thymic cellularity when the host was Cx43$^{+/-}$, regardless of donor genotype. This suggests that expression of Cx43 in the thymic stroma, rather than in the thymocytes themselves, is the determinant factor in thymic cellularity. Nonetheless, these experiments uphold the importance of Cx43 expression in the thymus.

Given the results of these studies, we hypothesized that a complete absence of Cx43 expression would further skew thymocyte development. To address this question, we created chimeric mice in which greater than 99% of thymocytes were derived from the donor CD45.2 population. Surprisingly, there seemed to be no effect of the absence of Cx43 expression on the ability of thymocytes generated from fetal liver-derived stem cells to repopulate the thymus. Cx43$^{+/+, +/-}$, and $^{-/-}$ fetal liver cells were equally effective when assessed by cell number, by a proportion of CD45.2 positive cells, as well as by markers of thymocyte differentiation (CD4 and CD8 expression) (Nguyen and Taffet, 2009). In the end, we could find no significant differences in the thymus populations of either Cx43$^{+/-}$ or Cx43$^{-/-}$ radiation chimeras compared to chimeras created using donor cells expressing wild-type levels of Cx43.

Owing to this lack of difference in the radiation chimeras, we performed studies similar to those done by Montecino-Rodriquez et al., looking for differences in the thymocyte profiles of nonchimeric Cx43$^{+/-}$ mice (Montecino-Rodriguez and Dorshkind, 2001; Montecino-Rodriguez et al., 2000). In contrast to their results, we did not find any difference in the proportion of double-positive thymocytes. However, when thymocytes were analyzed by first gating on CD3+ cells, significant differences in the single-positive thymocyte populations were observed between Cx43$^{+/-}$ mice and their wild-type littermates (Nguyen and Taffet, 2009). Specifically, there was a decrease in the percentage of CD4+ cells and a subsequent increase in CD8+ thymocytes. These results support the general findings of Montecino-Rodriquez et al.—in our chimeric mice, since recipient animals express wild-type levels of Cx43 independently of donor cell genotype, results from both groups imply that the Cx43 content of the thymic stroma is more important than that of thymocytes in the development of T cells.

We also assessed the T cell populations in the periphery. As shown in Figure 3.2, both Cx43$^{+/+}$ and Cx43$^{-/-}$ cells effectively reconstituted lymph nodes; however, at 80%, the reconstitution was not as complete as that of the thymus or spleen, probably due to the presence of longer-lived, radio-resistant, memory populations in the lymph nodes (Nguyen and Taffet, 2009). However, the proportion of CD4+ and CD8+ cells in the lymph nodes was similar in chimeras generated from both Cx43$^{+/+}$ and Cx43$^{-/-}$ donors. As with the spleen (Nguyen and Taffet, 2009), B cell repopulation in the

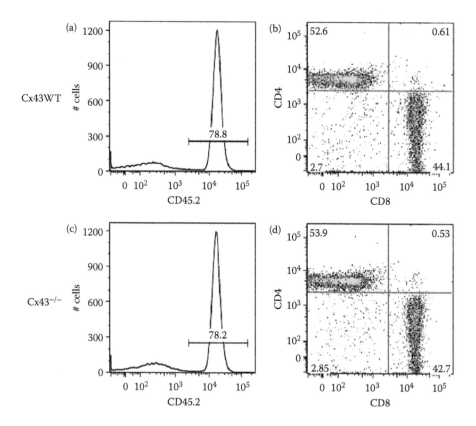

FIGURE 3.2 Analysis of lymph node cells from radiation chimeric mice. Mice were irradiated and reconstituted with fetal liver cells from wild-type and Cx43 knockout mice. After 8 weeks, lymph nodes were isolated and the cells analyzed by flow cytometry. Panels (a) and (c) show the proportion of cells in the lymph node from the donor mice (CD45.2+). There was no significant difference in reconstitution due to the Cx43 content of the donor cells. Panels (b) and (d) show the proportion of CD4+ and CD8+ cells within the CD45.2 positive cell population.

lymph nodes was unaffected by Cx43 depletion. Overall, both T and B lymphocyte populations could be successfully reconstituted in the absence of Cx43.

T cells were apparently capable of effective migration from the bone marrow to the thymus and then to the spleen and lymph nodes, while B cells could move from the bone marrow to the spleen and lymph nodes. This result was unexpected, given that Cx43 has been suggested to be critical for leukocyte migration (Oviedo-Orta et al., 2002).

Although Cx43 expression in immune cells seems to be dispensable for lymphocyte development, Cx43 has been reported to have roles in lymphocyte immune effector functions. In terms of B cells, it was shown that blocking gap junction communication in mixed lymphocyte cultures significantly reduced the secretion of immunoglobulins and IL-10. In particular, production of IgG, IgM, and IgA was greatly reduced following exposure to connexin mimetic peptides (Evans and Boitano, 2001; Oviedo-Orta et al., 2001). In our own studies, we analyzed antibody production by immunization with the T-dependent antigen KLH, which requires T cell interaction with antigen-presenting cells as well as B cells, and did not see a defect in antibody production.

Similar to lymphocytes, we did not detect a difference in macrophage production from Cx43$^{-/-}$ fetal liver cells (Nguyen and Taffet, 2009). We used a standard model of peritoneal inflammation to induce a substantial macrophage response. Animals that were irradiated and allowed to reconstitute were injected intraperitoneally with Brewer's thioglycollate broth. Peritoneal lavage was performed after 72 h and the populations of inflammatory cells were analyzed. No significant differences were observed in the counts of responding cells between chimeric animals reconstituted from Cx43$^{+/+,\ +/-}$, or $^{-/-}$ donors. In all cases, approximately 97% of the recovered cells were CD45.2 (donor) positive and greater than 60% of these cells were CD11b positive macrophages (Nguyen and Taffet, 2009). This result implied that leukocyte transmigration was not dependent on Cx43 expression, as had been predicted by *in vitro* models (Zahler et al., 2003).

The radiation chimera model is an interesting model in which to study the role of Cx43 in immune function. We have demonstrated the successful reconstitution of T and B lymphocytes, monocytes, and neutrophils in these animals. Basic immune functions are preserved as attested by the ability to produce antibody and respond to inflammatory stimulation (Nguyen and Taffet, 2009). Additionally, we have maintained chimeric animals of all Cx43 genotypes for several months after reconstitution, indicating that basic immune function persists. Further studies on T cell regulation (Bopp et al., 2007), immunity to bacteria (Anand et al., 2008), and antigen cross-presentation (Neijssen et al., 2005, 2007; Saccheri et al., 2010) will be of great interest in the future.

3.3   CULTURE OF CELLS DIRECTLY FROM FETAL LIVER

It is widely accepted that hematopoietic stem cells can be cultured in the presence of growth factors to produce a variety of myeloid cell types. Originally, media conditioned by the fibroblast cell line L-929 was used to stimulate the production of macrophages from bone marrow progenitor cells as well as the proliferation of peritoneal macrophages elicited using thioglycollate media (Hara and Ogawa, 1978; Stewart et al., 1975). The culture of bone marrow cells with L-cell conditioned media, and more recently, recombinant M-CSF, results in a population that is essentially 100% macrophages based on flow cytometry. Therefore, this is a technique that can yield a large number of essentially pure macrophages from a limited number of donors. One limitation of bone marrow derivation of macrophages is that marrow is most easily harvested from adult mice, limiting the technique's application to genes that do not result in pre- or neonatal lethality, such as Cx43. As the murine fetal liver has long been recognized as a potent source of granulocyte and macrophage progenitor cells (Cline and Moore, 1972; Moore and Williams, 1973), it is possible to culture macrophages from a fetal liver in cases where the deleted gene, such as Cx43, is essential for postnatal viability (Crowley et al., 1997; Underhill et al., 1998; Levine et al., 2005; Kanazawa and Kudo, 2005; Soni et al., 2006; Ogawa et al., 2004). We have recently been employing this technique to study the function of murine macrophages with a Cx43$^{-/-}$ genotype.

In Figure 3.3, we show the flow cytometry from cultures of fetal liver-derived cells from Cx43$^{+/+,\,+/-}$, and $^{-/-}$ mice. For this type of study, mice heterozygous for the gene in question were mated, and fetuses were removed at approximately 12–15 days of gestation. It is at this time that the liver has the highest content of HSC (Morrison et al., 1995). Once the fetal livers had been removed and placed in culture, a sample of the remaining fetal material was used for genotyping. After 24 h, nonadherent cells of identical genotypes were pooled and cultured in M-CSF supplemented media for 1 week. The adherent cells contain fibroblasts that may overgrow the culture. This technique provides (assuming a Mendelian inheritance) macrophages that are homozygously and heterozygously knocked out of the gene of interest, as well as wild-type macrophages for comparison from within the same litter. As shown in Figure 3.3,

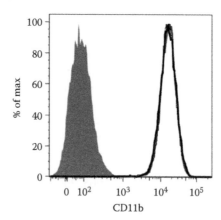

FIGURE 3.3 Analysis of macrophages cultured from fetal liver cells. Cx43$^{+/+}$, Cx43$^{+/-}$, and Cx43$^{-/-}$ fetal liver cells were cultured in M-CSF for 7 days and assessed for the proportion of cells expressing the macrophage marker CD11b. The unstained wild-type cells are represented by the gray filled area. Cells stained for CD11b are Cx43$^{+/+}$ (gray line), Cx43$^{+/-}$ (light black line), and Cx43$^{-/-}$ (heavy black line). There was no Cx43 dependence to *in vitro* macrophage production.

nearly all the cells (>99%) from Cx43$^{+/+, +/-}$, and $^{-/-}$ fetal livers are positive for the macrophage marker CD11b. We could not detect any difference in cell yield from the different genotypes. Thus, large numbers of nearly pure macrophages can be produced from Cx43 KO mice.

We have also analyzed the functional activity of the macrophages derived from these cultures. In Figure 3.4, we show the phagocytic potential of the macrophages from one such culture. Fetal liver macrophages cultured from Cx43$^{+/+, +/-}$, and $^{-/-}$ mice were exposed to zymosan particles labeled with Texas Red for 1 h and opsonized

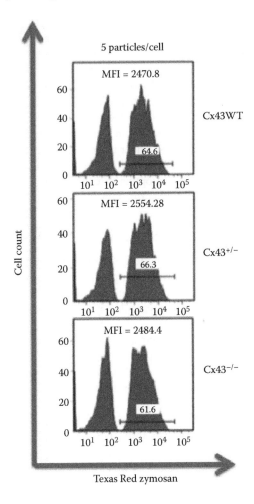

FIGURE 3.4 Phagocytosis of labeled zymosan by cultured macrophages. Macrophages were cultured from Cx43$^{+/+}$, Cx43$^{+/-}$, and Cx43$^{-/-}$ fetal liver cells. These cells were then incubated with Texas Red-conjugated zymosan particles and analyzed for phagocytosis by flow cytometry. The panels show that cultured macrophages efficiently take up zymosan particles. There was no difference in either the proportion of cells that take up zymosan or the relative amount of particle uptake (mean fluorescence intensity) between macrophages cultured from Cx43$^{+/+}$, Cx43$^{+/-}$, and Cx43$^{-/-}$ fetal liver cells.

with normal mouse serum. After washing, we assessed the capability of these cells to phagocytose zymosan. Microscopic analysis indicated that most of the zymosan appeared to be internalized (data not shown). Flow cytometric analysis was used to determine the percentage of cells that had taken up zymosan and the relative amount of internalized particles was assessed as mean fluorescence intensity (MFI). We did not detect any significant difference in either the percentage of cells that had taken up particles or the amount of zymosan particles ingested. We have performed similar studies with other types of particles such as bacteria and antibody-coated erythrocytes but have not detected a defect in phagocytosis among macrophages lacking Cx43.

In a recent publication, it was reported that macrophages from Cx43 KO fetal livers were not capable of phagocytosis (Anand et al., 2008). These investigators assessed macrophages isolated directly from fetal livers. These cells were isolated by adherence and were not generated in culture. Thus, the function of these macrophages may represent a state that is induced by changes in the liver due to Cx43 depletion. In our study, the adherent cells were removed and the nonadherent cells were cultured. Therefore, the environment in which the macrophages were cultured was identical for all three genotypes. Under these conditions, we found that phagocytosis was not dependent on Cx43 expression. The culturing of macrophages from Cx43 KO fetal liver cells is a useful option in exploring the impact of reduced or absent Cx43 expression in macrophage function.

3.4   CONDITIONAL AND TISSUE-SPECIFIC KNOCKOUTS

In the past, the lack of viable adult models of Cx43 deletion due to neonatal lethality has hindered the study of Cx43's role in physiological systems at later stages of murine development. Transgenic models that delete Cx43 in specific tissues while maintaining expression in the cardiovascular system (cre/loxP technology) would be extremely useful in overcoming these obstacles. Some examples of using the cre/loxP system to study the role of Cx43 in specific nonimmune tissues are endothelial cells (Liao et al., 2001), heart (Gutstein et al., 2001), Sertoli cells (Sridharan et al., 2007), smooth muscle (Doring et al., 2007), oocytes (Gershon et al., 2008), and astrocytes (Requardt et al., 2009). Interestingly, there are only limited studies of cre/loxP depletion of Cx43 in immune cells.

Experiments done by Presley et al. (2005) deleted Cx43 by cre/loxP technology using the Mx1-promoter. The goal of this study was to examine the contribution of Cx43 expression within the stromal compartment of the bone marrow. These studies showed that depletion of Cx43 in bone marrow stromal cells significantly decreased the short-term hematopoietic recovery as assessed by measurements of peripheral blood cell count. In this case, Mx1 expression is activated in response to exposure to poly IC. Mice were analyzed for hematopoietic regenerative ability after 5-fluorouracil treatment. The investigators noted a 90% reduction of Cx43 in the bone marrow in the induced Mx-Cre/Cx43$^{flox/flox}$ mice. This depletion resulted in a defect in short-term repopulation ability but not a long-term defect (Presley et al., 2005). In this study, there was no note of an impact on immune function in the Cx43 depleted animals. In a second study, Cx43 was depleted throughout the mouse using an estrogen receptor responsive cre (Eltzschig et al., 2006). In this cre system, treatment of the animals with 4-hydroxytamoxifen results in Cx43 depletion in multiple tissues. In this study, it was shown that ATP release by neutrophils is dependent on Cx43. Although this was a useful method to produce Cx43-deficient neutrophils, the depletion was not specific to neutrophils alone.

In 2011, Kuczma et al. (2011) produced a mouse that utilized cre under the CD4 promoter to deplete Cx43 in T lymphocytes. This study showed normal development of conventional T cells. Cx43-depleted CD4+ and CD8+ cells developed normally and migrated to lymph nodes and spleen. The primary difference noted was a greatly decreased proportion of Foxp3+ cells, which was accompanied by increased populations of activated T cells. The authors concluded that the expression of FoxP3 was dependent on Cx43 expression.

It should be noted that CD4-cre is not specific to CD4+ lymphocytes alone. Since all T cells pass through a double-positive (CD4+, CD8+) stage during thymic development, the cre recombinase is expressed, excising and deleting the floxed Cx43 in CD8+ T cells as well (Lee et al., 2001). It is interesting to note that CD4 is also expressed in some macrophages, and that transgenes expressing CD4 promoter sequence are active in those cells (Hanna et al., 1994). It is not clear whether there is any expression of cre recombinase in macrophages of CD4-cre transgenic animals.

Given the extraordinary benefits of tissue-specific cre/loxP technology, and ready availability of Cx43 "floxed" animals, this model system may represent the future of the study of connexins in immune function. Currently, animals are available that express cre recombinase under several immune cell-specific promoters. For macrophages, mice have been developed to express cre under the macrophage lysozyme (m-lyz) promoter (Clausen et al., 1999), as well as the CD11b promoter (Ferron and Vacher, 2005). Other immune cell types can be studied using mice expressing cre under the promoters for CD11c (dendritic cells) (Melillo et al., 2010), CD19 (B cells) (Rickert et al., 1997), and lck (T cells) (Zhang et al., 2005; Lee et al., 2001). Of course, this list is not exhaustive and new cre/loxP immune models are being developed continuously.

Both the cre/loxP technology and the radiation chimera mouse models described here are similar in that the overall goal is to bypass the lethal cardiovascular abnormalities in Cx43 KO mice. Cre/loxP technology has the advantage of being a less invasive and "cleaner" transgenic model in which to delete Cx43 in specific immune cell types versus our Cx43 radiation chimera model (Nguyen and Taffet, 2009) where all immune cells have Cx43 depleted.

When embarking upon studies of the functional role of connexins in the immune system, investigators today have the benefit of multitudinous approaches. In contrast with early investigations, blocking endogenous channels using chemical inhibitors and transfection of immune cell lines for expression of normal, tagged, or mutant connexins are no longer the only available approaches to identifying how these proteins contribute to immune cell function. Such approaches remain powerful tools, just as the studies that employed them remain critically relevant to the field. Improvements to these approaches, and the development of entirely new methods such as more specific inhibitors, Nucleofection, improved transduction, chimeric mice, and cre/loxP technology, have offered, and will continue to offer, new perspectives. We look forward to the next two decades, and beyond, of research into the role(s) of connexins in immunity. With the foundations that have been laid, they will undoubtedly be even more fruitful than those of the past.

REFERENCES

Akinc, A., Zumbuehl, A., Goldberg, M., Leshchiner, E. S., Busini, V., Hossain, N., Bacallado, S. A. et al. 2008. A combinatorial library of lipid-like materials for delivery of RNAi therapeutics. *Nat Biotechnol,* 26, 561–9.

Albina, J. E., Cui, S., Mateo, R. B., and Reichner, J. S. 1993. Nitric oxide-mediated apoptosis in murine peritoneal macrophages. *J Immunol,* 150, 5080–5.

Amarzguioui, M., Lundberg, P., Cantin, E., Hagstrom, J., Behlke, M. A., and Rossi, J. J. 2006. Rational design and *in vitro* and *in vivo* delivery of Dicer substrate siRNA. *Nat Protoc,* 1, 508–17.

Anand, R. J., Dai, S., Gribar, S. C., Richardson, W., Kohler, J. W., Hoffman, R. A., Branca, M. F. et al. 2008. A role for connexin43 in macrophage phagocytosis and host survival after bacterial peritoneal infection. *J Immunol,* 181, 8534–43.

Bermudez-Fajardo, A., Yliharsila, M., Evans, W. H., Newby, A. C., and Oviedo-Orta, E. 2007. CD4+ T lymphocyte subsets express connexin 43 and establish gap junction channel communication with macrophages in vitro. *J Leukoc Biol,* 82, 608–12.

Bodi, E., Hurtado, S. P., Carvalho, M. A., Borojevic, R., and Carvalho, A. C. 2004. Gap junctions in hematopoietic stroma control proliferation and differentiation of blood cell precursors. *An Acad Bras Cienc,* 76, 743–56.

Boitano, S. and Evans, W. H. 2000. Connexin mimetic peptides reversibly inhibit Ca(2+) signaling through gap junctions in airway cells. *Am J Physiol Lung Cell Mol Physiol,* 279, L623–30.

Bopp, T., Becker, C., Klein, M., Klein-Hessling, S., Palmetshofer, A., Serfling, E., Heib, V. et al. 2007. Cyclic adenosine monophosphate is a key component of regulatory T cell-mediated suppression. *J Exp Med,* 204, 1303–10.

Bruzzone, R., Barbe, M. T., Jakob, N. J., and Monyer, H. 2005. Pharmacological properties of homomeric and heteromeric pannexin hemichannels expressed in *Xenopus* oocytes. *J Neurochem,* 92, 1033–43.

Burt, J. M. and Spray, D. C. 1989. Volatile anesthetics block intercellular communication between neonatal rat myocardial cells. *Circ Res,* 65, 829–37.

Cancelas, J. A., Koevoet, W. L., De Koning, A. E., Mayen, A. E., Rombouts, E. J., and Ploemacher, R. E. 2000. Connexin-43 gap junctions are involved in multiconnexin-expressing stromal support of hemopoietic progenitors and stem cells. *Blood,* 96, 498–505.

Chadjichristos, C. E., Scheckenbach, K. E., Van Veen, T. A., Richani Sarieddine, M. Z., De Wit, C., Yang, Z., Roth, I. et al. 2010. Endothelial-specific deletion of connexin40 promotes atherosclerosis by increasing Cd73-dependent leukocyte adhesion. *Circulation,* 121, 123–31.

Chanson, M., Derouette, J. P., Roth, I., Foglia, B., Scerri, I., Dudez, T., and Kwak, B. R. 2005. Gap junctional communication in tissue inflammation and repair. *Biochim Biophys Acta,* 1711, 197–207.

Clausen, B. E., Burkhardt, C., Reith, W., Renkawitz, R., and Forster, I. 1999. Conditional gene targeting in macrophages and granulocytes using LysMcre mice. *Transgenic Res,* 8, 265–77.

Cline, M. J. and Moore, M. A. 1972. Embryonic origin of the mouse macrophage. *Blood,* 39, 842–9.

Crowley, M. T., Costello, P. S., Fitzer-Attas, C. J., Turner, M., Meng, F., Lowell, C., Tybulewicz, V. L., and Defranco, A. L. 1997. A critical role for Syk in signal transduction and phagocytosis mediated by Fcgamma receptors on macrophages. *J Exp Med*, 186, 1027–39.

Dahl, G., Nonner, W., and Werner, R. 1994. Attempts to define functional domains of gap junction proteins with synthetic peptides. *Biophys J*, 67, 1816–22.

Davidson, J. S., Baumgarten, I. M., and Harley, E. H. 1986. Reversible inhibition of intercellular junctional communication by glycyrrhetinic acid. *Biochem Biophys Res Commun*, 134, 29–36.

Diaz, J., Tornel, P. L., Jara, P., Canizares, F., Egea, J. M., and Martinez, P. 1995. The value of polymorphonuclear elastase in adult respiratory distress syndrome. *Clin Chim Acta*, 236, 119–27.

Dokka, S., Toledo, D., Shi, X., Ye, J., and Rojanasakul, Y. 2000a. High-efficiency gene transfection of macrophages by lipoplexes. *Int J Pharm*, 206, 97–104.

Dokka, S., Toledo, D., Wang, L., Shi, X., Huang, C., Leonard, S., and Rojanasakul, Y. 2000b. Free radical-mediated transgene inactivation of macrophages by endotoxin. *Am J Physiol Lung Cell Mol Physiol*, 279, L878–83.

Doring, B., Pfitzer, G., Adam, B., Liebregts, T., Eckardt, D., Holtmann, G., Hofmann, F., Feil, S., Feil, R., and Willecke, K. 2007. Ablation of connexin43 in smooth muscle cells of the mouse intestine: Functional insights into physiology and morphology. *Cell Tissue Res*, 327, 333–42.

Dos Reis, G. A. and De Oliveira-Castro, G. M. 1978. [Mechanism of slow hyperpolarization in activated macrophages (proceedings)]. *An Acad Bras Cienc*, 50, 127.

Duran-Struuck, R. and Dysko, R. C. 2009. Principles of bone marrow transplantation (BMT): Providing optimal veterinary and husbandry care to irradiated mice in BMT studies. *J Am Assoc Lab Anim Sci*, 48, 11–22.

Ebert, O., Finke, S., Salahi, A., Herrmann, M., Trojaneck, B., Lefterova, P., Wagner, E. et al. 1997. Lymphocyte apoptosis: Induction by gene transfer techniques. *Gene Ther*, 4, 296–302.

Eckardt, S. and Mclaughlin, K. J. 2008. Transplantation of chimeric fetal liver to study hematopoiesis. *Methods Mol Biol*, 430, 195–211.

Eltzschig, H. K., Eckle, T., Mager, A., Kuper, N., Karcher, C., Weissmuller, T., Boengler, K., Schulz, R., Robson, S. C., and Colgan, S. P. 2006. ATP release from activated neutrophils occurs via connexin 43 and modulates adenosine-dependent endothelial cell function. *Circ Res*, 99, 1100–8.

Escher, G., Hoang, A., Georges, S., Tchoua, U., El-Osta, A., Krozowski, Z., and Sviridov, D. 2005. Demethylation using the epigenetic modifier, 5-azacytidine, increases the efficiency of transient transfection of macrophages. *J Lipid Res*, 46, 356–65.

Eugenin, E. A., Branes, M. C., Berman, J. W., and Saez, J. C. 2003. TNF-alpha plus IFN-gamma induce connexin43 expression and formation of gap junctions between human monocytes/macrophages that enhance physiological responses. *J Immunol*, 170, 1320–8.

Evans, W. H. and Boitano, S. 2001. Connexin mimetic peptides: Specific inhibitors of gap-junctional intercellular communication. *Biochem Soc Trans*, 29, 606–12.

Ferron, M. and Vacher, J. 2005. Targeted expression of Cre recombinase in macrophages and osteoclasts in transgenic mice. *Genesis*, 41, 138–45.

Fonseca, P. C., Nihei, O. K., Urban-Maldonado, M., Abreu, S., De Carvalho, A. C., Spray, D. C., Savino, W., and Alves, L. A. 2004. Characterization of connexin 30.3 and 43 in thymocytes. *Immunol Lett*, 94, 65–75.

Foster, D. A., Taffet, S. M., Ruffolo, R. R., Poste, G., and Metcalf, B. W. 1997. Complex regulation of NF-kappa B transcription factor complex. NF-kB in inhibitable by tyrosine kinase inhibitors. *Cell Cycle Regulation*. Ruffolo, R. R., Poste, G., and Metcalf, B. W. (Eds). Amsterdam: Harwood Academic Publishers.

Gaziri, I. F., Oliveira-Castro, G. M., Machado, R. D., and Barcinski, M. A. 1975. Structure and permeability of junctions in phytohemagglutinin stimulated human lymphocytes. *Experientia*, 31, 172–4.

Gershon, E., Plaks, V., Aharon, I., Galiani, D., Reizel, Y., Sela-Abramovich, S., Granot, I., Winterhager, E., and Dekel, N. 2008. Oocyte-directed depletion of connexin43 using the Cre-LoxP system leads to subfertility in female mice. *Dev Biol*, 313, 1–12.

Goldberg, G. S., Moreno, A. P., Bechberger, J. F., Hearn, S. S., Shivers, R. R., Macphee, D. J., Zhang, Y. C., and Naus, C. C. 1996. Evidence that disruption of connexon particle arrangements in gap junction plaques is associated with inhibition of gap junctional communication by a glycyrrhetinic acid derivative. *Exp Cell Res*, 222, 48–53.

Guan, X., Wilson, S., Schlender, K., K. and Ruch, R. J. 1996. Gap-junction disassembly and connexin 43 dephosphorylation induced by 18 beta-glycyrrhetinic acid. *Mol Carcinog*, 16, 157–164.

Gutstein, D. E., Morley, G. E., Tamaddon, H., Vaidya, D., Schneider, M. D., Chen, J., Chien, K. R., Stuhlmann, H., and Fishman, G. I. 2001. Conduction slowing and sudden arrhythmic death in mice with cardiac-restricted inactivation of connexin43. *Circ Res*, 88, 333–9.

Haeryfar, S. M. and Hoskin, D. W. 2004. Thy-1: More than a mouse pan-T cell marker. *J Immunol*, 173, 3581–8.

Hanna, Z., Simard, C., Laperriere, A., and Jolicoeur, P. 1994. Specific expression of the human CD4 gene in mature CD4+ CD8- and immature CD4+ CD8+ T cells and in macrophages of transgenic mice. *Mol Cell Biol*, 14, 1084–94.

Hara, H. and Ogawa, M. 1978. Murine hemopoietic colonies in culture containing normoblasts, macrophages, and megakaryocytes. *Am J Hematol*, 4, 23–34.

Herve, J. C. and Sarrouilhe, D. 2002. Modulation of junctional communication by phosphorylation: Protein phosphatases, the missing link in the chain. *Biol Cell*, 94, 423–432.

Iglesias, R., Locovei, S., Roque, A., Alberto, A. P., Dahl, G., Spray, D. C., and Scemes, E. 2008. P2X7 receptor-Pannexin1 complex: Pharmacology and signaling. *Am J Physiol Cell Physiol*, 295, C752–60.

Iwasaki, A. 2006. The use of bone marrow-chimeric mice in elucidating immune mechanisms. *Methods Mol Med*, 127, 281–92.

Janssen, W. J., Muldrow, A., Kearns, M. T., Barthel, L., and Henson, P. M. 2010. Development and characterization of a lung-protective method of bone marrow transplantation in the mouse. *J Immunol Methods*, 357, 1–9.

Jara, P. I., Boric, M. P., and Saez, J. C. 1995. Leukocytes express connexin 43 after activation with lipopolysaccharide and appear to form gap junctions with endothelial cells after ischemia-reperfusion. *Proc Natl Acad Sci U S A*, 92, 7011–5.

Kanazawa, K. and Kudo, A. 2005. TRAF2 is essential for TNF-alpha-induced osteoclastogenesis. *J Bone Miner Res*, 20, 840–7.

Kaner, R. J., Worgall, S., Leopold, P. L., Stolze, E., Milano, E., Hidaka, C., Ramalingam, R. et al. 1999. Modification of the genetic program of human alveolar macrophages by adenovirus vectors *in vitro* is feasible but inefficient, limited in part by the low level of expression of the coxsackie/adenovirus receptor. *Am J Respir Cell Mol Biol*, 20, 361–70.

Kiel, M. J., Yilmaz, O. H., Iwashita, T., Terhorst, C., and Morrison, S. J. 2005. SLAM family receptors distinguish hematopoietic stem and progenitor cells and reveal endothelial niches for stem cells. *Cell*, 121, 1109–21.

Kuczma, M., Lee, J. R., and Kraj, P. 2011. Connexin 43 signaling enhances the generation of Foxp3+ regulatory T cells. *J Immunol*, 187, 248–57.

Kwak, B. R. and Jongsma, H. J. 1999. Selective inhibition of gap junction channel activity by synthetic peptides. *J Physiol*, 516(Pt 3), 679–85.

Kwak, B. R., Veillard, N., Pelli, G., Mulhaupt, F., James, R. W., Chanson, M., and Mach, F. 2003. Reduced connexin43 expression inhibits atherosclerotic lesion formation in low-density lipoprotein receptor-deficient mice. *Circulation*, 107, 1033–9.

Lee, P. P., Fitzpatrick, D. R., Beard, C., Jessup, H. K., Lehar, S., Makar, K. W., Perez-Melgosa, M. et al. 2001. A critical role for Dnmt1 and DNA methylation in T cell development, function, and survival. *Immunity,* 15, 763–74.

Leon-Ponte, M., Kirchhof, M. G., Sun, T., Stephens, T., Singh, B., Sandhu, S., and Madrenas, J. 2005. Polycationic lipids inhibit the pro-inflammatory response to LPS. *Immunol Lett,* 96, 73–83.

Levine, R. F., Derks, R., and Beaman, K. 2005. Regeneration and tolerance factor. A vacuolar ATPase with consequences. *Chem Immunol Allergy,* 89, 126–34.

Leyva, F. J., Anzinger, J. J., Mccoy, J. P., Jr., and Kruth, H. S. 2011. Evaluation of transduction efficiency in macrophage colony-stimulating factor differentiated human macrophages using HIV-1 based lentiviral vectors. *BMC Biotechnol,* 11, 13.

Liao, Y., Day, K. H., Damon, D. N., and Duling, B. R. 2001. Endothelial cell-specific knockout of connexin 43 causes hypotension and bradycardia in mice. *Proc Natl Acad Sci U S A,* 98, 9989–94.

Locovei, S., Bao, L., and Dahl, G. 2006. Pannexin 1 in erythrocytes: Function without a gap. *Proc Natl Acad Sci U S A,* 103, 7655–9.

Mack, K. D., Wei, R., Elbagarri, A., Abbey, N., and Mcgrath, M. S. 1998. A novel method for DEAE-dextran mediated transfection of adherent primary cultured human macrophages. *J Immunol Methods,* 211, 79–86.

Marina-Garcia, N., Franchi, L., Kim, Y. G., Miller, D., Mcdonald, C., Boons, G. J., and Nunez, G. 2008. Pannexin-1-mediated intracellular delivery of muramyl dipeptide induces caspase-1 activation via cryopyrin/NLRP3 independently of Nod2. *J Immunol,* 180, 4050–7.

Martin, C. A., Homaidan, F. R., Palaia, T., Burakoff, R., and EL-Sabban, M. E. 1998. Gap junctional communication between murine macrophages and intestinal epithelial cell lines. *Cell Adhes Commun,* 5, 437–49.

Matsue, H., Yao, J., Matsue, K., Nagasaka, A., Sugiyama, H., Aoki, R., Kitamura, M., and Shimada, S. 2006. Gap junction-mediated intercellular communication between dendritic cells (DCs) is required for effective activation of DCs. *J Immunol,* 176, 181–90.

Mattern, T., Flad, H. D., Brade, L., Rietschel, E. T., and Ulmer, A. J. 1998. Stimulation of human T lymphocytes by LPS is MHC unrestricted, but strongly dependent on B7 interactions. *J Immunol,* 160, 3412–8.

Melillo, J. A., Song, L., Bhagat, G., Blazquez, A. B., Plumlee, C. R., Lee, C., Berin, C., Reizis, B., and Schindler, C. 2010. Dendritic cell (DC)-specific targeting reveals Stat3 as a negative regulator of DC function. *J Immunol,* 184, 2638–45.

Mendoza-Naranjo, A., Saez, P. J., Johansson, C. C., Ramirez, M., Mandakovic, D., Pereda, C., Lopez, M. N., Kiessling, R., Saez, J. C., and Salazar-Onfray, F. 2007. Functional gap junctions facilitate melanoma antigen transfer and cross-presentation between human dendritic cells. *J Immunol,* 178, 6949–57.

Montecino-Rodriguez, E. and Dorshkind, K. 2001. Regulation of hematopoiesis by gap junction-mediated intercellular communication. *J Leukoc Biol,* 70, 341–7.

Montecino-Rodriguez, E., Leathers, H., and Dorshkind, K. 2000. Expression of connexin 43 (Cx43) is critical for normal hematopoiesis. *Blood,* 96, 917–24.

Moore, J. C., Tsang, S. Y., Rushing, S. N., Lin, D., Tse, H. F., Chan, C. W., and Li, R. A. 2008. Functional consequences of overexpressing the gap junction Cx43 in the cardiogenic potential of pluripotent human embryonic stem cells. *Biochem Biophys Res Commun,* 377, 46–51.

Moore, M. A. and Williams, N. 1973. Analysis of proliferation and differentiation of foetal granulocyte-macrophage progenitor cells in haemopoietic tissue. *Cell Tissue Kinet,* 6, 461–76.

Morrison, S. J., Hemmati, H. D., Wandycz, A. M., and Weissman, I. L. 1995. The purification and characterization of fetal liver hematopoietic stem cells. *Proc Natl Acad Sci U S A,* 92, 10302–6.

Navarro, J. and Touraine, J. L. 1989. Promotion of fetal liver engraftment by T cells in a murine semiallogeneic model without graft-versus-host reaction. *Transplantation,* 47, 871–6.

Neijssen, J., Herberts, C., Drijfhout, J. W., Reits, E., Janssen, L., and Neefjes, J. 2005. Cross-presentation by intercellular peptide transfer through gap junctions. *Nature,* 434, 83–8.

Neijssen, J., Pang, B., and Neefjes, J. 2007. Gap junction-mediated intercellular communication in the immune system. *Prog Biophys Mol Biol,* 94, 207–18.

Nguyen, T. D. and Taffet, S. M. 2009. A model system to study Connexin 43 in the immune system. *Mol Immunol,* 46, 2938–46.

O'Donnell, P. M. and Taffet, S. M. 2002. The proximal promoter region is essential for lipopolysaccharide induction and cyclic AMP inhibition of mouse tumor necrosis factor-alpha. *J Interferon Cytokine Res,* 22, 539–48.

Ogawa, S., Lozach, J., Jepsen, K., Sawka-Verhelle, D., Perissi, V., Sasik, R., Rose, D. W., Johnson, R. S., Rosenfeld, M. G., and Glass, C. K. 2004. A nuclear receptor corepressor transcriptional checkpoint controlling activator protein 1-dependent gene networks required for macrophage activation. *Proc Natl Acad Sci U S A,* 101, 14461–6.

Ogawa, T., Shimauchi, H., Uchida, H., and Mori, Y. 1996. Stimulation of splenocytes in C3H/HeJ mice with *Porphyromonas gingivalis* lipid A in comparison with enterobacterial lipid A. *Immunobiology,* 196, 399–414.

Ojielo, C. I., Cooke, K., Mancuso, P., Standiford, T. J., Olkiewicz, K. M., Clouthier, S., Corrion, L. et al. 2003. Defective phagocytosis and clearance of *Pseudomonas aeruginosa* in the lung following bone marrow transplantation. *J Immunol,* 171, 4416–24.

Oliveira-Castro, G. M., Barcinski, M. A., and Cukierman, S. 1973. Intercellular communication in stimulated human lymphocytes. *J Immunol,* 111, 1616–9.

Oloris, S. C., Mesnil, M., Reis, V. N., Sakai, M., Matsuzaki, P., Fonseca EDE, S., Da Silva, T. C. et al. 2007. Hepatic granulomas induced by *Schistosoma mansoni* in mice deficient for connexin 43 present lower cell proliferation and higher collagen content. *Life Sci,* 80, 1228–35.

Oviedo-Orta, E., Errington, R. J., and Evans, W. H. 2002. Gap junction intercellular communication during lymphocyte transendothelial migration. *Cell Biol Int,* 26, 253–63.

Oviedo-Orta, E., Gasque, P., and Evans, W. H. 2001. Immunoglobulin and cytokine expression in mixed lymphocyte cultures is reduced by disruption of gap junction intercellular communication. *FASEB J,* 15, 768–74.

Oviedo-Orta, E. and Howard Evans, W. 2004. Gap junctions and connexin-mediated communication in the immune system. *Biochim Biophys Acta,* 1662, 102–12.

Oviedo-Orta, E., Hoy, T., and Evans, W. H. 2000. Intercellular communication in the immune system: Differential expression of connexin 40 and 43, and perturbation of gap junction channel functions in peripheral blood and tonsil human lymphocyte subpopulations. *Immunology,* 99, 578–90.

Pan, H., Mostoslavsky, G., Eruslanov, E., Kotton, D. N., and Kramnik, I. 2008. Dual-promoter lentiviral system allows inducible expression of noxious proteins in macrophages. *J Immunol Methods,* 329, 31–44.

Pelegrin, P., Barroso-Gutierrez, C., and Surprenant, A. 2008. P2X7 receptor differentially couples to distinct release pathways for IL-1beta in mouse macrophage. *J Immunol,* 180, 7147–57.

Pelegrin, P. and Surprenant, A. 2006. Pannexin-1 mediates large pore formation and interleukin-1beta release by the ATP-gated P2X7 receptor. *Embo J,* 25, 5071–82.

Pelegrin, P. and Surprenant, A. 2007. Pannexin-1 couples to maitotoxin- and nigericin-induced interleukin-1beta release through a dye uptake-independent pathway. *J Biol Chem,* 282, 2386–94.

Philpott, N. J., Nociari, M., Elkon, K. B., and Falck-Pedersen, E. 2004. Adenovirus-induced maturation of dendritic cells through a PI3 kinase-mediated TNF-alpha induction pathway. *Proc Natl Acad Sci U S A,* 101, 6200–5.

Plotkin, L. I., Lezcano, V., Thostenson, J., Weinstein, R. S., Manolagas, S. C., and Bellido, T. 2008. Connexin 43 is required for the anti-apoptotic effect of bisphosphonates on osteocytes and osteoblasts in vivo. *J Bone Miner Res,* 23, 1712–21.

Polacek, D., Lal, R., Volin, M. V., and Davies, P. F. 1993. Gap junctional communication between vascular cells. Induction of connexin43 messenger RNA in macrophage foam cells of atherosclerotic lesions. *Am J Pathol,* 142, 593–606.

Presley, C. A., Lee, A. W., Kastl, B., Igbinosa, I., Yamada, Y., Fishman, G. I., Gutstein, D. E., and Cancelas, J. A. 2005. Bone marrow connexin-43 expression is critical for hematopoietic regeneration after chemotherapy. *Cell Commun Adhes,* 12, 307–17.

Reaume, A. G., De Sousa, P. A., Kulkarni, S., Langille, B. L., Zhu, D., Davies, T. C., Juneja, S. C., Kidder, G. M., and Rossant, J. 1995. Cardiac malformation in neonatal mice lacking connexin43. *Science,* 267, 1831–4.

Reinecke, H., Minami, E., Virag, J. I., and Murry, C. E. 2004. Gene transfer of connexin43 into skeletal muscle. *Hum Gene Ther,* 15, 627–36.

Requardt, R. P., Kaczmarczyk, L., Dublin, P., Wallraff-Beck, A., Mikeska, T., Degen, J., Waha, A., Steinhauser, C., Willecke, K., and Theis, M. 2009. Quality control of astrocyte-directed Cre transgenic mice: The benefits of a direct link between loss of gene expression and reporter activation. *Glia,* 57, 680–92.

Rickert, R. C., Roes, J., and Rajewsky, K. 1997. B lymphocyte-specific, Cre-mediated mutagenesis in mice. *Nucleic Acids Res,* 25, 1317–8.

Rosendaal, M. and Stone, M. 2003. Enhancement of repopulation haemopoiesis by heterozygous connexin 43 stem cells seeded on wild-type connexin 43 stroma. *Clin Sci (Lond.),* 105, 561–8.

Saccheri, F., Pozzi, C., Avogadri, F., Barozzi, S., Faretta, M., Fusi, P., and Rescigno, M. 2010. Bacteria-induced gap junctions in tumors favor antigen cross-presentation and antitumor immunity. *Sci Transl Med,* 2, 44ra57.

Sakurai, H., Kawabata, K., Sakurai, F., Nakagawa, S., and Mizuguchi, H. 2008. Innate immune response induced by gene delivery vectors. *Int J Pharm,* 354, 9–15.

Sarieddine, M. Z., Scheckenbach, K. E., Foglia, B., Maass, K., Garcia, I., Kwak, B. R., and Chanson, M. 2009. Connexin43 modulates neutrophil recruitment to the lung. *J Cell Mol Med,* 13, 4560–70.

Shao, Q., Wang, H., Mclachlan, E., Veitch, G. I., and Laird, D. W. 2005. Down-regulation of Cx43 by retroviral delivery of small interfering RNA promotes an aggressive breast cancer cell phenotype. *Cancer Res,* 65, 2705–11.

Soni, S., Bala, S., Gwynn, B., Sahr, K. E., Peters, L. L., and Hanspal, M. 2006. Absence of erythroblast macrophage protein (Emp) leads to failure of erythroblast nuclear extrusion. *J Biol Chem,* 281, 20181–9.

Sridharan, S., Brehm, R., Bergmann, M., and Cooke, P. S. 2007. Role of connexin 43 in Sertoli cells of testis. *Ann N Y Acad Sci,* 1120, 131–43.

Srinivas, M., Duffy, H. S., Delmar, M., Spray, D. C., Zipes, D. Z., and Jalife, J. 2004. Prospects for pharmacological targeting of gap junction channels. *Cardiac Electrophysiology: From Cell to Bedside.* Zipes, D. P. and Jalife, J. J. (Eds). Philadelphia: Saunders.

Srinivas, M., Hopperstad, M. G., and Spray, D. C. 2001. Quinine blocks specific gap junction channel subtypes. *Proc Natl Acad Sci USA,* 98, 10942–7.

Srinivas, M. and Spray, D. C. 2003. Closure of gap junction channels by arylaminobenzoates. *Mol Pharmacol,* 63, 1389–97.

Stewart, C. C., Lin, H. S., and Adles, C. 1975. Proliferation and colony-forming ability of peritoneal exudate cells in liquid culture. *J Exp Med,* 141, 1114–32.

Stewart, S. A., Dykxhoorn, D. M., Palliser, D., Mizuno, H., Yu, E. Y., An, D. S., Sabatini, D. M. et al. 2003. Lentivirus-delivered stable gene silencing by RNAi in primary cells. *RNA,* 9, 493–501.

Stohlman, S. A., Hinton, D. R., Parra, B., Atkinson, R., and Bergmann, C. C. 2008. CD4 T cells contribute to virus control and pathology following central nervous system infection with neurotropic mouse hepatitis virus. *J Virol*, 82, 2130–9.

Taffet, S. M., Shurtleff, S. A., Powanda, M. C., Oppenheim, J. J., Kluger, M. J., and Dinarello, C. A. 1990. Identification of multiple regulatory elements in the TNFalpha gene. *Molecular and Cellular Biology of Cytokines*. New York: Alan R. Liss, Inc.

Teshima, T., Ordemann, R., Reddy, P., Gagin, S., Liu, C., Cooke, K. R., and Ferrara, J. L. 2002. Acute graft-versus-host disease does not require alloantigen expression on host epithelium. *Nat Med*, 8, 575–81.

Thompson, C. D., Frazier-Jessen, M. R., Rawat, R., Nordan, R. P., and Brown, R. T. 1999. Evaluation of methods for transient transfection of a murine macrophage cell line, RAW 264.7. *Biotechniques*, 27, 824–6, 828–30, 832.

Tolmachov, O., Ma, Y. L., Themis, M., Patel, P., Spohr, H., Macleod, K. T., Ullrich, N. D., Kienast, Y., Coutelle, C., and Peters, N. S. 2006. Overexpression of connexin 43 using a retroviral vector improves electrical coupling of skeletal myoblasts with cardiac myocytes in vitro. *BMC Cardiovasc Disord*, 6, 25.

Tomaras, G. D., Foster, D. A., Burrer, C. M., and Taffet, S. M. 1999. ETS transcription factors regulate an enhancer activity in the third intron of TNF-alpha. *J Leukoc Biol*, 66, 183–93.

Underhill, D. M., Chen, J., Allen, L. A., and Aderem, A. 1998. MacMARCKS is not essential for phagocytosis in macrophages. *J Biol Chem*, 273, 33619–23.

Van De Parre, T. J., Martinet, W., Schrijvers, D. M., Herman, A. G., and De Meyer, G. R. 2005. mRNA but not plasmid DNA is efficiently transfected in murine J774A.1 macrophages. *Biochem Biophys Res Commun*, 327, 356–60.

Van Os, R., Sheridan, T. M., Robinson, S., Drukteinis, D., Ferrara, J. L., and Mauch, P. M. 2001. Immunogenicity of Ly5 (CD45)-antigens hampers long-term engraftment following minimal conditioning in a murine bone marrow transplantation model. *Stem Cells*, 19, 80–7.

Van Tendeloo, V. F., Ponsaerts, P., Lardon, F., Nijs, G., Lenjou, M., Van Broeckhoven, C., Van Bockstaele, D. R., and Berneman, Z. N. 2001. Highly efficient gene delivery by mRNA electroporation in human hematopoietic cells: Superiority to lipofection and passive pulsing of mRNA and to electroporation of plasmid cDNA for tumor antigen loading of dendritic cells. *Blood*, 98, 49–56.

Vliagoftis, H., Hutson, A. M., Mahmudi-Azer, S., Kim, H., Rumsaeng, V., Oh, C. K., Moqbel, R., and Metcalfe, D. D. 1999. Mast cells express connexins on their cytoplasmic membrane. *J Allergy Clin Immunol*, 103, 656–62.

Wang, J., MA, M., Locovei, S., Keane, R. W., and Dahl, G. 2007. Modulation of membrane channel currents by gap junction protein mimetic peptides: Size matters. *Am J Physiol Cell Physiol*, 293, C1112–9.

Woehrle, T., Yip, L., Elkhal, A., Sumi, Y., Chen, Y., Yao, Y., Insel, P. A., and Junger, W. G. 2010a. Pannexin-1 hemichannel-mediated ATP release together with P2X1 and P2X4 receptors regulate T-cell activation at the immune synapse. *Blood*, 116, 3475–84.

Woehrle, T., Yip, L., Manohar, M., Sumi, Y., Yao, Y., Chen, Y., and Junger, W. G. 2010b. Hypertonic stress regulates T cell function via pannexin-1 hemichannels and P2X receptors. *J Leukoc Biol*, 88, 1181–9.

Xaus, J., Comalada, M., Valledor, A. F., Lloberas, J., Lopez-Soriano, F., Argiles, J. M., Bogdan, C., and Celada, A. 2000. LPS induces apoptosis in macrophages mostly through the autocrine production of TNF-alpha. *Blood*, 95, 3823–31.

Zahler, S., Hoffmann, A., Gloe, T., and Pohl, U. 2003. Gap-junctional coupling between neutrophils and endothelial cells: A novel modulator of transendothelial migration. *J Leukoc Biol*, 73, 118–26.

Zeng, L., Yang, S., Wu, C., Ye, L., and Lu, Y. 2006. Effective transduction of primary mouse blood- and bone marrow-derived monocytes/macrophages by HIV-based defective lentiviral vectors. *J Virol Methods,* 134, 66–73.

Zhang, D. J., Wang, Q., Wei, J., Baimukanova, G., Buchholz, F., Stewart, A. F., Mao, X., and Killeen, N. 2005. Selective expression of the Cre recombinase in late-stage thymocytes using the distal promoter of the Lck gene. *J Immunol,* 174, 6725–31.

Zhang, X., Edwards, J. P., and Mosser, D. M. 2009. The expression of exogenous genes in macrophages: Obstacles and opportunities. *Methods Mol Biol,* 531, 123–43.

Zhou, X., Cui, Y., Huang, X., Yu, Z., Thomas, A. M., Ye, Z., Pardoll, D. M., Jaffee, E. M., and Cheng, L. 2003. Lentivirus-mediated gene transfer and expression in established human tumor antigen-specific cytotoxic T cells and primary unstimulated T cells. *Hum Gene Ther,* 14, 1089–105.

Zsengeller, Z., Otake, K., Hossain, S. A., Berclaz, P. Y., and Trapnell, B. C. 2000. Internalization of adenovirus by alveolar macrophages initiates early proinflammatory signaling during acute respiratory tract infection. *J Virol,* 74, 9655–67.

4 Gap Junctions in Antigen-Presenting Cells

Pablo J. Sáez, Kenji F. Shoji, and Juan Carlos Sáez

CONTENTS

4.1 INTRODUCTION

Dendritic cells (DCs) constitute a heterogeneous cell population that emerges in the bone marrow (BM) from a macrophage and DC precursor, generating the following: (1) a common DC precursor and (2) a monocyte precursor (Liu and Nussenzweig 2010). In addition, DCs can also emerge from monocytes under inflammatory conditions (Shortman and Naik 2007). Two major categories of DCs have been established as follows: (1) conventional (cDCs) and (2) inflammatory DCs. DC differentiation is reached by the former during resting steady-state conditions, while the latter do so during inflammatory conditions (Merad and Manz 2009). In addition, cDCs can be classified into several subtypes that share the ability to pick up, process, and present antigens to T cells. DC subtypes can be recognized through the expression of different cell surface markers, pattern cytokine secretions, migration pathways, locations, and functions (Shortman and Naik 2007). Although plasmacytoid DCs (pDCs) are crucial in the interferon I-mediated immune responses (Shortman and Naik 2007), they are outside the scope of this review. Immature DCs recognize "danger signals," which includes pathogen-associated molecular patterns (PAMPs) and damage-associated molecular patterns (DAMPs). They activate DCs, which in turn activate T cells after antigen presentation (Kono and Rock 2008).

DCs are highly specialized antigen-presenting cells (APCs), named "professional" APCs. Other immune cells, such as macrophages and B cells, are also APCs but less efficient than DCs (Janeway et al. 1987; Kleindienst and Brocker 2005; Kurt-Jones et al. 1988; Mosser and Edwards 2008). After antigen uptake, DCs expose it at the cell surface in the major histocompatibility complex class II (MHC II), which is recognized by T cell receptors (TCR) located at the cell surface of T cells (Krummel and Cahalan 2010). For antigen presentation, DCs migrate through lymphatic vessels from periphery to lymph nodes (LNs), or other secondary lymphoid organs (SLOs), where naïve T cells reside (Randolph et al. 2005). Antigen recognition induces a highly regulated maturation process in DCs that includes upregulation of molecules associated with antigen presentation, cell adhesion, migration, and cellular communication (Corvalán et al. 2007; Reis e Sousa 2006; Saccheri et al. 2010).

Once antigen-bearing DCs reach lymph nodes, they show slow motility but also frequently show shape changes and cellular process extensions. This cell dynamic allows transient contacts (minutes) with T cells, also observed in other SLOs (Bousso 2008). Later on, DCs lose their motility and establish long-lasting contacts (hours) with T cells, allowing the formation of a very specialized contact zone named immunological synapse, required for T cell activation and proliferation (Bousso 2008; Krummel and Cahalan 2010). Changes in protein distribution occur during the formation of the immunological synapse. Such changes allow the formation of supramolecular activation clusters (SMACs) that contain a centralized accumulation of TCR–MHC II complexes (cSMAC), surrounded by a peripheral zone (pSMAC) where transmembrane interactions occur (Krummel and Cahalan 2010). The immunological synapse formation provides proper cell adhesion, which allows gap junction channels (GJCs) to form and stabilize (Sáez et al. 2000). This is an important mechanism for cellular cross talk.

The following two protein families can constitute GJCs: connexins (Cxs) and pannexins (Panxs), although Panx GJCs have only been demonstrated in osteoclasts (Ishikawa et al. 2011). Each GJC is formed by two hemichannels (HCs), which are oligohexamers of Cxs or Panxs (although Panx2 might form octamers) provided by adjacent cells (Figure 4.1) (Ambrosi et al. 2010; Sáez et al. 2003). They allow direct intercellular transfer of ions and small molecules, including cyclic nucleotides, inositol triphosphate (IP_3), micro ribonucleic acid (RNA), and antigenic peptides up to ~1800 Da (Harris 2007; Katakowski et al. 2010; Neijssen et al. 2005; Orellana et al. 2009). Although, Cxs and Panxs are tetraspanning membrane proteins with similar membrane topology. They differ in several other features, including lack of significant homology in primary structure and number of extracellular cysteines (six in Cxs and four in Panxs, although Cx29 also has four), number of family members (21 for Cxs and 3 for Panxs), and glycosylation state (Cxs are proteins and Panxs are glycoproteins) (Figure 4.1) (D'Hondt et al. 2009).

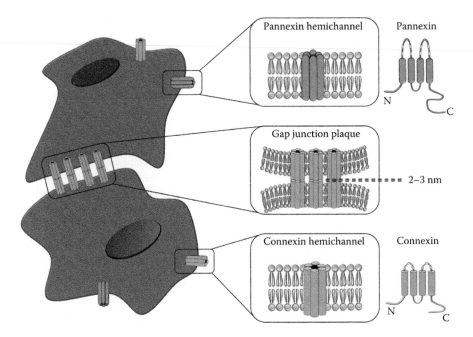

FIGURE 4.1 Scheme of gap junction channels, hemichannels, connexins, and pannexins. Left: two adjoining cells forming a gap junction channel (GJC) plaque at the cell interface. Each cell presents hemichannels (HCs) formed by subunits termed connexins (Cxs) or pannexins (Panxs). Middle top: Panx HCs formed by six Panxs subunits. Middle center: GJC plaque formed by an aggregation of many GJCs that leave a virtual space of 2–3 nm at the adjoining region. Middle bottom: Cx HCs constituted by six Cxs subunits. Right: membrane topology of Panxs (top) or Cxs (bottom). The white squares denote extracellular cysteine residues. N and C represent amino and carboxyl termini, respectively.

Several Cxs have been detected in immune system cells, Cx43 being the most ubiquitously expressed even by DCs (Corvalán et al. 2007; Matsue et al. 2006; Mendoza-Naranjo et al. 2007; Neijssen et al. 2007; Saccheri et al. 2010; Sáez et al. 2000). GJC-mediated intercellular communication has been demonstrated *in vitro* between stimulated, but not resting, murine and human DCs (Concha et al. 1988; Matsue et al. 2006; Mendoza-Naranjo et al. 2007). In addition, gap junctional communication between DCs and T cells has been observed after antigen exposition, suggesting their involvement in immunological synapse and local signaling (Elgueta et al. 2009). Other immune cells express GJCs under inflammatory conditions, which allow the rapid spread of activating or inhibitory signals between neighboring cells (Bopp et al. 2007; Ring et al. 2010). Because GJCs might contribute to the activation of the immune response and induction of T cell-mediated suppression, a rising interest of immunologists in GJC-mediated intercellular communication has occurred during the last decade.

Here, we summarize the current knowledge on GJC expression in APCs during inflammation or antigen presentation, and we discuss their possible role in amplifying, upregulating, or downregulating the immune responses.

4.2 EXPRESSION OF CONNEXINS BY DENDRITIC CELLS AND T CELLS

4.2.1 CONNEXINS IN DENDRITIC CELLS

The expression of GJCs by immune system cells has been studied for around four decades, starting in the early 1970s when Hülser and Peters worked on T cells (Hülser and Peters 1971, 1972). Later on, studies done in macrophages were reported and the possibility of GJC formation between DCs, and DCs and T cells was suggested (Concha et al. 1988, 1993; Levy et al. 1976; Porvaznik and MacVittie 1979). During the last decade, a rising interest on DCs has occurred and expression of Cx GJCs has been proposed to play a relevant role in cross-presentation, amplification, and also downregulation of immune responses (Table 4.1) (Neijssen et al. 2005, 2007).

Langerhans cells (LCs). Immature LCs reside in the epidermal layer of the skin and after antigen uptake they migrate to draining LNs and present antigens to T cells (Merad et al. 2008). The first ultrastructural study done in DCs showed GJC-like membrane specializations (Concha et al. 1988). The presence of GJC-like structures at cellular interfaces of LCs isolated from skin and T cells was also described in both murine and human samples (Concha et al. 1988, 1993). Similar observations were obtained in human samples from peripheral lymph after induction of contact dermatitis (Brand et al. 1995).

Recently, Cx43 immunoreactivity in MHC II-positive cells (probably LCs) at the human epidermis and human appendix was described (Neijssen et al. 2005). In addition, the expression of Cxs has been suggested through functional analyses in cultures of XS52 cells, which is a cell line derived from the murine epidermis with LCs-like phenotype (Matsue et al. 2006). However, Zimmerli et al. (2007) showed the absence of Cx43 in LCs of normal human skin, suggesting that resting LCs do not form GJCs. The latter is consistent with previous observations obtained in other DC subtypes, including cell lines derived from murine DCs and primary cultures of murine or human DCs that do not express functional GJCs under control conditions but show gap junctional communication and upregulation of Cx43 levels upon activation with proinflammatory agents (e.g., cytokines or bacterial products) (Corvalán et al. 2007; Krenacs et al. 1997; Matsue et al. 2006; Mendoza-Naranjo et al. 2007; Zimmerli et al. 2007). The role of Cxs in LCs will be described in detail in Section 6.2 of Chapter 6.

Follicular dendritic cells (FDCs). SLOs such as LNs, spleen, and tonsils contain B cell-rich zones called follicles, where FDCs are located. At SLO follicles FDCs present antigens to B cells and perform antigen retention (Batista and Harwood 2009). After antigen challenge, an immune response is developed and B cell proliferation occurs within the follicles, leading to the formation of germinal centers (GCs) (Klein and Dalla-Favera 2008). Using *in situ* hybridization, Cx43 mRNA was detected in cells morphologically identified as FDCs in reactive human tonsil (Krenacs et al. 1997). Immunofluorescence analysis revealed Cx43 in human tonsil GCs, where Cx43 reactivity colocalized with CD21 and CD35, which are molecular markers of FDCs (Krenacs and Rosendaal 1995, 1998; Krenacs et al. 1997). The ultrastructural analyses of freeze-fracture replicas showed the hallmark of

TABLE 4.1
Gap Junction Channels between DCs and between DCs and Lymphocytes

APC	Connexins	Species	Coupled Cells	Techniques	References
LCs	Cx43	Human, mouse	LCs–T cells	EM, IF	Brand et al. (1995); Concha et al. (1988, 1993); Neijssen et al. (2005)
XS52 (cell line derived from LCs)	ND	Mouse	LCs–LCs	DT	Matsue et al. (2006)
FDCs	Cx43	Human, mouse	FDCs–FDCs FDCs–B cell	DT, EM, IF	Krenacs and Rosendaal (1995, 1998); Krenacs et al. (1997)
BMDCs	Cx43	Mouse	DCs–DCs DCs–CD4+ DCs–CD8+ DCs–T reg	DT, RT-PCR, IF, WB	Elgueta et al. (2009); Matsue et al. (2006); Ring et al. (2010)
tsDC (cell line derived from BMDC)	Cx43, Cx45	Mouse	DCs–DCs	DT, RT-PCR, WB	Corvalán et al. (2007)
MoDCs	Cx43, Cx45	Human	DCs–DCs	DT, WB, FC	Mendoza-Naranjo et al. (2007)
sDC	Cx43	Mouse	sDC–T cells	DT, IF	Elgueta et al. (2009)
DC1 (cell line derived from sDC)	Cx43	Mouse	NE	WB	Winzler et al. (1997)
CD11c DCs at LNs	Cx43	Mouse	DCs–T reg	DT, IF	Ring et al. (2010)
DEC205+ DCs at LNs	Cx43, Cx45	Mouse	NE	IF, β-gal	Corvalán et al. (2007)
CD4+ and CD8+ DCs at LNs	Cx43	Mouse	NE	FC	Saccheri et al. (2010)

Note: BMDCs, bone marrow-derived DCs; Cxs, connexins; DCs, dendritic cells; DT, dye transfer; EM, electron microscopy; FC, flow cytometry; FDCs, follicular dendritic cells; FF, freeze fracture; LCs, Langerhans cells; IF, immunofluorescence; LNs, lymph nodes; MoDCs, monocyte-derived DCs; NE, not evaluated; pDCs, plasmacytoid DCs; T reg, regulatory T cells; RT-PCR, reverse transcription polymerase chain reaction; sDCs, splenic DCs; tsDC, thermo-sensitive DCs; Wb, Western blot.

GJs associated with FDCs in human tonsil GCs (Krenacs et al. 1997). In reactive spleen samples, Cx43 mRNA was found by Northern blot analyses (Krenacs et al. 1997). Accordingly, immunofluorescence analyses showed Cx43 reactivity in FDCs in reactive human spleen (Krenacs and Rosendaal 1995). However, neither Cx26 nor Cx32 were detected in LNs, spleen, or tonsils (Krenacs and Rosendaal 1995). Immature splenic DCs (sDCs) reside in the spleen, the organ that filters the blood and is the main site for immune (innate and adaptive) responses to blood-borne antigens (Mebius and Kraal 2005). Isolated sDCs from mice show immunoreactivity for

Cx43, which is redistributed to sDC–T cell interfaces in the presence of antigenic peptides (Elgueta et al. 2009). In agreement with the previous report, DC1, which is a cell line derived from murine sDC, showed an increase in its relative protein levels of Cx43 after lipopolysaccharide (LPS) stimulation (Saccheri et al. 2010). For more details, see Section 2.1 of Chapter 2.

Monocyte and bone marrow-derived dendritic cells. Monocytes and DCs emerge from a common precursor in the BM. Then, they populate different organs or circulate in the blood (see above). The *in vitro* differentiation of BM precursors into DCs facilitates the study of DC biology (Inaba et al. 2009). Relative levels of Cx43 mRNA and protein were detected in BM-derived DCs (BMDCs), and they were upregulated by LPS treatment, but were unaffected by interferon-γ (IFN-γ) (Matsue et al. 2006). Thermo-sensitive DCs (tsDC), an immortalized cell line derived from murine BMDC, express detectable Cx43 and Cx45 mRNA and protein levels evaluated through RT-PCR and Western blot analyses, respectively (Corvalán et al. 2007; Volkmann et al. 1996). Moreover, treatment with tumor necrosis factor-α (TNF-α) plus interleukin-1β (IL-1β) increased Cxs 43 and 45 proteins and mRNA levels in tsDCs. In addition, a synergic effect with this cytokines was observed after IFN-γ treatment (Corvalán et al. 2007).

Monocyte-derived DCs (MoDCs) represent a good model to expand and grow DCs *in vitro* (Sallusto and Lanzavecchia 1994). Human MoDCs are used for immunotherapy in patients with cancer such as melanoma. Thus, the antigen delivery and amplification of immune responses play a pivotal role in the immune response to tumors (Mendoza-Naranjo et al. 2007). Human MoDCs under resting conditions express Cxs 43 and 45 proteins, and Cx43 levels increased slightly after treatment with a melanoma cell lysate (MCL). However, human MoDCs exposed to MCL plus TNF-α showed a significant increase in Cx43 levels, but not in Cx45 levels (Mendoza-Naranjo et al. 2007). Furthermore, in MoDCs treated with MCL plus TNF-α, immunofluorescence analyses revealed Cx43 recruitment to the cell membrane where it formed GJC plaques (Mendoza-Naranjo et al. 2007).

Migratory dendritic cells in lymphoid organs. So far, few *in vivo* approaches using murine models after tissue injury or tumor challenge have been described. An increase in Cx43 reactivity occurs in mice LNs after immune challenge, suggesting that GJCs might also be upregulated (Krenacs et al. 1997). Under resting conditions, both Cx43 and Cx45 were detected in LNs, but Cx43 colocalized with the DC marker DEC205 (DEC205$^+$Cx43$^-$) in a small population of paracortical DCs, whereas Cx45 was present in a larger population of DCs (DEC205$^+$Cx45$^+$) (Corvalán et al. 2007). After inducing striate muscle injury with BaCl$_2$ used as a myotoxin, a significant increase in the number of DEC205$^+$Cx43$^+$ cells at draining LNs was observed. Similar results were obtained in transgenic mice that express the lacZ reporter gene encoding β-galactosidase in one Cx43 allele. On the other hand, the population of DEC205$^+$Cx45$^+$ did not show an increase in Cx45 levels, but showed a redistribution of the protein after muscle damage; DEC205$^+$Cx45$^+$ cells formed tight aggregates. Cx32 was not detected in LNs under resting conditions or after muscle injury (Corvalán et al. 2007). These studies show the expression of Cxs 43 and 45 in cDCs under resting conditions and the increase expression or changes in distribution of these Cxs in cDCs or inflammatory DCs (Figure 4.2).

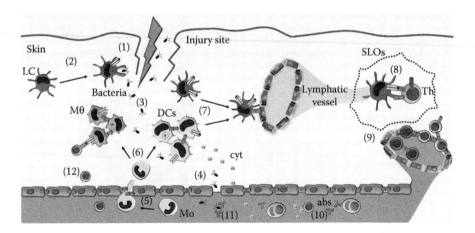

FIGURE 4.2 Illustration of homocellular and heterocellular gap junctional communication at specific stages of immune responses. (1) After skin injury or infection, several events occur. (2) Immune cells located at areas surrounding the wounded site are recruited; in this picture, skin resident Langerhans cells (LCs) are recruited by danger signals (e.g., PAMPs and DAMPs) released by local cells. LCs phagocytose bacteria and expose their antigens in MHC II molecules. (3) The tissue is invaded by bacterial spreading. (4) Bacteria infect endothelial cells (ECs), which lose their cellular junctions and produce cyotkines (cyt), leading to an increase in vascular permeability, and bacteria enter the bloodstream. (5) After upregulation of cell adhesion molecules (CAMs), ECs allow leukocyte rolling and extravasation. (6) After extravasation, monocyte (Mo) differentiate into macrophages (Mθ) or dendritic cells (DCs). (7) DCs (LCs or inflammatory) phagocytose and expose bacteria antigens in MHC II at their cell surface and enter lymph vessels. (8) DCs migrate into draining secondary lymphoid organs (SLOs) and present antigens to naïve T helper cells (Th), inducing their proliferation. (9) Clonal expansion of antigen-specific LT and B cells (LBs) occurs and both cell types enter the blood. (10) LBs differentiate into plasma B cells, which produce antibodies (abs), and Th cells secrete cytokines, which also stimulate LBs. (11) Blood bacteria are opsonized by circulating abs. (12) Some Th cells enter the tissue and interact with Mθ, favoring the elimination of the pathogen.

Downregulation of Cxs, including Cx43, or gap junctional communication occurs in several tumor cells during tumorigenesis (Naus and Laird 2010). However, the intratumor injection of *Salmonella typhimurium* upregulates the relative levels of Cx43 in B16 cells and makes the tumor more immunogenic and thus allows the activation of DCs and consequently activation of CD8+ T cells (Figure 4.3), favoring tumor rejection (Saccheri et al. 2010). The antitumor immune response will be discussed in detail in Section 5.1 of Chapter 5.

4.2.2 Connexins in T Cells

T cells comprise the second major class of lymphocytes. Basically, T cells are the key elements in the immune response because they can function both as effectors that can mediate cytolitic responses and as helpers coordinating humoral and cell-mediated responses. In T cells, Cx43 has been attributed to be the main GJC protein.

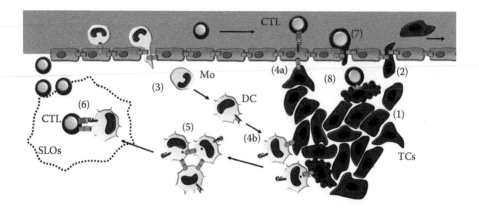

FIGURE 4.3 Gap junctional communication during tumorigenesis. (1) At early stages during tumorigenesis, tumor cells (TCs) proliferate within a tissue and downregulate gap junctional communication. (2) During metastasis, TCs could reexpress some Cxs for intravasation and invade other tissues. (3) Danger signals (e.g., ATP) are released at the tumor site, recruiting monocytes (Mo) that might differentiate into inflammatory dendritic cells (DCs). (4a) GJC-mediated cross-presentation of tumor antigens from TCs to endothelial cells (ECs). Then, ECs might present antigens in MHC I to antigen-specific cytotoxic T cells (CTLs). (4b) Through phagocytosis or cross-presentation, DCs acquire the antigen from TCs. (5) After taking up antigens, DCs might amplify the number of antigen-bearing DCs through GJC-mediated cross-presentation. (6) DCs migrate into draining secondary lymph organs (SLOs) where they present antigens to specific CTLs and induce clonal expansion. (7) CTLs homing at tumor site and after extravasation crawl toward TCs. (8) Through granzyme reactions, CTLs kill TCs by apoptosis.

In human, only Cx43 has been identified from blood-derived T cells, and Cx40 as well as Cx43 were identified in T cells derived from human tonsils (Oviedo-Orta et al. 2000). Moreover, during blastic transformation, CD4+ but not CD8+ T cells increased the surface expression of Cx43, but it did not affect the extent of dye transfer (Oviedo-Orta et al. 2000). On the other hand, it was recently related that Cx43 upregulation favors T cell proliferation via HC-mediated cysteine uptake (Figure 4.2) (Oviedo-Orta et al. 2010).

In addition, both Th1 and Th2 effector cells express Cx43 (Bermudez-Fajardo et al. 2007). On the other hand, CD4+ naïve T cells and naturally occurring regulatory T cells (T reg) from a murine background expressed the mRNA of Cxs 31.1, 32, 43, 45, and 46 (Bermudez-Fajardo et al. 2007). Homocellular GJCs between T cells and heterocellular GJCs between T cells and other cell types, including B cells, endothelial cells (ECs), macrophages, and DCs have been documented (Bermudez-Fajardo et al. 2007; Elgueta et al. 2009; Oviedo-Orta and Evans 2002; Oviedo-Orta et al. 2000).

It has been proposed that the role of GJC formation in T cells is to participate in B cell maturation, T cell activation by DCs, and T cell suppression between naïve T cells and T reg (Figure 4.2) (Bopp et al. 2007; Elgueta et al. 2009; Oviedo-Orta et al. 2001).

4.3 WHY CELL CONTACTS BETWEEN DENDRITIC CELLS AND T-CELLS MIGHT PRESENT GAP JUNCTIONS?

Although antigen presentation might occur in a contact-independent manner (see Section 4.4.2), the importance of cellular contacts during this process is clear. Interestingly, before the formation of heterocellular contacts between DCs and T cells, homocellular interactions occur between DCs. These contacts might be relevant to amplifying the immune response by increasing the number of responsive antigen-bearing DCs. Under resting steady-state conditions, many homocellular contacts are established at DC clusters at LNs, which increase after antigen recognition (Bousso 2008; Lindquist et al. 2004). Furthermore, DC maturation increases the expression of cell adhesion molecules (CAM), favoring cell–cell contacts and formation of GJCs (Imhof and Aurrand-Lions 2004; Sáez et al. 2000). In several cellular systems, appropriate cell adhesion is required for GJC formation, and clustering of DCs increases the possibility of establishing this intercellular communication pathway (Meyer et al. 1992; Musil et al. 1990).

4.3.1 Importance of Cell–Cell Contacts between Dendritic Cells

Cell adhesion is crucial for leukocyte extravasation at infection or injury sites. This process is principally mediated by selectins and integrins expressed in the vasculature surface of the inflamed site (Abram and Lowell 2009; Ley et al. 2007). The importance of cell adhesion is unveiled in leukocyte adhesion deficiencies, which might be due to integrin mutations that lead to several syndromes such as increased susceptibility to pathogen infection, decreased chemotaxis, and impaired wound healing; notably, the lack of integrins might be lethal (Abram and Lowell 2009; Ley et al. 2007). Extravasation from blood vessels might be important for inflammatory DCs, but not for tissue resident DCs already present in the site of infection. After antigen encounter, DCs have to invade afferent lymphatic vessels in a process also mediated by CAM. The antibody blockade of both ICAM-1 and VCAM-1 impaired the lymphatic transmigration of antigen-bearing DCs and thus prevented the development of virus-specific CTL immune responses (Teoh et al. 2009). Recently, a role for GJCs in cell adhesion has been proposed, which is compatible with the fact that GJC blockers avoid extravasation of monocytes and neutrophils (Elias et al. 2007; Eugenín et al. 2003; Sarieddine et al. 2009; Véliz et al. 2008). Because the impairment of cell adhesion through L-CAM or N-cadherin immunoneutralization prevents GJ formation and changes the phosphorylation state of Cx43 expressed by DCs, the upregulation of CAMs in activated DCs might favor GJC-mediated communication between DCs-lymphatic vessels and BM DC-sinusoidal wall (Figure 4.2) (Abram and Lowell 2009; Campbell 1982; Meyer et al. 1992; Musil et al. 1990).

In the skin, both LCs and dermal DCs (dDCs) are PAMP and DAMP sensors, and intravital observations using two-photon microscopy in a murine model showed that activation with LPS or protozoan parasites leads to morphological changes, motility loss, and interaction with surrounding DCs (Ng et al. 2008). DCs exposed to *Escherichia coli* or their supernatants show Ca^{2+} fluxes that propagate between neighboring DCs through long membrane connections named tunneling nanotubes

(TNT; see Section 4.4.2) (Salter and Watkins 2009; Watkins and Salter 2005). Another mechanism for transmission of Ca^{2+} waves is through GJCs, which might be present at TNTs (Sáez et al. 2003; Wang et al. 2010).

4.3.2   GAP JUNCTIONS BETWEEN DENDRITIC CELLS

The expression of Cxs in mammals is ubiquitous and gap junctional communication is normally present in different organs such as the brain, heart, liver, lung, and other tissues under resting conditions (Sáez et al. 2003). In most tissues, the exposure to proinflammatory agents such as β-amyloid peptide, PAMPs (e.g., LPS or peptidoglycan (PGN)), and cytokines (e.g., TNF-α, IL-1β, and IFN-γ) induces a drastic reduction in gap junctional communication (González et al. 2002; Hu and Xie 1994; Leaphart et al. 2007; Martin and Prince 2008; Orellana et al. 2009; Simon et al. 2004). Inversely, immune cells under control conditions do not show communication via GJCs, but stimulation with proinflammatory agents as those mentioned above induces changes in Cx distribution, switch or increase in Cxs expression, and in some cases GJC formation (Neijssen et al. 2007). *In vitro* studies have shown induction of GJC-mediated communication between APCs exposed to a particular combination of proinflammatory agents (Table 4.1).

The epidermis is the outer layer of the skin and is the first physical barrier to avoid the entry of pathogens into the organism (Nestle et al. 2009). The main cells of the epidermis, the keratinocytes, produce cytokines, including TNF-α, IL-1β, IL-6, IL-10, and IL-18, which modulate the activity of DCs and other skin resident cells (Nestle et al. 2009). The stimulation of tsDCs for 7 h with a conditioned medium obtained from keratinocytes induced their maturation and transient expression of functional GJCs, which were partially abolished with TNF-α and IL-1β neutralizing antibodies and completely blocked with GJC inhibitors (Corvalán et al. 2007). Stimulation of tsDC, D2SC/1 (a cell line derived from spleen DCs) or BMDCs with TNF-α plus IL-1β (for 5–8 h) mimics the induction of gap junctional communication induced by keratinocyte conditioned medium. In addition, other cytokines modulate the effect of TNF-α plus IL-1β on gap junctional communication; while IFN-γ increases the extent and duration, IL-6 prevents the effect of TNF-α plus IL-1β (Figure 4.2) (Corvalán et al. 2007).

PAMPs such as LPS or CpG oligodeoxynucleotide (ODN) activate toll-like receptor (TLR) 4 and TLR9, respectively, and are classical stimuli to induce the maturation of APCs (Kono and Rock 2008). However, LPS did not induce gap junctional communication in APCs (monocytes, macrophages, and microglia), but when coapplied with IFN-γ the functional expression of GJCs occurred (Eugenín et al. 2001, 2003, 2007). Similar results were obtained with cultures of BMDCs and XS52 cells that became coupled after treatment for 24 h with LPS or CpG ODN plus IFN-γ (Matsue et al. 2006). Both treatments that induced gap junctional communication in DCs depend on the autocrine effect of TNF-α and were mimicked with TNF-α plus IFN-γ, and blocked with GJC inhibitors. In addition, the blockade of DC GJCs prevented the upregulation of surface markers related with antigen presentation, such as CD40, CD80, CD86, and MHC II, and reduced their capacity to induce proliferation of $CD4^+$ T cells after alloantigenic presentation (Matsue et al. 2006).

FDCs form a meshwork within the follicles, where Cx43 reactivity has been detected (see Section 4.2.1) (Krenacs and Rosendaal 1995, 1998; Krenacs et al. 1997). In addition, dye transfer experiments have revealed intercellular coupling between cultured FDCs under resting conditions. Moreover, coupled FDCs were observed in tonsil samples of children with recurrent inflammation (Krenacs et al. 1997), suggesting that repetitive antigen challenge evokes a microenvironment that induces gap junctional communication.

MHC II antigen presentation occurs when extracellular antigens are taken up by APCs, loaded in MHC II at the cell surface and presented to CD4+ T cells (Figure 4.4) (Vyas et al. 2008). On the other hand, MHC I antigen presentation corresponds to antigen peptides derived from cytosolic proteins and occurs in almost all cells of the body that express MHC I. During MHC I antigen presentation, endogenous or viral peptides are presented to CD8+ T cells inducing an immune response to viruses and thus killing the infected cells (Vyas et al. 2008). Cross-presentation is a mechanism through which exogenous peptides are acquired by APCs and presented in MHC I to CD8+ T cells, allowing the development of immune responses against viruses or tumors from noninfected APCs (Vyas et al. 2008). Recently, Neijssen et al. (2005) showed gap junctional communication as a new pathway for cross-presentation in monocytes, which via GJCs acquired influenza peptides from infected cells and then induced CTL activation (Neijssen et al. 2005). This mechanism suggests a "bystander effect," by which uninfected cells acquire virus peptides and can be eliminated by the

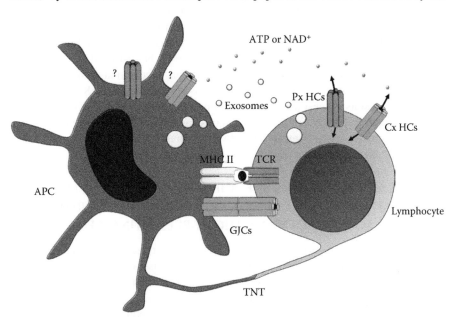

FIGURE 4.4 Gap junctional and hemichannel communication during antigen presentation. An antigen-presenting cell (APC) exposes antigens in MHC II and forms GJCs with a lymphocyte (a T or B cell). Other mechanisms of cellular communication are depicted, including the opening of hemichannels (HCs), exosomes, and tunneling nanotubes (TNT). ATP, NAD+, and other small molecules could be released to the extracellular milieu via HCs.

CTL after MHC I antigen presentation. During tumor development, a similar mechanism might occur (Handel et al. 2007; Neijssen et al. 2005, 2007).

As mentioned before, MoDCs are used in melanoma immunotherapy, so the amplification of immune responses plays a pivotal role against tumor. Melanoma antigens are presented by DCs to the CTL, so MCLs induce DC maturation, which is potentiated with TNF-α. Indeed, 24 h treatment with MCL plus TNF-α induced transient gap junctional communication between MoDCs favorin cross-presentation, and was blocked with GJC inhibitors (Mendoza-Naranjo et al. 2007). Similar results were obtained with apoptotic cells. A431 epithelial carcinoma and B16 melanoma cells transfected with Cx43 efficiently transferred apoptotic peptides to BMDCs, which activate CTLs (Pang et al. 2009). However, in A431 apoptotic cells, cross-presentation still occurs although less efficiently in the absence of Cx43, suggesting the existence of two mechanisms for peptide delivery to BMDCs as follows: (1) dependent on Cx43 GJCs and (2) independent of GJCs possibly provided by direct uptake of apoptotic peptides (Figure 4.3) (Pang et al. 2009). Otherwise, in B16 cells, the cross-presentation was completely dependent on GJCs (Pang et al. 2009).

Recently, Saccheri et al. (2010) demonstrated GJC-mediated cross-presentation using melanoma-derived B16 cells. B16 cells showed downregulation of Cxs. In this study, Cx43 expression was induced by infecting B16 cells *in vitro* or directly by infecting the tumor with *Salmonella typhimurium*. Cx43 upregulation was associated with dye transfer via GJCs between B16 and BMDCs, and was required for cross-presentation of ovalbumin (OVA) peptides from B16 to BMDCs, which after cross-presentation activate OVA-specific CTLs (Saccheri et al. 2010). Interestingly, a lower progression of the tumor was observed *in vivo* when B16 cells infected with *Salmonella typhimurium* expressed Cx43. In addition, vaccination with DCs loaded with cell lysate from infected B16 cells increased mice survival (Saccheri et al. 2010). These results suggest that *in vivo* cross-presentation through GJCs is an effective mechanism to develop an immune response against tumors (Figure 4.3). Since melanoma peptides can be transferred between human MoDCs, GJCs might constitute a molecular target to develop new therapies to treat cancer.

4.4 ARE GAP JUNCTIONS SUFFICIENT TO EXPLAIN THE NEED FOR DIRECT PHYSICAL CONTACT BETWEEN DENDRITIC CELLS SO TO BE GOOD ANTIGEN-PRESENTING CELLS?

The heterocellular contacts between DCs and T cells generated during antigen presentation are required for either induction or tolerance responses of the immune response (Bousso 2008; Hugues et al. 2004; Shakhar et al. 2005). During DC-induced T cell activation, both cell contacts and local signaling play a pivotal role in determining changes in free intracellular Ca^{2+} concentration ($[Ca^{2+}]_i$) required for T cell activation (Krummel and Cahalan 2010; Vig and Kinet 2009; Wei et al. 2007). Because GJCs are permeable to several second messengers (e.g., Ca^{2+}, IP_3, and cAMP), they permit diffusional transfer or propagation of second messenger waves (e.g., different types of Ca^{2+} waves), and thus their participation in antigen presentation is relevant (Harris 2007; Tour et al. 2007).

In this section, we highlight the possible role of DC–T cell GJCs and their contribution to antigen presentation. Other possible pathways for antigen presentation are also briefly described.

4.4.1 GAP JUNCTIONS BETWEEN DENDRITIC CELLS AND T CELLS AS PART OF IMMUNE SYNAPSE

Primary cultures of sDCs and T cells obtained from an antigen-specific system (OT I and OT II) express Cx43, but they do not form functional GJCs under resting conditions (without antigen peptides) (Elgueta et al. 2009). However, the exposure to antigen peptide induced gap junctional communication between DCs and T cells after 6–24 h. Maximal coupling was observed around 8 h after antigen challenge and was abolished by GJC blockers (Elgueta et al. 2009). Functional GJCs between BMDCs and $CD4^+$ or $CD8^+$ was shown, and the blockade of DC–T cell GJCs prevents IL-2 secretion in both $CD4^+$ and $CD8^+$ T cells (Elgueta et al. 2009). The GJC blockade also prevented activation and delayed polyclonal expansion of $CD4^+$ T cells, but did not affect the polyclonal expansion (Elgueta et al. 2009). Since the used GJC inhibitors are likely to be effective during a limited period of time, the above findings dissect the importance of heterocellular (DC–T cells) and homocellular (T cell-T cell) GJCs present in DC–T cell co-cultures exposed to antigen peptides (Figure 4.4).

An important subset of $CD4^+$ T cells is constituted by T reg cells. They are involved in the suppression of immune responses to a specific antigen thorough a process named *tolerance*. A good model to study T reg cells *in vitro* and *in vivo* is the hapten-induced contact hypersensitivity reaction, where a hapten is taken up by DCs and then presented to T cells inducing tolerance (Green et al. 2009). Haptenized BMDCs showed functional GJC-mediated communication with T reg cells, evaluated with the dye transfer technique, which was abolished with GJC blockers (Ring et al. 2010). T reg cells loaded with the cell tracker calcein were injected into mice before sensitization, and 48 h later were found in draining LNs. In confocal images of LN cross sections, a close association between T reg cells and DCs was identified using CD11c as a DC marker. Reactive plaques of Cx43 were detected at DC–T reg interfaces (Ring et al. 2010). DCs sensitized with the hapten and isolated from LNs induced proliferation of hapten-specific $CD8^+$ T cells. But when T reg cells were injected into mice, the isolated DCs were unable to induce $CD8^+$ T cell proliferation. However, blocking T reg cell GJCs before injection restored the ability of isolated DCs to stimulate $CD8^+$ T cells *in vitro* (Ring et al. 2010). This finding was corroborated *in vivo* by using isolated $CD8^+$ T cells that respond to sensitization under control conditions. Additionally, T reg cell GJCs were inhibited, but not when untreated T reg cells were injected into the mice. Furthermore, *in vivo* experiments showed that ear-swelling produced with sensitization is inhibited by T reg cells in a GJC-dependent manner (Ring et al. 2010).

4.4.2 INTERCELLULAR COMMUNICATION PATHWAYS DURING ANTIGEN PRESENTATION AND CROSS-PRESENTATION

A "two-signal" model was proposed for antigen presentation. In this model, the proper activation of T cells requires the following two signals: (1) one delivered by

TCRs after appropriate engagement with the peptide–MHC II complex, and (2) the right costimulation through engagement of T cell CD28 and APC CD80 or CD86, and many other receptors. Signal 1 alone would lead to a tolerogenic response, and no immune response would occur. However, the requirement of a third signal corresponding to soluble molecules released by APCs that might determine the full activation or tolerance of naïve T cells has been described (Figure 4.4) (Reise Sousa 2006). From the many mechanism that has been proposed to participate in the release of these soluble molecules, hemichannels and exosomes play a pivotal role in the antigen presentation. In addition to GJCs, other mechanisms, such as tunneling nanotubes, has been proposed to participate in the antigen presentation through connecting adjacent cells (Gousset and Zurzolo 2009; Thery et al. 2009; Woehrle et al. 2010).

Hemichannels. HCs allow the direct transfer of molecules between the cytoplasm and extracellular milieu, including ATP, cAMP, NAD^+, and other small molecules that are known to modulate antigen presentation, and can be activating or tolerogenic signals (Bopp et al. 2007; Harris 2007; Hubert et al. 2010; la Sala et al. 2002; Oviedo-Orta and Evans 2002).

In cultured cells under resting conditions, HCs present a low open probability, but under particular physiological and pathophysiological conditions, their activity can be enhanced (Schalper et al. 2008). $CD4^+$ T cells express Cx and Panx HCs, which mediate the release of ATP for efficient TCR activation and proliferation of T cells (Oviedo-Orta et al. 2010; Schenk et al. 2008; Woehrle et al. 2010). Because macrophages express Cx and Panx HCs, it is possible that DCs also do so, thus being relevant to determine the fate of the immune response (Figure 4.4).

Exosomes. Exocytosed vesicles of endosomal origin with 40–100 nm in diameter are named exosomes and differ from other secreted vesicles (e.g., microvesicles and membrane particles). They have cholesterol-enriched membranes and contain lipid rafts and proteins involved in membrane transport and fusion (Simons and Raposo 2009; Thery et al. 2009). The secretion of exosomes is a constitutive or induced mechanism depending on the cell type; cultured DCs secrete them constitutively although the extent of secretion varies depending on the activation state of DCs (Thery et al. 2009). $CD4^+$ T cell stimulation with exosomes released by DCs has been demonstrated. *In vivo*, exosomes containing MHC II at FDC–T cell interfaces have been detected (Denzer et al. 2000; Thery et al. 2002). Transfer of MHC I and MHC II loaded exosomes between DCs also occurs and *in vivo* studies unveiled that DCs are required for exosome stimulation of $CD4^+$ T cells (Viaud et al. 2010). In addition, PAMPs and DAMPs modulate the secretion of exosomes. For example, ATP (which is a DAMP) activation of the $P2X_7$ receptor increases the extent of exosome secretion by DCs, which is a response synergized by LPS treatment (Ramachandra et al. 2010). Because ATP is released into the synaptic cleft during antigen presentation, exosomes might serve as pathways to provide the signal 3 (Figure 4.4).

Exosomes can be loaded with inflammatory or anti-inflammatory agents for their use in immunotherapy against different infections and cancer treatment, or to avoid graft rejection and autoimmune disease (Ramachandra et al. 2010). Clearly, this new field in intercellular communication will be a focus of therapeutic interest in the following years.

Tunneling nanotubes. Various membranous connections are known to mediate intercellular communication of mammal cells, including TNTs that are thin open-ended membranous tunnels between cells (Davis 2009). TNTs allow the transfer of organelles, proteins, and organic molecules (e.g., dyes), and also propagate Ca^{2+} signals (Dorban et al. 2010; Gousset and Zurzolo 2009; Langevin et al. 2010; Rustom et al. 2004; Salter and Watkins 2009). TNTs have been visualized in cultured murine BMDCs treated with LPS or challenged with *Escherichia coli.* In both situations, Ca^{2+} waves were observed between surrounding DCs (Salter and Watkins 2009; Watkins and Salter 2005). *In vivo*, TNTs have been visualized between MHC II$^+$ cells (probably DCs) in the corneal stroma; in the LPS-inflamed cornea, the number and length of TNTs between MHC II$^+$ cells increased (Chinnery et al. 2008). Thus, DCs might direct cross-presentation under inflammatory conditions or during antigen presentation via TNTs.

TNTs has been involved in several diseases, including prion diseases and human immunodeficiency virus (HIV). Prion diseases are fatal neurodegenerative disorders that affect several species (Gousset and Zurzolo 2009). Prions enter into the organism through the gut and then invade the SLOs, particularly macrophages and FDCs, where their amplification occurs (Gousset and Zurzolo 2009). However, if FDCs are not migratory DCs, how does invasion of the nervous system occur? It is known that the first invaded neural tissues belong to the enteric nervous system but the mechanism responsible for infection remained unknown (Gousset and Zurzolo 2009). Recent *in vitro* studies showed TNT transfer of prion protein (PrP^{sc}) from DCs to peripheral or central neurons (Dorban et al. 2010; Langevin et al. 2010).

Macrophages can also connect to each other through TNTs, which are increased in number after infection with HIV and allow the transfer of HIV particles (Eugenin et al. 2009; Onfelt et al. 2006), DCs might use a similar mechanism to amplify the immune response. However, because Cx43 has been found along TNTs and electrical communication through GJCs at TNTs has been demonstrated, further studies will be required to fully understand this way of cell–cell communication (Figure 4.4) (Wang et al. 2010).

4.5 CONNEXIN GAP JUNCTIONS IN OTHER ANTIGEN-PRESENTING CELLS

Monocytes/macrophages. Macrophages represent a widely distributed family of mononuclear leukocytes. They participate in the production, activation, mobilization, and regulation of all immune effector cells. Mounting evidence supports the claim that monocytes/macrophages form GJCs between themselves and with other cell types, including lymphocytes and endothelial cells (Bermudez-Fajardo et al. 2007; Eugenín et al. 2003; Jara et al. 1995; Levy et al. 1976). Although Cx37 is present in monocytes/macrophages recruited from early atheromas (Wong et al. 2006), Cx43 is considered to be the main connexin expressed by macrophages.

Cx43 was detected in foam cells (macrophages filled with lipid vacuoles) of atherosclerotic plaques, murine peritoneal macrophages as well as in the macrophage cell lines such as J774 and RAW264.7 (Anand et al. 2008; Fortes et al. 2004; Polacek et al. 1993). Nevertheless, neither Cx43 mRNA nor protein expression was detected in nonactivated human monocytes/macrophages (Eugenín et al. 2003; Jara et al. 1995; Polacek et al. 1993). However, Cx43 emerges in many proinflammatory conditions. For instance, treatment with a combination of either LPS or TNF-α plus IFN-γ increased Cx43 mRNA and protein levels, as well as the number of dye coupled cells (Eugenín et al. 2003). In agreement, Cx43 has been detected in macrophages at inflammatory foci (Polacek et al. 1993). Gap junctional communication seems to play a relevant role in coordinating monocyte/macrophage extravasation, since GJC blockers reduced the numbers of monocytes/macrophages that migrate across a blood–brain barrier model (Eugenín et al. 2003) due to a reduction in the secretion of matrix metalloproteinase-2 (MMP-2) (Eugenín et al. 2003).

Moreover, Cx43 HCs were shown to participate in the mechanism of phagocytosis. For instance, blockade of Cx43 by pharmacological agents, siRNA, and Cx43 mutant mice reduced phagocytosis in macrophages. In this work, Cx43 inhibition was shown to stop FcR-induced activation of RhoA and reduced the extent of actin cup formation, which is involved in particle internalization (Anand et al. 2008). Most strikingly, by using an *in vivo* adoptive transfer, Wong et al. (2006) showed that Cx37 elimination enhances monocyte/macrophage recruitment to the endothelium. Opening of Cx37 HC in primary monocytes/macrophages as well as in a macrophage cell line inhibits leukocyte adhesion through ATP release. Thus, Cx37 HCs were suggested to control the initiation of atherosclerotic plaque formation by regulating monocyte adhesion (Figure 4.2) (Wong et al. 2006).

Microglia. They are APCs and their activation is a hallmark in neurodegenerative diseases. Under resting conditions, microglia have a ramified morphology and monitor their local microenvironment. However, they can rapidly become activated in response to DAMPs or PAMPs (Inoue 2008; Ransohoff and Perry 2009). Activated microglia secrete several cytokines and other proinflammatory molecules with autocrine properties (Hanisch 2002; Orellana et al. 2009; Ransohoff and Perry 2009). Microglia express Cxs 32, 36, and 43 and microglia activation enhances their relative levels and/or activity (Orellana et al. 2009). After treatment with TNF-α plus IFN-γ or PGN, microglia showed an increase in Cx43 levels and a transient induction in GJC-mediated communication (Eugenín et al. 2001; Garg et al. 2005). Both cytokines and PAMPs increase $[Ca^{2+}]_i$ and might upregulate Cx43 GJCs, since microglia treated with a Ca^{2+} ionophore exhibit upregulation of Cx43 and GJC formation (Farber and Kettenmann 2006; Martínez et al. 2002). Eugenín et al. (2001) showed that under proinflammatory conditions, Cx43$^{-/-}$ microglia do not form GJCs, suggesting that homocellular GJCs are constituted mainly by Cx43 (Dobrenis et al. 2005; Eugenín et al. 2001).

A provocative hypothesis is that heterocellular gap junctional communication between microglia and neurons through Cx36-based GJCs might mediate intercellular transfer of deleterious signals for neurons (Dobrenis et al. 2005). Up to now, the heterocellular GJCs between microglia and T cells have not been described but they might occur because they are able to present antigens in MHC II, and a recent report showed cross-presentation into MHC I to CD8$^+$ T cells (Beauvillain et al. 2008).

Microglia recruitment to brain stab wounds occurs between 24 and 48 h after applying the injury. At the stab wound, microglia formed clusters and showed Cx43 reactivity at homocellular interfaces, suggesting the formation of GJCs *in vivo* (Eugenín et al. 2001). Moreover, real-time measurements using two-photon microscopy demonstrated ATP- and HC-dependent recruitment of microglia to the injured sites (Davalos et al. 2005). Other proinflammatory agents such as TNF-α also increased the activity of HCs in microglia, leading to glutamate release and subsequent neuronal death (Takeuchi et al. 2006).

Kupffer cells (KCs). They represent the largest population of resident macrophages in the body. They are active phagocytes and secrete many key inflammatory cytokines such as TNF-α, IL-1β, and IL-6, as well as GM-CSF and chemokines like MIP-1α and RANTES (Racanelli and Rehermann 2006). Surveillance, uptake, and degradation of intravascular cell debris are among the main functional roles of these cells. By releasing IL-6 onto hepatocytes, KCs also regulate the secretion of acute phase proteins such as C-reactive protein, anti-α1-antitrypsin, ceruloplasmin, or haptoglobin, thereby controlling systemic and local inflammatory reactions (Ishibashi et al. 2009).

Gap junction communication between KCs has been suggested to occur both *in vivo* and *in vitro* under inflammatory conditions. Under confocal microscopy, large aggregates of KCs expressing Cx43 were observed in liver sections from 6 h LPS-treated rats. Moreover, Cx43 was preferentially localized at KC–KC interfaces, suggesting the *in vivo* formation of GJCs. In support of this notion, treatment with LPS plus IFN-γ promoted the formation of GJCs between cultured KCs, which is a response associated with an increase in Cx43 mRNA and protein levels (Eugenín et al. 2007). These results suggest that KCs might contribute to develop the liver diseases.

Recently, *in vivo* experiments show the cross-presentation of *Leishmania donovani* peptides from KCs to CD8+ T cells at infection foci in a murine model (Beattie et al. 2010). In this study, KCs were shown to be the most efficient and numerous APCs in the generation of the CD8+ T cell-mediated immune response against *Leishmania* (Beattie et al. 2010). This study suggests that GJCs between KCs and T cells might be established during *in vivo* contacts and cross-presentation.

Mast cells (MCs). They are recognized as critical effectors in allergic disorders and other immunoglobulin E-associated immune responses. In an *in vitro* model for collagen organization, MCs have been observed to enhance contraction by a cell–cell contact-dependent mechanism. MCs express Cxs 32 and 43, the latter being the main Cx located at the cell membrane (Vliagoftis et al. 1999). In addition, gap junctional communication between MCs forming monolayers was demonstrated by scrape loading technique. Pretreatment with an inhibitor of the fatty acid hydrolase, that is, an enzyme that degrades oleamide (an endogenous inhibitor of GJC), decreased gap junctional communication (Moyer et al. 2004). In addition, demonstration of gap junctional communication between MCs and fibroblast was obtained using the calcein intercellular transfer assay (Au et al. 2007; Moyer et al. 2004). Interestingly, cocultures of MCs and fibroblasts showed a high rate and degree of fibroblast collagen lattice contraction in a process dependent on Cx43 GJCs between fibroblasts and MCs (Au et al. 2007; Moyer et al. 2004). However, heterocellular interaction between MCs and T cells were also shown.

Under control conditions or after PAMPs (LPS, PGN, and lipoteichoic acid) exposition, a very small subpopulation of MCs expressed MHC II, but after IFN-γ plus

IL-4 priming the expression of MHC II increased in 50% of MCs (Gaudenzio et al. 2009). After IFN-γ plus IL-4 treatment, both freshly isolated and cultured peritoneal cell-derived MC efficiently presented OVA peptides at MHC II to effector CD4+ T cells in an OT II murine model (Gaudenzio et al. 2009). Although CD4+ T cell activation by MCs was heterogeneous and less efficient than DCs, probably due to MHC II heterogeneous expression, they establish functional immunological synapse with a molecular and functional hallmark (Gaudenzio et al. 2009). Similar to other DCs and macrophages, the establishment of GJCs between MCs and T cells might be occurring, but because PAMPs did not increase the expression of MHC II, this MC–T cell interaction might occur after cytokine secretion during T cell polarization and not at the initiation of the immune response.

Neutrophils. The most abundant leukocytes in the blood are neutrophils, which represent 50–70% of the circulating white bloods cells. They are the first cell types to respond to most infections (particularly bacterial and fungal infections), but unlike resident macrophages, MCs, and immature DCs, neutrophils do not reside in peripheral tissues prior to infections. Rather, they are recruited from the circulation to the inflamed foci by cytokines and chemokines produced by resident macrophages and MCs (Nathan 2006). Most studies indicate that human neutrophils under resting conditions do not present Cxs, since detections for most Cxs failed at the mRNA and protein levels (Brañes et al. 2002; Scerri et al. 2006). Cx expression appears to be controlled by different soluble factors, including proinflammatory agents. For instance, dye transfer occurs in aggregates of freshly isolated human polymorphonuclear cells (PMN) treated with endothelial cell-conditioned medium and LPS or TNF-α. Neutrophils present Cxs 37, 40, and 43, all of which were detected at the protein level by both Western blot and immunofluorescence (Brañes et al. 2002; Eltzschig et al. 2006; Zahler et al. 2003).

Using flow cytometry analyses, bidirectional heterocellular dye coupling between PMNs and ECs, but not to HeLa cells that do not express Cxs, was found under basal conditions and was prevented by specific GJC blockers (Zahler et al. 2003). The increase in intercellular coupling between ECs and neutrophils resulted in a 50% reduction in transmigration, suggesting that coupled cells migrate to an unstimulated endothelium in a lesser extent than uncoupled PMNs (Zahler et al. 2003). In contrast, leukocyte adhesion induced by TNF-α in the microcirculation of the hamster cheek pouch was reduced or prevented by GJC blockers, acting through a mechanism that did not affect cell adhesion (Véliz et al. 2008). A similar function has been observed for Cx43 and Panx1 in neutrophils, as both proteins form HCs permeable to ATP and their opening during inflammation or hypoxia could contribute to increase the extracellular ATP concentration. In fact, the ATP release via Cx43 HCs by fMLP (*N*-formyl Met–Leu–Phe)-activated PMNs was blocked by GJC inhibitors and did not occur in PMNs from Cx43$^{-/-}$ mice (Zahler et al. 2003). In addition, it has also been proposed that Panx1 HCs allow ATP release from activated PMNs. In fMLP-treated PMNs, translocation of Panx1 to polarized regions and HC activation was observed. Additionally, ATP release was inhibited by Panx1 HC blockers such as carbenoxolone and [10]Panx1 (Chen et al. 2010). Thus, opening of Cx and Panx HCs seems to be a fundamental mechanism for ATP release from activated PMNs similar to monocyte/macrophages.

As we proposed above, HCs might constitute a source for the "third signal" during antigen presentation, which already has been reported in neutrophils. Murine and human neutrophils express MHC II after activation, and after peptide exposition become efficient APCs to CD8$^+$ T cells, similar to macrophages but less efficient than DCs (Beauvillain et al. 2007). In the murine model, the cross-presentation to CD8$^+$ T cells was shown both *in vitro* and *in vivo*, which induced the activation (not tolerance) of CD8$^+$ T cells differentiating into effector cells (Beauvillain et al. 2007). Because neutrophils are abundant in the blood, they are the first cell type to migrate at the site of infection and might migrate to draining lymph nodes. Thus, the formation of heterocellular GJCs between neutrophils and T cells might be possible.

B cells. They are responsible for humoral immunity. Induction of a humoral immune response against foreign proteins begins with antigen recognition between antigens primed B cells and Th cells. Th cells activate B cells via a mechanism similar to immune synapse between DC–T cell. In addition, GJCs between B and T cells as well as between B cells have been demonstrated (Oviedo-Orta et al. 2000). The formation of functional GJCs was demonstrated in freshly isolated human B cells, suggesting that these cells carry a preformed pool of GJCs subunits mostly formed by Cxs 40 and 43, which are upregulated during inflammation and promotes IgM synthesis (Figure 4.2) (Oviedo-Orta et al. 2000).

Endothelial cells. Virtually all the microvascular and small vessel ECs are positive for MHC II, but they do not express costimulatory signals such as CD80 or CD86 molecules. Therefore, they are unable to fully stimulate T cells (Rose 1998). ECs are believed to recruit Ag-specific cells into the inflammatory sites by displaying cognate MHC peptide complexes and controlling extravasations by regulating many intracellular and signaling pathways (Marelli-Berg and Jarmin 2004). In addition, the *in vitro* transfer of antigenic peptides from melanoma cells to ECs, which through MHC I-mediated cross-presentation to CTLs allow the elimination of tumor cells, was recently reported (Benlalam et al. 2009). This mechanism *in vivo* might significantly support T cells homing at tumor site (Figure 4.3).

Electrical coupling mediated by GJCs was proven to be the basis of different vascular responses (Scheckenbach et al. 2010). Cxs 37, 40, 43, and 45 were consistently found in the vascular wall (Chanson and Kwak 2007). Although the Cx expression profiles have not yet been fully described for all vascular territories, it is clear that Cx expression is not uniform in all blood vessels (Chanson and Kwak 2007; Hill et al. 2001). ECs most commonly express Cx37 and Cx40, but Cx43 has also been detected in ECs near vascular branch points as well as in ECs of capillaries (Chanson and Kwak 2007). Cx-dependent Ca^{2+} signaling was proposed to provide proinflammatory signaling mechanisms that promote spatial expansion of inflammation. For instance, Ca^{2+} waves have been reported to promote exocitosis of P selectin, thereby promoting leukocyte rolling to the vascular surface (Scheckenbach et al. 2010). In this context, GJCs between ECs and many different leukocytes have been reported, which appears to differentially regulate transendothelial migration (Eugenín et al. 2003; Guinan et al. 1988; Oviedo-Orta and Evans 2002; Zahler et al. 2003). For example, inhibition of GJCs decreased monocyte migration and produced a modest reduction of lymphocytes (Eugenín et al. 2003; Oviedo-Orta and Evans 2002). In addition, a high ratio of Cx40 versus Cx43 delays neutrophil adhesion to ECs, whereas a low ratio favoring Cx43 promotes transmigration across the endothelial barrier (Scheckenbach et al. 2010). In the lung, the proinflammatory role of Cx43 was confirmed *in vivo* by using Cx43$^{+/-}$ mice, in which almost a 50% reduction in neutrophil recruitment to the alveolar space during lung inflammation occurred (Sarieddine et al. 2009). In agreement, leukocyte adhesion and extravasation induced by TNF-α was reduced or prevented by GJC blockers in preparations of hamster microcirculation (Figure 4.2) (Véliz et al. 2008).

4.6 CONCLUDING REMARKS

Cellular interactions dependent or independent of cellular contacts constitute an important mechanism in coordination and might determine the fate of immune responses. The establishment of electrical synapse is well reported in innate and adaptive immune cells. However, ultrastuctural analyses will be required to identify the structure of GJCs between DCs and T cells, although it has already been suggested (Concha et al. 1988, 1993). In addition, the identification of GJC structures between DCs and T cells *in vivo* must still be demonstrated.

GJC formation is highly regulated by posttranslational modifications (e.g., phosphorylation, S-nitrosylation, ubiquitination) (Sáez et al. 2003). In this sense, further studies will be required to show the signaling involved in the regulation of formation, opening, and degradation of GJCs in immune cells. Regulation of GJCs with cytokines, which are key components in immune cell interactions, was reported for several immune cells, but the signaling involved was not described. Furthermore, the effect of important stimulatory and regulatory cytokines such as IL-4, IL-10, IL-12, IL-17, and IFN-β in Cx-mediated cell communication must still be elucidated. In addition, because several immune cells express Panxs, the formation of Panx GJCs became plausible.

Because contact-independent cell communication was recently discovered and T cell HCs were shown to be involved in proliferation and ATP release, the study of HCs as a new mechanism for delivery of a third signal in the immunological synapse must be studied. Woehrle et al. (2010) suggest the involvement of Panx HCs in the immunological synapse, but further studies will be required to show the structure and function of HCs at DC–T cell interfaces.

The immune system is constituted by different cell types that are divided into subsets that have different functions. So, the functional expression of GJCs and HCs might be differentially regulated between subsets. In the same way, the transfer of different signals might favor the activation or tolerance in the immune response. In such cases, permeability of GJCs constituted by different Cxs might be relevant in determining signal transfer, and thus the fate of the immune response.

Finally, the evidence and importance of GJCs and HCs in immune responses is increasingly greater and such mechanisms of cellular communication should be considered. Because defects in Cx- or Panx-mediated communication might result in delayed or inefficient immune response, they have to be evaluated in immune diseases such as hypersensitivity or autoimmune responses. The regulation of cell communication could be an important approach to modulate immune responses so as to generate, for example, tolerance to avoid graft rejection, or T cell activation to induce antitumoral responses.

ACKNOWLEDGMENTS

This work was partially supported by grants from CONICYT (24100062 to PJS), FONDECYT (1070591 to JCS), FONDEF (D07I1086 to JCS), and CONICYT-Anillo ACT71 (to JCS).

REFERENCES

Abram, C. L., and C. A. Lowell. 2009. The ins and outs of leukocyte integrin signaling. *Annu Rev Immunol* 27:339–62.

Ambrosi, C., O. Gassmann, J. N. Pranskevich, D. Boassa, A. Smock, J. Wang, G. Dahl, C. Steinem, and G. E. Sosinsky. 2010. Pannexin1 and Pannexin2 channels show quaternary similarities to connexons and different oligomerization numbers from each other. *J Biol Chem* 285(32):24420–31.

Anand, R. J., S. Dai, S. C. Gribar, W. Richardson, J. W. Kohler, R. A. Hoffman, M. F. Branca et al. 2008. A role for connexin43 in macrophage phagocytosis and host survival after bacterial peritoneal infection. *J Immunol* 181(12):8534–43.

Au, S. R., K. Au, G. C. Saggers, N. Karne, and H. P. Ehrlich. 2007. Rat mast cells communicate with fibroblasts via gap junction intercellular communications. *J Cell Biochem* 100(5):1170–7.

Batista, F. D., and N. E. Harwood. 2009. The who, how and where of antigen presentation to B cells. *Nat Rev Immunol* 9(1):15–27.

Beattie, L., A. Peltan, A. Maroof, A. Kirby, N. Brown, M. Coles, D. F. Smith, and P. M. Kaye. 2010. Dynamic imaging of experimental Leishmania donovani-induced hepatic granulomas detects Kupffer cell-restricted antigen presentation to antigen-specific CD8 T cells. *PLoS Pathog* 6(3):e1000805.

Beauvillain, C., Y. Delneste, M. Scotet, A. Peres, H. Gascan, P. Guermonprez, V. Barnaba, and P. Jeannin. 2007. Neutrophils efficiently cross-prime naive T cells in vivo. *Blood* 110(8):2965–73.

Beauvillain, C., S. Donnou, U. Jarry, M. Scotet, H. Gascan, Y. Delneste, P. Guermonprez, P. Jeannin, and D. Couez. 2008. Neonatal and adult microglia cross-present exogenous antigens. *Glia* 56(1):69–77.

Benlalam, H., A. Jalil, M. Hasmim, B. Pang, R. Tamouza, M. Mitterrand, Y. Godet et al. 2009. Gap junction communication between autologous endothelial and tumor cells induce cross-recognition and elimination by specific CTL. *J Immunol* 182(5):2654–64.

Bermudez-Fajardo, A., M. Yliharsila, W. H. Evans, A. C. Newby, and E. Oviedo-Orta. 2007. CD4+ T lymphocyte subsets express connexin 43 and establish gap junction channel communication with macrophages in vitro. *J Leukoc Biol* 82(3):608–12.

Bopp, T., C. Becker, M. Klein, S. Klein-Hessling, A. Palmetshofer, E. Serfling, V. Heib et al. 2007. Cyclic adenosine monophosphate is a key component of regulatory T cell-mediated suppression. *J Exp Med* 204(6):1303–10.

Bousso, P. 2008. T-cell activation by dendritic cells in the lymph node: Lessons from the movies. *Nat Rev Immunol* 8(9):675–84.

Brand, C. U., T. Hunziker, T. Schaffner, A. Limat, H. A. Gerber, and L. R. Braathen. 1995. Activated immunocompetent cells in human skin lymph derived from irritant contact dermatitis: An immunomorphological study. *Br J Dermatol* 132(1):39–45.

Brañes, M. C., J. E. Contreras, and J. C. Sáez. 2002. Activation of human polymorphonuclear cells induces formation of functional gap junctions and expression of connexins. *Med Sci Monit* 8(8):BR313–23.

Campbell, F. R. 1982. Intercellular contacts between migrating blood cells and cells of the sinusoidal wall of bone marrow. An ultrastructural study using tannic acid. *Anat Rec* 203(3):365–74.

Concha, M., C. D. Figueroa, and I. Caorsi. 1988. Ultrastructural characteristics of the contact zones between Langerhans cells and lymphocytes. *J Pathol* 156(1):29–36.

Concha, M., A. Vidal, G. Garces, C. D. Figueroa, and I. Caorsi. 1993. Physical interaction between Langerhans cells and T-lymphocytes during antigen presentation in vitro. *J Invest Dermatol* 100(4):429–34.

Corvalán, L. A., R. Araya, M. C. Brañes, P. J. Sáez, A. M. Kalergis, J. A. Tobar, M. Theis, K. Willecke, and J. C. Sáez. 2007. Injury of skeletal muscle and specific cytokines induce the expression of gap junction channels in mouse dendritic cells. *J Cell Physiol* 211(3):649–60.

Chanson, M., and B. R. Kwak. 2007. Connexin37: A potential modifier gene of inflammatory disease. *J Mol Med* 85(8):787–95.

Chen, Y., Y. Yao, Y. Sumi, A. Li, U. K. To, A. Elkhal, Y. Inoue et al. 2010. Purinergic signaling: A fundamental mechanism in neutrophil activation. *Sci Signal* 3(125):ra45.

Chinnery, H. R., E. Pearlman, and P. G. McMenamin. 2008. Cutting edge: Membrane nanotubes in vivo: A feature of MHC class II+ cells in the mouse cornea. *J Immunol* 180(9):5779–83.

D'Hondt, C., R. Ponsaerts, H. De Smedt, G. Bultynck, and B. Himpens. 2009. Pannexins, distant relatives of the connexin family with specific cellular functions? *Bioessays* 31(9):953–74.

Davalos, D., J. Grutzendler, G. Yang, J. V. Kim, Y. Zuo, S. Jung, D. R. Littman, M. L. Dustin, and W. B. Gan. 2005. ATP mediates rapid microglial response to local brain injury in vivo. *Nat Neurosci* 8(6):752–8.

Davis, D. M. 2009. Mechanisms and functions for the duration of intercellular contacts made by lymphocytes. *Nat Rev Immunol* 9(8):543–55.

Denzer, K., M. van Eijk, M. J. Kleijmeer, E. Jakobson, C. de Groot, and H. J. Geuze. 2000. Follicular dendritic cells carry MHC class II-expressing microvesicles at their surface. *J Immunol* 165(3):1259–65.

Dobrenis, K., H. Y. Chang, M. H. Pina-Benabou, A. Woodroffe, S. C. Lee, R. Rozental, D. C. Spray, and E. Scemes. 2005. Human and mouse microglia express connexin36, and functional gap junctions are formed between rodent microglia and neurons. *J Neurosci Res* 82(3):306–15.

Dorban, G., V. Defaweux, E. Heinen, and N. Antoine. 2010. Spreading of prions from the immune to the peripheral nervous system: A potential implication of dendritic cells. *Histochem Cell Biol* 133(5):493–504.

Elgueta, R., J. A. Tobar, K. F. Shoji, J. De Calisto, A. M. Kalergis, M. R. Bono, M. Rosemblatt, and J. C. Sáez. 2009. Gap junctions at the dendritic cell-T cell interface are key elements for antigen-dependent T cell activation. *J Immunol* 183(1):277–84.

Elias, L. A., D. D. Wang, and A. R. Kriegstein. 2007. Gap junction adhesion is necessary for radial migration in the neocortex. *Nature* 448(7156):901–7.

Eltzschig, H. K., T. Eckle, A. Mager, N. Kuper, C. Karcher, T. Weissmuller, K. Boengler, R. Schulz, S. C. Robson, and S. P. Colgan. 2006. ATP release from activated neutrophils occurs via connexin 43 and modulates adenosine-dependent endothelial cell function. *Circ Res* 99(10):1100–8.

Eugenín, E. A., M. C. Brañes, J. W. Berman, and J. C. Sáez. 2003. TNF-alpha plus IFN-gamma induce connexin43 expression and formation of gap junctions between human monocytes/macrophages that enhance physiological responses. *J Immunol* 170(3):1320–8.

Eugenín, E. A., D. Eckardt, M. Theis, K. Willecke, M. V. Bennett, and J. C. Sáez. 2001. Microglia at brain stab wounds express connexin 43 and *in vitro* form functional gap junctions after treatment with interferon-gamma and tumor necrosis factor-alpha. *Proc Natl Acad Sci USA* 98(7):4190–5.

Eugenín, E. A., P. J. Gaskill, and J. W. Berman. 2009. Tunneling nanotubes (TNT) are induced by HIV-infection of macrophages: A potential mechanism for intercellular HIV trafficking. *Cell Immunol* 254(2):142–8.

Eugenín, E. A., H. E. González, H. A. Sanchez, M. C. Brañes, and J. C. Sáez. 2007. Inflammatory conditions induce gap junctional communication between rat Kupffer cells both *in vivo* and in vitro. *Cell Immunol* 247(2):103–10.

Farber, K., and H. Kettenmann. 2006. Functional role of calcium signals for microglial function. *Glia* 54(7):656–65.

Fortes, F. S., I. L. Pecora, P. M. Persechini, S. Hurtado, V. Costa, R. Coutinho-Silva, M. B. Braga et al. 2004. Modulation of intercellular communication in macrophages: Possible interactions between GAP junctions and P2 receptors. *J Cell Sci* 117(Pt 20):4717–26.

Garg, S., M. Md Syed, and T. Kielian. 2005. *Staphylococcus aureus*-derived peptidoglycan induces Cx43 expression and functional gap junction intercellular communication in microglia. *J Neurochem* 95(2):475–83.

Gaudenzio, N., N. Espagnolle, L. T. Mars, R. Liblau, S. Valitutti, and E. Espinosa. 2009. Cell-cell cooperation at the T helper cell/mast cell immunological synapse. *Blood* 114(24):4979–88.

González, H. E., E. A. Eugenín, G. Garcés, N. Solis, M. Pizarro, L. Accatino, and J. C. Sáez. 2002. Regulation of hepatic connexins in cholestasis: Possible involvement of Kupffer cells and inflammatory mediators. *Am J Physiol Gastrointest Liver Physiol* 282(6):G991–1001.

Gousset, K., and C. Zurzolo. 2009. Tunnelling nanotubes: A highway for prion spreading? *Prion* 3(2):94–8.

Green, D. R., T. Ferguson, L. Zitvogel, and G. Kroemer. 2009. Immunogenic and tolerogenic cell death. *Nat Rev Immunol* 9(5):353–63.

Guinan, E. C., B. R. Smith, P. F. Davies, and J. S. Pober. 1988. Cytoplasmic transfer between endothelium and lymphocytes: Quantitation by flow cytometry. *Am J Pathol* 132(3):406–9.

Handel, A., A. Yates, S. S. Pilyugin, and R. Antia. 2007. Gap junction-mediated antigen transport in immune responses. *Trends Immunol* 28(11):463–6.

Hanisch, U. K. 2002. Microglia as a source and target of cytokines. *Glia* 40(2):140–55.

Harris, A. L. 2007. Connexin channel permeability to cytoplasmic molecules. *Prog Biophys Mol Biol* 94(1–2):120–43.

Hill, C. E., J. K. Phillips, and S. L. Sandow. 2001. Heterogeneous control of blood flow amongst different vascular beds. *Med Res Rev* 21(1):1–60.

Hu, V. W., and H. Q. Xie. 1994. Interleukin-1 alpha suppresses gap junction-mediated intercellular communication in human endothelial cells. *Exp Cell Res* 213(1):218–23.

Hubert, S., B. Rissiek, K. Klages, J. Huehn, T. Sparwasser, F. Haag, F. Koch-Nolte, O. Boyer, M. Seman, and S. Adriouch. 2010. Extracellular NAD+ shapes the Foxp3+ regulatory T cell compartment through the ART2-P2X7 pathway. *J Exp Med* 207(12):2561–8.

Hugues, S., L. Fetler, L. Bonifaz, J. Helft, F. Amblard, and S. Amigorena. 2004. Distinct T cell dynamics in lymph nodes during the induction of tolerance and immunity. *Nat Immunol* 5(12):1235–42.

Hülser, D. F., and J. H. Peters. 1972. Contact cooperation in stimulated lymphocytes. II. Electrophysiological investigations on intercellular communication. *Exp Cell Res* 74(2):319–26.

Hülser, D. F., and J. H. Peters. 1971. Intercellular communication in phytohemagglutinin-induced lymphocyte agglutinates. *Eur J Immunol* 1(6):494–5.

Imhof, B. A., and M. Aurrand-Lions. 2004. Adhesion mechanisms regulating the migration of monocytes. *Nat Rev Immunol* 4(6):432–44.

Inaba, K., W. J. Swiggard, R. M. Steinman, N. Romani, G. Schuler, and C. Brinster. 2009. Isolation of dendritic cells. *Curr Protoc Immunol* 86:3.7.1–3.7.19.

Inoue, K. 2008. Purinergic systems in microglia. *Cell Mol Life Sci* 65(19):3074–80.

Ishibashi, H., M. Nakamura, A. Komori, K. Migita, and S. Shimoda. 2009. Liver architecture, cell function, and disease. *Semin Immunopathol* 31(3):399–409.

Ishikawa, M., T. Iwamoto, T. Nakamura, A. Doyle, S. Fukumoto, and Y. Yamada. 2011. Pannexin 3 functions as an ER Ca(2+) channel, hemichannel, and gap junction to promote osteoblast differentiation. *J Cell Biol* 193(7):1257–74.

Janeway, C. A., Jr., J. Ron, and M. E. Katz. 1987. The B cell is the initiating antigen-presenting cell in peripheral lymph nodes. *J Immunol* 138(4):1051–5.

Jara, P. I., M. P. Boric, and J. C. Sáez. 1995. Leukocytes express connexin 43 after activation with lipopolysaccharide and appear to form gap junctions with endothelial cells after ischemia-reperfusion. *Proc Natl Acad Sci USA* 92(15):7011–5.

Katakowski, M., B. Buller, X. Wang, T. Rogers, and M. Chopp. 2010. Functional microRNA is transferred between glioma cells. *Cancer Res* 70(21):8259–63.

Klein, U., and R. Dalla-Favera. 2008. Germinal centres: Role in B-cell physiology and malignancy. *Nat Rev Immunol* 8(1):22–33.

Kleindienst, P., and T. Brocker. 2005. Concerted antigen presentation by dendritic cells and B cells is necessary for optimal CD4 T-cell immunity in vivo. *Immunology* 115(4):556–64.

Kono, H., and K. L. Rock. 2008. How dying cells alert the immune system to danger. *Nat Rev Immunol* 8(4):279–89.

Krenacs, T., and M. Rosendaal. 1998. Gap-junction communication pathways in germinal center reactions. *Dev Immunol* 6(1–2):111–8.

Krenacs, T., and M. Rosendaal. 1995. Immunohistological detection of gap junctions in human lymphoid tissue: Connexin43 in follicular dendritic and lymphoendothelial cells. *J Histochem Cytochem* 43(11):1125–37.

Krenacs, T., M. van Dartel, E. Lindhout, and M. Rosendaal. 1997. Direct cell/cell communication in the lymphoid germinal center: Connexin43 gap junctions functionally couple follicular dendritic cells to each other and to B lymphocytes. *Eur J Immunol* 27(6):1489–97.

Krummel, M. F., and M. D. Cahalan. 2010. The immunological synapse: A dynamic platform for local signaling. *J Clin Immunol* 30(3):364–72.

Kurt-Jones, E. A., D. Liano, K. A. HayGlass, B. Benacerraf, M. S. Sy, and A. K. Abbas. 1988. The role of antigen-presenting B cells in T cell priming in vivo. Studies of B cell-deficient mice. *J Immunol* 140(11):3773–8.

la Sala, A., S. Sebastiani, D. Ferrari, F. Di Virgilio, M. Idzko, J. Norgauer, and G. Girolomoni. 2002. Dendritic cells exposed to extracellular adenosine triphosphate acquire the migratory properties of mature cells and show a reduced capacity to attract type 1 T lymphocytes. *Blood* 99(5):1715–22.

Langevin, C., K. Gousset, M. Costanzo, O. Richard-Le Goff, and C. Zurzolo. 2010. Characterization of the role of dendritic cells in prion transfer to primary neurons. *Biochem J* 431(2):189–98.

Leaphart, C. L., F. Qureshi, S. Cetin, J. Li, T. Dubowski, C. Baty, D. Beer-Stolz, F. Guo, S. A. Murray, and D. J. Hackam. 2007. Interferon-gamma inhibits intestinal restitution by preventing gap junction communication between enterocytes. *Gastroenterology* 132(7):2395–411.

Levy, J. A., R. M. Weiss, E. R. Dirksen, and M. R. Rosen. 1976. Possible communication between murine macrophages oriented in linear chains in tissue culture. *Exp Cell Res* 103(2):375–85.

Ley, K., C. Laudanna, M. I. Cybulsky, and S. Nourshargh. 2007. Getting to the site of inflammation: The leukocyte adhesion cascade updated. *Nat Rev Immunol* 7(9):678–89.

Lindquist, R. L., G. Shakhar, D. Dudziak, H. Wardemann, T. Eisenreich, M. L. Dustin, and M. C. Nussenzweig. 2004. Visualizing dendritic cell networks in vivo. *Nat Immunol* 5(12):1243–50.

Liu, K., and M. C. Nussenzweig. 2010. Origin and development of dendritic cells. *Immunol Rev* 234(1):45–54.

Marelli-Berg, F. M., and S. J. Jarmin. 2004. Antigen presentation by the endothelium: A green light for antigen-specific T cell trafficking? *Immunol Lett* 93(2–3):109–13.

Martin, F. J., and A. S. Prince. 2008. TLR2 regulates gap junction intercellular communication in airway cells. *J Immunol* 180(7):4986–93.

Martínez, A. D., E. A. Eugenín, M. C. Brañes, M. V. Bennett, and J. C. Sáez. 2002. Identification of second messengers that induce expression of functional gap junctions in microglia cultured from newborn rats. *Brain Res* 943(2):191–201.

Matsue, H., J. Yao, K. Matsue, A. Nagasaka, H. Sugiyama, R. Aoki, M. Kitamura, and S. Shimada. 2006. Gap junction-mediated intercellular communication between dendritic cells (DCs) is required for effective activation of DCs. *J Immunol* 176(1):181–90.

Mebius, R. E., and G. Kraal. 2005. Structure and function of the spleen. *Nat Rev Immunol* 5(8):606–16.

Mendoza-Naranjo, A., P. J. Sáez, C. C. Johansson, M. Ramírez, D. Mandakovic, C. Pereda, M. N. López, R. Kiessling, J. C. Sáez, and F. Salazar-Onfray. 2007. Functional gap junctions facilitate melanoma antigen transfer and cross-presentation between human dendritic cells. *J Immunol* 178(11):6949–57.

Merad, M., and M. G. Manz. 2009. Dendritic cell homeostasis. *Blood* 113(15):3418–27.

Merad, M., F. Ginhoux, and M. Collin. 2008. Origin, homeostasis and function of Langerhans cells and other langerin-expressing dendritic cells. *Nat Rev Immunol* 8(12):935–47.

Meyer, R. A., D. W. Laird, J. P. Revel, and R. G. Johnson. 1992. Inhibition of gap junction and adherens junction assembly by connexin and A-CAM antibodies. *J Cell Biol* 119(1):179–89.

Mosser, D. M., and J. P. Edwards. 2008. Exploring the full spectrum of macrophage activation. *Nat Rev Immunol* 8(12):958–69.

Moyer, K. E., G. C. Saggers, and H. P. Ehrlich. 2004. Mast cells promote fibroblast populated collagen lattice contraction through gap junction intercellular communication. *Wound Repair Regen* 12(3):269–75.

Musil, L. S., B. A. Cunningham, G. M. Edelman, and D. A. Goodenough. 1990. Differential phosphorylation of the gap junction protein connexin43 in junctional communication-competent and -deficient cell lines. *J Cell Biol* 111(5 Pt 1):2077–88.

Nathan, C. 2006. Neutrophils and immunity: Challenges and opportunities. *Nat Rev Immunol* 6(3):173–82.

Naus, C. C., and D. W. Laird. 2010. Implications and challenges of connexin connections to cancer. *Nat Rev Cancer* 10(6):435–41.

Neijssen, J., C. Herberts, J. W. Drijfhout, E. Reits, L. Janssen, and J. Neefjes. 2005. Cross-presentation by intercellular peptide transfer through gap junctions. *Nature* 434(7029):83–8.

Neijssen, J., B. Pang, and J. Neefjes. 2007. Gap junction-mediated intercellular communication in the immune system. *Prog Biophys Mol Biol* 94(1–2):207–18.

Nestle, F. O., P. Di Meglio, J. Z. Qin, and B. J. Nickoloff. 2009. Skin immune sentinels in health and disease. *Nat Rev Immunol* 9(10):679–91.

Ng, L. G., A. Hsu, M. A. Mandell, B. Roediger, C. Hoeller, P. Mrass, A. Iparraguirre et al. 2008. Migratory dermal dendritic cells act as rapid sensors of protozoan parasites. *PLoS Pathog* 4(11):e1000222.

Onfelt, B., S. Nedvetzki, R. K. Benninger, M. A. Purbhoo, S. Sowinski, A. N. Hume, M. C. Seabra, M. A. Neil, P. M. French, and D. M. Davis. 2006. Structurally distinct membrane nanotubes between human macrophages support long-distance vesicular traffic or surfing of bacteria. *J Immunol* 177(12):8476–83.

Orellana, J. A., P. J. Sáez, K. F. Shoji, K. A. Schalper, N. Palacios-Prado, V. Velarde, C. Giaume, M. V. Bennett, and J. C. Sáez. 2009. Modulation of brain hemichannels and gap junction channels by pro-inflammatory agents and their possible role in neurodegeneration. *Antioxid Redox Signal* 11(2):369–99.

Oviedo-Orta, E., and W. H. Evans. 2002. Gap junctions and connexins: Potential contributors to the immunological synapse. *J Leukoc Biol* 72(4):636–42.

Oviedo-Orta, E., P. Gasque, and W. H. Evans. 2001. Immunoglobulin and cytokine expression in mixed lymphocyte cultures is reduced by disruption of gap junction intercellular communication. *FASEB J* 15(3):768–74.

Oviedo-Orta, E., T. Hoy, and W. H. Evans. 2000. Intercellular communication in the immune system: Differential expression of connexin40 and 43, and perturbation of gap junction channel functions in peripheral blood and tonsil human lymphocyte subpopulations. *Immunology* 99(4):578–90.

Oviedo-Orta, E., M. Perreau, W. H. Evans, and I. Potolicchio. 2010. Control of the proliferation of activated CD4+ T cells by connexins. *J Leukoc Biol* 88(1):79–86.

Pang, B., J. Neijssen, X. Qiao, L. Janssen, H. Janssen, C. Lippuner, and J. Neefjes. 2009. Direct antigen presentation and gap junction mediated cross-presentation during apoptosis. *J Immunol* 183(2):1083–90.

Polacek, D., R. Lal, M. V. Volin, and P. F. Davies. 1993. Gap junctional communication between vascular cells. Induction of connexin43 messenger RNA in macrophage foam cells of atherosclerotic lesions. *Am J Pathol* 142(2):593–606.

Porvaznik, M., and T. J. MacVittie. 1979. Detection of gap junctions between the progeny of a canine macrophage colony-forming cell in vitro. *J Cell Biol* 82(2):555–64.

Racanelli, V., and B. Rehermann. 2006. The liver as an immunological organ. *Hepatology* 43(2 Suppl 1):S54–62.

Ramachandra, L., Y. Qu, Y. Wang, C. J. Lewis, B. A. Cobb, K. Takatsu, W. H. Boom, G. R. Dubyak, and C. V. Harding. 2010. Mycobacterium tuberculosis synergizes with ATP to induce release of microvesicles and exosomes containing major histocompatibility complex class II molecules capable of antigen presentation. *Infect Immun* 78(12):5116–25.

Randolph, G. J., V. Angeli, and M. A. Swartz. 2005. Dendritic-cell trafficking to lymph nodes through lymphatic vessels. *Nat Rev Immunol* 5(8):617–28.

Ransohoff, R. M., and V. H. Perry. 2009. Microglial physiology: Unique stimuli, specialized responses. *Annu Rev Immunol* 27:119–45.

Reis e Sousa, C. 2006. Dendritic cells in a mature age. *Nat Rev Immunol* 6(6):476–83.

Ring, S., S. Karakhanova, T. Johnson, A. H. Enk, and K. Mahnke. 2010. Gap junctions between regulatory T cells and dendritic cells prevent sensitization of CD8(+) T cells. *J Allergy Clin Immunol* 125(1):237–46 e1–7.

Rose, M. L. 1998. Endothelial cells as antigen-presenting cells: Role in human transplant rejection. *Cell Mol Life Sci* 54(9):965–78.

Rustom, A., R. Saffrich, I. Markovic, P. Walther, and H. H. Gerdes. 2004. Nanotubular highways for intercellular organelle transport. *Science* 303(5660):1007–10.

Saccheri, F., C. Pozzi, F. Avogadri, S. Barozzi, M. Faretta, P. Fusi, and M. Rescigno. 2010. Bacteria-induced gap junctions in tumors favor antigen cross-presentation and antitumor immunity. *Sci Transl Med* 2(44):44–57.

Sáez, J. C., V. M. Berthoud, M. C. Brañes, A. D. Martínez, and E. C. Beyer. 2003. Plasma membrane channels formed by connexins: Their regulation and functions. *Physiol Rev* 83(4):1359–400.

Sáez, J. C., M. C. Brañes, L. A. Corvalán, E. A. Eugenín, H. González, A. D. Martínez, and F. Palisson. 2000. Gap junctions in cells of the immune system: Structure, regulation and possible functional roles. *Braz J Med Biol Res* 33(4):447–55.

Sallusto, F., and A. Lanzavecchia. 1994. Efficient presentation of soluble antigen by cultured human dendritic cells is maintained by granulocyte/macrophage colony-stimulating factor plus interleukin 4 and downregulated by tumor necrosis factor alpha. *J Exp Med* 179(4):1109–18.

Salter, R. D., and S. C. Watkins. 2009. Dendritic cell altered states: What role for calcium? *Immunol Rev* 231(1):278–88.

Sarieddine, M. Z., K. E. Scheckenbach, B. Foglia, K. Maass, I. Garcia, B. R. Kwak, and M. Chanson. 2009. Connexin43 modulates neutrophil recruitment to the lung. *J Cell Mol Med* 13(11–12):4560–70.

Scerri, I., O. Tabary, T. Dudez, J. Jacquot, B. Foglia, S. Suter, and M. Chanson. 2006. Gap junctional communication does not contribute to the interaction between neutrophils and airway epithelial cells. *Cell Commun Adhes* 13(1–2):1–12.

Schalper, K. A., N. Palacios-Prado, J. A. Orellana, and J. C. Sáez. 2008. Currently used methods for identification and characterization of hemichannels. *Cell Commun Adhes* 15(1):207–18.

Scheckenbach, K. E., S. Crespin, B. R. Kwak, and M. Chanson. 2010. Connexin channel-dependent signaling pathways in inflammation. *J Vasc Res* 48(2):91–103.

Schenk, U., A. M. Westendorf, E. Radaelli, A. Casati, M. Ferro, M. Fumagalli, C. Verderio, J. Buer, E. Scanziani, and F. Grassi. 2008. Purinergic control of T cell activation by ATP released through pannexin-1 hemichannels. *Sci Signal* 1(39):ra6.

Shakhar, G., R. L. Lindquist, D. Skokos, D. Dudziak, J. H. Huang, M. C. Nussenzweig, and M. L. Dustin. 2005. Stable T cell-dendritic cell interactions precede the development of both tolerance and immunity in vivo. *Nat Immunol* 6(7):707–14.

Shortman, K., and S. H. Naik. 2007. Steady-state and inflammatory dendritic-cell development. *Nat Rev Immunol* 7(1):19–30.

Simon, A. M., A. R. McWhorter, H. Chen, C. L. Jackson, and Y. Ouellette. 2004. Decreased intercellular communication and connexin expression in mouse aortic endothelium during lipopolysaccharide-induced inflammation. *J Vasc Res* 41(4):323–33.

Simons, M., and G. Raposo. 2009. Exosomes—Vesicular carriers for intercellular communication. *Curr Opin Cell Biol* 21(4):575–81.

Takeuchi, H., S. Jin, J. Wang, G. Zhang, J. Kawanokuchi, R. Kuno, Y. Sonobe, T. Mizuno, and A. Suzumura. 2006. Tumor necrosis factor-alpha induces neurotoxicity via glutamate release from hemichannels of activated microglia in an autocrine manner. *J Biol Chem* 281(30):21362–8.

Teoh, D., L. A. Johnson, T. Hanke, A. J. McMichael, and D. G. Jackson. 2009. Blocking development of a CD8+ T cell response by targeting lymphatic recruitment of APC. *J Immunol* 182(4):2425–31.

Thery, C., L. Duban, E. Segura, P. Veron, O. Lantz, and S. Amigorena. 2002. Indirect activation of naive CD4+ T cells by dendritic cell-derived exosomes. *Nat Immunol* 3(12):1156–62.

Thery, C., M. Ostrowski, and E. Segura. 2009. Membrane vesicles as conveyors of immune responses. *Nat Rev Immunol* 9(8):581–93.

Tour, O., S. R. Adams, R. A. Kerr, R. M. Meijer, T. J. Sejnowski, R. W. Tsien, and R. Y. Tsien. 2007. Calcium Green FlAsH as a genetically targeted small-molecule calcium indicator. *Nat Chem Biol* 3(7):423–31.

Véliz, L. P., F. G. González, B. R. Duling, J. C. Sáez, and M. P. Boric. 2008. Functional role of gap junctions in cytokine-induced leukocyte adhesion to endothelium in vivo. *Am J Physiol Heart Circ Physiol* 295(3):H1056–66.

Viaud, S., C. Thery, S. Ploix, T. Tursz, V. Lapierre, O. Lantz, L. Zitvogel, and N. Chaput. 2010. Dendritic cell-derived exosomes for cancer immunotherapy: What's next? *Cancer Res* 70(4):1281–5.

Vig, M., and J. P. Kinet. 2009. Calcium signaling in immune cells. *Nat Immunol* 10(1):21–7.

Vliagoftis, H., A. M. Hutson, S. Mahmudi-Azer, H. Kim, V. Rumsaeng, C. K. Oh, R. Moqbel, and D. D. Metcalfe. 1999. Mast cells express connexins on their cytoplasmic membrane. *J Allergy Clin Immunol* 103(4):656–62.

Volkmann, A., J. Neefjes, and B. Stockinger. 1996. A conditionally immortalized dendritic cell line which differentiates in contact with T cells or T cell-derived cytokines. *Eur J Immunol* 26(11):2565–72.

Vyas, J. M., A. G. Van der Veen, and H. L. Ploegh. 2008. The known unknowns of antigen processing and presentation. *Nat Rev Immunol* 8(8):607–18.

Wang, X., M. L. Veruki, N. V. Bukoreshtliev, E. Hartveit, and H. H. Gerdes. 2010. Animal cells connected by nanotubes can be electrically coupled through interposed gap-junction channels. *Proc Natl Acad Sci USA* 107(40):17194–9.

Watkins, S. C., and R. D. Salter. 2005. Functional connectivity between immune cells mediated by tunneling nanotubules. *Immunity* 23(3):309–18.

Wei, S. H., O. Safrina, Y. Yu, K. R. Garrod, M. D. Cahalan, and I. Parker. 2007. Ca2+ signals in CD4+ T cells during early contacts with antigen-bearing dendritic cells in lymph node. *J Immunol* 179(3):1586–94.

Woehrle, T., L. Yip, A. Elkhal, Y. Sumi, Y. Chen, Y. Yao, P. A. Insel, and W. G. Junger. 2010. Pannexin-1 hemichannel-mediated ATP release together with P2X1 and P2X4 receptors regulate T-cell activation at the immune synapse. *Blood* 116(18):3475–84.

Wong, C. W., T. Christen, I. Roth, C. E. Chadjichristos, J. P. Derouette, B. F. Foglia, M. Chanson, D. A. Goodenough, and B. R. Kwak. 2006. Connexin37 protects against atherosclerosis by regulating monocyte adhesion. *Nat Med* 12(8):950–4.

Zahler, S., A. Hoffmann, T. Gloe, and U. Pohl. 2003. Gap-junctional coupling between neutrophils and endothelial cells: A novel modulator of transendothelial migration. *J Leukoc Biol* 73(1):118–26.

Zimmerli, S. C., F. Masson, J. Cancela, P. Meda, and C. Hauser. 2007. Cutting edge: Lack of evidence for connexin-43 expression in human epidermal Langerhans cells. *J Immunol* 179(7):4318–21.

5 Connect the Immune System

Roles of Gap Junctions in Antigen Presentation and T Cell Activation

Baoxu Pang and Jacques Neefjes

CONTENTS

Major histocompatibility complex (MHC) class I molecules are expressed on all mammalian cells. They present peptide fragments of intracellular pathogens at the cell surface for consideration by the immune system. The so-called cytotoxic T cells (CTLs) can recognize these and then decide to kill the infected cell. The process of antigen presentation by MHC class I molecules is understood in detail. It involves a number of steps dedicated to ensuring that antigenic fragments are only presented to the immune system by infected cells and not by innocent bystanders. This can be achieved by retaining antigenic fragments in the infected cell, which prevents the elimination of noninfected neighboring cells by immune cells, or—in immunology language—prevents innocent bystander kill. However, activation of T cells also requires presentation of peptide fragments by dendritic cells (DCs) that act as master regulators in the immune system. The fact that these DCs present fragments that are derived from other cells means that they are able to break the dogma of classical MHC class I antigen presentation where antigenic peptides are presented only by the infected cells. Recently, peptides have been shown to diffuse through gap junctions. This allows peptides to be transferred from an infected cell to its neighbor for presentation by the noninfected cell and then allows innocent bystander kill. Connexins are expressed in the immune system under various conditions, but were not really considered to be important by the immunological community, hence receiving little attention. However, the involvement of peptide and second messenger transfer through gap junctions in various immune responses is now becoming evident. Here, we will discuss the current state of knowledge and speculate about its importance in the control of autoimmune diseases, through the modulation of regulatory T cell activity and its role in cancer immunotherapy, and through the control of proper T cell responses against tumor tissue.

5.1 STANDARD MODEL OF ANTIGEN PRESENTATION BY MHC CLASS I MOLECULES

MHC class I molecules present peptide fragments of around nine amino acids that are made inside cells to the outside immune world. CTLs may then recognize the peptide fragments bound to the MHC class I molecules and eliminate the poisoned cells. The CTLs are selected to recognize "everything-minus-self-peptides," including fragments from viruses and mutated (cancer) proteins. CTLs thus eliminate infected cells, transformed cells, and other cells expressing nonself intracellular proteins. To make these fragments, proteins, in the form of old proteins (Rock et al., 1994), misfolded proteins, or defective ribosomal products (DRiPs) (Schubert et al., 2000; Reits et al., 2000), are degraded by cytosolic or nuclear proteasomes into fragments that may be further trimmed or destroyed by cytosolic aminopeptidases (Reits et al., 2003). In fact, more than 99% of peptides will be destroyed before consideration by MHC class I molecules. To reach the peptide-binding site of MHC class I molecules, cytosolic peptides have to be translocated over a membrane, a task performed by the transporter associated with antigen processing (TAP), residing in the endoplasmic reticulum (ER) (Neefjes et al., 1993). TAP also acts as a scaffold for association with many, but not all MHC class I molecules that are waiting for peptide loading (Neisig et al., 1996) via the dedicated chaperone tapasin (Ortmann et al., 1997). In the ER, peptides may be further trimmed or rapidly removed unless they bind to MHC class I molecules (Saric et al., 2002; York et al., 2002). Peptide binding to an MHC class I molecule is also a signal to leave the ER for transport to the plasma membrane for antigen presentation (Figure 5.1). This pathway ensures that antigen presentation of viral peptides only occurs on the surface of infected cells and not the innocent bystanders, protecting uninfected cells from elimination during the ensuing immune response.

5.2 AN EXCEPTION: ANTIGEN CROSS-PRESENTATION VIA GAP JUNCTIONS

Before specific CTLs can act to clear the infected or tumor cells, they have to be activated and expanded with the help of a particular class of immune cells, the DCs. For specific activation, the DCs have to present the exact same foreign or mutated peptide on their MHC class I molecules on the cell surface as found on the infected/ transformed cells. The DCs have to acquire this antigen from other cells, which implies that the normal cellular boundaries are lost, as is the dogma of MHC class

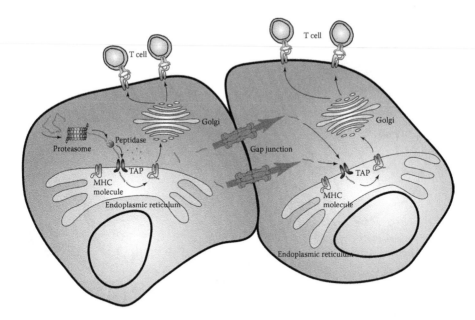

FIGURE 5.1 Classical MHC class I antigen presentation and gap junction-mediated antigen cross-presentation. In the classical MHC class I antigen presentation pathway, endogenous old proteins and DRiPs are degraded by nuclear or cytosolic proteasomes. Although the majority of the peptides are degraded by peptidases into free amino acids, a fraction may only be trimmed. The survivors can be transported into the ER by TAP, the transporter associated with antigen processing, where long peptides can be further trimmed into smaller peptides. With the help of other chaperones, peptides can bind to MHC class I molecules. These peptide–MHC complexes then leave the ER for transport to the cell surface and presentation to T cells. For gap junction-mediated antigen cross-presentation, proteasomes are not needed for antigen processing in the recipient cells, since only small peptides can diffuse via gap junctions between cells. These exogenous peptides then access the normal antigen presentation pathway for presentation by MHC class I molecules on the neighboring cells. As a result, exogenous antigens are cross-presented by the recipient cells.

I antigen presentation by the infected cell only. How do peptides from other cells manage to enter the MHC class I antigen presentation pathways of DCs? Multiple models have been proposed (Groothuis et al., 2005). One model involves the transfer of peptides through gap junctions between infected cells and DCs (Neijssen et al., 2005, 2007) (Figure 5.2). These gap junctions would solve the topological problems faced for cross-presentation by the DCs, where peptides have to pass multiple membranes from the infected cells to the DCs. When gap junctions are involved, cytosolic peptides from the infected cell simply enter the cytosol of the associated DC. Immunohistochemistry analysis of human tissues has already identified gap junctions in different subtypes of DCs, such as Langerhans cells (Concha et al., 1988) and follicular DCs (Krenács and Rosendaal, 1995; Krenacs et al., 1997). Monocytes and DCs usually express gap junction subunits poorly but are able to upregulate connexin 43 (Cx43), in response to "danger" signals and pathogen factors, such as

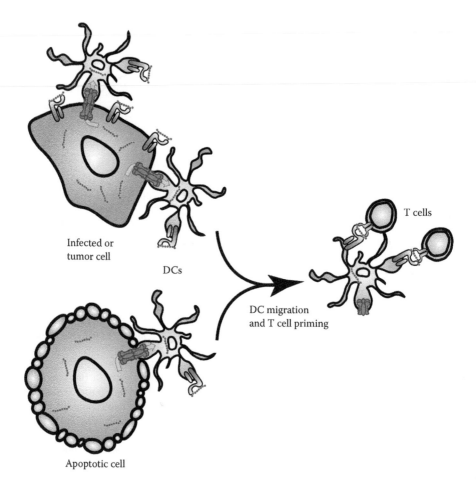

FIGURE 5.2 Antigen cross-presentation by DCs. Functional gap junctions can form between DCs and infected, tumor, or apoptotic cells. As a result, antigenic peptides can diffuse into the cytosol of DCs and be cross-presented by DCs. Activated DCs bearing the non-self peptides then migrate to lymph nodes and activate specific T cells, which proliferate and eliminate the infected or tumor cells.

interferon γ, TNF-α, and lipopolysaccharide (LPS) (Matsue et al., 2006; Mendoza-Naranjo et al., 2007; Saccheri et al., 2010). Most connexins have a highly restricted tissue distribution with the exception of Cx43, which is expressed by most tissues and a series of immune cells, including B cells, Langerhans cells, and other cells (Neijssen et al., 2007). These cells are subsequently able to communicate with other cells via Cx43-containing gap junctions. Once DCs form functional gap junctions with transformed cells or cells infected by viruses, antigenic peptides can be transferred into the cytosol of the DCs and presented on the cell surface of DCs by MHC class I molecules (Neijssen et al., 2005). In principle, DCs may make multiple gap junction-mediated contacts with the tissue to sense the state of the cell's cytosol and continuously sample fractions of cytosol for cross-presentation. When peptides from viral proteins or mutated proteins are transferred, a proper T cell response will be stimulated and the related disease is controlled.

Peptide exchange via gap junctions also ensures that the recipient DCs present exactly the same peptides as those that are made and presented by the original infected cells or transformed cells. Endogenous proteins are degraded by proteasomes in the form of DRiPs or misfolded proteins. The degradation products are then further trimmed by various aminopeptidases before they become the proper size for fitting MHC class I molecules. Immune cells, such as DCs, contain a specialized type of proteasomes called the immunoproteasome, with different cleavage specificities as compared to standard resident proteasomes present in most tissues. Since DCs mainly contain immunoproteasomes (Macagno et al., 1999), they may present different peptide fragments of a protein, which will then not activate T cells responding to peptides made in tissues by the standard proteasomes (Chapatte et al., 2006). Indeed, immunoproteasomes failed to generate proper antigenic peptides from several melanoma antigens such as Melan-A^{mart} (Morel et al., 2000). Incorrect peptide generation for cross-presentation due to different proteasome types would not be an issue when antigenic peptides are acquired by DCs via gap junctions because the peptides are from exactly the same pool of antigens as those in the original cells and are also prepared in the original cells (Figure 5.1). Manipulating tumor cells to upregulate connexins and establish functional gap junctions with DCs can result in a better antitumor immune response. It has been shown in mice that *Salmonella* localizes to tumors as these are more oxygen-poor regions (which *Salmonella* prefers). As a result, *Salmonella* induces a local inflammation and induces Cx43 expression in the melanoma tumor. This then supports cross-presentation and induces successful antitumor immune responses, due to gap junction-mediated contact between DCs and the tumor cells and the resulting transfer of peptides from tumor cells to DCs for cross-presentation (Saccheri et al., 2010). This result is important as it suggests that the induction of gap junction contact between the tumor tissue and DCs may be essential for strong antitumor immune responses. As bacterial infection in tumors may not always be a means of treatment, alternative methods to induce Cx43 expression in tumors, monocytes, and DCs have to be defined. It can help the immune system to surveil the cytosol of dangerous cells such as infected and transformed cells.

5.3 INNOCENT CELLS DO NOT DIE IN VAIN

Many cells die in response to anticancer treatment or infection. Notably, cell debris and apoptosis induction may result in stronger immune responses (Albert et al., 1998; Rovere et al., 1998). The question then is whether gap junctions are involved in the exchange of information—in the form of antigenic peptides—when cells undergo apoptosis. Using Caspase 9 constructs that can be activated by small compounds (also allowing controlled initiation of apoptosis), functional gap junction-mediated contacts were observed during the apoptosis phase and these contacts persisted until the apoptotic cell formed blebs and fully decomposed. During the initial phase of apoptosis, peptide transfer between cells continued and novel peptides generated by caspases were able to enter other associated cells, including DCs (Pang et al., 2009)

(Figure 5.2). Cells undergoing apoptosis apparently still allow the DCs to sample cytosolic contents for cross-presentation. This peptide repertoire will differ in detail from normal cells as peptides generated by caspases are also included.

5.4   GAP JUNCTIONS IN T CELL ACTIVATION

T cells are the effectors in the adaptive immune system. One subtype of T cells, CD8+ CTLs use their specific T cell receptors to survey the peptides presented by MHC class I molecules on the surface of resident cells. Once they recognize foreign or mutated peptides, they will trigger apoptosis of the target cells, and thus protect the body from infection and cancer. Compared to innate immune responses, CTL-mediated adaptive immune responses are more accurate and fine-tuned. Each T cell only expresses one type of T cell receptor and specifically recognizes one unique peptide presented by a defined MHC class I molecule. To respond to a broad range of unforeseen peptides, the adaptive immune system maintains its large T cell army in an economical way. The few CTLs specific for a new antigen are present in low numbers but are efficiently expanded when stimulated by professional antigen-presenting cells (APC), such as DCs and macrophages. These APCs continuously patrol through tissues to evaluate the health status of the resident cells. Once they recognize foreign or mutated antigens, the resulting peptide fragments can be presented to the T cells, resulting in the activation and expansion of the latter and the specific control of the disease. This cascade of events implies that the T cell-mediated specific immune response is slow. Acute responses to infections are made by resident antibodies (although expansion of their producer B cells also requires T cells and is slow) and the innate immune response where pathogen patterns are recognized by toll-like receptors (TLRs) and other proteins.

Multiple signaling pathways are activated in T cells via the T cell receptor and various costimulatory molecules that together induce proper T cell activation and proliferation (Smith-Garvin et al., 2009). During the activation process, T cells are in close contact with the APCs. Since gap junction subunits are widely expressed in T cells, functional gap junctions may support communication between various subtypes of T cells such as CD8+ CTLs, CD4+ T helper cells and regulatory T cells (Treg), or T cells and other contacting cells such as DCs. Although the presence of gap junctions in T cells has been suggested for some decades (Wekerle et al., 1980), this did not get much attention as their function was unclear. Only recently, it was found that functional gap junctions can form between various subtypes of T cells and APCs such as DCs and macrophages (Bermudez-Fajardo et al., 2007; Elgueta et al., 2009), allowing exchange of small metabolites and peptides, and playing important roles in T cell activation. The formation of gap junctions between DCs and CD4+ T cells might promote the T cell receptor–peptide–MHC class II-induced activation signaling pathway, since blocking functional gap junctions will attenuate subsequent T cell activation (Elgueta et al., 2009). However, the result of such interactions may be different under various conditions. For example, the formation of gap junctions between DCs and Tregs can also attenuate the activation of specific CD8+ CTLs by DCs (Ring et al., 2010). How it is regulated is still unclear. Possibly, the second messenger cyclic adenosine monophosphate (cAMP) is transferred between DC and

CD8⁺ CTLs via gap junctions to attenuate immune responses, as cAMP exchange between Tregs and the conventional T cells impairs T cell activation (Bopp et al., 2007). Ca^{2+} ions may be another relevant second messenger transferred through gap junctions between different immune cell types to induce T cell activation. Of note, the role of gap junctions in the immune system remains poorly studied and the timing of the gap junction formation, the regulation of gap junction closure/opening, and the type of molecules exchanged for a biological effect are mostly unknown.

T cell activity is essential for proper immune responses. Controlling T cell responses may prevent autoimmune activity that is the result of "uncontrolled activation" of these cells. Tregs can control conventional T cell activity through suppression of DCs (Ring et al., 2010). But suppression of autoimmune responses is more often the result of a direct interaction between Tregs and conventional T cells (Nakamura et al., 2001). It was originally believed that TNF-β on the cell surface of Tregs could suppress other T cells. But this model has not been supported by others (Randolph and Fathman, 2006). Later, it was found that a direct interaction between Tregs and conventional T cells may be essential to form gap junctions to transfer cAMP from Tregs into conventional CD4⁺ T helper cells (Bopp et al., 2007), as a result preventing activation of the latter (Figure 5.3). This may explain why the suppression of immune responses by Tregs is cell contact dependent. Tregs infiltrate tumors, possibly to suppress the antitumor response of conventional T cells (Wang

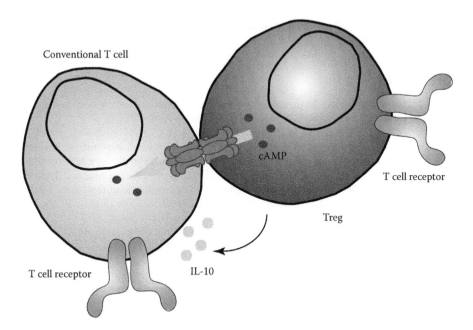

FIGURE 5.3 Gap junctions in T cells. Functional gap junctions can form between conventional T cells and Tregs. Through gap junctions, small molecules such as cAMP can diffuse from Tregs into conventional T cells. In combination with cytokines, they can result in suppression of conventional T cells and induction of immune tolerance. The full span of immune responses resulting from these contacts is poorly evaluated.

et al., 2004). In this concept, tumors attract Tregs that contact and couple to tissue-resident T cells and suppress T cell-mediated apoptosis. In this context, inhibiting gap junction formation between Tregs and antitumor T cells might result in better antitumor responses.

5.5 SUMMARY

Although cell–cell contact has long been known to occur between immune cells, immunologists have mainly considered these as interactions between various trans-membrane receptors that mediate different signaling pathways. Only recently, the contribution of gap junctions to the regulation of immune responses is being appreciated. Connexins are widely expressed by immune cells and their expression can be under the control of "danger signals." Small messenger molecules indicating the health status of a tissue can diffuse via gap junctions into immune cells that then may induce immune responses in a cell surface receptor-independent manner. This issue has not yet been studied extensively by immunologists. Besides second messengers, peptides can also diffuse through gap junctions between resident cells and DCs, resulting in antigen cross-presentation by the MHC class I molecules on DCs and activation of T cells. Activating gap junction formation can contribute to better antitumor and antivirus responses, while simultaneously supporting the activity of Tregs. Gap junctions support the exchange of cytosolic information between immune cells and the tissue. Yet, a detailed understanding of the wealth of small molecules exchanged between the tissue and immune cells and the resulting responses is only beginning to become understood.

REFERENCES

Albert, M. L., Sauter, B., and Bhardwaj, N. 1998. Dendritic cells acquire antigen from apoptotic cells and induce class I-restricted CTLs. *Nature,* 392, 86–89.

Bermudez-Fajardo, A., Ylihärsilä, M., Evans, W. H., Newby, A. C., and Oviedo-Orta, E. 2007. CD4+ T lymphocyte subsets express connexin 43 and establish gap junction channel communication with macrophages in vitro. *Journal of Leukocyte Biology,* 82, 608–612.

Bopp, T., Becker, C., Klein, M., Klein-Hebling, S., Palmetshofer, A., Serfling, E., Heib, V. et al. 2007. Cyclic adenosine monophosphate is a key component of regulatory T cell-mediated suppression. *The Journal of Experimental Medicine,* 204, 1303–1310.

Chapatte, L., Ayyoub, M., Morel, S., Peitrequin, A.L., Lévy, N., Servis, C., van den Eynde, B. J., Valmori, D., and Lévy, F. 2006. Processing of tumor-associated antigen by the proteasomes of dendritic cell controls *in vivo* T-cell responses. *Cancer Research,* 66, 5461–5468.

Concha, M., Figueroa, C. D., and Caorsi, I. 1988. Ultrastructural characteristics of the contact zones between langerhans cells and lymphocytes. *The Journal of Pathology,* 156, 29–36.

Elgueta, R., Tobar, J. A., Shoji, K. F., de Calisto, J., Kalergis, A. M., Bono, M. R., Rosemblatt, M., and Sáez, J. C. 2009. Gap junctions at the dendritic cell–T cell interface are key elements for antigen-dependent T cell activation. *The Journal of Immunology,* 183, 277–284.

Groothuis, T. A. M., Griekspoor, A. C., Neijssen, J. J., Herberts, C. A., and Neefjes, J. J. 2005. MHC class I alleles and their exploration of the antigen-processing machinery. *Immunological Reviews,* 207, 60–76.

Krenács, T. and Rosendaal, M. 1995. Immunohistological detection of gap junctions in human lymphoid tissue: Connexin43 in follicular dendritic and lymphoendothelial cells. *Journal of Histochemistry & Cytochemistry,* 43, 1125–1137.

Krenacs, T., van Dartel, M., Lindhout, E., and Rosendaal, M. 1997. Direct cell/cell communication in the lymphoid germinal center: Connexin43 gap junctions functionally couple follicular dendritic cells to each other and to B lymphocytes. *European Journal of Immunology,* 27, 1489–1497.

Macagno, A., Gilliet, M., Sallusto, F., Lanzavecchia, A., Nestle, F. O., and Groettrup, M. 1999. Dendritic cells up-regulate immunoproteasomes and the proteasome regulator Pa28 during maturation. *European Journal of Immunology,* 29, 4037–4042.

Matsue, H., Yao, J., Matsue, K., Nagasaka, A., Sugiyama, H., Aoki, R., Kitamura, M., and Shimada, S. 2006. Gap junction-mediated intercellular communication between dendritic cells (DCs) is required for effective activation of DCs. *The Journal of Immunology,* 176, 181–190.

Mendoza-Naranjo, A., Saéz, P. J., Johansson, C. C., Ramírez, M., Mandakovic, D., Pereda, C., López, M. N., Kiessling, R., Sáez, J. C., and Salazar-Onfray, F. 2007. Functional gap junctions facilitate melanoma antigen transfer and cross-presentation between human dendritic cells. *The Journal of Immunology,* 178, 6949–6957.

Morel, S., Lévy, F., Burlet-Schiltz, O., Brasseur, F., Probst-Kepper, M., Peitrequin, A.L., Monsarrat, B. et al. 2000. Processing of some antigens by the standard proteasome but not by the immunoproteasome results in poor presentation by dendritic cells. *Immunity,* 12, 107–117.

Nakamura, K., Kitani, A., and Strober, W. 2001. Cell contact-dependent immunosuppression by Cd4 + Cd25 + regulatory T cells is mediated by cell surface-bound transforming growth factor β. *The Journal of Experimental Medicine,* 194, 629–644.

Neefjes, J. J., Momburg, F., and Hammerling, G. J. 1993. Selective and ATP-dependent translocation of peptides by the MHC-encoded transporter. *Science,* 261, 769–771.

Neijssen, J., Herberts, C., Drijfhout, J. W., Reits, E., Janssen, L., and Neefjes, J. 2005. Cross-presentation by intercellular peptide transfer through gap junctions. *Nature,* 434, 83–88.

Neijssen, J., Pang, B., and Neefjes, J. 2007. Gap junction-mediated intercellular communication in the immune system. *Progress in Biophysics and Molecular Biology,* 94, 207–218.

Neisig, A., Wubbolts, R., Zang, X., Melief, C., and Neefjes, J. 1996. Allele-specific differences in the interaction of MHC class I molecules with transporters associated with antigen processing. *The Journal of Immunology,* 156, 3196–3206.

Ortmann, B., Copeman, J., Lehner, P. J., Sadasivan, B., Herberg, J. A., Grandea, A. G., Riddell, S. R. et al. 1997. A critical role for tapasin in the assembly and function of multimeric MHC class I-TAP complexes. *Science,* 277, 1306–1309.

Pang, B., Neijssen, J., Qiao, X., Janssen, L., Janssen, H., Lippuner, C., and Neefjes, J. 2009. Direct antigen presentation and gap junction mediated cross-presentation during apoptosis. *The Journal of Immunology,* 183, 1083–1090.

Randolph, D. A. and Fathman, C. G. 2006. CD4 + CD25+ regulatory T cells and their therapeutic potential. *Annual Review of Medicine,* 57, 381–402.

Reits, E., Griekspoor, A., Neijssen, J., Groothuis, T., Jalink, K., Van Veelen, P., Janssen, H., Calafat, J., Drijfhout, J. W., and Neefjes, J. 2003. Peptide diffusion, protection, and degradation in nuclear and cytoplasmic compartments before antigen presentation by MHC class I. *Immunity,* 18, 97–108.

Reits, E. A. J., Vos, J. C., Gromme, M., and Neefjes, J. 2000. The major substrates for TAP in vivo are derived from newly synthesized proteins. *Nature,* 404, 774–778.

Ring, S., Karakhanova, S., Johnson, T., Enk, A. H., and Mahnke, K. 2010. Gap junctions between regulatory T cells and dendritic cells prevent sensitization of Cd8+ T cells. *Journal of Allergy and Clinical Immunology,* 125, 237–246.e7.

Rock, K. L., Gramm, C., Rothstein, L., Clark, K., Stein, R., Dick, L., Hwang, D., and Goldberg, A. L. 1994. Inhibitors of the proteasome block the degradation of most cell proteins and the generation of peptides presented on MHC class I molecules. *Cell,* 78, 761–771.

Rovere, P., Vallinoto, C., Bondanza, A., Crosti, M. C., Rescigno, M., Ricciardi-Castagnoli, P., Rugarli, C., and Manfredi, A. A. 1998. Cutting edge: Bystander apoptosis triggers dendritic cell maturation and antigen-presenting function. *The Journal of Immunology,* 161, 4467–4471.

Saccheri, F., Pozzi, C., Avogadri, F., Barozzi, S., Faretta, M., Fusi, P., and Rescigno, M. 2010. Bacteria-induced gap junctions in tumors favor antigen cross-presentation and antitumor immunity. *Science Translational Medicine,* 2, 44ra57.

Saric, T., Chang, S.-C., Hattori, A., York, I. A., Markant, S., Rock, K. L., Tsujimoto, M., and Goldberg, A. L. 2002. An IFN-[gamma]-induced aminopeptidase in the ER, ERAP1, trims precursors to MHC class I-presented peptides. *Nature Immunology,* 3, 1169–1176.

Schubert, U., Anton, L. C., Gibbs, J., Norbury, C. C., Yewdell, J. W., and Bennink, J. R. 2000. Rapid degradation of a large fraction of newly synthesized proteins by proteasomes. *Nature,* 404, 770–774.

Smith-Garvin, J. E., Koretzky, G. A., and Jordan, M. S. 2009. T cell activation. *Annual Review of Immunology,* 27, 591–619.

Wang, H. Y., Lee, D. A., Peng, G., Guo, Z., Li, Y., Kiniwa, Y., Shevach, E. M., and Wang, R.-F. 2004. Tumor-specific human CD4+ regulatory T cells and their ligands: Implications for immunotherapy. *Immunity,* 20, 107–118.

Wekerle, H., Ketelsen, U.P., and Ernst, M. 1980. Thymic nurse cells. Lymphoepithelial cell complexes in murine thymuses: Morphological and serological characterization. *The Journal of Experimental Medicine,* 151, 925–944.

York, I. A., Chang, S.-C., Saric, T., Keys, J. A., Favreau, J. M., Goldberg, A. L., and Rock, K. L. 2002. The ER aminopeptidase ERAP1 enhances or limits antigen presentation by trimming epitopes to 8–9 residues. *Nature Immunology,* 3, 1177–1184.

6 Gap Junctions and Connexins in the Immune Defense Against Tumors

Flavio A. Salazar-Onfray

CONTENTS

6.1 GJIC AND CANCER

Gap junction intercellular communications (GJIC) are fundamental mechanisms for the maintenance of the cellular homeostatic balance. GJIC are involved in almost every aspect of cellular life, particularly those related to the control of proliferation, cell differentiation, and cell death. Other important biological processes such as regulation of gene expression or antigen cross-presentation also require the participation of Cxs and gap junctions (GJs) (Krysko et al., 2005; Vinken et al., 2006; Rodríguez-Sinovas et al., 2007; Neijssen et al., 2007). Variations of normal levels of Cxs have been observed in many human diseases, which have common feature alterations in intercellular communications, apoptosis regulation and cellular proliferation. Typical examples of such disease are cancer, Alzheimer's disease, atherosclerosis, and ischemia (Lin et al., 1998; Trosko and Chang, 2000).

In this respect, a reduction or complete loss of the GJIC, usually associated with changes in the expression levels of Cxs, has been observed in several types of human cancers. The kind of Cx that is lost during the stage of tumor progression varies along with the type of tumor. Because Cx43 is the most widely expressed Cx in human cells, it is not surprising that in many types of cancer, this Cx constitutes a potential therapeutic target (Kandouz et al., 2010). In fact, very low or nonexisting levels of Cxs expression has been observed in prostate cancer (Tsai et al., 1996; Wang et al., 2007), glioma (Huang et al., 1999), lung cancer (Jinn et al., 1998), skin cancer (Tada and Hashimoto, 1997), breast cancer (Hirschi et al., 1996), osteosarcoma (Zhang et al., 2003), and melanoma (Haass et al., 2004). There are various mechanisms described underlying the loss of Cxs, functional hemichannels (Hch), and/or GJIC in carcinogenesis. However, it has been frequently observed that epigenetic modifications, such as DNA hypermethylation and histone deacetylation are one of the triggering mechanisms of Cxs gene expression silencing (Vinken et al., 2006; Hattori et al., 2007). Further, it is common to observe an aberrant localization of Cxs in tumor cells due to inappropriate phosphorylations of the C-terminal domain (Pointis et al., 2007). In general, Cx43 is considered a type II suppressor–suppressor gene.

On the other hand, in a couple of different studies, Cxs are also considered membrane proteins with adhesive properties (Lin et al., 2002; Cotrina et al., 2008). The attachment of tumor cells in transition from a primary site to a secondary organ site requires the attachment as well as the migration of tumor cells through the vascular endothelium, a process known as tumor cell diapedesis. It has been shown that immediately following adhesion to the endothelium, tumor cells establish GJIC with the endothelial cells. In fact, several studies have found that the communication between tumor cells and endothelial cells is mediated by Cxs, which are critical to tumor cell extravasations to the metastatic site (el-Sabban et al., 1991, 1994; Pollmann et al., 2005). It has been reported that Cx43-mediated GJIC enhances breast tumor cell diapedesis (Pollmann et al., 2005). Furthermore, the diminished Cx43 expression reduces adhesion of breast cancer cells to the pulmonary endothelium (Pollmann et al., 2005). Likewise, upregulation of Cx43 was seen in tumor cell–endothelial cell contact areas both *in vitro* and *in vivo* (Elzarrad et al., 2008). Accumulated evidences indicate that Cx26, another Cx family member, is overexpressed in carcinomas of the pancreas, head and neck, colon, and prostate, as well as in keratinocyte-derived skin tumors (Villaret et al., 2000; Pfeffer et al., 2004; Kanczuga-Koda et al., 2005; Tate et al., 2006; Haass et al., 2006). Furthermore, recent reports showed that high Cx26 expression was associated with poor prognosis of lung squamous cell carcinoma and breast carcinoma (Ito et al., 2006; Naoi et al., 2007), and with colorectal tumor cell metastasis to the lung (Ezumi et al., 2008). Together, these lines of evidence seem to be contrary to the conventional role of Cxs as tumor suppressors and instead suggest that Cx26 may play a role in tumorigenesis.

A key initial event required for metastasis is the tumor cell detachment from the primary site. Several studies have shown that adenocarcinoma progression is accompanied by the release of single cells through an epithelial–mesenchymal transition (EMT). Interestingly, Cx43 repression has been observed in the EMT process in a different kind of cells, for example, embryonic carcinoma cells (de Boer et al., 2007), hepatic stellate cells (Lim et al., 2009), colorectal cancer cells (Lee et al., 2012), and breast tumor cells (McLachlan et al., 2006). Moreover, the reexpression of Cx43 partially reverts the EMT on breast cancer cells (McLachlan et al., 2006). It has been shown that Cx43 repression following EMT requires the Snail transcription factor through the binding to Snail consensus sequence in the Cx43 promoter (de Boer et al., 2007; Lim et al., 2009).

In particular, the role of GJIC in tumor progression has been studied, mainly through the exogenous expression of Cxs in derived–derived cell lines, normally associated with an *in vitro* and *in vivo* inhibition of the tumor growth (Hattori et al.,

2007; Fukushima et al., 2007; Pointis et al., 2007). Usually, cell proliferation inhibition is dependent on the amount of established GJIC and/or the presence of functional Hch, but inhibition of the tumor growth has also been observed independently of intercellular communication levels, suggesting a role for Cx proteins (Zhang et al., 2003). The mechanisms involved in the inhibition of the tumor growth due to overexpression of the Cx genes in tumor cell lines are diverse. It has been reported, for example, that Cxs may enhance the sensitivity of tumor cells to cellular death, propagating death signals, or controlling cell proliferation through the regulation of specific kinase phosphorylations (Krutovskikh et al., 2002; Huang et al., 2002; Zhang et al., 2003; Fujimoto et al., 2005; Vinken et al., 2006). It is well documented that Cx43 Hch and GJs play an important role in the communication of cell death messages between cells (Krysko et al., 2005; Rodríguez-Sinovas et al., 2007; Decrock et al., 2011a). The exogenous introduction of Cxs in various experimental models were found to facilitate apoptotic cell death (Decrock et al., 2009). The chemical nature of the signals that mediate these effects are largely unknown, but some evidence suggests that Ca^{2+} and IP_3 are the major cell death messengers passing through Cx43 Hch and GJs (Decrock et al., 2011a,b). In this respect, very few studies have been performed in human melanoma cell lines. In a recently published work, it was observed that transfection of a malignant melanoma line with the gene encoding for Cx43 resulted in the suppression of the growth independently of anchorage, which indicated that Cx43 also may act like a suppressor gene in melanoma (Su et al., 2000). Our own unpublished results suggest that in human melanoma cells, Cx43 downregulates cellular proliferation, possibly increasing the cellular death rate *in vitro* and in immunodeficient mice (Tittarelli et al., 2012). Particularly, we were able to observe a direct correlation between the Cx43 expression and TNF-α-mediated apoptosis of melanoma cells.

Besides the direct role played by Cxs and GJs in the proliferation and metastatic capacity of tumor cells, an important issue is related to the participation of GJIC between cells from the immune system and their capacity to recognize and destroy tumor cells.

6.2 GJ AND IMMUNE SYSTEM ANTITUMOR RESPONSE

As reviewed in the Prequel and Chapter 1 of this book, one of the main forms of communication between immune system cells is through cell–cell contacts. This form of communication involves the participation of Cxs and GJs in the majority of immune responses (Figure 6.1).

Regarding the immune system ability to activate specific CTL against tumor cells *in vivo*, it has been demonstrated that it requires a prior antigen presentation in a particular context, given by professional APC, among which the most important are the DCs (Steinman, 2007; Steinman and Banchereau, 2007). Recent studies demonstrated the central role that DCs have in the direct interaction with tumor cells, and their specific action on the T cell-mediated immune response. DCs are located as residents in the areas of interaction between the environment and the organism. The DCs capture particles from the extracellular environment and receive signals from the inflamed tissue induced by pathogen or tumor invasion. DCs capture antigens,

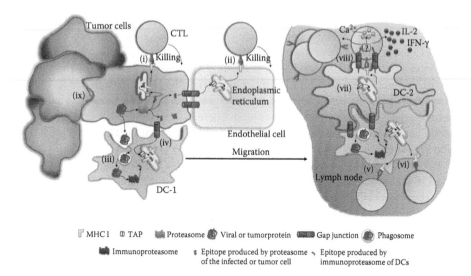

FIGURE 6.1 Principal roles of gap junctions in the immune defense against tumors. (i) Tumor-associated antigenic peptides derived from cytoplasmatic proteins are processed in the proteasome, transported into the endoplasmic reticulum (ER) by the transporter associated with antigen processing (TAP), assembled with major histocompatibility complex I (MHC I) molecules, and then moved to the cell surface. Cytotoxic T lymphocytes (CTLs) detect tumor antigens and destroy the tumor cell. (ii) Processed antigenic peptides can also be transferred from one cell to its neighbor (other tumor cells or endothelial cells) through gap junctions activating CTL-mediated killing. (iii) DCs can obtain exogenous antigenic proteins by phagosomes internalization and transference to cytosol and immunoproteasomes antigen processing and cross-presention to CD8+ T cells (v). (iv) Gap junctions facilitate the aquisition of tumor-derived peptides by DCs and antigen cross-presentation to CD8+ T cell in the lymph nodes (vi). Therefore, DCs are able to present not only antigens generated by the immunoproteasome in their own cytoplasm but also peptides generated by tumor cells proteasome, increasing the chance of activated tumor-specific CTLs. (vii) DCs transfer antigenic peptides to adjacent DCs (DC-2) through gap junctions in the periphery and lymph nodes, extending the local immune response and amplifying the cross-presentation of tumor-associated antigens. (viii) GJ accumulates at the immunological synapse during antigen-specific T cell priming, mediating the cross talk between DCs and T cells regulating Ca^{2+} signals and controlling T cell cytokine release, and proliferation. (ix) Reduction or complete loss of gap junctions in tumor cells, usually associated with changes in the level of expression of Cxs, has been observed in several types of human cancers. As a result, they will not receive growth inhibitory signals, differentiation stop signals, and death molecule signals from the surrounding cells, leading to tumor growth.

and mature and migrate into secondary lymphoid organs, where they deliver signals to T cells for their activation, proliferation, and transference to the peripheral tissues to perform their effector functions (Steinman and Banchereau, 2007).

During the maturation, the DC's phenotype change, starting a rapid increase in the surface expression of antigen presentation molecules MHC class I and class II, costimulatory molecules such as CD40, CD80, and CD86 (B7.1 and B7.2, respectively), CD83 (a specific activation marker), adhesion molecules to T lymphocytes as CD58 and CD48 (Blast-1 and LFA-3, respectively), and the receptor CCR-7, which recognizes the chemokines CCL19 and CCL21, responsible for the migration of DC toward secondary lymphoid organs, where antigen-specific naive T cells activation occurs (Mellman and Steinman, 2001; Delamarre et al., 2003). Antigens are presented by the DC in the context of MHC class I or class II, depending on the compartment in which they have been originated. When antigens are internalized from the extracellular compartment, they are presented in the context of MHC class II that can be recognized by CD4+ T cells (Mellman and Steinman, 2001). Another route of antigen presentation is the so-called endogenous pathway. In this pathway, peptides derived from intracellular endogenous proteins or proteins encoded from intracellular infectious agents are presented associated with the MHC class I molecules (Zinkernagel, 2002). The MHC class I–peptides complex is recognized by CD8+ T cells (Steinman and Banchereau, 2007), which are activated to become CTLs, which have the ability to recognize and eliminate tumor cells (Mellman and Steinman, 2001). Another alternative pathway is named cross-presentation, in which exogenous antigens are presented through MHC class I molecules (Burgdorf et al., 2008). This cross-presentation pathway has proved to be essential for generating CTL against viruses, transplanted cells, and tumor antigens (Amigorena and Savina, 2010; Kurts et al., 2010). It has been described that DC maturation regulates the cross-presentation process in murine DCs (Delamarre et al., 2003). Microbial products, cytokines, and physical stimuli are capable of activating antigen processing and presentation on MHC II molecules, but only a subset of these stimuli also facilitates MHC I cross-presentation of the same antigen, thus indicating that the exogenous MHC I and MHC II pathways are differentially regulated during DC maturation (Delamarre et al., 2003). Additionally, it has been described that the presence of a combination of cytokines may enhance GJ-mediated transfer between DCs (Matsue et al., 2006). GJIC between DCs can occur in response to specific combinations of a defined stimulus. In fact, GJIC between DCs was recently described in a murine model, where the long-term DC line XS52 and also murine bone marrow-derived DC became effectively dye-coupled when activated with LPS or TNF-α plus IFN-γ (Matsue et al., 2006). These observations point to GJ formation as an event associated with DC maturation.

Regarding DC maturation and Cx expression, our own results show that functional GJIC formation between human (h) DC was markedly increased by a melanoma cell lysate, concomitant with the induction of a more evident mature phenotype of DCs (Mendoza-Naranjo et al., 2007). Moreover, we found that Cx43 expression was significantly increased further upon melanoma cell lysate stimulation, where a systematic increase of Cx43 GJ plaques formation between the human dendritic cells (hDC) was also observed. In fact, it has been reported that Cx43–GJs functionally couple follicular DC to each other and to B lymphocytes, in this way participating in direct cell–cell communication in lymphoid germinal centers (Krenacs et al., 1997). We also found that inhibition of GJ formation by using a specific Cx mimetic peptide that binds to Cx43 extracellular loop-1 at the plasma membrane surface diminished the capability of hDC to acquire melanoma antigens from adjacent cells and inhibited melanoma-associated antigen (MAA)-specific T cell activation as assessed by IFN-γ production, suggesting a close relationship between Cx43 membrane expression and Ag transfer (Mendoza-Naranjo et al., 2007). In our model, the DC can transfer antigenic peptides to DC adjacent through GJ, extending the local immune response (Mendoza-Naranjo et al., 2007).

6.3   GJ AND T CELL ACTIVATION

Cxs are expressed on T lymphocytes. Moreover, an increased Cx43 expression in activated T lymphocytes has been reported, particularly in its phosphorylated form. The use of Cx43 blockers reduced the proliferation of these cells in a dose-dependent manner, suggesting the involvement of Cx43 Hch in the maintenance of clonal expansion of T lymphocytes (Oviedo-Orta et al., 2010). The GJs have also been involved in the regulation of peripheral immune response by regulatory T cells. It has been demonstrated that regulatory T cells mediate their suppressive capacity on effector T cells by the transfer of cAMP through GJ channels (Bopp et al., 2007). Additionally, there is also evidence of GJIC between CD4+ T cells and macrophages, boosting in this model a Th1 immune response, the most relevant type of response against tumors (Bermudez-Fajardo et al., 2007).

The initiation of a tumor-associated antigen-specific immune response requires a productive engagement of T cell receptors (TCRs) by MHC–peptide complexes (pMHC) on the APC (Babbitt et al., 1985). This TCR engagement by cognate pMHC results in the formation of a highly organized protein network known as the immunological synapse (IS), which is required for T cell activation and proliferation (Grakoui et al., 1999). The mature IS is characterized by the assembly of specific proteins on the T cell and APC membranes into supramolecular activation clusters (SMACs). These structures consist of a centralized accumulation of TCRs and pMHC (cSMAC), surrounded by a peripheral ring (pSMAC) containing the integrin LFA-1 and its receptor ICAM-1 (Monks et al., 1998; Grakoui et al., 1999). The IS comprises a multitude of structures, many of which are mediators of intercellular communication (Trautmann and Valitutti, 2003). Multiple surface molecules spatially segregated at the IS mediate intercellular communication and activate intracellular signaling pathways, resulting in T cell activation and proliferation. Antigen-dependent T cell activation is a cell–cell contact-dependent process, suggesting that mediators of intercellular communication are directly involved. We have recently shown that Cx43 accumulates at the IS during antigen-specific T cell priming as both GJs and stand-alone Hchs (Mendoza-Naranjo et al., 2011). Redistribution of Cx43 to the IS was antigen specific and time dependent, with maximal accumulation occurring as early as 30 min after DC–T cell conjugate formation, at which time a mature IS has been formed (Lee et al., 2002). We found that Cx43 accumulated at the pSMAC in T cells and colocalized with LFA-1, which is essential for adhesion and signaling within the IS.

Cell surface molecules from all over the T cell membrane are transported to the IS through a mechanism involving the cell cytoskeleton and motor proteins (Wulfing and Davis, 1998). We have shown that Cx43 recruitment to the synapse required an intact actin's cytoskeleton, as the inhibitors of actin polymerization abolished Cx43 accumulation (Mendoza-Naranjo et al., 2011). In our hands, the inhibition of microtubules did not affect the recruitment of Cx43 to the synapse. Therefore, the Cx43 pool relocated to the IS is not likely to be a newly synthesized Cx43, but Cx43 already allocated in the plasma membrane that redistributed to the synapse (Figure 6.2).

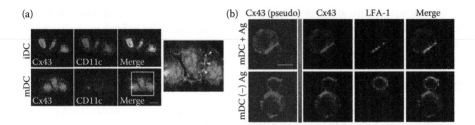

FIGURE 6.2 **(See color insert.)** Gap junctions in the immune defense against tumors. (a) Cx43 is localized at cell–cell contacts between human mature DCs (mDCs). Immature and mature DCs were costained for Cx43 (green) and CD11c (red). High-power confocal magnification shows Cx43 at cell–cell contacts, colocalizing with CD11c (arrowheads). Scale bar = 20 mm. (b) Cx43 accumulates at the immunological synapse in an antigen (Ag)-specific way. Representative images of Cx43 and LFA-1 distribution after incubation of oval-bumin (OVA)–DCs (DC + Ag) or LPS–DCs (DC (–) Ag) with OT-II T cells are shown. Scale bar = 5 mm.

We demonstrated the specific involvement of Cx43 GJ channels in mediating bidirectional communication between DCs and T cells at the IS, and the use of a GJ drug inhibitor, a specific mimetic peptide, or gene silencing resulted in the blockage of intercellular communication between these cells (Mendoza-Naranjo et al., 2011). Intercellular communication has been previously described between macrophages and T lymphocytes, in particular the Th1 cell subset (Bermudez-Fajardo et al., 2007), as well as unidirectional communication from DCs to T cells (Elgueta et al., 2009), which further reinforces our observations. The nature of the intracellular signals exchanged through GJs at the synapse is presently unknown, and further studies are required to identify the molecules that travel through GJs formed between DCs and T cells. Antigen-specific activation of T lymphocytes via stimulation of the TCR complex is marked by a rapid and sustained increase of intracellular Ca^{2+}, which is required for gene transcription, cellular proliferation, and differentiation (Lewis, 2001). A sustained Ca^{2+} signal for many hours is also necessary to stimulate the nuclear factor of activated T cells (NFAT), a transcription factor that regulates the expression of various cytokine genes, including IL-2 (Lewis, 2001). Different studies have described Cx43 participating in Ca^{2+} influx in various cell types (Lin et al., 2004). We presented evidence supporting the participation of Cx43 in regulating Ca^{2+} oscillations in the IS. We demonstrated that a specific blockade of Cx43 prevents the sustained rise of intracellular Ca^{2+} that was seen in T cells forming conjugates with antigen-pulsed DCs, in both murine and human models (Mendoza-Naranjo et al., 2011). The increase in intracellular Ca^{2+} is recognized as an obligatory step in the cascade of signals that finally results in T cell proliferation (Jensen et al., 1999). A role for GJs in T cell activation was recently described, and inhibition of GJIC was responsible for reduced IL-2 secretion and cell proliferation (Elgueta et al., 2009). The impaired Ca^{2+} signals we observed as a result of blocking GJs and Cx43 are likely to account for the reduced lymphocyte activation seen after inhibiting GJIC, suggesting that Ca^{2+} signals regulated by GJs may possibly be one of the mechanisms controlling T cell activation.

In summary, there are many open questions related to the Cxs and GJ role in the antitumor immune response that probably will be promptly solved in the coming years.

REFERENCES

Amigorena S, and Savina A. Intracellular mechanisms of antigen cross presentation in dendritic cells. *Curr Opin Immunol.* 2010; 22(1): 109–17.

Babbitt BP, Allen PM, Matsueda G, Haber E, and Unanue ER. Binding of immunogenic peptides to Ia histocompatibility molecules. *Nature.* 1985; 317: 359–61.

Bermudez-Fajardo A, Ylihärsilä M, Evans WH, Newby AC, and Oviedo-Orta E. CD4+ T lymphocyte subsets express connexin 43 and establish gap junction channel communication with macrophages in vitro. *J Leukoc Biol.* 2007; 82(3): 608–12.

Bopp T, Becker C, Klein M, Klein-Hessling S, Palmetshofer A, Serfling E, Heib V et al. Cyclic adenosine monophosphate is a key component of regulatory T cell-mediated suppression. *J Exp Med.* 2007; 204: 1303–10.

Burgdorf S, Schölz C, Kautz A, Tampé R, and Kurts C. Spatial and mechanistic separation of cross presentation and endogenous antigen presentation. *Nat Immunol.* 2008; 9(5): 558–66.

Cotrina ML, Lin JH, and Nedergaard M. Adhesive properties of connexin hemichannels. *Glia.* 2008; 56: 1791–8.

de Boer TP, van Veen TAB, Bierhuizen MFA et al. Connexin 43 repression following epithelium-to-mesenchyme transition in embryonal carcinoma cells requires Snail1 transcription factor. *Differentiation.* 2007; 75: 208–18.

Decrock E, Krysko DV, Vinken M et al. Transfer of IP$_3$ through gap junctions is critical, but not sufficient, for the spread of apoptosis. *Cell Death Differ.* 2011a. doi:10.1038/cdd.2011.176

Decrock E, Vinken M, Bol M et al. Calcium and connexin-based intercellular communication, a deadly catch? *Cell Calcium.* 2011b; 50:310–21.

Decrock E, Vinken M, and De Vuyst E. Connexin-related signaling in cell death: To live or let die? *Cell Death Differ.* 2009; 16: 524–36.

Delamarre L, Holcombe H, Giodini A, and Mellman I. Presentation of exogenous antigens on MHC class I and MHC class II molecules is differentially regulated during dendritic cell maturation. *J Exp Med.* 2003; 198: 111–22.

el-Sabban ME, and Pauli BU. Adhesion-mediated gap junctional communication between lung-metastatic cancer cells and endothelium. *Invasion Metastasis.* 1994; 14:164–76.

el-Sabban ME, and Pauli BU. Cytoplasmic dye transfer between metastatic tumour cells and vascular endothelium. *J Cell Biol.* 1991; 115: 1375–82.

Elgueta R, Tobar JA, Shoji KF, De Calisto J, Kalergis AM, Bono MR, Rosemblatt M, and Saez JC. Gap junctions at the dendritic cell–T cell interface are key elements for antigen-dependent T cell activation. *J Immunol.* 2009; 183: 277–84.

Elzarrad MK, Haroon A, Willecke K, Dobrowolski R, Gillespie MN, and Al-Mehdi AB. Connexin-43 upregulation in micrometastases and tumour vasculature and its role in tumour cell attachment to pulmonary endothelium. *BMC Med.* 2008; 6: 20.

Ezumi K, Yamamoto H, Murata K, Higashiyama M, Damdinsuren B, Nakamura Y, Kyo N et al. Aberrant expression of connexin 26 is associated with lung metastasis of colorectal cancer. *Clin Cancer Res.* 2008; 14(3): 677–84.

Fujimoto E, Sato H, Shirai S, Nagashima Y, Fukumoto K, Hagiwara H, Negishi E et al. Connexin 32 as a tumour suppressor gene in a metastatic renal cell carcinoma cell line. *Oncogeneology.* 2005; 24: 3684–90.

Fukushima M, Hattori Y, Yoshizawa T, and Maitani Y. Combination of non-viral connexin 43 therapy and docetaxel inhibits the growth of human prostate cancer in mice. *Int J Oncol.* 2007; 30: 225–31.

Grakoui A, Bromley SK, Sumen C, Davis MM, Shaw AS, Allen PM, and Dustin ML. The immunological synapse: A molecular machine controlling T cell activation. *Science.* 1999; 285: 221–7.

Haass N, Smalley K, and Herlyn M. The role of altered cell-cell communication in melanoma progression. *J Mol Histol.* 2004; 35: 309–18.

Haass NK, Wladykowski E, Kief S, Moll I, and Brandner JM. Differential induction of connexins 26 and 30 in skin tumors and their adjacent epidermis. *J Histochem Cytochem.* 2006; 54: 171–82.

Hattori Y, Fukushima M, and Maitani Y. Non-viral delivery of the connexin 43 gene with histone deacetylase inhibitor to human nasopharyngeal tumour cells enhances gene expression and inhibits *in vivo* tumour growth. *Int J Oncol.* 2007; 30: 1427–39.

Hirschi K, Xu C, Tsukamoto T, and Sager R. Gap junction genes Cx26 and Cx43 individually suppress the cancer phenotype of human mammary carcinoma cells and restore differentiation potential. *Cell Growth Differ.* 1996; 7: 861–70.

Huang R, Hossain M, Sehgal A, and Boynton A. Reduced connexin 43 expression in high-grade human brain glioma cells. *J Surg Oncol.* 1999; 70: 21–4.

Huang R, Lin Y, Wang C, Gano J, Lin B, Shi Q, Boynton A, Burke J, and Huang R. Connexin 43 suppresses human glioblastoma cell growth by down-regulation of monocyte chemotactic protein 1, as discovered using protein array technology. *Cancer Res.* 2002; 62: 2806–12.

Ito A, Koma Y, Uchino K et al. Increased expression of connexin 26 in the invasive component of lung squamous cell carcinoma: Significant correlation with poor prognosis. *Cancer Lett.* 2006; 234: 239–48.

Jensen BS, Odum N, Jorgensen NK, Christophersen P, and Olesen SP. Inhibition of T cell proliferation by selective block of Ca(2+)-activated K(+) channels. *Proc Natl Acad Sci USA.* 1999; 96: 10917–21.

Jinn Y, Ichioka M, and Marumo F. Expression of connexin32 and connexin43 gap junction proteins and E-cadherin in human lung cancer. *Cancer Lett.* 1998; 127: 161–9.

Kanczuga-Koda L, Sulkowski S, Koda M, and Sulkowska M. Alterations in connexin26 expression during colorectal carcinogenesis. *Oncology.* 2005; 68: 217–22.

Kandouz M, and Batist G. Gap junctions and connexins as therapeutic targets in cancer. *Expert Opin Ther Targets.* 2010;14(7): 681–92.

Krenacs T, van Dartel M, Lindhout E, and Rosendaal M. Direct cell/cell communication in the lymphoid germinal center: Connexin43 gap junctions functionally couple follicular dendritic cells to each other and to B lymphocytes. *Eur J Immunol.* 1997; 27(6): 1489–97.

Krutovskikh V, Piccoli C, and Yamasaki H. Gap junction intercellular communication propagates cell death in cancerous cells. *Oncogeneology.* 2002; 21: 1989–99.

Krysko DV, Leybaert L, Vandenabeele P, and D'Herde K. Gap junctions and the propagation of cell survival and cell death signals. *Apoptosis.* 2005; 10: 459–69.

Kurts C, Robinson BW, and Knolle PA. Cross-priming in health and disease. *Nat Rev Immunol.* 2010; 10(6): 403–14.

Lee CC, Chen WS, Chen CC, Chen LL, Lin YS, Fan CS, and Huang TS. TCF12 protein functions as transcriptional repressor of E-cadherin, and its overexpression is correlated with metastasis of colorectal cancer. *J Biol Chem.* 2012; 287: 2798–809.

Lee KH, Holdorf AD, Dustin ML, Chan AC, Allen PM, and Shaw AS. T cell receptor signaling precedes immunological synapse formation. *Science.* 2002; 295: 1539–42.

Lewis RS. Calcium signaling mechanisms in T lymphocytes. *Annu Rev Immunol.* 2001; 19: 497–521.

Lim MC, Maubach G, and Zhuo L. TGF-β1 down-regulates connexin43 expression and gap junction intercellular communication in rat hepatic stellate cells. *Eur J Cell Biol.* 2009; 88: 719–30.

Lin GC, Rurangirwa JK, Koval M, and Steinberg TH. Gap junctional communication modulates agonist-induced calcium oscillations in transfected HeLa cells. *J Cell Sci.* 2004; 117: 881–7.

Lin J, Weigel H, Cotrina M, Liu S, Bueno E, Hansen A, Hansen T, Goldman S, and Nedergaard M. Gap-junction-mediated propagation and amplification of cell injury. *Nat Neurosci.* 1998; 1: 494–500.

Lin JH, Takano T, Cotrina ML et al. Connexin 43 enhances the adhesivity and mediates the invasion of malignant glioma cells. *J Neurosci.* 2002; 22: 4302–11.

Matsue H, Yao J, Matsue K, Nagasaka A, Sugiyama H, Aoki R, Kitamura M, and Shimada S. Gap junction-mediated intercellular communication between dendritic cells (DCs) is required for effective activation of DCs. *J Immunol.* 2006; 176: 181–190.

McLachlan E, Shao Q, Wang HL, Langlois S, and Laird DW. Connexins act as tumour suppressors in three-dimensional mammary cell organoids by regulating differentiation and angiogenesis. *Cancer Res.* 2006; 66: 9886–94.

Mellman I, and Steinman RM. Dendritic cells: Specialized and regulated antigen processing machines. *Cell.* 2001; 106(3): 255–8.

Mendoza-Naranjo A, Bouma G, Pereda C, Ramírez M, Webb KF, Tittarelli A, López MN et al. Functional gap junctions accumulate at the immunological synapse and regulate calcium signaling in T cells. *J Immunol.* 2011; 187(6): 3121–32.

Mendoza-Naranjo A, Saéz PJ, Johansson CC, Ramírez M, Mandakovic D, Pereda C, López MN, Kiessling R, Sáez JC, and Salazar-Onfray F. Functional gap junctions facilitate melanoma antigen transfer and cross-presentation between human dendritic cells. *J Immunol.* 2007; 178(11): 6949–57.

Monks CR, Freiberg BA, Kupfer H, Sciaky N, and Kupfer A. Three-dimensional segregation of supramolecular activation clusters in T cells. *Nature.* 1998; 395: 82–6.

Naoi Y, Miyoshi Y, Taguchi T et al. Connexin 26 expression is associated with lymphatic vessel invasion and poor prognosis in human breast cancer. *Breast Cancer Res Treat.* 2007; 106: 11–7.

Oviedo-Orta E, Perreau M, Evans WH, and Potolicchio I. Control of the proliferation of activated CD4+ T cells by connexins. *J Leukoc Biol.* 2010; 88: 79–86.

Pfeffer F, Koczan D, Adam U et al. Expression of connexin26 in islets of Langerhans is associated with impaired glucose tolerance in patients with pancreatic adenocarcinoma. *Pancreas.* 2004; 29: 284–90.

Pointis G, Fiorini C, Gilleron J, Carette D, and Segretain D. Connexins as precocious markers and molecular targets for chemical and molecular agents in carcinogenesis. *Curr Med Chem.* 2007; 14: 2288–303.

Pollmann MA, Shao Q, Laird DW et al. Connexin 43 mediated gap junctional communication enhances breast tumour cell diapedesis in culture. *Breast Cancer Res.* 2005; 7: R522–34.

Rodríguez-Sinovas A, Cabestrero A, López D, Torre I, Morente M, Abellán A, Miró E, Ruiz-Meana M, and García-Dorado D. The modulatory effects of connexin 43 on cell death/survival beyond cell coupling. *Prog Biophys Mol Biol.* 2007; 94(1–2): 219–32.

Steinman RM. Dendritic cells: Understanding immunogenicity. *Eur J Immunol.* 2007; 37(Suppl 1): S53–60.

Steinman RM, and Banchereau J. Taking dendritic cells into medicine. *Nature.* 2007; 449(7161): 419–26.

Su YA, Bittner ML, Chen Y, Tao L, Jiang Y, Zhang Y, Stephan DA, and Trent JM. Identification of suppressor–suppressor genes using human melanoma cell lines UACC903, UACC903(+6), and SRS3 by comparison of expression profiles. *Mol Carcinog.* 2000; 28(2): 119–27.

Tada J, and Hashimoto K. Ultrastructural localization of gap junction protein connexin 43 in normal human skin, basal cell carcinoma, and squamous cell carcinoma. *J Cutan Pathol.* 1997; 24(10): 628–35.

Tate AW, Lung T, Radhakrishnan A, Lim SD, Lin X, and Edlund M. Changes in gap junctional connexin isoforms during prostate cancer progression. *Prostate.* 2006; 66: 19–31.

Tittarelli A, Corvalán F, Mendoza-Naranjo A, López MN, and Salazar-Onfray F. Connexin 43 expression regulates proliferation and cell death of human melanoma cells. 2012. (Manuscript in preparation).

Trautmann A, and Valitutti S. The diversity of immunological synapses. *Curr Opin Immunol.* 2003; 15: 249–54.

Trosko E, and Chang C. Modulation of cell–cell communication in the cause and chemoprevention/chemotherapy of cancer. *Biofactors.* 2000; 12: 259–63.

Tsai H, Werber J, Davia M, Edelman M, Tanaka K, Melman A, Christ G, and Geliebter J. Reduced connexin 43 expression in high-grade, human prostatic adenocarcinoma cells. *Biochem Biophys Res Commun.* 1996; 227: 64–9.

Villaret DB, Wang T, Dillon D et al. Identification of genes overexpressed in head and neck squamous cell carcinoma using a combination of complementary DNA subtraction and microarray analysis. *Laryngo-scope.* 2000; 110: 374–81.

Vinken M, Vanhaecke T, Papeleu P, Snykers S, Henkens T, and Rogiers V. Connexins and their channels in cell growth and cell death. *Cell Signal.* 2006; 18: 592–600.

Wang M, Berthoud V, and Beyer E. Connexin 43 increases the sensitivity of prostate cancer cells to TNF-induced apoptosis. *J Cell Sci.* 2007; 120: 320–9.

Wulfing C, and Davis MM. A receptor/cytoskeletal movement triggered by costimulation during T cell activation. *Science.* 1998; 282: 2266–9.

Zhang Y, Kaneda M, and Morita I. The gap junction-independent suppressing–suppressing effect of connexin 43. *J Biol Chem.* 2003; 278: 44852–6.

Zinkernagel RM. On cross-priming of MHC class I-specific CTL: Rule or exception? *Eur J Immunol.* 2002; 32(9): 2385–92.

exchange of electrical and chemical signals. Gap junctions (GJ) are essential for this signal exchange. GJ channels are composed of six connexins forming a hemi-channel in the cell membrane (see Chapter 1). This pathway is involved in signaling and exchange between intracellular and extracellular spaces. Additionally, two hemi-channels on adjacent cells can form a full GJ channel, allowing the direct exchange of ions, small metabolites, and other second messenger molecules between cells in contact (Saez et al., 2003). GJ communication contributes to synchronization of cells within a tissue and coherent or conducted cellular responses. More than 20 different mammalian connexins have been described (Sohl and Willecke, 2004). Many combinations of connexins are possible, which may assemble into channels, and each type of GJ channel has its proper gating and permeability properties. In addition, the expression pattern of connexins is complex with multiple connexins in one cell type. The very short half-life of connexins (about 1–5 h for Cx43) indicates that channels are renewed several times per day and participate in that complexity (Saez et al., 2003).

Four connexins have been described in the vascular system, connecting ECs and SMCs in the blood vessel wall (Hill et al., 2001, 2002; de Wit et al., 2006). Connexins have important functions in vascular physiology, such as conduction of vasomotor responses and tone among SMCs (Christ et al., 1996; Figueroa et al., 2004; de Wit et al., 2006), capillary sprouting and endothelial repair (Kwak et al., 2001). Deletion of connexins is thus expected to affect the vasculature. Indeed, homozygous Cx43-deficient mice die at birth because of severe malformations in the cardiac outflow tract (Reaume et al., 1995). Homozygous Cx40-deficient mice display impaired conduction of vasodilation along arterioles and suffer from hypertension (de Wit et al., 2000). In contrast, mice with an endothelial-specific deletion of Cx43 suffer from bradycardia and hypotension (Liao et al., 2001), although contradictory results have also been described (Theis et al., 2001). Cx45-deficient embryos exhibit striking abnormalities in vascular development and die between embryonic day (E) 9.5 and 10.5 (Kruger et al., 2000). Although homozygous Cx37-deficient mice do not display a particular vascular phenotype (Figueroa et al., 2006), mice deficient in Cx37 and Cx40 die perinatally showing severe vascular abnormalities, such as local hemorrhages and hemangioma (Simon and McWhorter, 2002). Apart from blood vessels, Cx37 and Cx43 are also expressed in lymphatic vessels and genetic deletion of these connexins induces lymphedema (Kanady et al., 2011). Thus, connexins are important proteins that ensure correct blood vessel formation and homeostasis.

7.3 PATHOGENESIS OF ATHEROSCLEROSIS

The pathogenesis of atherosclerosis can be divided in sequential phases due to the progressive form of the disease. During these different phases, many cell types change their regular fate pointing to a pro-inflammatory phenotype, deleterious for the patient (Kanady et al., 2011; Sabine et al., 2012). Interestingly, the expression pattern of connexins varies between blood vessel type (Hill et al., 2001) and species (de Wit et al., 2006). Moreover, connexin expression patterns are additionally modified during the formation of the atherosclerotic plaque (Kwak et al., 2002), which further complicates the analysis. We will therefore picture the different phases in order and discuss the expression and roles of connexins.

7.4 INITIATION OF THE DISEASE IS DUE TO ENDOTHELIAL DYSFUNCTION

ECs are the first cells implicated in the pathogenesis of atherosclerosis. They form a monolayer covering the interior side of the blood vessel wall, forming a barrier between blood and the pro-coagulant sub-endothelial matrix proteins. Healthy ECs express a typical array of proteins allowing for an undisturbed blood flow and for an impairment of the coagulation cascade, such as tissue factor (TF) pathway inhibitor (TFPI). Autocrine secretion of nitric oxide (NO) to regulate vascular tone (preserving a relative vasodilated state) and normal blood flow are essential in maintaining a healthy endothelium.

Endothelial dysfunction is the characteristic of the earliest phase of atherosclerosis and is induced by many factors. Many potentially modifiable risk factors described in the INTERHEART study, such as smoking, hypertension, diabetes, obesity, hypercholesterolemia (Yusuf et al., 2004), push ECs towards a pro-inflammatory phenotype, leading to a decrease in NO production and antithrombotic factors, and to an increase of vasoconstrictor-prothrombotic products and vascular permeability. Thus, an imbalance from a relative vasodilated phenotype toward a more vaso-constrictive state occurs. Altogether this phenomenon promotes the development of atherosclerosis.

The endothelium, where Cx37 and Cx40 are abundantly expressed (Yeh et al., 1998) (Figure 7.1), is a well-coupled tissue, as shown by the efficient diffusion of a GJ permeable dye in ECs from the aorta (Ebong et al., 2006). Cx43 is also detected in few arterial ECs and in capillaries (Cowan et al., 1998; Gabriels and Paul, 1998; Theis et al., 2001). Atherosclerotic lesions are found typically in large- and medium-sized arteries where a high laminar shear stress (the tangential force of the flowing blood on the endothelial surface) is observed. In contrast, arterial bifurcations experience turbulent flow (oscillatory shear stress), a condition that has been often associated with the development of atherosclerosis. Interestingly, while Cx43 is mostly absent in murine aortic endothelium, this connexin is abundant in ECs localized at the downstream edge of the ostia of branching vessels and at flow dividers (Gabriels and Paul, 1998). These observations have been confirmed by *in vitro* studies, demonstrating a positive correlation between Cx43 expression and disturbed flow patterns (Cowan et al., 1998; DePaola et al., 1999; Davies et al., 2001). In fact, the expression of endothelial Cx43 increases in response to oscillatory shear stress *in vitro*, whereas pressure itself does not affect endothelial Cx43 expression (Kwak et al., 2005). Finally, oscillatory shear stress-induced expression of Cx43 is thought to be involved in sensitivity to mechanic stimuli, in particular in ECs of cardiac valves (DePaola et al., 1999; Inai et al., 2004).

Other factors might also influence the expression of connexins in the vasculature. Among the risk factors of atherosclerosis, aging induces a general decrease in connexin expression in rat aortic ECs; however, Cx40 seems to remain undisturbed over long periods (Yeh et al., 2000). Nicotine also induces a decrease in Cx43 expression in human umbilical vein ECs (HUVECs), due to enhanced protein degradation (Tsai et al., 2004). Hypertension is not only induced by the absence of Cx40 expression, but hypertension itself is also modifying expression of connexins in the vascular

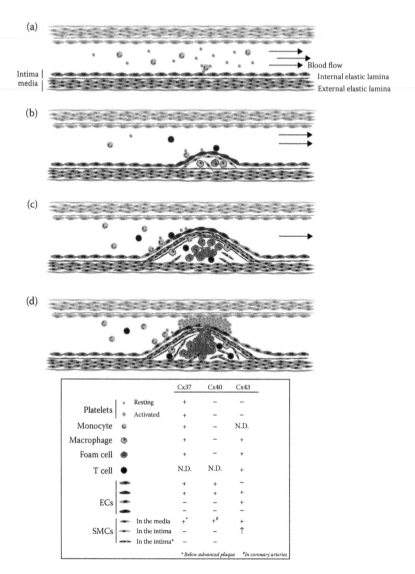

FIGURE 7.1 Atherosclerotic plaque formation occurs in distinct phases and changes in the pattern of connexin expression are observed during each step. (a) Normal blood vessel, with healthy ECs and SMCs, circulating monocytes and platelets express low level of connexin37 (Cx37). (b) In areas of turbulent flow or where the endothelium is dysfunctional, increased leukocyte adhesion and augmented endothelial permeability leads to leukocyte transmigration and formation of an early atheromatous lesion. (c) In the advanced plaque, SMCs have been recruited, proliferated, and secreted ECM, this covering the atherosclerotic plaque with a fibrous cap. A necrotic core is visible, composed mostly of apoptotic debris. Foam cells in the vicinity of the core express Cx37 and Cx43. Cx43 is expressed by ECs in the shoulder of the plaque. (d) Ruptured plaque. This life-threatening event occurs when degrading enzymes and hemodynamic stress weakened the fibrous cap of vulnerable plaque. Coagulation and activation of circulating platelets occurs immediately, leading to blood flow stop, embolisms, and ischemia of tissue.

wall. In ECs from hypertensive rats, the expression of Cx37 and Cx43 is reduced, while Cx40 expression is not modified (Yeh et al., 2006a). Furthermore, carvedilol, a β-blocker used to treat hypertension, directly upregulates endothelial Cx43, independently of its antioxidant activity (Yeh et al., 2006a). Moreover, the endogenous inhibitor of NO synthase decreases Cx43 expression and GJ communication in HUVECs (Jia et al., 2009). Oxidation products of lipoprotein-derived phospholipids are significant for the development of atherosclerosis. These products upregulate Cx43, downregulate Cx37, but do not affect, *in vivo* and in culture, Cx40 expression in ECs isolated from murine carotid arteries (Isakson et al., 2006). Treatment with omega-3 polyunsaturated fatty acids (the so-called "good" fat) increases the expression of Cx43 in the endothelium of spontaneously hypertensive rats. Moreover, this treatment increases the phosphorylation of this connexin (Dlugosova et al., 2009). ECs are also sensitive to cholesterol titers. Lowering low-density lipoprotein (LDL)-cholesterol levels is considered to reduce the risk of atherosclerosis events, despite the fact that not all patients under cholesterol-lowering therapy are protected against the disease (Sacks et al., 1996). Interestingly, increasing the cholesterol level in the culture medium, or treatment with LDL, increases GJ assembly between hepatoma cells (Meyer et al., 1990, 1991). Besides the effect of circulating level of cholesterol, membrane cholesterol content also affects the intrinsic properties of GJ channels. Indeed, heptanol, a well-known GJ blocker (Spear et al., 1990), decreases the fluidity of cholesterol-rich domains, leading to a decrease of GJ communication in neonatal rat cardiomyocytes, thus indicating the importance of cholesterol-rich domain fluidity for proper and efficient GJ coupling (Bastiaanse et al., 1993).

High-glucose levels also downregulate Cx43 expression in the microvasculature (Li and Roy, 2009). Diabetes mellitus, another risk factor of atherosclerosis (Goldberg, 2004; White and Chew, 2008), influences connexin expression as well. Indeed, diabetes mellitus markedly decreases Cx43 GJs in cardiomyocytes. Statins, lipid-lowering drugs with multiple pleiotropic effects, reverse this diabetes-induced consequence (Sheu et al., 2007). In contrast, endothelial Cx37 and Cx40 are downregulated in diabetic ApoE$^{-/-}$ mice; and simvastatin exacerbates this effect (Hou et al., 2008). In a similar fashion, ECs from coronary artery of diabetic mice show lower protein levels of Cx37 and Cx40, but not Cx43, and a reduction in GJ communication (Makino et al., 2008).

Other factors also influence the endothelial connexin biology. Ischemia–reperfusion-induced reoxygenation increases the production of superoxide, leading to vascular disorders. Hypoxia and subsequent reoxygenation of ECs inhibits GJ communication, by modulating Cx43, in HUVECs (Zhang et al., 2000). Moreover, rapid reoxygenation reduces electrical coupling and protein kinase A activity in ECs (Bolon et al., 2005). This effect is not observed in Cx40$^{-/-}$ mice, demonstrating that the abrupt reoxygenation might target Cx40.

As mentioned above, ECs play an important role at the interface between inflammation and coagulation. Many diseases critically involve this interface, particularly sepsis, a life-threatening state defined by an infection associated with a systemic inflammatory response syndrome (SIRS). This leads to simultaneous bleeding and organ failure due to an impaired perfusion. In a rat model of peritonitis induced by cecal ligation and puncture, the expression of Cx40 increases in aortic ECs (Rignault et al., 2005). In contrast, connexin expression and intercellular communication decrease in a model of endotoxemia induced by lipopolysaccharides (LPS) (Simon et al., 2004). Thus, changes in connexin expression depend apparently on the source of infection, or at least on the inflammatory model used. The integrity of the endothelium, maintained by NO, is essential in sepsis; sepsis seems to impair vasoconstriction by acting on Cx37 in arterioles (McKinnon et al., 2009). C-reactive protein (CRP) is a conserved protein present in low level in humans. But in response to many pathologies, such as infection, cancer or tissue injury, CRP titer increases rapidly up to 1000-fold (Pepys and Hirschfield, 2003). CRP is associated with coronary atherosclerotic disease (Alvarez Garcia et al., 2003; Sampietro et al., 2002), and is used as a predictive marker for cardiovascular risk (Ikonomidis et al., 2008). The actual question is to determine whether CRP is only a risk marker or rather a functional risk mediator (Verma and Yeh, 2003). Nonetheless, CRP influences the gene-expression profile of human vascular ECs (venous and arterial) and enhances monocytes adhesion to ECs (Wang et al., 2005b) by activating extracellular signal-regulated kinases (ERK)1/2. Moreover, CRP upregulates Cx43 mRNA expression, as shown by microarray analysis (Wang et al., 2005b). In contrast, another study showed that CRP has no effect on Cx43 GJ in human aortic ECs (Wang et al., 2010). Activation of ECs (Hu and Xie, 1994, van Rijen et al., 1998) or SMCs (Mensink et al., 1995) with LPS, tumor necrosis factor (TNF)-α, IL-1α, or IL-1β reduce GJ communication. In fact, ECs are very sensitive to TNF-α that activates cells by promoting expression of adhesion molecules. Hence, TNF-α decreases the expression of Cx37 and Cx40 in ECs, but does not modify Cx43 expression (van Rijen et al., 1998). Finally, TNF-α induces closure of myoendothelial GJs in co-culture of human ECs and SMCs (Hu and Cotgreave, 1997) thus modifying both gating and permeability of GJs.

Growth factors are also involved in connexin biology. Thus, vascular endothelial growth factor (VEGF) impairs GJ communication in ECs (Suarez and Ballmer-Hofer, 2001). This disruption is due to Cx43 internalization and phosphorylation (Thuringer, 2004). Moreover, epidermal growth factor decreases the GJ coupling in early passage HUVECs (Xie and Hu, 1994). In contrast, transforming growth factor (TGF)-β increases Cx43 synthesis (Larson et al., 1997, 2001) but decreases Cx37 expression in ECs (Larson et al., 1997). Increased Cx43 expression occurs via p38 and phosphatidylinositol 3-kinase (PI3K)/Akt signaling pathways in mammary gland epithelial cells (Tacheau et al., 2008). Finally, basic fibroblast growth factor (bFGF) increases Cx43 expression and GJ communication, but the opposite occurs after incubation with antibodies against bFGF, that is, decreased Cx43 expression during wound healing (Pepper and Meda, 1992). In summary, a very large range of atherosclerosis-associated stimuli that induce endothelial dysfunction also modify connexin expression and GJ communication in vascular wall.

7.5 CELLULAR RECRUITMENT BY ECs AND LEUKOCYTES INFILTRATION IN THE ARTERIAL INTIMA

The second phase of the development of atherosclerosis is the initiation of inflammatory cell recruitment by the activated vessel wall, and the migration or diapedesis into the intima. This occurs by the appearance of adhesion molecules and production of chemokines. Circulating cells, mainly monocytes, adhere to the dysfunctional endothelium that expresses specific adhesion molecules in response to the accumulation of cholesterol in the intima (Cybulsky and Gimbrone, 1991), such as vascular cell-adhesion molecule-1 (VCAM-1), an integrin receptor. Monocytes and lymphocytes express integrin very late antigen-4 (VLA-4), and bind thereafter firmly to the endothelium leading, from a rolling phenotype, toward firm adhesion. Moreover, chemokines such as monocyte chemoattractant protein 1 (MCP-1), stimulate recruited monocytes to enter the intima. This sequence of events determines the development of atherosclerosis, occurring very early in the disease.

As described above, TNF-α induces modification of connexins in ECs that affect the migration of leukocytes in different inflammatory pathologies. A possible role for connexins in leukocyte migration has been studied in different *in vitro* transmigration models and the results are conflicting (De Maio et al., 2002; Chanson et al., 2005). Indeed, diapedesis of neutrophils can be increased or not by using connexin-mimetic peptides or pharmacological channel blockers (Zahler et al., 2003; Chanson et al., 2005; Scerri et al., 2006). Using the same modulators, monocytes transmigration is decreased (Eugenin et al., 2003), but lymphocytes transmigration is only modestly affected (Oviedo-Orta et al., 2002).

As there is a lack of noninvasive methods to accurately detect and characterize atherosclerotic lesions in humans, mouse models have been widely used to study the mechanisms involved in atherogenesis. The two most commonly employed mouse models of atherosclerosis are the apolipoprotein E-deficient mouse (ApoE$^{-/-}$) (Zhang et al., 1992) and the LDL receptor-deficient mouse (LDLR$^{-/-}$) (Ishibashi et al., 1993, 1994). Both mouse models rapidly develop atherosclerosis on a high-fat, high cholesterol diet. As in humans, atherosclerosis in mice develops in regions of the vasculature subjected to low or oscillatory wall shear stress (Suo et al., 2007). Preference sites in the mouse are the aortic root, the lesser curvature of the aortic arch and branch points of the brachiocephalic, left carotid and subclavian arteries. In contrast to the human situation, the first segment and the first branch of all major coronary arteries are usually protected from the disease (Hu et al., 2005). Although mouse models have proven very useful for investigations toward mechanisms of disease initiation and early plaque growth, another limitation of the mouse models is that plaque rupture with superimposed thrombosis, the most common complication of human atherosclerosis, is rarely observed in mice (Plump and Lum, 2009). Nevertheless, these mouse models have helped to reveal that the expression of connexins is modulated during atherogenesis. Endothelial Cx37 and Cx40 expression is considerably reduced in LDLR$^{-/-}$ mice on a cholesterol-rich diet for several months (Kwak et al., 2002) but also in wild-type mice (Chu et al., 2010). ECs covering advanced atherosclerotic plaques no longer express Cx37 or Cx40 (Kwak et al., 2002) (Figure 7.1). We recently generated ApoE$^{-/-}$ mice with endothelial-specific deletion of Cx40. In mice on a cholesterol-rich diet, we observed increased progression of atherosclerosis. Moreover, spontaneous lesions were even observed in the aortic sinuses of young mice without such a diet. These lesions showed monocyte infiltration into the intima, increased expression of VCAM-1, and decreased expression of the ecto-enzyme CD73 in the endothelium. Endothelial CD73 is known to induce antiadhesion signaling via the production of adenosine. *In vitro* experiments on the mouse endothelial cell line bEnd.3 revealed that the reduction of Cx40 expression using siRNA or antisense decreased CD73 expression and activity and increased leukocyte adhesion to bEnd.3 cells (Chadjichristos et al., 2010). An adenosine receptor agonist reversed these effects. Thus, Cx40-mediated gap junctional communication contributes to a quiescent nonactivated endothelium by propagating adenosine-evoked anti-inflammatory signals between ECs. Disrupting this mechanism by targeting Cx40 promotes leukocyte adhesion to the endothelium, thus accelerating atherosclerosis.

Cx37 is not only present in ECs but also in monocytes and macrophages in early and late atheromas (Kwak et al., 2002) (Figure 7.1). In addition, Cx37 expression is induced in medial SMCs beneath advanced atherosclerotic lesions (Kwak et al., 2002) (Figure 7.1). Atherosclerosis is exacerbated in Cx37$^{-/-}$ApoE$^{-/-}$ mice fed a high-cholesterol diet (Wong et al., 2006), both in descending aorta and aortic sinuses. The transmigration of monocytic cells into atherosclerotic lesions is due to the presence of Cx37 on these cells, and not to the expression of Cx37 on ECs, as shown by adoptive transfer. Furthermore, *in vitro* adhesion of monocytic cells to an activated ECs monolayer is increased when Cx37 is absent in monocytic cells. The antiadhesive effect is due to extracellular release by cells of ATP through Cx37 hemi-channels. A genetic polymorphism in the human gene encoding Cx37 (GJA4) has been reported as a potential prognostic marker for atherosclerosis (Chanson and Kwak, 2007). Interestingly, this Cx37 polymorphism affects the adhesiveness of monocytic cells (Wong et al., 2006).

After arrival in the intima, monocytes mature into macrophages. Macrophages express scavenger receptors, such as CD36, that allow engulfment of lipids giving cells the typical foam cell appearance as seen by microscopy. Oxidized LDL cholesterol enters cells and encourages a positive feedback loop to increase CD36 expression on the cell surface. Moreover, this signaling implies cellular cytoskeleton rearrangement leading to trapping of the cells in plaque (Park et al., 2009). Platelets also express CD36 that can promote thrombosis (Valiyaveettil and Podrez, 2009).

Circulating cells expressing connexins may form GJ and thus allow for communication between cells. This communication can be demonstrated by dye-coupling or dual voltage-clamp experiments. Thus, murine macrophages are electrically and dye-coupled in adherent cultures (Levy et al., 1976; Martin et al., 1998). Interestingly, inflammatory conditions (cellular activation) seems to increase GJ coupling *in vitro*, as seen in microglia (Eugenin et al., 2001) as well as in human monocytes (Eugenin et al., 2003). In contrast, other studies have not been able to detect dye transfer between human or murine monocytic cells (Polacek et al., 1993; Alves et al., 1996), between human monocytic cells and ECs, and between human monocytic cells and SMCs (Polacek et al., 1993; Eugenin et al., 2003). Various connexins have been found in monocytic cells, and connexin expression seems dependent on the activation of these cells: Cx43 is present in mouse macrophage cell lines (Beyer and Steinberg, 1991; Alves et al., 1996; Wong et al., 2006), activated peritoneal hamster and mouse macrophages (Jara et al., 1995; Alves et al., 1996), and in monocytic cells stimulated by TNF-α or interferon (INF)-γ, and Cx37 has been detected in human and mouse monocytes (Wong et al., 2006).

In addition to macrophages, T-cells are the second most important cell population in the atherosclerotic plaque (comprising about 10%) (Jonasson et al., 1986).

They govern the transition from a latent plaque to a vulnerable plaque (Hansson and Libby, 2006). CD4[+] clones of T-cells isolated from atherosclerotic plaques recognize oxidized LDL suggesting a local stimulation of T-cells by monocytes/macrophages in the plaque involving major histo-compatibility antigen II (Stemme et al., 1995). These T-cells express a pro-inflammatory (T_H1) phenotype suggesting that athero-sclerosis is a T_H1-driven pathology (Stemme et al., 1995; Zhou et al., 2001). T_H1-type cytokines are predominant in plaques but IL-4, a T_H2-type cytokine, is also pro-duced by a few cells in the plaque (Frostegard et al., 1999).

Lymphocytes also participate in GJ communication. Cx43 is found in human blood-derived T-, B-cells, and CD56[+] (natural killer [NK]) cells (Oviedo-Orta et al., 2000). Tonsil-derived T- and B-cells express both Cx40 and Cx43 (Oviedo-Orta et al., 2000). Moreover, electrical coupling has been observed after phytohemagglu-tinin stimulation (Hulser and Peters, 1971). Dye coupling between lymphocytes and ECs has been described (Oviedo-Orta et al., 2002). This coupling was bidirectional and when blocked, it increased neutrophil transmigration through the EC barrier. It has been recently demonstrated that neutrophils also contribute to the pathogenesis of atherosclerosis (Paulsson et al., 2007). Neutrophils express Cx37, Cx40, and Cx43, but not Cx32 (Branes et al., 2002; Zahler et al., 2003). After activation by LPS or TNF-α, connexins in human neutrophils cluster into GJs; however, no dye coupling was observed between neutrophils (Branes et al., 2002). In contrast, functional GJs have been reported between ECs and neutrophils (Zahler et al., 2003). This coupling was decreased by TNF-α (Zahler et al., 2003).

Activated ECs recruit mainly monocytes and T-cells (Hansson and Libby, 2006), as well as also platelets (Theilmeier et al., 2002). Platelet rolling is initiated by selec-tins that are present on activated endothelial and platelet surfaces (Frenette et al., 2000). Indeed, mice lacking P-selectin on their platelets develop smaller atheroscle-rotic lesions than mice with wild-type platelets (Burger and Wagner, 2003). This protection is enhanced when both P- and E-selectin are also lacking (Dong et al., 1998). The lack of the integrin α_{IIb} (a subunit of the fibrinogen receptor $\alpha_{IIb}\beta_3$ pres-ent on platelets) attenuates atherosclerotic lesion formation (Massberg et al., 2005). Interestingly, expression of Cx37 was recently discovered in platelets and in their precursor cells, the megakaryocyte (Angelillo-Scherrer et al., 2011). Deletion of the Cx37 gene in mice shortened bleeding time and increased thrombus propen-sity. Moreover, aggregation was increased in murine Cx37[−/−] platelets, or in murine Cx37[+/+] and human platelets treated with GJ blockers. Intracellular microinjection of neurobiotin, a Cx37-permeant tracer, revealed functional GJs in platelet aggre-gates. Further experiments revealed that the establishment of GJ communication between Cx37 expressing platelets provided a mechanism to limit thrombus pro-pensity (Angelillo-Scherrer et al., 2011). Moreover, healthy subjects homozygous for Cx37-1019C, a prognostic marker for atherosclerosis, display increased platelet responses as compared to subjects carrying the Cx37-1019T allele. Expression of these polymorphic channels in communication-deficient cells revealed a decreased permeability of Cx37-1019C channels for neurobiotin. Together, these observations are consistent with the hypothesis that an antiaggregating signal may spread more efficiently between platelets through Cx37-319S GJ channels, which in turn results in decreased platelet reactivity.

Thus, connexins are expressed and GJ channels may be functional in blood cells. This reveals an additional possibility for a communication network between circulating cells and vascular cells, a key event during pathogenesis.

7.6 SMC MIGRATION DURING ATHEROSCLEROSIS DEVELOPMENT

Medial SMCs are coupled to each other by connexins (Straub et al., 2010). In vascular SMCs, the GJ communication is homocellular and bidirectional (Figueroa and Duling, 2009). Moreover, SMCs and ECs are coupled by GJs (Straub et al., 2010) in microvasculature, but probably not in large vessels where atherosclerosis occurs. Normally, SMCs are contained within the media, but, in the expansion phase of the atherosclerosis, SMCs transmigrate from the media of the vessels to the intima. These intimal SMCs have unique properties: they are less differentiated, have a higher proliferative index and display an increased production of ECM components, proteases, and cytokines. This leads to a strong fibrous cap covering the atherosclerotic plaque, thus safely isolating the plaque from the blood. SMCs in the advanced plaque may also take up lipids and acquire a foam cell appearance. As with macrophage foam cells, they eventually die and release their lipids. This process generates the typical lipid core of atherosclerotic lesions.

With respect to the expression of connexins in SMCs, Cx37, Cx40, and Cx45 have been shown in SMCs of small elastic or resistance arteries (Little et al., 1995; van Kempen and Jongsma, 1999; Nakamura et al., 1999; Haefliger et al., 2000) but Cx43 is considered as the predominant connexin in the media (Severs et al., 2001). This has been reported mostly for aorta, a large elastic artery (Severs et al., 2001; van Kempen and Jongsma, 1999). However, Cx43 can barely be detected in the media of muscular arteries (Hong and Hill, 1998) and of coronary arteries (Bruzzone et al., 1993; Gros et al., 1994; Verheule et al., 1997; Hong and Hill, 1998; Yeh et al., 1998). Thus, the presence of connexins in SMCs may lead to a further complexity in communication exchanges within the developing atherosclerotic lesion.

Cx43 expression differs in intimal SMCs compared to media SMCs and depends on the disease stage. Thus, intimal Cx43 expression increased at early stages but was decreased in advanced human plaques (Blackburn et al., 1995) (Figure 7.1). Intimal Cx43 expression is also decreased in advanced lesions of hypercholesterolemic rabbit (Polacek et al., 1997) as well as in LDLR$^{-/-}$ mice (Kwak et al., 2002). Furthermore, Cx43 mRNA has been found in foam cells of human and rabbit atherosclerotic carotid arteries (Polacek et al., 1997, 1993). In mice, the reduction of Cx43 expression inhibits the formation of atherosclerotic lesions. On a cholesterol-rich diet, Cx43$^{+/-}$LDLR$^{-/-}$ mice display reduced atherosclerotic lesions, as shown by smaller lipid cores and fewer macrophages and T-cells despite normal white blood cells counts (Kwak et al., 2003). In addition, lesions have thicker fibrous caps containing more SMCs and interstitial collagen.

The phosphorylation state of connexins is important with respect to channel gating, and the potential exchange of solutes (see Chapter 1). Thus, the phosphorylation of a specific serine in Cx43 (ser368) induced a reduction of dye transfer and calcium exchange between SMCs and ECs in the vessel wall (Straub et al., 2010).

The phosphorylation state of Cx43 participates also to the proliferation of SMCs (Johnstone et al., 2009) as well as in their differentiation (Chadjichristos et al., 2008). Moreover, a variation of Cx37 phosphorylation modifies the proliferation of an endothelial tumor cell line (Morel et al., 2010). Finally, Losartan, an angiotensin-converting enzyme inhibitor, reduces the expression of Cx43 but not Cx40, both overexpressed in the intima of a rabbit model of atherosclerosis (Ruan et al., 2010).

Nevertheless, connexins connecting ECs and SMCs seem to influence each other, adding another level of complexity atherosclerosis studies. Indeed, deletion of Cx43 in SMCs reduces the expression of Cx43 in endothelium from the aorta (Liao et al., 2007), and deletion of Cx43 in ECs reduces Cx43 transcription in the adjacent vascular smooth muscle layer (Liao et al., 2001).

7.7 ATHEROSCLEROTIC PLAQUE RUPTURE ACCOUNTS FOR THE MAJORITY OF SUDDEN CARDIAC DEATH

Sudden luminal thrombosis may result from different pathological processes: plaque rupture or plaque erosion (Virmani et al., 2002). This event is responsible for acute coronary syndrome (ACS) (Burke et al., 1997). Hence, following the rupture of a plaque, a coronary artery can become occluded by a thrombus (Figure 7.1d), causing interruption of blood flow and therefore oxygen supply resulting in infarction of the tissue normally irrigated by this coronary artery. The resulting necrosis may lead to various complications such as myocardial rupture, arrhythmias, or finally heart failure. Clinical outcome may be improved by different treatments such as fibrinolysis, coronary artery bypass grafting, or percutaneous coronary interventions with a balloon catheter followed by placement of a stent.

At the advanced phase of atherosclerosis, plaques may be stable or unstable. Unstable plaques, prone to rupture, are characterized by a thin fibrous cap, a large lipid core and are relatively rich in macrophages and other inflammatory cells. Plaque erosion is associated with interruption of the endothelial monolayer that may also lead to luminal thrombosis. However, these lesions are often characterized by a small necrotic core or a thick fibrous cap that isolates the necrotic core from the blood vessel lumen (Virmani et al., 2006). Plaque quality underpins the evolution of the pathology.

The intrinsic activity state of macrophages appeared a key determinant of the plaque quality. Indeed, macrophage foam cells produce pro-inflammatory cytokines that will, as a positive drawback, stimulate macrophage proliferation and lipid uptake. Moreover, activated macrophages produce matrix metalloproteinases (MMPs) that participate in the degradation of the ECM. On the other hand, collagen fibers contained in the fibrous cap are essential to stabilize the plaque (Shah et al., 1995). IFN-γ, a T_H1 cytokine, inhibits both collagen production and proliferation of SMCs, the main source of this ECM (Amento et al., 1991). All these processes result in weakening of the fibrous cap. Stabilization of the plaque is therefore a new and potent therapeutic goal. Interestingly, activation of monocytic cells by combination of TNF-α and IFN-γ, which are associated with the release of MMP-2 and MMP-3, induces Cx43-dependent GJ communication (Eugenin et al., 2003). Enhanced MMP release may also favor plaque rupture. T_H1 cells secrete more MMPs than T_H0 or

T$_H$2 (Bermudez-Fajardo et al., 2007). Interestingly, Cx43 expression and dye coupling between macrophages and T-cells is increased in T$_H$1 CD4$^+$ T-cells compared to Th0 and Th2 (Oviedo-Orta et al., 2008), pointing to a potential role of connexins in atherosclerotic plaque quality. Accordingly, Cx43$^{+/-}$LDLR$^{-/-}$ mice display more stable atherosclerotic lesions (Wong et al., 2003).

Lastly, rupture of the fibrous cap induces a cascade of thrombogenic events, thus allowing activation of platelets and leukocytes by TF, contained in the necrotic core. The main source of TF are monocytes, responsible for the acute thrombus propagation overlying an unstable plaque (Giesen et al., 1999). TF can be found in the acellular lipid core and is associated with macrophages and SMCs (Tremoli et al., 1999). T-cells appear not to be able to synthesize TF, but they can induce the production of TF by macrophages in the plaque (Mach et al., 1997; Buchner et al., 2003). Interestingly, statins have not only been found to affect connexin expression (Kwak et al., 2003; Yeh et al., 2003) but have also been found to decrease TF transcription (Colli et al., 1997). Hence, once the plaque has been formed, many cells and processes are in charge of its quality and its proclivity to rupture, leading to the life-threatening thrombosis and myocardial infarction (MI).

7.8 CONNEXINS ARE CARDIOVASCULAR RISK MARKERS AND POTENT TARGETS FOR THERAPY

Attention to mutations and polymorphisms in connexin genes is increasing, particularly in relation to cardiovascular disease. As described earlier, the C allele from the Cx37-C1019T polymorphism is a potential prognostic marker for carotid atherosclerotic plaque development as reported in a Swedish population (Boerma et al., 1999) and for CAD as reported in Taiwanese, Swiss, and Chinese populations (Yeh et al., 2001; Wong et al., 2007; Pitha et al., 2010). In an evaluation between the C1019T polymorphism and ankle/brachial blood pressure index in diabetic patients, a subclinical atherosclerotic state, the C allele indicates higher risk (Pitha et al., 2010). In contrast, the Cx37-1019T allele is associated with acute MI in Japanese and Sicilian populations (Yamada et al., 2002; Listi et al., 2005, 2007; Leu et al., 2011). Moreover, the T allele is related to intima-medial thickening and associated with a more elevated rate of ischemic stroke (Leu et al., 2011). Finally, the Cx37-C1019T polymorphism appears as a prognostic marker of death after ACS in a prospective study (Lanfear et al., 2007). However, the association of this polymorphism with CAD or MI was not replicated by an Irish study (Horan et al., 2006). Moreover, the Cx37-C1019T polymorphism is not associated with carotid artery intima-media thickness, carotid artery compliance or brachial artery flow-mediated dilatation (all markers of subclinical atherosclerosis) in young adults in a Finish study (Collings et al., 2007). Finally, polymorphisms in the promoter of Cx40 gene has been linked to hypertension in men (Firouzi et al., 2006).

Altogether, these data demonstrate the clinical importance of connexins in atherosclerosis and hypertension and their promising role as potential prognostic markers. However, the need for more prospective studies in larger populations and for well-defined clinical end points remains.

Statins are the regular treatment of atherosclerotic disease and are widely used. Therapeutic efficacy of statins is certainly linked to their role in lowering plasma cholesterol (Vaughan et al., 2000), but in addition, statins have multiple cholesterol-independent pleiotropic effects that also reduce atherosclerosis-related morbidity and mortality (Libby and Sasiela, 2006). Interestingly, statins inhibit the expression of Cx43 in human vascular cells and decreased GJ communication between ECs or SMCs, within the pharmacological range of regular therapeutic prescription for its cholesterol-lowering action (Kwak et al., 2003). Mice are relatively resistant to the lipid-lowering effect of statins, allowing the discrimination between pleiotropic and cholesterol-lowering effects *in vivo* (Libby and Aikawa, 2002). Statins reduce Cx43 plaque expression in LDLR$^{-/-}$ mice (Kwak et al., 2003). Furthermore, statin-treated mice display similar changes with respect to the characteristics of plaque stability, as in Cx43$^{+/-}$LDLR$^{-/-}$ mice (Kwak et al., 2003). Moreover, statins reverse the decreased expression of endothelial Cx37 after long-term cholesterol-rich diet (Yeh et al., 2003). Statins inhibited also GJ communication in cultured SMCs (Shen et al., 2010). Thus, statins emerge as potent modulators of connexins, which in turn might contribute to the pleiotropic effects of these compounds in reduction of atherosclerosis mortality.

7.9 RESTENOSIS IS A LIFE-THREATENING SIDE EFFECT OF CURRENT THERAPY DURING CARDIAC ISCHEMIA

Along with pharmacological therapy, direct intervention on atherosclerotic arteries is often necessary. Despite the use of state-of-art drug eluting stents (DES) to impair SMC proliferation in response to the mechanical manipulation, the reocclusion of the artery, caused by neointimal hyperplasia (restenosis) or thrombosis, still remains an important problem (Newsome et al., 2008). To prevent thrombosis reendothelialization is essential but often compromised by the proliferation-inhibiting compounds on the DES.

To study restenosis, various animal models have been exposed to carotid balloon distension injury or stenting. In the rat balloon injury model, an increased expression of Cx43 is measured in SMCs within the media first, and later in the neointima as well (Yeh et al., 1997). Thus, Cx43 expression seems enhanced in intimal thickening after acute vascular injury (Yeh et al., 1997; Plenz et al., 2004; Wong et al., 2006; Liao et al., 2007). We should, however, be careful in the interpretation of these data, because, in contrast to the human situation, the rat model is known to display very little inflammatory response after balloon injury. Other connexins may play an additional role in the restenotic process. In a porcine model, Cx40 expression is decreased and Cx43 expression markedly increased in stent-induced intimal thickening (Chadjichristos et al., 2008). Cx43 has been shown to participate in the proliferation and differentiation of SMCs *in vitro*. Thus, SMCs isolated from pig coronary arteries show an heterogeneous phenotype, with both spindle-shaped SMCs (S-SMCs) and rhomboid SMCs (R-SMCs) (Hao et al., 2002). S-SMCs are predominant in the normal media, whereas R-SMCs are present in intimal thickening induced by stent injury. R-SMCs show an increased proliferative index, increased migratory and proteolytic phenotype. S-SMCs express Cx40 and Cx43. R-SMCs, devoid of Cx40, display higher Cx43 expression and increased cell-to-cell coupling than S-SMCs. Moreover, S-SMCs treated with platelet-derived growth factor-BB acquire an R phenotype, with an upregulation of Cx43, but a loss of Cx40. Importantly, this can be prevented by a reduction of Cx43 expression using an antisense oligonucleotide (Chadjichristos et al., 2008).

Using knockout mouse models, the progression of restenosis changes when Cx43 is reduced (Wang et al., 2005a; Chadjichristos et al., 2006; Liao et al., 2007). However, when Cx43 expression is ubiquitously reduced neointima formation is decreased (Chadjichristos et al., 2006). In contrast, Cx43 deletion in SMCs only leads to an increased neointimal formation (Liao et al., 2007). Moreover, systemic application of the GJ blocker carbenoxolone inhibits neointimal formation in rats (Song et al., 2009). Thus, targeting Cx43 in SMCs, ECs and leukocytes is important to reduce restenosis. Indeed, macrophages with a reduced Cx43 expression (Cx43$^{+/-}$LDLR$^{-/-}$) infiltrate less in tissue 7 days after balloon injury and Cx43$^{+/-}$LDLR$^{-/-}$ peritoneal macrophages display decreased chemotactic activity for SMCs. These events, together with reduced SMC infiltration and proliferation, result in a reduced neointimal thickening in Cx43$^{+/-}$LDLR$^{-/-}$ mice subjected to balloon-induced vascular injury.

Finally, DES are potent players in the prevention of restenosis. However, by reducing reendothelization, DES appear to have a life-threatening side effect by enhancing the risk of thrombosis (Sheiban et al., 2002). Indeed, when ECs are grown on metallic stents, they display a decreased expression of von Willebrand factor (vWf), endothelial nitric synthase (eNOS) and Cx43, all signs of endothelial dysfunction (Yeh et al., 2006b). Thus, the ideal DES would not only reduce inflammation and SMC proliferation but also enhance reendothelization. Interestingly, immunolabeling with anti-vWf antibodies on serial cross sections from balloon-injured carotids in Cx43$^{+/+}$LDLR$^{-/-}$ and Cx43$^{+/-}$LDLR$^{-/-}$ mice revealed complete endothelial repair 14 days postinjury in Cx43$^{+/-}$LDLR$^{-/-}$ mice. In contrast, vWf immunostaining was absent at this time in all Cx43$^{+/+}$LDLR$^{-/-}$ mice (Chadjichristos et al., 2006). Thus, a reduction in Cx43 expression restricted neointima formation after acute balloon injury by limiting the inflammatory response as well as the proliferation and migration of SMCs toward the damaged site. Moreover, these events resulted in accelerated endothelial repair. Altogether, these findings suggest that Cx43 might be a promising target for local delivery strategies aimed at reducing restenosis after percutaneous coronary interventions. In this respect, recent applications of connexin-specific blocking peptides on SMCs and isolated blood vessels are of particular interest. Thus, strategies targeting the GJ protein might not only be of interest to prevent the burden of atherosclerosis in patients, but might be of interest in combination with other therapeutic strategies for the disease as well.

ACKNOWLEDGMENTS

This work was supported by the Swiss National Science Foundation (310030-127551 to BRK) and the Leenaards Foundation.

REFERENCES

Alvarez Garcia, B., Ruiz, C., Chacon, P., Sabin, J. A., and Matas, M. 2003. High-sensitivity C-reactive protein in high-grade carotid stenosis: Risk marker for unstable carotid plaque. *J Vasc Surg*, 38, 1018–24.

Alves, L. A., Coutinho-Silva, R., Persechini, P. M. et al. 1996. Are there functional gap junctions or junctional hemichannels in macrophages? *Blood,* 88, 328–34.

Amento, E. P., Ehsani, N., Palmer, H., and Libby, P. 1991. Cytokines and growth factors positively and negatively regulate interstitial collagen gene expression in human vascular smooth muscle cells. *Arterioscler Thromb,* 11, 1223–30.

Angelillo-Scherrer, A., Fontana, P., Burnier, L. et al. 2011. Connexin 37 limits thrombus propensity by downregulating platelet reactivity. *Circulation,* 124, 930–9.

Bastiaanse, E. M., Jongsma, H. J., Van Der Laarse, A., and Takens-Kwak, B. R. 1993. Heptanol-induced decrease in cardiac gap junctional conductance is mediated by a decrease in the fluidity of membranous cholesterol-rich domains. *J Membr Biol,* 136, 135–45.

Bermudez-Fajardo, A., Yliharsila, M., Evans, W. H., Newby, A. C., and Oviedo-Orta, E. 2007. CD4+ T lymphocyte subsets express connexin 43 and establish gap junction channel communication with macrophages in vitro. *J Leukoc Biol,* 82, 608–12.

Beyer, E. C. and Steinberg, T. H. 1991. Evidence that the gap junction protein connexin-43 is the ATP-induced pore of mouse macrophages. *J Biol Chem,* 266, 7971–4.

Blackburn, J. P., Peters, N. S., Yeh, H. I. et al. 1995. Upregulation of connexin43 gap junctions during early stages of human coronary atherosclerosis. *Arterioscler Thromb Vasc Biol,* 15, 1219–28.

Boerma, M., Forsberg, L., Van Zeijl, L. et al. 1999. A genetic polymorphism in connexin 37 as a prognostic marker for atherosclerotic plaque development. *J Intern Med,* 246, 211–8.

Bolon, M. L., Ouellette, Y., Li, F., and Tyml, K. 2005. Abrupt reoxygenation following hypoxia reduces electrical coupling between endothelial cells of wild-type but not connexin40 null mice in oxidant- and PKA-dependent manner. *Faseb J,* 19, 1725–7.

Branes, M. C., Contreras, J. E., and Saez, J. C. 2002. Activation of human polymorphonuclear cells induces formation of functional gap junctions and expression of connexins. *Med Sci Monit,* 8, BR313–23.

Bruzzone, R., Haefliger, J. A., Gimlich, R. L., and Paul, D. L. 1993. Connexin40, a component of gap junctions in vascular endothelium, is restricted in its ability to interact with other connexins. *Mol Biol Cell,* 4, 7–20.

Buchner, K., Henn, V., Grafe, M. et al. 2003. CD40 ligand is selectively expressed on CD4+ T cells and platelets: Implications for CD40-CD40 L signalling in atherosclerosis. *J Pathol,* 201, 288–95.

Burger, P. C. and Wagner, D. D. 2003. Platelet P-selectin facilitates atherosclerotic lesion development. *Blood,* 101, 2661–6.

Burke, A. P., Farb, A., Malcom, G. T. et al. 1997. Coronary risk factors and plaque morphology in men with coronary disease who died suddenly. *N Engl J Med,* 336, 1276–82.

Chadjichristos, C. E., Matter, C. M., Roth, I. et al. 2006. Reduced connexin43 expression limits neointima formation after balloon distension injury in hypercholesterolemic mice. *Circulation,* 113, 2835–43.

Chadjichristos, C. E., Morel, S., Derouette, J. P. et al. 2008. Targeting connexin 43 prevents platelet-derived growth factor-BB-induced phenotypic change in porcine coronary artery smooth muscle cells. *Circ Res,* 102, 653–60.

Chadjichristos, C. E., Scheckenbach, K. E., Van Veen, T. A. et al. 2010. Endothelial-specific deletion of connexin40 promotes atherosclerosis by increasing CD73-dependent leukocyte adhesion. *Circulation,* 121, 123–31.

Chanson, M., Derouette, J. P., Roth, I. et al. 2005. Gap junctional communication in tissue inflammation and repair. *Biochim Biophys Acta,* 1711, 197–207.

Chanson, M. and Kwak, B. R. 2007. Connexin37: A potential modifier gene of inflammatory disease. *J Mol Med,* 85, 787–95.

Christ, G. J., Spray, D. C., El-Sabban, M., Moore, L. K., and Brink, P. R. 1996. Gap junctions in vascular tissues. Evaluating the role of intercellular communication in the modulation of vasomotor tone. *Circ Res,* 79, 631–46.

Chu, P. H., Yeh, H. I., Wu, H. H. et al. 2010. Deletion of the FHL2 gene attenuates the formation of atherosclerotic lesions after a cholesterol-enriched diet. *Life Sci*, 86, 365–71.

Colli, S., Eligini, S., Lalli, M. et al. 1997. Vastatins inhibit tissue factor in cultured human macrophages. A novel mechanism of protection against atherothrombosis. *Arterioscler Thromb Vasc Biol*, 17, 265–72.

Collings, A., Islam, M. S., Juonala, M. et al. 2007. Associations between connexin37 gene polymorphism and markers of subclinical atherosclerosis: The Cardiovascular Risk in Young Finns study. *Atherosclerosis*, 195, 379–84.

Cowan, D. B., Lye, S. J., and Langille, B. L. 1998. Regulation of vascular connexin43 gene expression by mechanical loads. *Circ Res*, 82, 786–93.

Cybulsky, M. I. and Gimbrone, M. A., Jr. 1991. Endothelial expression of a mononuclear leukocyte adhesion molecule during atherogenesis. *Science*, 251, 788–91.

Dahlof, B. 2010. Cardiovascular disease risk factors: Epidemiology and risk assessment. *The American Journal of Cardiology*, 105, 3A–9A.

Davies, P. F., Shi, C., Depaola, N., Helmke, B. P., and Polacek, D. C. 2001. Hemodynamics and the focal origin of atherosclerosis: A spatial approach to endothelial structure, gene expression, and function. *Ann N Y Acad Sci*, 947, 7–16; discussion 16–7.

De Maio, A., Vega, V. L., and Contreras, J. E. 2002. Gap junctions, homeostasis, and injury. *J Cell Physiol*, 191, 269–82.

De Wit, C., Hoepfl, B., and Wolfle, S. E. 2006. Endothelial mediators and communication through vascular gap junctions. *Biol Chem*, 387, 3–9.

De Wit, C., Roos, F., Bolz, S. S. et al. 2000. Impaired conduction of vasodilation along arterioles in connexin40-deficient mice. *Circ Res*, 86, 649–55.

Depaola, N., Davies, P. F., Pritchard, W. F., Jr. et al. 1999. Spatial and temporal regulation of gap junction connexin43 in vascular endothelial cells exposed to controlled disturbed flows in vitro. *Proc Natl Acad Sci USA*, 96, 3154–9.

Dlugosova, K., Okruhlicova, L., Mitasikova, M. et al. 2009. Modulation of connexin-43 by omega-3 fatty acids in the aorta of old spontaneously hypertensive rats. *J Physiol Pharmacol*, 60, 63–9.

Dong, Z. M., Chapman, S. M., Brown, A. A. et al. 1998. The combined role of P- and E-selectins in atherosclerosis. *J Clin Invest*, 102, 145–52.

Ebong, E. E., Kim, S., and Depaola, N. 2006. Flow regulates intercellular communication in HAEC by assembling functional Cx40 and Cx37 gap junctional channels. *Am J Physiol Heart Circ Physiol*, 290, H2015–23.

Eugenin, E. A., Branes, M. C., Berman, J. W., and Saez, J. C. 2003. TNF-alpha plus IFN-gamma induce connexin43 expression and formation of gap junctions between human monocytes/macrophages that enhance physiological responses. *J Immunol*, 170, 1320–8.

Eugenin, E. A., Eckardt, D., Theis, M. et al. 2001. Microglia at brain stab wounds express connexin 43 and *in vitro* form functional gap junctions after treatment with interferon-gamma and tumor necrosis factor-alpha. *Proc Natl Acad Sci USA*, 98, 4190–5.

Figueroa, X. F. and Duling, B. R. 2009. Gap junctions in the control of vascular function. *Antioxid Redox Signal*, 11, 251–66.

Figueroa, X. F., Isakson, B. E., and Duling, B. R. 2004. Connexins: Gaps in our knowledge of vascular function. *Physiology (Bethesda)*, 19, 277–84.

Figueroa, X. F., Isakson, B. E., and Duling, B. R. 2006. Vascular gap junctions in hypertension. *Hypertension*, 48, 804–11.

Firouzi, M., Kok, B., Spiering, W. et al. 2006. Polymorphisms in human connexin40 gene promoter are associated with increased risk of hypertension in men. *J Hypertens*, 24, 325–30.

Frenette, P. S., Denis, C. V., Weiss, L. et al. 2000. P-Selectin glycoprotein ligand 1 (PSGL-1) is expressed on platelets and can mediate platelet–endothelial interactions in vivo. *J Exp Med*, 191, 1413–22.

Frostegard, J., Ulfgren, A. K., Nyberg, P. et al. 1999. Cytokine expression in advanced human atherosclerotic plaques: Dominance of pro-inflammatory (Th1) and macrophage-stimulating cytokines. *Atherosclerosis*, 145, 33–43.

Gabriels, J. E. and Paul, D. L. 1998. Connexin43 is highly localized to sites of disturbed flow in rat aortic endothelium but connexin37 and connexin40 are more uniformly distributed. *Circ Res,* 83, 636–43.

Giesen, P. L., Rauch, U., Bohrmann, B. et al. 1999. Blood-borne tissue factor: Another view of thrombosis. *Proc Natl Acad Sci USA,* 96, 2311–5.

Goldberg, I. J. 2004. Why does diabetes increase atherosclerosis? I don't know! *J Clin Invest,* 114, 613–5.

Gros, D., Jarry-Guichard, T., Ten Velde, I. et al. 1994. Restricted distribution of connexin40, a gap junctional protein, in mammalian heart. *Circ Res,* 74, 839–51.

Haefliger, J. A., Polikar, R., Schnyder, G. et al. 2000. Connexin37 in normal and pathological development of mouse heart and great arteries. *Dev Dyn,* 218, 331–44.

Hansson, G. K. and Libby, P. 2006. The immune response in atherosclerosis: A double-edged sword. *Nat Rev Immunol,* 6, 508–19.

Hao, H., Ropraz, P., Verin, V. et al. 2002. Heterogeneity of smooth muscle cell populations cultured from pig coronary artery. *Arterioscler Thromb Vasc Biol,* 22, 1093–9.

Hill, C. E., Phillips, J. K., and Sandow, S. L. 2001. Heterogeneous control of blood flow amongst different vascular beds. *Med Res Rev,* 21, 1–60.

Hill, C. E., Rummery, N., Hickey, H., and Sandow, S. L. 2002. Heterogeneity in the distribution of vascular gap junctions and connexins: Implications for function. *Clin Exp Pharmacol Physiol,* 29, 620–5.

Hong, T. and Hill, C. E. 1998. Restricted expression of the gap junctional protein connexin 43 in the arterial system of the rat. *J Anat,* 192 (Pt 4), 583–93.

Horan, P. G., Allen, A. R., Patterson, C. C. et al. 2006. The connexin 37 gene polymorphism and coronary artery disease in Ireland. *Heart,* 92, 395–6.

Hou, C. J., Tsai, C. H., Su, C. H. et al. 2008. Diabetes reduces aortic endothelial gap junctions in ApoE-deficient mice: Simvastatin exacerbates the reduction. *J Histochem Cytochem,* 56, 745–52.

Hu, J. and Cotgreave, I. A. 1997. Differential regulation of gap junctions by proinflammatory mediators in vitro. *J Clin Invest,* 99, 2312–6.

Hu, W., Polinsky, P., Sadoun, E., Rosenfeld, M. E., and Schwartz, S. M. 2005. Atherosclerotic lesions in the common coronary arteries of ApoE knockout mice. *Cardiovascular Pathology: The Official Journal of the Society for Cardiovascular Pathology,* 14, 120–5.

Hu, V. W. and Xie, H. Q. 1994. Interleukin-1 alpha suppresses gap junction-mediated intercellular communication in human endothelial cells. *Exp Cell Res,* 213, 218–23.

Hulser, D. F. and Peters, J. H. 1971. Intercellular communication in phytohemagglutinin-induced lymphocyte agglutinates. *Eur J Immunol,* 1, 494–5.

Ikonomidis, I., Stamatelopoulos, K., Lekakis, J., Vamvakou, G. D., and Kremastinos, D. T. 2008. Inflammatory and non-invasive vascular markers: The multimarker approach for risk stratification in coronary artery disease. *Atherosclerosis,* 199, 3–11.

Inai, T., Mancuso, M. R., Mcdonald, D. M. et al. 2004. Shear stress-induced upregulation of connexin 43 expression in endothelial cells on upstream surfaces of rat cardiac valves. *Histochem Cell Biol,* 122, 477–83.

Isakson, B. E., Kronke, G., Kadl, A., Leitinger, N., and Duling, B. R. 2006. Oxidized phospholipids alter vascular connexin expression, phosphorylation, and heterocellular communication. *Arterioscler Thromb Vasc Biol,* 26, 2216–21.

Ishibashi, S., Brown, M. S., Goldstein, J. L. et al. 1993. Hypercholesterolemia in low density lipoprotein receptor knockout mice and its reversal by adenovirus-mediated gene delivery. *The Journal of Clinical Investigation,* 92, 883–93.

Ishibashi, S., Goldstein, J. L., Brown, M. S., Herz, J., and Burns, D. K. 1994. Massive xanthomatosis and atherosclerosis in cholesterol-fed low density lipoprotein receptor-negative mice. *The Journal of Clinical Investigation,* 93, 1885–93.

Jara, P. I., Boric, M. P., and Saez, J. C. 1995. Leukocytes express connexin 43 after activation with lipopolysaccharide and appear to form gap junctions with endothelial cells after ischemia–reperfusion. *Proc Natl Acad Sci USA,* 92, 7011–5.

Jia, S. J., Zhou, Z., Zhang, B. K. et al. 2009. Asymmetric dimethylarginine damages connexin43-mediated endothelial gap junction intercellular communication. *Biochem Cell Biol,* 87, 867–74.

Johnstone, S. R., Ross, J., Rizzo, M. J. et al. 2009. Oxidized phospholipid species promote *in vivo* differential cx43 phosphorylation and vascular smooth muscle cell proliferation. *Am J Pathol,* 175, 916–24.

Jonasson, L., Holm, J., Skalli, O., Bondjers, G., and Hansson, G. K. 1986. Regional accumulations of T cells, macrophages, and smooth muscle cells in the human atherosclerotic plaque. *Arteriosclerosis,* 6, 131–8.

Kanady, J. D., Dellinger, M. T., Munger, S. J., Witte, M. H., and Simon, A. M. 2011. Connexin37 and Connexin43 deficiencies in mice disrupt lymphatic valve development and result in lymphatic disorders including lymphedema and chylothorax. *Dev Biol,* 354, 253–66.

Kruger, O., Plum, A., Kim, J. S. et al. 2000. Defective vascular development in connexin 45-deficient mice. *Development,* 127, 4179–93.

Kwak, B. R., Mulhaupt, F., Veillard, N., Gros, D. B., and Mach, F. 2002. Altered pattern of vascular connexin expression in atherosclerotic plaques. *Arterioscler Thromb Vasc Biol,* 22, 225–30.

Kwak, B. R., Pepper, M. S., Gros, D. B., and Meda, P. 2001. Inhibition of endothelial wound repair by dominant negative connexin inhibitors. *Mol Biol Cell,* 12, 831–45.

Kwak, B. R., Silacci, P., Stergiopulos, N., Hayoz, D., and Meda, P. 2005. Shear stress and cyclic circumferential stretch, but not pressure, alter connexin43 expression in endothelial cells. *Cell Commun Adhes,* 12, 261–70.

Kwak, B. R., Veillard, N., Pelli, G. et al. 2003. Reduced connexin43 expression inhibits atherosclerotic lesion formation in low-density lipoprotein receptor-deficient mice. *Circulation,* 107, 1033–9.

Lanfear, D. E., Jones, P. G., Marsh, S. et al. 2007. Connexin37 (GJA4) genotype predicts survival after an acute coronary syndrome. *Am Heart J,* 154, 561–6.

Larson, D. M., Christensen, T. G., Sagar, G. D., and Beyer, E. C. 2001. TGF-beta1 induces an accumulation of connexin43 in a lysosomal compartment in endothelial cells. *Endothelium,* 8, 255–60.

Larson, D. M., Wrobleski, M. J., Sagar, G. D., Westphale, E. M., and Beyer, E. C. 1997. Differential regulation of connexin43 and connexin37 in endothelial cells by cell density, growth, and TGF-beta1. *Am J Physiol,* 272, C405–15.

Leu, H. B., Chung, C. M., Chuang, S. Y. et al. 2011. Genetic variants of connexin37 are associated with carotid intima-medial thickness and future onset of ischemic stroke. *Atherosclerosis,* 214, 101–6.

Levy, J. A., Weiss, R. M., Dirksen, E. R., and Rosen, M. R. 1976. Possible communication between murine macrophages oriented in linear chains in tissue culture. *Exp Cell Res,* 103, 375–85.

Li, A. F. and Roy, S. 2009. High glucose-induced downregulation of connexin 43 expression promotes apoptosis in microvascular endothelial cells. *Invest Ophthalmol Vis Sci,* 50, 1400–7.

Liao, Y., Day, K. H., Damon, D. N., and Duling, B. R. 2001. Endothelial cell-specific knockout of connexin 43 causes hypotension and bradycardia in mice. *Proc Natl Acad Sci USA,* 98, 9989–94.

Liao, Y., Regan, C. P., Manabe, I. et al. 2007. Smooth muscle-targeted knockout of connexin43 enhances neointimal formation in response to vascular injury. *Arterioscler Thromb Vasc Biol,* 27, 1037–42.

Libby, P. and Aikawa, M. 2002. Stabilization of atherosclerotic plaques: New mechanisms and clinical targets. *Nat Med,* 8, 1257–62.

Libby, P., Ridker, P. M., and Hansson, G. K. 2011. Progress and challenges in translating the biology of atherosclerosis. *Nature,* 473, 317–25.

Libby, P. and Sasiela, W. 2006. Plaque stabilization: Can we turn theory into evidence? *Am J Cardiol,* 98, 26P–33P.

Listi, F., Candore, G., Balistreri, C. R. et al. 2007. Connexin37 1019 gene polymorphism in myocardial infarction patients and centenarians. *Atherosclerosis,* 191, 460–1.

Listi, F., Candore, G., Lio, D. et al. 2005. Association between C1019T polymorphism of connexin37 and acute myocardial infarction: A study in patients from Sicily. *Int J Cardiol,* 102, 269–71.

Little, T. L., Beyer, E. C., and Duling, B. R. 1995. Connexin 43 and connexin 40 gap junctional proteins are present in arteriolar smooth muscle and endothelium in vivo. *Am J Physiol,* 268, H729–39.

Lloyd-Jones, D. M. 2010. Cardiovascular risk prediction: Basic concepts, current status, and future directions. *Circulation,* 121, 1768–77.

Mach, F., Schonbeck, U., Bonnefoy, J. Y., Pober, J. S., and Libby, P. 1997. Activation of monocyte/macrophage functions related to acute atheroma complication by ligation of CD40: Induction of collagenase, stromelysin, and tissue factor. *Circulation,* 96, 396–9.

Makino, A., Platoshyn, O., Suarez, J., Yuan, J. X., and Dillmann, W. H. 2008. Downregulation of connexin40 is associated with coronary endothelial cell dysfunction in streptozotocin-induced diabetic mice. *Am J Physiol Cell Physiol,* 295, C221–30.

Martin, C. A., El-Sabban, M. E., Zhao, L., Burakoff, R., and Homaidan, F. R. 1998. Adhesion and cytosolic dye transfer between macrophages and intestinal epithelial cells. *Cell Adhes Commun,* 5, 83–95.

Massberg, S., Schurzinger, K., Lorenz, M. et al. 2005. Platelet adhesion via glycoprotein IIb integrin is critical for atheroprogression and focal cerebral ischemia: An *in vivo* study in mice lacking glycoprotein IIb. *Circulation,* 112, 1180–8.

Mckinnon, R. L., Bolon, M. L., Wang, H. X. et al. 2009. Reduction of electrical coupling between microvascular endothelial cells by NO depends on connexin37. *Am J Physiol Heart Circ Physiol,* 297, H93–H101.

Mensink, A., De Haan, L. H., Lakemond, C. M., Koelman, C. A., and Koeman, J. H. 1995. Inhibition of gap junctional intercellular communication between primary human smooth muscle cells by tumor necrosis factor alpha. *Carcinogenesis,* 16, 2063–7.

Meyer, R. A., Lampe, P. D., Malewicz, B., Baumann, W. J., and Johnson, R. G. 1991. Enhanced gap junction formation with LDL and apolipoprotein B. *Exp Cell Res,* 196, 72–81.

Meyer, R., Malewicz, B., Baumann, W. J., and Johnson, R. G. 1990. Increased gap junction assembly between cultured cells upon cholesterol supplementation. *J Cell Sci,* 96(Pt 2), 231–8.

Morel, S., Burnier, L., Roatti, A. et al. 2010. Unexpected role for the human Cx37 C1019T polymorphism in tumour cell proliferation. *Carcinogenesis,* 31, 1922–31.

Nakamura, K., Inai, T., Nakamura, K., and Shibata, Y. 1999. Distribution of gap junction protein connexin 37 in smooth muscle cells of the rat trachea and pulmonary artery. *Arch Histol Cytol,* 62, 27–37.

Newsome, L. T., Kutcher, M. A., and Royster, R. L. 2008. Coronary artery stents: Part I. Evolution of percutaneous coronary intervention. *Anesth Analg,* 107, 552–69.

Oviedo-Orta, E., Bermudez-Fajardo, A., Karanam, S., Benbow, U., and Newby, A. C. 2008. Comparison of MMP-2 and MMP-9 secretion from T helper 0, 1 and 2 lymphocytes alone and in coculture with macrophages. *Immunology,* 124, 42–50.

Oviedo-Orta, E., Errington, R. J., and Evans, W. H. 2002. Gap junction intercellular communication during lymphocyte transendothelial migration. *Cell Biol Int,* 26, 253–63.

Oviedo-Orta, E., Hoy, T., and Evans, W. H. 2000. Intercellular communication in the immune system: Differential expression of connexin40 and 43, and perturbation of gap junction channel functions in peripheral blood and tonsil human lymphocyte subpopulations. *Immunology,* 99, 578–90.

Park, Y. M., Febbraio, M., and Silverstein, R. L. 2009. CD36 modulates migration of mouse and human macrophages in response to oxidized LDL and may contribute to macrophage trapping in the arterial intima. *J Clin Invest*, 119, 136–45.

Paulsson, J., Dadfar, E., Held, C., Jacobson, S. H., and Lundahl, J. 2007. Activation of peripheral and *in vivo* transmigrated neutrophils in patients with stable coronary artery disease. *Atherosclerosis*, 192, 328–34.

Pepper, M. S. and Meda, P. 1992. Basic fibroblast growth factor increases junctional communication and connexin 43 expression in microvascular endothelial cells. *J Cell Physiol*, 153, 196–205.

Pepys, M. B. and Hirschfield, G. M. 2003. C-reactive protein: A critical update. *J Clin Invest*, 111, 1805–12.

Pitha, J., Hubacek, J. A., and Pithova, P. 2010. The connexin 37 (1019C > T) gene polymorphism is associated with subclinical atherosclerosis in women with type 1 and 2 diabetes and in women with central obesity. *Physiol Res*, 59, 1029–32.

Plenz, G., Ko, Y. S., Yeh, H. I. et al. 2004. Upregulation of connexin43 gap junctions between neointimal smooth muscle cells. *Eur J Cell Biol*, 83, 521–30.

Plump, A. S. and Lum, P. Y. 2009. Genomics and cardiovascular drug development. *Journal of the American College of Cardiology*, 53, 1089–100.

Polacek, D., Bech, F., Mckinsey, J. F., and Davies, P. F. 1997. Connexin43 gene expression in the rabbit arterial wall: Effects of hypercholesterolemia, balloon injury and their combination. *J Vasc Res*, 34, 19–30.

Polacek, D., Lal, R., Volin, M. V., and Davies, P. F. 1993. Gap junctional communication between vascular cells. Induction of connexin43 messenger RNA in macrophage foam cells of atherosclerotic lesions. *Am J Pathol*, 142, 593–606.

Reaume, A. G., De Sousa, P. A., Kulkarni, S. et al. 1995. Cardiac malformation in neonatal mice lacking connexin43. *Science*, 267, 1831–4.

Rignault, S., Haefliger, J. A., Gasser, D. et al. 2005. Sepsis up-regulates the expression of connexin 40 in rat aortic endothelium. *Crit Care Med*, 33, 1302–10.

Ruan, L. M., Cai, W., Chen, J. Z., and Duan, J. F. 2010. Effects of Losartan on expression of connexins at the early stage of atherosclerosis in rabbits. *Int J Med Sci*, 7, 82–9.

Sabine, A., Agaralov, Y., Maby-El Hajjami et al. 2012. Mechanotransduction PROX1, and FOXC2 cooperate to control connexin37 and calcineurin during lymphatic-valve formation. *Dev Cell*, 22, 430–45.

Sacks, F. M., Pfeffer, M. A., Moye, L. A. et al. 1996. The effect of pravastatin on coronary events after myocardial infarction in patients with average cholesterol levels. Cholesterol and Recurrent Events Trial investigators. *N Engl J Med*, 335, 1001–9.

Saez, J. C., Berthoud, V. M., Branes, M. C., Martinez, A. D., and Beyer, E. C. 2003. Plasma membrane channels formed by connexins: Their regulation and functions. *Physiol Rev*, 83, 1359–400.

Sampietro, T., Bigazzi, F., Dal Pino, B. et al. 2002. Increased plasma C-reactive protein in familial hypoalphalipoproteinemia: A proinflammatory condition? *Circulation*, 105, 11–4.

Scerri, I., Tabary, O., Dudez, T. et al. 2006. Gap junctional communication does not contribute to the interaction between neutrophils and airway epithelial cells. *Cell Commun Adhes*, 13, 1–12.

Severs, N. J., Rothery, S., Dupont, E. et al. 2001. Immunocytochemical analysis of connexin expression in the healthy and diseased cardiovascular system. *Microsc Res Tech*, 52, 301–22.

Shah, P. K., Falk, E., Badimon, J. J. et al. 1995. Human monocyte-derived macrophages induce collagen breakdown in fibrous caps of atherosclerotic plaques. Potential role of matrix-degrading metalloproteinases and implications for plaque rupture. *Circulation*, 92, 1565–9.

Sheiban, I., Carrieri, L., Catuzzo, B. et al. 2002. Drug-eluting stent: The emerging technique for the prevention of restenosis. *Minerva Cardioangiol*, 50, 443–53.

Shen, J., Wang, L. H., Zheng, L. R., Zhu, J. H., and Hu, S. J. 2010. Lovastatin inhibits gap junctional communication in cultured aortic smooth muscle cells. *J Cardiovasc Pharmacol Ther*, 15, 296–302.

Sheu, J. J., Chang, L. T., Chiang, C. H. et al. 2007. Impact of diabetes on cardiomyocyte apoptosis and connexin43 gap junction integrity: Role of pharmacological modulation. *Int Heart J*, 48, 233–45.

Simon, A. M. and Mcwhorter, A. R. 2002. Vascular abnormalities in mice lacking the endothelial gap junction proteins connexin37 and connexin40. *Dev Biol*, 251, 206–20.

Simon, A. M., Mcwhorter, A. R., Chen, H., Jackson, C. L., and Ouellette, Y. 2004. Decreased intercellular communication and connexin expression in mouse aortic endothelium during lipopolysaccharide-induced inflammation. *J Vasc Res*, 41, 323–33.

Sohl, G. and Willecke, K. 2004. Gap junctions and the connexin protein family. *Cardiovasc Res*, 62, 228–32.

Song, M., Yu, X., Cui, X. et al. 2009. Blockade of connexin 43 hemichannels reduces neointima formation after vascular injury by inhibiting proliferation and phenotypic modulation of smooth muscle cells. *Exp Biol Med (Maywood)*, 234, 1192–200.

Spear, J. F., Balke, C. W., Lesh, M. D. et al. 1990. Effect of cellular uncoupling by heptanol on conduction in infarcted myocardium. *Circ Res*, 66, 202–17.

Stemme, S., Faber, B., Holm, J. et al. 1995. T lymphocytes from human atherosclerotic plaques recognize oxidized low density lipoprotein. *Proc Natl Acad Sci USA*, 92, 3893–7.

Straub, A. C., Johnstone, S. R., Heberlein, K. R. et al. 2010. Site-specific connexin phosphorylation is associated with reduced heterocellular communication between smooth muscle and endothelium. *J Vasc Res*, 47, 277–86.

Suarez, S. and Ballmer-Hofer, K. 2001. VEGF transiently disrupts gap junctional communication in endothelial cells. *J Cell Sci*, 114, 1229–35.

Suo, J., Ferrara, D. E., Sorescu, D. et al. 2007. Hemodynamic shear stresses in mouse aortas: Implications for atherogenesis. *Arteriosclerosis, Thrombosis, and Vascular Biology*, 27, 346–51.

Tacheau, C., Fontaine, J., Loy, J., Mauviel, A., and Verrecchia, F. 2008. TGF-beta induces connexin43 gene expression in normal murine mammary gland epithelial cells via activation of p38 and PI3K/AKT signaling pathways. *J Cell Physiol*, 217, 759–68.

Theilmeier, G., Michiels, C., Spaepen, E. et al. 2002. Endothelial von Willebrand factor recruits platelets to atherosclerosis-prone sites in response to hypercholesterolemia. *Blood*, 99, 4486–93.

Theis, M., De Wit, C., Schlaeger, T. M. et al. 2001. Endothelium-specific replacement of the connexin43 coding region by a lacZ reporter gene. *Genesis*, 29, 1–13.

Thuringer, D. 2004. The vascular endothelial growth factor-induced disruption of gap junctions is relayed by an autocrine communication via ATP release in coronary capillary endothelium. *Ann N Y Acad Sci*, 1030, 14–27.

Tremoli, E., Camera, M., Toschi, V., and Colli, S. 1999. Tissue factor in atherosclerosis. *Atherosclerosis*, 144, 273–83.

Tsai, C. H., Yeh, H. I., Tian, T. Y. et al. 2004. Down-regulating effect of nicotine on connexin43 gap junctions in human umbilical vein endothelial cells is attenuated by statins. *Eur J Cell Biol*, 82, 589–95.

Valiyaveettil, M. and Podrez, E. A. 2009. Platelet hyperreactivity, scavenger receptors and atherothrombosis. *J Thromb Haemost*, 7(Suppl 1), 218–21.

Van Kempen, M. J. and Jongsma, H. J. 1999. Distribution of connexin37, connexin40 and connexin43 in the aorta and coronary artery of several mammals. *Histochem Cell Biol*, 112, 479–86.

Van Rijen, H. V., Van Kempen, M. J., Postma, S., and Jongsma, H. J. 1998. Tumour necrosis factor alpha alters the expression of connexin43, connexin40, and connexin37 in human umbilical vein endothelial cells. *Cytokine*, 10, 258–64.

Vaughan, C. J., Gotto, A. M., Jr., and Basson, C. T. 2000. The evolving role of statins in the management of atherosclerosis. *J Am Coll Cardiol*, 35, 1–10.

Verheule, S., Van Kempen, M. J., Te Welscher, P. H., Kwak, B. R., and Jongsma, H. J. 1997. Characterization of gap junction channels in adult rabbit atrial and ventricular myocardium. *Circ Res*, 80, 673–81.

Verma, S. and Yeh, E. T. 2003. C-reactive protein and atherothrombosis—Beyond a biomarker: An actual partaker of lesion formation. *Am J Physiol Regul Integr Comp Physiol*, 285, R1253–6; discussion R1257–8.

Virmani, R., Burke, A. P., Farb, A., and Kolodgie, F. D. 2006. Pathology of the vulnerable plaque. *J Am Coll Cardiol*, 47, C13–8.

Virmani, R., Burke, A. P., Kolodgie, F. D., and Farb, A. 2002. Vulnerable plaque: The pathology of unstable coronary lesions. *J Interv Cardiol*, 15, 439–46.

Wang, L., Chen, J., Sun, Y. et al. 2005a. Regulation of connexin expression after balloon injury: Possible mechanisms for antiproliferative effect of statins. *Am J Hypertens*, 18, 1146–53.

Wang, H. H., Yeh, H. I., Wang, C. Y. et al. 2010. C-reactive protein, sodium azide, and endothelial connexin43 gap junctions. *Cell Biol Toxicol*, 26, 153–63.

Wang, Q., Zhu, X., Xu, Q. et al. 2005b. Effect of C-reactive protein on gene expression in vascular endothelial cells. *Am J Physiol Heart Circ Physiol*, 288, H1539–45.

White, H. D. and Chew, D. P. 2008. Acute myocardial infarction. *Lancet*, 372, 570–84.

Wong, C. W., Burger, F., Pelli, G., Mach, F., and Kwak, B. R. 2003. Dual benefit of reduced Cx43 on atherosclerosis in LDL receptor-deficient mice. *Cell Commun Adhes*, 10, 395–400.

Wong, C. W., Christen, T., Pfenniger, A., James, R. W., and Kwak, B. R. 2007. Do allelic variants of the connexin37 1019 gene polymorhism differentially predict for coronary artery disease and myocardial infarction? *Atherosclerosis*, 191, 355–61.

Wong, C. W., Christen, T., Roth, I. et al. 2006. Connexin37 protects against atherosclerosis by regulating monocyte adhesion. *Nat Med*, 12, 950–4.

Xie, H. Q. and Hu, V. W. 1994. Modulation of gap junctions in senescent endothelial cells. *Exp Cell Res*, 214, 172–6.

Yamada, Y., Izawa, H., Ichihara, S. et al. 2002. Prediction of the risk of myocardial infarction from polymorphisms in candidate genes. *N Engl J Med*, 347, 1916–23.

Yeh, H. I., Chang, H. M., Lu, W. W. et al. 2000. Age-related alteration of gap junction distribution and connexin expression in rat aortic endothelium. *J Histochem Cytochem*, 48, 1377–89.

Yeh, H. I., Chou, Y., Liu, H. F., Chang, S. C., and Tsai, C. H. 2001. Connexin37 gene polymorphism and coronary artery disease in Taiwan. *Int J Cardiol*, 81, 251–5.

Yeh, H. I., Lee, P. Y., Su, C. H. et al. 2006a. Reduced expression of endothelial connexins 43 and 37 in hypertensive rats is rectified after 7-day carvedilol treatment. *Am J Hypertens*, 19, 129–35.

Yeh, H. I., Lu, S. K., Tian, T. Y. et al. 2006b. Comparison of endothelial cells grown on different stent materials. *J Biomed Mater Res A*, 76, 835–41.

Yeh, H. I., Lu, C. S., Wu, Y. J. et al. 2003. Reduced expression of endothelial connexin37 and connexin40 in hyperlipidemic mice: Recovery of connexin37 after 7-day simvastatin treatment. *Arterioscler Thromb Vasc Biol*, 23, 1391–7.

Yeh, H. I., Lupu, F., Dupont, E., and Severs, N. J. 1997. Upregulation of connexin43 gap junctions between smooth muscle cells after balloon catheter injury in the rat carotid artery. *Arterioscler Thromb Vasc Biol*, 17, 3174–84.

Yeh, H. I., Rothery, S., Dupont, E., Coppen, S. R., and Severs, N. J. 1998. Individual gap junction plaques contain multiple connexins in arterial endothelium. *Circ Res*, 83, 1248–63.

Yla-Herttuala, S., Bentzon, J. F., Daemen, M. et al. 2011. Stabilisation of atherosclerotic plaques. Position paper of the European Society of Cardiology (ESC) Working Group on atherosclerosis and vascular biology. *Thromb Haemost*, 106, 1–19.

Yusuf, S., Hawken, S., Ounpuu, S. et al. 2004. Effect of potentially modifiable risk factors associated with myocardial infarction in 52 countries (the INTERHEART study): Case – control study. *Lancet*, 364, 937–52.

Zahler, S., Hoffmann, A., Gloe, T., and Pohl, U. 2003. Gap-junctional coupling between neutrophils and endothelial cells: A novel modulator of transendothelial migration. *J Leukoc Biol*, 73, 118–26.

Zhang, Y. W., Morita, I., Zhang, L. et al. 2000. Screening of anti-hypoxia/reoxygenation agents by an *in vitro* method. Part 2: Inhibition of tyrosine kinase activation prevented hypoxia/reoxygenation-induced injury in endothelial gap junctional intercellular communication. *Planta Med*, 66, 119–23.

Zhang, S. H., Reddick, R. L., Piedrahita, J. A., and Maeda, N. 1992. Spontaneous hypercholesterolemia and arterial lesions in mice lacking apolipoprotein E. *Science*, 258, 468–71.

Zhou, X., Caligiuri, G., Hamsten, A., Lefvert, A. K. and Hansson, G. K. 2001. LDL immunization induces T-cell-dependent antibody formation and protection against atherosclerosis. *Arterioscler Thromb Vasc Biol*, 21, 108–14.

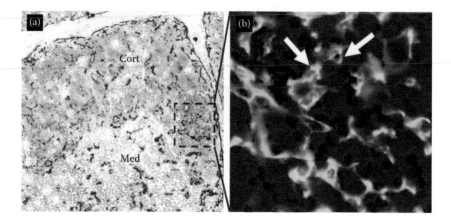

FIGURE 2.2 See the text for the figure caption.

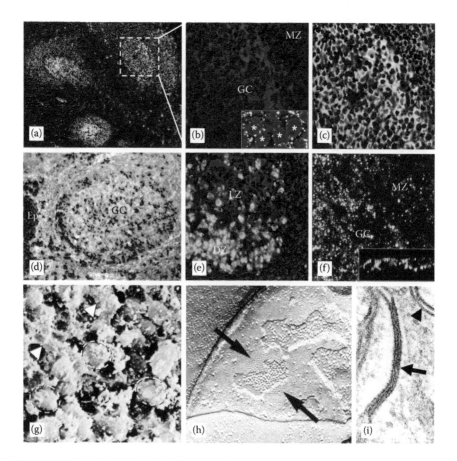

FIGURE 2.3 See the text for the figure caption.

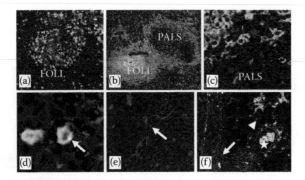

FIGURE 2.4 See the text for the figure caption.

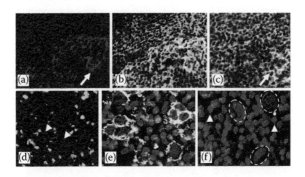

FIGURE 2.5 See the text for the figure caption.

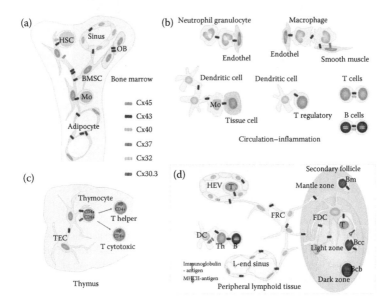

FIGURE 2.6 See the text for the figure caption.

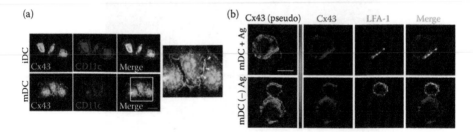

FIGURE 6.2 Gap junctions in the immune defense against tumors. (a) Cx43 is localized at cell–cell contacts between human mature DCs (mDCs). Immature and mature DCs were costained for Cx43 (green) and CD11c (red). High-power confocal magnification shows Cx43 at cell–cell contacts, colocalizing with CD11c (arrowheads). Scale bar = 20 mm. (b) Cx43 accumulates at the immunological synapse in an antigen (Ag)-specific way. Representative images of Cx43 and LFA-1 distribution after incubation of OVA–DCs (DC + Ag) or LPS–DCs (DC (–) Ag) with OT-II T cells are shown. Scale bar = 5 mm.

FIGURE 13.1 Highly ordered patterns of Cx43 GJs (green) at intercalated disks between cardio-myoctes in the healthy rat ventricle. Following myocardial infarction, Cx43 redistributes to myocyte lateral surfaces at the injury border zone—that is, the cell edges marked by wheat germ agglutinin (WGA) labeling (red) in the image shown. Nuclei are marked by blue DAPI labeling.

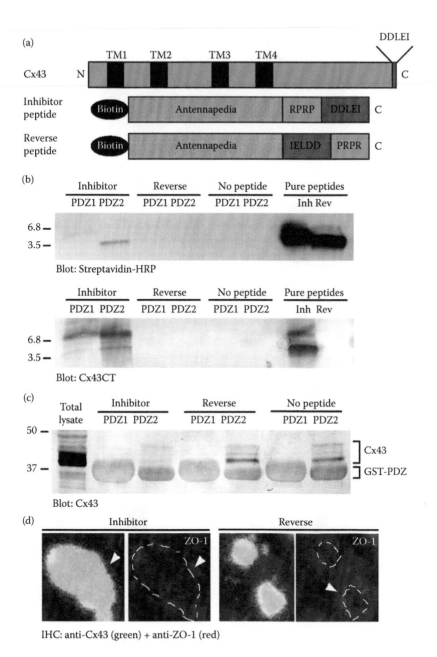

FIGURE 13.2 αCT1 binds to the PDZ2 domain of ZO-1 and blocks interaction with Cx43 carboxyl terminus. (a) Schemetic of domain structure of Cx43 and αCT1 and reverse control (REV) peptides. (b) Blots αCT1 and REV (~3.5 kDa) detected by streptavidin-HRP (top) and Ab recognizing Ct of Cx43 (bottom). αCT1 was pulled down by GST-PDZ2, but not GST-PDZ1. (c) Endogenous Cx43 pulled from HeLa lysates by GST-PDZ2 reduced in presence of αCT1. (d) αCT1 increases GJ size (green) and decreases ZO-1 (red) localization at the GJ edge compared to REV control peptide. (From Hunter, A. W. et al. 2006. *Mol Biol Cell* 16: 5686–5698.)

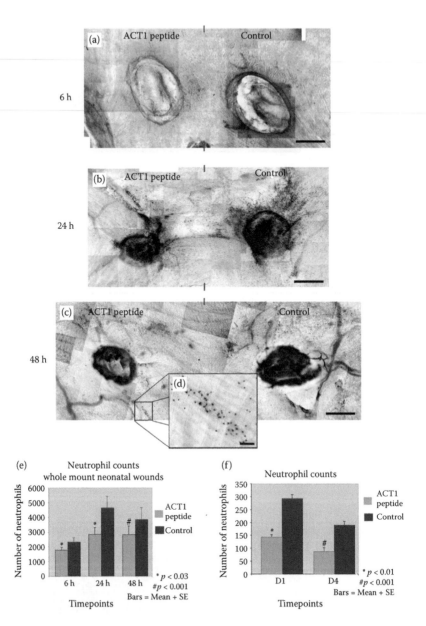

FIGURE 13.3 αCT1 reduces neutrophil recruitment to skin wounds (a–c). Neutrophil recruitment in paired excisional wounds on the back of neonatal mice over a 6, 24, and 48 h time course. The density of neutrophils at αCT1-treated wounds was visibly lower than the matching control wound on the same animal—particularly at 24 h (b). Inset (d) shows higher magnification of individually labeled neutrophils associated with a vascular structure adjacent to a peptide-treated wound. (e and f) Bar graphs showing significant reduction in the density of neutrophils at αCT1-peptide treated wounds (blue bars) relative to the paired control wounds (purple bars) in neonates (e) and adults (f). Scale a–c = 250 μm, inset = 50 μm. (From Ghatnekar, G. S. et al. 2009. *Regen Med* 4(2): 205–223.)

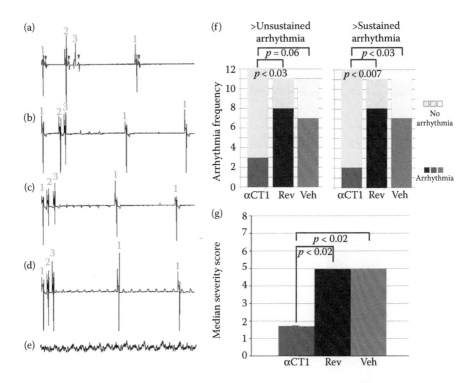

FIGURE 13.4 αCT1 reduces Cx43 GJ remodeling and arrhythmic propensity. Tracings from pacing protocols on isolated perfused ventricles illustrate no arrhythmia (a), three spontaneous premature beats (b), resolving tachycardia (c), sustained tachycardia (d), and fibrillation (e). The green numbers in figures a–d label the s1, s2, and s3 stimuli. The blue arrows in (a) show the stimulated ventricular action potentials. (f) Numbers of hearts displaying arrhythmias (dark red and blue colors) that were unsustained (left-hand bar graph) or sustained (right-hand bar graph) in αCT1, Rev, and Veh groups following pacing. Lighter colors within bars indicate hearts within groups in which arrhythmia was not induced by pacing. (g) Graphical representation of the median severity of arrhythmia for the three treatment groups ($p < 0.02$ αCT1/Rev, $p < 0.02$ αCT1/Veh). $n \geq 11$ (mice/group). (From O'Quinn, M. P. et al. 2011. *Circ Res* 108(6): 704–715.)

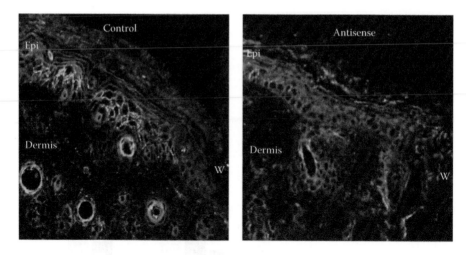

FIGURE 15.1 See the text for the figure caption.

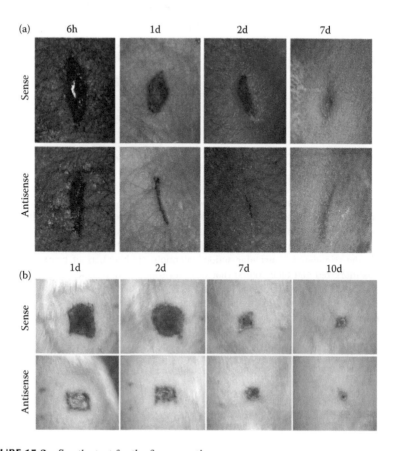

FIGURE 15.2 See the text for the figure caption.

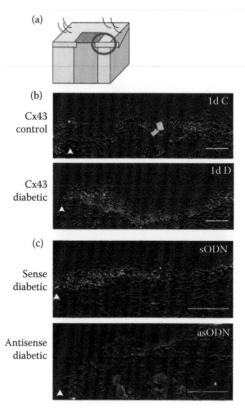

FIGURE 15.3 See the text for the figure caption.

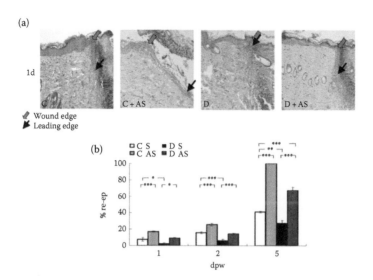

FIGURE 15.4 See the text for the figure caption.

8 Connexins in Lung Function and Inflammation

Marc Chanson and Michael Koval

CONTENTS

8.1 PATHWAYS FOR INTERCELLULAR COMMUNICATION IN LUNG

The lung is a heterogeneous organ consisting of more than 40 distinct cell types (Crapo et al. 1982). Essential to pulmonary function are two major subcompartments, the epithelial-lined airspace and the pulmonary circulation, which act in concert to promote gas exchange between the extracellular environment and bloodstream. Connective tissue composed of fibroblasts provides additional structural integrity. Recruitment of leukocytes from the circulation into the airspaces enhances host defense and is regulated by a finely tuned balance between the pulmonary endothelium, respiratory epithelium, and cells of the immune system (Doerschuk 2001). While there are several pathways that enable these distinct cellular compartments to communicate, here, we focus on roles for connexins and pannexins in facilitating lung function and inflammation.

8.1.1 Connexins

Gap junctions play several functional roles in the lung (Chatterjee et al. 2007; Foglia et al. 2009; Johnson and Koval 2009). Morphologic electron microscopic evidence from human lung suggests multiple classes of intercellular contacts containing gap junctions (Sirianni et al. 2003). This not only includes homotypic contacts between adjacent cells in epithelial and endothelial monolayers but also includes heterotypic contacts that enable cross-communication between the epithelium and endothelium. Intriguingly, heterocellular contacts in the alveolar/vascular unit frequently involve lung wall fibroblasts, suggesting a role for fibroblasts in mediating communication between subcompartments in the alveolar/vascular unit (Sirianni et al. 2003). A comparable morphologic analysis of lungs from chronic obstructive pulmonary dysplasia (COPD) patients showed that heterocellular gap junction contacts involving fibroblasts were disrupted, although whether this is a cause or effect of COPD is less clear at present (Sirianni et al. 2006).

Of the 20 mammalian connexins (Sohl and Willecke 2004), several are differentially expressed throughout the lung (Table 8.1). The conductive airway epithelium is comprised of mucus-secreting cells such as goblet cells and submucosal glands, which function together with ciliated cells as essential components of airway defense

TABLE 8.1
Connexin Expression by Lung Cells

Cell Type	Connexin Expression	References
Airway epithelium	Cx26, Cx30, Cx31, Cx32, Cx37, Cx43	Boitano et al. (1998), Huang et al. (2003a), Kojima et al. (2007), Ruch et al. (2001), Traub et al. (1998), Wiszniewski et al. (2007)
Trachea	Cx26, Cx43, Cx46	Boitano and Evans (2000), Carson et al. (1998), Isakson et al. (2006)
Alveolar epithelium		Abraham et al. (1999, 2001), Avanzo
Type II	Cx26, Cx32, Cx37, Cx43, Cx46	et al. (2007), Guo et al. (2001), Isakson
Type I	Cx26, Cx37, Cx40, Cx43, Cx46	et al. (2001a,b), Kasper et al. (1996), Lee et al. (1997), Traub et al. (1998)
Pulmonary endothelium	Cx37, Cx40, Cx43	Chatterjee et al. (2007), Parthasarathi et al. (2006), Rignault et al. (2007), Traub et al. (1998), Yeh et al. (1998)
Smooth muscle	Cx37, Cx40, Cx43	Beyer et al. (1992), Kasper et al. (1996), Nakamura et al. (1999)
Lung fibroblasts	Cx43, Cx45	Banoub et al. (1996), Kasper et al. (1996), Trovato-Salinaro et al. (2006), Zhang et al. (2004)
Neutrophils	Cx37, Cx40, Cx43	Eltzschig et al. (2006), Sarieddine et al. (2009), Zahler et al. (2003)
Monocytes/ macrophages	Cx37, Cx43	Eugenin et al. (2003), Kwak et al. (2002), Wong et al. (2006)

by promoting mucus clearance. The pattern of connexin expression by human respiratory epithelium depends on cell phenotype and stage of differentiation. While several connexin isoforms are detected in the undifferentiated airway epithelium (Table 8.1), Cx26 and Cx43 rapidly disappear upon differentiation whereas Cx30 and Cx31 remain detectable in the well-polarized epithelium. Primary cultures of well-differentiated human airway epithelial cells have also been found to express mRNAs for Cx30.3 and Cx31.1 (Wiszniewski et al. 2006, 2007).

On the contrary, the distal airspaces, which are sites of gas exchange, are lined by a thin epithelium composed of a heterogeneous mixture of type I (ATI) and II (ATII) alveolar epithelial cells. In normal lungs, ATI and ATII alveolar epithelial cells express Cx26, Cx32, Cx43, and Cx46. Cx43 is fairly ubiquitous and is the major connexin functionally interconnecting ATI and ATII cells (Abraham et al. 2001). In contrast, Cx32 is expressed exclusively by ATII alveolar epithelial cells in normal adult rat lung; however, ATI cells are unable to form functional gap junctions with Cx32. Although Cx32 has the potential to interconnect two ATII cells, ATII cells are seldom in direct contact, suggesting an alternative role for Cx32 in ATII cell function. For example, Cx32 may function by forming high-conductance hemichannels (De Vuyst et al. 2006). Cx32 has the potential to regulate lung epithelial cell growth since Cx32-deficient mice are more susceptible to benzene-induced lung toxicity and have a higher incidence of lung tumors (King et al. 2005; King and Lampe 2004; Yoon et al. 2004). A role of Cx32 in modulating acute lung inflammation is supported by the extensive and long-lasting response to intratracheal instillation of lipopolysaccharide (LPS) in Cx32-deficient mice (Marc Chanson, unpublished results).

A recent analysis of Cx43-deficient mice suggests a crucial role for Cx43 in lung development (Nagata et al. 2009). Lungs of Cx43-deficient mice appear to be arrested in the pseudoglandular or canalicular stage of development, are poorly branched, and are enriched for type II cells relative to type I cells. Although Cx43-deficient fetal lungs are morphologically distinct from normal fetal lungs, the mechanism by which Cx43 regulates lung differentiation is not known at present. However, since endothelial cell-targeted Cx43-deficient mice have apparently normal lung development (Liao et al. 2001; Parthasarathi et al. 2006), roles for connexins are more likely to reflect proliferation and/or differentiation of epithelia or mesenchyma, as opposed to an effect on alveolarization downstream from pulmonary vessel development (Jakkula et al. 2000). In lung tissue, the pulmonary microvascular endothelium is of particular interest as margination and transmigration of inflammatory cells occur predominantly in the alveolar capillary network, in contrast to the systemic circulation (DiStasi and Ley 2009). The major vascular connexins are Cx37, Cx40, and Cx43, which are expressed to differing extents depending upon tissue type and location (Figueroa et al. 2004). These connexins, when expressed by the same cell, have been shown to form heteromeric (mixed) gap junction channels, including Cx40/Cx43, Cx37/Cx43, and Cx37/Cx40 channels. The ability of connexins to intermix fine tunes the permeability and gating characteristics of gap junction channels. For instance, an increase in the amount of Cx43 relative to Cx40 augments intercellular communication through heteromeric Cx40/Cx43 channels as assessed by dye coupling (Burt et al. 2001). Moreover, lung inflammation is critically dependent on the relative expression of Cx40 versus Cx43 in the pulmonary vasculature, as described in Section 8.3.1.

Other examples of heteromeric gap junction channels with unique permeability characteristics include the following connexin pairings: Cx37/Cx43, Cx40/Cx43, and Cx43/Cx45. Given that different classes of vascular cells express different ratios of these connexins, this suggests that intercellular communication at different points in the vascular tree is regulated through differential connexin expression (Cottrell and Burt 2005).

8.1.2 PANNEXINS

There is evidence in support of (Jiang and Gu 2005) and evidence critical of (Spray et al. 2006) connexin hemichannels playing physiologic roles. Thus, it was not surprising that a second class of proteins, pannexins, also form high conductance plasma membrane channels (Scemes et al. 2007). Pannexins have been shown to mediate calcium currents and ATP release, including in airway epithelial cells (Ransford et al. 2009), where pannexin-1 (Px1) was shown to be a critical component of the P2X7 receptor complex (Pelegrin and Surprenant 2006). P2X7 receptors are found throughout lung epithelia, particularly ATI cells (Barth et al. 2007; Chen et al. 2004), as well as alveolar macrophages (Kolliputi et al. 2010) and neutrophils (Chen et al. 2010). Moreover, P2X7 receptor activity is also required for optimum processing and secretion of several proinflammatory hormones such as IL-1β and IL-6 (Kolliputi et al. 2010).

Underscoring a role for Px1 in lung injury and inflammation, P2X7-deficient mice are resistant to cigarette smoke-induced inflammation, asthma, and pulmonary fibrosis (Lucattelli et al. 2011; Muller et al. 2011; Riteau et al. 2010). For transgenic models of inflammation and asthma, bone marrow chimeric mice were used to confirm that it was leukocyte P2X7 that was critical, since P2X7-deficient mice reconstituted with normal bone marrow had a comparable pathology as P2X7-deficient mice alone (Lucattelli et al. 2011; Muller et al. 2011; Riteau et al. 2010). In contrast, bleomycin-induced pulmonary fibrosis in mouse models were used to demonstrate that epithelial P2X7 and Px1 were critical for fibrosis to develop following injury. This is consistent with a likely role for ATP released by P2X7 receptors/Px1 channels acting as a homing signal for cells undergoing apoptosis (Chekeni et al. 2010). Underscoring the clinical significance of ATP release in human pathology, patients with idiopathic pulmonary fibrosis have significantly higher levels of ATP in bronchoalveolar lavage fluid as compared with control patients (Riteau et al. 2010).

8.2 INTRAEPITHELIAL SIGNALING IN PROTECTING THE LUNG

8.2.1 CONTROL OF MUCOCILIARY CLEARANCE

The respiratory tract is chronically exposed to environmental pathogens and pollutants. Lung epithelia provide a primary defense against pathogens, since it forms a physical barrier that isolates the airspace from other tissues. However, the sterility of pulmonary tissues is also maintained by the concerted effects of innate and acquired host defense systems, which recognize, localize, kill, and remove pathogens. The intercellular propagation of calcium waves coordinates several aspects of lung physiology, including ciliary beating, inflammatory cytokine production, surfactant production, and host defense (Figures 8.1 and 8.2). The synchronization of ciliary beating is important to clear the airways from toxicants and microorganisms by ensuring the directional flow of mucus out of lungs (Knowles and Boucher 2002; Travis et al. 2001). Two mechanisms of propagation, which are not mutually exclusive, have been proposed (Figure 8.1). One involves the transmission of inositol trisphosphate from one cell to another via gap junction channels (Boitano et al. 1992). A second mechanism is the ATP release to the extracellular space, which, in turn, stimulates purinergic receptors to mobilize intracellular calcium in surrounding cells (Homolya et al. 2000; Ransford et al. 2009). In addition to providing parallel pathways to transmit calcium transients between cells, connexins and pannexins may also regulate each other, although the mechanisms of interaction between these proteins have not yet been fully elucidated (Cotrina et al. 1998).

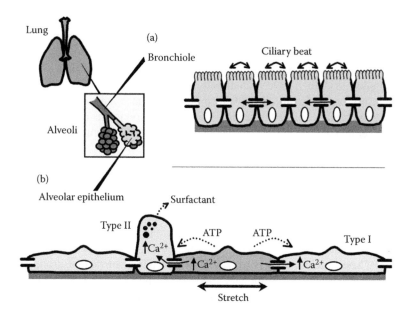

FIGURE 8.1 Intercellular communication in the lung. The lung consists of several distinct functional compartments. Shown in the inset are the terminal airspaces, alveoli, and bronchioles. (a) In the airways, including bronchioles, diffusion of IP3 through gap junctions enables the propagation of calcium waves, which help synchronize ciliary beating to allow directional transport of mucus. (b) The alveolar epithelium is a heterogeneous monolayer consisting of type II (AT2) cells and type I (AT1) cells. The alveolus acts as an integrated system where ATI cells respond to mechanical stimulation with an increase in intracellular calcium, which, in turn, is transmitted to AT2 cells via gap junctions to induce lamellar body fusion and secretion of pulmonary surfactant. Also shown is the alternative pathway, mediated by ATP secretion and paracrine stimulation via purinergic receptors.

8.2.2 CONNEXINS AND TISSUE BARRIER FUNCTION

Lung barrier function is controlled by tight junctions (Koval 2006; Matter et al. 2005; Schneeberger and Lynch 2004; Turner 2006; Van Itallie and Anderson 2006). Although connexins themselves do not provide the physical paracellular barrier, there is increasing evidence for a role of connexins in optimizing tight junctions. For instance, Cx26 was found to contribute to tight junction formation in airway epithelial cells and to interact with occludin (Kojima et al. 2007; Nusrat et al. 2000). Also, the gap junction inhibitors glycyrrhetinic acid and oleamide have been found to decrease barrier function of rat lung endothelial cells (Nagasawa et al. 2006).

In endothelium, Cx40 and Cx43 have been shown to bind to several tight junction proteins, including occludin, claudin-5, and ZO-1, based on coimmunoprecipitation analysis (Nagasawa et al. 2006). Moreover, expression of transfected Cx32 induced an ~25% increase in the barrier function of immortalized hepatocytes derived from Cx32-deficient mice (Kojima et al. 2002). In this system, Cx32 expression enhanced the localization of ZO-1 and JAM-A to the plasma membrane, suggesting an increase in tight junction formation. Since Cx40 and Cx43 interact with ZO-1 (Giepmans 2006), a protein that also directly interacts with claudins and occludin (Denker and Nigam 1998), these connexins are likely to interact with tight junctions via ZO-1. In contrast, while Cx32 can also coimmunoprecipitate with tight junction proteins (Kojima et al. 2001), it does not directly bind to ZO-1. Instead, interactions between Cx32 and tight junctions may be mediated by another scaffold protein, such as Discs Large homolog 1 (Dlgh1), which can directly bind to Cx32 (Duffy et al. 2007). Since several classes of connexin-deficient mice appear to have sufficient tissue barrier function, these studies suggest potential roles for most connexins in fine tuning tight junction assembly and function, rather than as indispensable components.

8.2.3 PULMONARY SURFACTANT SECRETION

Gap junctions also contribute to the mechanical regulation of surfactant secretion. The alveolus functions as an integrated system in which ATI cells act as mechanical sensors that transmit calcium transients to ATII cells via gap junctions (Ichimura et al. 2006; Isakson et al. 2003) and purinergic receptors (Patel et al. 2005) during breathing. These calcium signals may also help to adjust surfactant production to stimuli such as changes in pulmonary blood pressure (Ashino et al. 2000; Wang et al. 2002). Furthermore, intercellular communication also enables calcium wave propagation from one alveolus to the next (Ichimura et al. 2006) as demonstrated by *in situ* fluorescence microscopy analysis of the intact lung. One implication of interalveolar calcium signaling is that partial lung collapse (atelectasis) or fibrosis can have downstream effects on surfactant secretion by neighboring alveoli, which are otherwise apparently normal by morphologic criteria.

The major protein constituents of pulmonary surfactant are surfactant protein-A (SP-A), SP-B, SP-C, and SP-D. SP-B and SP-C are hydrophobic and directly contribute to the biophysical properties of surfactant (Beers and Mulugeta 2005; Whitsett and Weaver 2002). In contrast, SP-A and SP-D are members of the collectin protein family and are largely hydrophilic (Crouch and Wright 2001; Hawgood and Poulain

2001). SP-A and SP-D play key roles in regulating lung inflammation, which can have downstream effects on oxidant stress and alveolar damage (Wright 2005). Consistent with an immunoregulatory role for SP-A and SP-D, these proteins are required for efficient clearance of bacterial infections (Giannoni et al. 2006). Both SP-A and SP-D have carbohydrate recognition domains that recognize bacterial polysaccharides while the collagenous stalk region of the proteins bind to neutrophils, macrophages, and type II cells (Hartshorn et al. 1998; O'Riordan et al. 1995; Sano et al. 1998). SP-A and SP-D also bind viruses (Hartshorn et al. 2000; LeVine et al. 1999). Thus, the collectin surfactant proteins act as coreceptors, or opsinizing agents, by coating pathogens and enabling them to be recognized and destroyed by the innate immune system in the lung (LeVine et al. 2000; Wert et al. 2000). A net decrease in gap junctional communication will have a negative impact on surfactant and lamellar body secretion (Ichimura et al. 2006; Parthasarathi et al. 2006). While a complete lack of surfactant production and secretion is clearly deleterious, an imbalance in the regulation of surfactant secretion can also compromise lung function. Whether this is due to decreased intercellular communication in the alveolus during injury or infection remains to be determined. However, a role for disrupted cell–cell communication in regulating surfactant release is potentially underscored by surfactant abnormalities associated with diseases such as acute respiratory distress syndrome (ARDS) and ventilator-induced injury, where cell–cell contacts are disrupted.

8.3 ACUTE LUNG INJURY, CALCIUM SIGNALING, AND INFLAMMATION

Calcium signaling within lung tissue is also important in the pathophysiology of the acute inflammatory response. Acute lung injury (ALI) and ARDS are characterized by massive invasion of cells and formation of protein-rich edema fluid in the interstitial and intra-alveolar spaces. These alterations develop rapidly over the entire lung, thus reflecting the rapid spread of inflammation (Ware and Matthay 2000).

During the acute phase of lung injury, connexin expression in the alveolus is altered, where Cx43 and Cx46 expression is elevated (Abraham et al. 2001; Kasper et al. 1996). Conversely, Cx40 expression at the whole-lung level decreases during the acute phase of injury (Rignault et al. 2007). Some Cx46-expressing alveolar epithelial cells do not express typical ATII cell markers and thus may represent a distinct subtype of cells proliferating in response to injury. Cx46 has relatively limited permeability, as compared with Cx32 and Cx43, suggesting a possible role for Cx46 in restricting metabolic depletion or intercellular transmission of toxic agents (Koval 2002).

8.3.1 CONTROL OF NEUTROPHIL RECRUITMENT

Propagation of calcium waves has been directly observed in the pulmonary circulation by imaging the intact perfused lung (Ying et al. 1996). The requirement for Cx43 in pulmonary endothelial calcium waves was confirmed by using an endothelium-specific Cx43-deficient mouse model in which these waves were no longer present (Parthasarathi et al. 2006). These calcium waves were generated from a distinct subset of pacemaker endothelial cells that are localized at the pulmonary branch points and that are particularly sensitive to mechanical stress. Interestingly, calcium waves induced by mechanical stimulation have been shown to increase pulmonary endothelial P-selectin expression at the cell surface, suggesting a link to the inflammatory response (Parthasarathi et al. 2006).

Given the role of Cx43-mediated calcium waves in promoting exocytosis of P-selectin in postcapillary venules, Cx43 has been proposed to provide a proinflammatory signaling pathway contributing to the spatial expansion of inflammation (Figure 8.2). Consistent with this model, specific Cx43 blocking peptides reduced adhesion of neutrophils to the surface of mouse pneumocyte and endothelial cell lines *in vitro* (Sarieddine et al. 2009). The proinflammatory role of Cx43 was further confirmed *in vivo* using Cx43$^{+/-}$ mice, since lungs from these mice have half the Cx43 protein content than wild-type mice. In contrast with control mice, Cx43$^{+/-}$ mice showed nearly 50% fewer neutrophils recruited to the alveolar space 24 h after induction of lung inflammation by intratracheal instillation of *Pseudomonas aeruginosa* LPS. Conversely, mice expressing a truncated form of Cx43 lacking most of the C-terminus (Cx43^{K258stop}) had increased neutrophil recruitment in response to instillation of LPS (Sarieddine et al. 2009). Since gap junctions containing Cx43^{K258stop} have increased open probability and decreased gating (Moreno et al. 2002), this suggests a role for channel function in neutrophil recruitment rather than a channel-independent role for Cx43, such as has been observed in studies linking Cx43 to directed cell migration (Jiang and Gu 2005).

In contrast to Cx43, which increases, lung Cx40 content has been shown to decrease in mouse and rabbit models of ALI (Rignault et al. 2007; Zhang et al. 2010). This phenotype was associated with an overload of intracellular calcium in endothelial cells, suggesting the potential for an interruption of calcium wave propagation. Consistent with a role for Cx40 in modulating inflammation, mice with an endothelial-specific Cx40 deletion showed increased neutrophil recruitment to the alveolar space in response to intratracheal LPS instillation (Chadjichristos et al. 2010).

Interestingly, pulmonary microvessels of mice lacking endothelial Cx40 also showed decreased expression of 5'-ecto-nucleotidase (CD73), an ectoenzyme that hydrolyzes extracellular nucleotides, for example, converts ATP to adenosine. There is a considerable body of evidence demonstrating that CD73 protects against ALI by preventing leukocyte adhesion to the endothelium (Eckle et al. 2007; Thompson et al. 2004; Völmer et al. 2006). Mechanistically, adenosine generated by CD73 stimulates A$_{2B}$ receptors to induce a cAMP-mediated signaling pathway, which increases transport of CD73 to the plasma membrane, creating a positive feedback loop that enhances the effect (Figure 8.2). This pathway was confirmed using siRNA to decrease Cx40 expression by mouse endothelial bEnd.3 cells, which also decreased the expression and activity of CD73 and enhanced neutrophil adhesion. Importantly, this effect was confirmed in *in vivo* transgenic mice deficient of endothelial Cx40 (Chadjichristos et al. 2010).

It is interesting to note the opposite regulation of Cx40 and Cx43 during the course of an inflammatory response in lungs (Figure 8.2). Whereas a high ratio for Cx40 (Cx40 > Cx43) delays the adhesion of neutrophils to the endothelium, a high ratio for Cx43 (Cx43 > Cx40) promotes their transmigration across the endothelial barrier. Given that gap junctional communication is critically dependent on the stoichiometry of heteromeric Cx40/Cx43 channels (Burt et al. 2001), this provides one of the best examples where acutely altered intercellular communication through gap

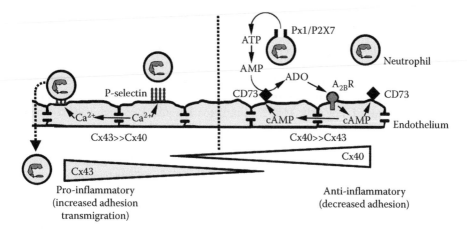

FIGURE 8.2 Potential roles of connexins in regulating neutrophil adhesion to the endothelium. As an initial stimulus, circulating neutrophils release ATP via hemichannels most likely composed of Px1. ATP is converted to adenosine (ADO) by the successive activity of CD39 (located at the surface of neutrophils and endothelial cells) and CD73 (located at the surface of endothelial cells). The conversion rate favors the production of ADO, which activates adenosine receptors ($A_{2B}R$) on the endothelium. $A_{2B}R$ triggers a cAMP-dependent signal cascade, which further activates CD73 and inhibits expression of endothelial proteins required for leukocyte adhesion. $A_{2B}R$ engagement also enhances Cx40-mediated gap junctional communication, which propagates intercellular transfer of cAMP to propagate an anti-inflammatory response. In contrast, in the presence of a proinflammatory stimulus, expression of Cx40 decreases while Cx43 increases. Increased Cx43 and decreased Cx40 expression by endothelium alters gap junction permeability and may facilitate transmission of calcium transients over cAMP. The calcium-dependent signals propagate through gap junctions, which, in turn, increase expression of endothelial cell adhesion such as P-selectin, which promotes neutrophil adhesion, thus providing a proinflammatory response.

junction channels has the potential to influence a physiologic response, in this case, modulation of neutrophil–endothelial interactions.

Alternatively, specific connexins may also provide for a bridge between inflammatory circulating cells and inflamed tissue cells, where Cx43 could promote neutrophil binding to endothelial cells, whereas Cx40 could be inhibitory. Although whether connexins are expressed by neutrophils is controversial (Sarieddine et al. 2009; Scerri et al. 2006), it was recently reported that neutrophils may release ATP via hemichannels made of Cx43 or Px1, which can contribute to a proinflammatory response (Eltzschig et al. 2006; Chen et al. 2010).

8.3.2 Airway Epithelial Inflammation

Recent findings highlight the importance of calcium waves within the conducting airway epithelium to activate the innate immune defense in response to inhaled bacteria, such as the opportunistic *Pseudomonas aeruginosa*. For instance, activation of Toll-like receptor 2 (TLR-2) on the apical surface of airway epithelial cell lines by *Pseudomonas* triggers an immediate calcium flux through gap junctions that is necessary and sufficient to stimulate NF-κB and MAP kinase signaling (Martin and Prince 2008). Both NF-κB and MAP kinases contribute to the synthesis and release of proinflammatory cytokines, including IL-8, which is a potent chemotactic factor for neutrophils. Calcium waves transmitted through Cx43 were shown to transiently amplify the proinflammatory response by recruitment of adjacent but nonstimulated cells (Martin and Prince 2008). This amplification, however, is limited in time. Indeed, not only TLR-2 stimulation but also LPS and TNF-α markedly decreased gap junctional communication within 30–60 min, an effect mediated by c-src-dependent tyrosine phosphorylation of Cx43, which decreases intercellular communication (Chanson et al. 2001; Huang et al. 2003a; Martin and Prince 2008).

A defect in negative regulation of Cx43 channels may contribute to an exaggerated inflammatory response. For example, cystic fibrosis (CF), a genetic disease caused by mutations in the CF transmembrane conductance regulator (*CFTR*) gene, is frequently associated with excessive neutrophil recruitment to the airways. Interestingly, airway epithelial cells isolated from CF patients fail to downregulate gap junctional communication likely due to several mechanisms, including inadequate c-src recruitment to membrane microdomains, or lipid rafts, in the absence of functional CFTR (Chanson et al. 2001; Dudez et al. 2008; Huang et al. 2003b). Although several factors are likely to contribute to the proinflammatory pathology of CF in lungs, these studies identify misregulation of Cx43 as one of these factors.

Although most studies examining roles for connexins in lung injury and inflammation have focused on Cx40 and Cx43, other connexins are also likely to play roles as well. For instance, using an ovalbumin-induced model for airway hypersensitivity-induced asthma, airway epithelial Cx37 expression was found to be reduced, which correlated with the inflammatory response and Th2 cytokine production (Park et al. 2007). Although this study did not characterize changes in other connexins, these findings are intriguing, given that Cx37 heteromerically interacts with Cx43 to attenuate gap junctional communication (Brink et al. 1997; Larson et al. 2000). Moreover, tyrosine kinases have been implicated in the control of the hypersensitivity response (Kamata et al. 2003). This raises the potential for gap junctional communication to be simultaneously modulated by multiple mechanisms to control inflammation, namely increased Cx43 phosphorylation and increased formation of Cx37/Cx43 heteromers. Further work is needed to determine whether this is the case.

8.4   ALCOHOLIC LUNG SYNDROME

Although alcohol abuse is classically associated with liver disease (Reuben 2007), recent evidence has confirmed that chronic alcohol abuse is also a major risk factor contributing to the severity of ARDS (Joshi and Guidot 2007; Moss et al. 1996, 2003). In a study of ICU patients, it was found that after adjusting for smoking and hepatic dysfunction, patients with a history of alcohol abuse were more than twice as susceptible for ARDS than nonalcoholic patients (Moss et al. 2003). In large part, the increased susceptibility to ARDS caused by chronic ethanol ingestion is due to a fundamental defect in lung barrier function as a result of impaired tight junction formation between ATI cells (Fernandez et al. 2007; Guidot et al. 2000). As described above, decreased barrier function (e.g., a leaky lung) contributes to the severity of ARDS (Ware and Matthay 2000).

8.4.1 Oxidant Stress and Connexins in Alcoholic Lung

Dietary ethanol causes oxidant stress in the lung (Moss et al. 2000). The lung is particularly sensitive to oxidant stress, which is minimized by the antioxidant gluta-thione in the alveolar airspaces (Jean et al. 2002). Thus, one source of oxidant stress is from ethanol metabolism to acetaldehyde, which directly depletes the reduced glu-tathione pool (Brown et al. 2004; Moss et al. 2000). The prominent role of oxidant stress and reactive oxygen species (ROS) in alcoholic lung suggests that antioxidant therapy could be a useful therapeutic approach. In fact, in animal models of alco-hol ingestion, a diet enriched in the glutathione precursor procystine prevents the alcoholic lung phenotype (Brown et al. 2001; Guidot and Brown 2000). However, complete reversal of the alcoholic lung phenotype requires several weeks of treat-ment and is not a suitable approach for an acute treatment regimen for alcoholic lung.

Three different studies have demonstrated that ethanol treatment of cells *in vitro* inhibits gap junctional communication (Abou Hashieh et al. 1996; Bokkala et al. 2001; Wentlandt et al. 2004). The ability of ethanol to inhibit gap junctions could be due to direct partitioning into cell membranes, analogous to the inhibitory effect of long-chain alcohols on connexins (Johnston et al. 1980; Chanson et al. 1989). Ethanol-induced depletion of the glutathione pool and increased oxidant stress can also inhibit gap junctional communication (Upham et al. 1997). The effect of etha-nol on connexin expression is more variable, where ethanol was shown to inhibit Cx43 expression (Bokkala et al. 2001; Wentlandt et al. 2004), but had little effect on Cx26 or Cx32 (Abou Hashieh et al. 1996; Wentlandt et al. 2004). Whether ethanol has a comparable effect on other connexins or cell types remains to be determined. In fact, specific alterations in gap junctional communication in response to oxidant stress may be a mechanism to decrease the intercellular transmission of toxic agents (Azzam et al. 2001; Elshami et al. 1996), while also maintaining intercellular trans-fer of antioxidant compounds, including glutathione (Frossard et al. 2003; Goldberg et al. 2004). Alternatively, complete shutdown of gap junctional communication can help preserve the bulk of the tissue at the expense of more extensive damage to iso-lated individual cells.

8.4.2 Hormone Signaling and Connexins in Alcoholic Lung

In addition to its metabolic effects on the antioxidant glutathione pool, ethanol also induces cell signaling pathways that contribute to oxidant stress. In particular, etha-nol stimulates angiotensin II activity (Bechara et al. 2005), which, in turn, upregu-lates NADPH oxidase (Nox) (Seshiah et al. 2002). Interestingly, angiotensin II has also been shown to upregulate cardiovascular and epithelial Cx43 expression and function (Bokkala et al. 2001; Dhein et al. 2002; Kansui et al. 2004; Kasi et al. 2007) and can antagonize the effect of ethanol on gap junctional communication (Bokkala et al. 2001). In contrast, Cx40 appears to be less affected by angiotensin II (Dhein et al. 2002). Given the role of endothelial Cx43 in inflammation, this is consistent with the notion that angiotensin II is proinflammatory as well (Boos and Lip 2006).

Clearly, inflammation and the concomitant infiltration of neutrophils and acti-vation of alveolar macrophages contribute to oxidant stress in response to ALI

(Christofidou-Solomidou and Muzykantov 2006). Also, the intense oxidant load on alcoholic lung provides a condition where the alveolar epithelium is prone to injury and apoptosis (Brown et al. 2001). This has influences on alveolar epithelial function by promoting the cells to undergo an epithelial-to-mesenchyme transition (EMT) (Kasai et al. 2005; Kim et al. 2006; Willis and Borok 2007). In essence, the alcoholic lung is primed to have an exaggerated response to the effects of subsequent insults (Araya et al. 2006; Kang et al. 2007).

In addition to impairing epithelial cell phenotype and compromising alveolar barrier function, EMT has been shown to decrease expression of Cx43 by embryonic carcinoma cells downstream of increased Snail expression and decreased cadherin expression (de Boer et al. 2007). Whether this is the case for alveolar epithelial cell junctions remains to be determined. In addition to ROS, reactive nitrogen species, including peroxynitrite, are also generated during ALI (Haddad et al. 1994; Sittipunt et al. 2001), which can inhibit gap junctional communication (Sharov et al. 1999).

8.5 CONCLUSION

By regulating extracellular and intercellular signaling, controlling the flow of metabolites and restricting the flow of toxic agents, connexins, and pannexins enable the cells of the lung to act as integrated systems. Although to date no human respiratory disease has been directly attributable to a connexin deficiency or mutation, there is considerable evidence from animal and *in vitro* models that connexins have the capacity to control lung inflammation. However, targeting connexins as a therapeutic approach to control lung injury and inflammation will pose several challenges. For instance, it seems unlikely that targeting connexins will prove beneficial in the advanced stages of a lung disease, such as ARDS or COPD. Tissue-specific heterogeneity in connexin expression adds an additional complication, as does the differing roles for connexins in gap junctional intercellular communication as opposed to connexin- and pannexin-mediated hemichannel activity. Moreover, approaches using gap junction inhibitors run the risk of having an adverse effect on processes where connexins have anti-inflammatory effects (such as vascular Cx40) or promote lung function, where connexins increase the ability of cells to secrete pulmonary surfactant in response to mechanical stress. In fact, the ability of Px1-deficient mice to resist lung injury and inflammation suggests that inhibitors of hemichannels may be the best therapeutic target as opposed to more global gap junction inhibitors. However, inhibition of inflammation may have deleterious consequences and connexin inhibitors may be more valuable in attenuating the inflammatory response.

Thus, it may prove that identifying pharmacologic agents that promote specific subclasses of gap junctional communication will provide a more fruitful approach to target inflammation. To be successful, this strategy will require targeting specific connexin subclasses or connexins expressed by a specific subset of cells, perhaps in combination with a hemichannel inhibitor. Continuing to identify specific roles for specific connexins and pannexins will provide an important foundation to determine whether targeting these intercellular communication pathways represents a feasible approach to the treatment of inflammatory lung disease.

ACKNOWLEDGMENTS

This work was supported by grants from the Swiss National Science Foundation (310000–119739) and Vaincre la Mucoviscidose (to MC), and the National Institutes of Health grants HL-083120 and AA-013757 (to MK).

REFERENCES

Abou Hashieh, I., S. Mathieu, F. Besson et al. 1996. Inhibition of gap junction intercellular communications of cultured rat hepatocytes by ethanol: Role of ethanol metabolism. *J Hepatol.* 24:360–7.

Abraham, V., M.L. Chou, K.M. DeBolt, M. Koval. 1999. Phenotypic control of gap junctional communication by cultured alveolar epithelial cells. *Am J Physiol.* 276:L825–34.

Abraham, V., M.L. Chou, P. George et al. 2001. Heterocellular gap junctional communication between alveolar epithelial cells. *Am J Physiol Lung Cell Mol Physiol.* 280:L1085–93.

Araya, J., S. Cambier, A. Morris et al. 2006. Integrin-mediated transforming growth factor-beta activation regulates homeostasis of the pulmonary epithelial-mesenchymal trophic unit. *Am J Pathol.* 169:405–15.

Ashino, Y., X. Ying, L.G. Dobbs et al. 2000. $[Ca^{2+}]_i$ oscillations regulate type II cell exocytosis in the pulmonary alveolus. *Am J Physiol.* 279:L5–13.

Avanzo, J.L., G. Mennecier, M. Mesnil, F.J. et al. 2007. Deletion of a single allele of Cx43 is associated with a reduction in the gap junctional intercellular communication and increased cell proliferation of mouse lung pneumocytes type II. *Cell Prolif.* 40:411–21.

Azzam, E.I., S.M. de Toledo, and J.B. Little. 2001. Direct evidence for the participation of gap junction-mediated intercellular communication in the transmission of damage signals from alpha-particle irradiated to nonirradiated cells. *Proc Natl Acad Sci USA.* 98:473–8.

Banoub, R.W., M. Fernstrom, and R.J. Ruch. 1996. Lack of growth inhibition or enhancement of gap junctional intercellular communication and connexin43 expression by beta-carotene in murine lung epithelial cells *in vitro. Cancer Lett.* 108:35–40.

Barth, K., K. Weinhold, A. Guenther et al. 2007. Caveolin-1 influences P2X7 receptor expression and localization in mouse lung alveolar epithelial cells. *FEBS J.* 274:3021–33.

Bechara, R.I., A. Pelaez, A. Palacio et al. 2005. Angiotensin II mediates glutathione depletion, transforming growth factor-beta1 expression, and epithelial barrier dysfunction in the alcoholic rat lung. *Am J Physiol Lung Cell Mol Physiol.* 289:L363–70.

Beers, M.F., and S. Mulugeta. 2005. Surfactant protein C biosynthesis and its emerging role in conformational lung disease. *Annu Rev Physiol.* 67:663–96.

Beyer, E.C., K.E. Reed, E.M. Westphale et al. 1992. Molecular cloning and expression of rat connexin40, a gap junction protein expressed in vascular smooth muscle. *J Membr Biol.* 127:69–76.

Boitano, S., E.R. Dirksen, and W.H. Evans. 1998. Sequence-specific antibodies to connexins block intercellular calcium signaling through gap junctions. *Cell Calcium.* 23:1–9.

Boitano, S., E.R. Dirksen, and M.J. Sandersen. 1992. Intracellular propagation of calcium waves mediated by inositol trisphosphate. *Science.* 258:292–5.

Boitano, S., and W.H. Evans. 2000. Connexin mimetic peptides reversibly inhibit Ca(2+) signaling through gap junctions in airway cells. *Am J Physiol Lung Cell Mol Physiol.* 279:L623–30.

Bokkala, S., H.M. Reis, E. Rubin et al. 2001. Effect of angiotensin II and ethanol on the expression of connexin 43 in WB rat liver epithelial cells. *Biochem J.* 357:769–77.

Boos, C.J., and G.Y. Lip. 2006. Is hypertension an inflammatory process? *Curr Pharm Des.* 12:1623–35.

Brink, P.R., K. Cronin, K. Banach et al. 1997. Evidence for heteromeric gap junction channels formed from rat connexin43 and human connexin37. *Am J Physiol.* 273(Pt 1):C1386–96.

Brown, L.A., F.L. Harris, and D.M. Guidot. 2001. Chronic ethanol ingestion potentiates TNF-alpha-mediated oxidative stress and apoptosis in rat type II cells. *Am J Physiol Lung Cell Mol Physiol.* 281:L377–86.

Brown, L.A., F.L. Harris, X.D. Ping et al. 2004. Chronic ethanol ingestion and the risk of acute lung injury: A role for glutathione availability? *Alcohol.* 33:191–7.

Burt, J.M., A.M. Fletcher, T.D. Steele et al. 2001. Alteration of Cx43:Cx40 expression ratio in A7r5 cells. *Am J Physiol Cell Physiol.* 280:C500–8.

Carson, J.L., W. Reed, B.M. Moats-Staats et al. 1998. Connexin 26 expression in human and ferret airways and lung during development. *Am J Respir Cell Mol Biol.* 18:111–9.

Chadjichristos, C.E., K.E. Scheckenbach, T.A. van Veen et al. 2010. Endothelial-specific deletion of connexin40 promotes atherosclerosis by increasing CD73-dependent leukocyte adhesion. *Circulation.* 121:123–31.

Chanson, M., P.Y. Berclaz, I. Scerri et al. 2001. Regulation of gap junctional communication by a pro-inflammatory cytokine in cystic fibrosis transmembrane conductance regulator-expressing but not cystic fibrosis airway cells. *Am J Pathol.* 158:1775–84.

Chanson, M., R. Bruzzone, D. Bosco et al. 1989. Effects of *n*-alcohols on junctional coupling and amylase secretion of pancreatic acinar cells. *J Cell Physiol.* 139:147–56.

Chatterjee, S., S. Baeter, and J. Bhattacharya. 2007. Endothelial and epithelial signaling in the lung. *Am J Physiol Lung Cell Mol Physiol.* 293:L517–9.

Chekeni, F.B., M.R. Elliott, J.K. Sandilos et al. 2010. Pannexin 1 channels mediate "find-me" signal release and membrane permeability during apoptosis. *Nature.* 467:863–7.

Chen, Y., Y. Yao, Y. Sumi et al. 2010. Purinergic signaling: A fundamental mechanism in neutrophil activation. *Sci Signal.* 3:ra45.

Chen, Z., N. Jin, T. Narasaraju et al. 2004. Identification of two novel markers for alveolar epithelial type I and II cells. *Biochem Biophys Res Commun.* 319:774–80.

Christofidou-Solomidou, M., and V.R. Muzykantov. 2006. Antioxidant strategies in respiratory medicine. *Treat Respir Med.* 5:47–78.

Cotrina, M.L., J.H. Lin, A. Alves-Rodrigues et al. 1998. Connexins regulate calcium signaling by controlling ATP release. *Proc Natl Acad Sci USA.* 95:15735–40.

Cottrell, G.T., and J.M. Burt. 2005. Functional consequences of heterogeneous gap junction channel formation and its influence in health and disease. *Biochim Biophys Acta.* 1711:126–41.

Crapo, J.D., B.E. Barry, P. Gehr et al. 1982. Cell number and cell characteristics of the normal human lung. *Am Rev Respir Dis.* 126:332–7.

Crouch, E., and J.R. Wright. 2001. Surfactant proteins a and d and pulmonary host defense. *Annu Rev Physiol.* 63:521–54.

de Boer, T.P., T.A. van Veen, M.F. Bierhuizen et al. 2007. Connexin43 repression following epithelium-to-mesenchyme transition in embryonal carcinoma cells requires Snail1 transcription factor. *Differentiation.* 75:208–18.

De Vuyst, E., E. Decrock, L. Cabooter et al. 2006. Intracellular calcium changes trigger connexin 32 hemichannel opening. *EMBO J.* 25:34–44.

Denker, B.M., and S.K. Nigam. 1998. Molecular structure and assembly of the tight junction. *Am J Physiol.* 274:F1–9.

Dhein, S., L. Polontchouk, A. Salameh et al. 2002. Pharmacological modulation and differential regulation of the cardiac gap junction proteins connexin 43 and connexin 40. *Biol Cell.* 94:409–22.

DiStasi, M.R., and K. Ley. 2009. Opening the flood-gates: How neutrophil-endothelial interactions regulate permeability. *Trends Immunol.* 30:547–56.

Doerschuk, C.M. 2001. Mechanisms of leukocyte sequestration in inflamed lungs. *Microcirculation.* 8:71–88.

Dudez, T., F. Borot, S. Huang et al. 2008. CFTR in a lipid raft-TNFR1 complex modulates gap junctional intercellular communication and IL-8 secretion. *Biochim Biophys Acta.* 1783:779–88.

Duffy, H.S., I. Iacobas, K. Hotchkiss et al. 2007. The gap junction protein connexin32 interacts with the Src homology 3/hook domain of discs large homolog 1. *J Biol Chem.* 282:9789–96.

Eckle, T., L. Fullbier, M. Wehrmann et al. 2007. Identification of ectonucleotidases CD39 and CD73 in innate protection during acute lung injury. *J Immunol.* 178:8127–37.

Elshami, A.A., A. Saavedra, H. Zhang et al. 1996. Gap junctions play a role in the "bystander effect" of the herpes simplex virus thymidine kinase/ganciclovir system in vitro. *Gene Ther.* 3:85–92.

Eltzschig, H.K., T. Eckle, A. Mager et al. 2006. ATP release from activated neutrophils occurs via connexin 43 and modulates adenosine-dependent endothelial cell function. *Circ Res.* 99:1100–8.

Eugenín, E.A., M.C. Brañes, J.W. Berman, and J.C. Sáez. 2003. TNF-alpha plus IFN-gamma induce connexin43 expression and formation of gap junctions between human monocytes/macrophages that enhance physiological responses. *J Immunol.* 170:1320–8.

Fernandez, A.L., M. Koval, X. Fan et al. 2007. Chronic alcohol ingestion alters claudin expression in the alveolar epithelium of rats. *Alcohol.* 41:371–9.

Figueroa, X.F., B.E. Isakson, and B.R. Duling. 2004. Connexins: Gaps in our knowledge of vascular function. *Physiology (Bethesda).* 19:277–84.

Foglia, B., I. Scerri, T. Dudez et al. 2009. The role of connexins in the respiratory epithelium. In *Connexin: A Guide.* A.L. Harris and D. Locke, editors. Humana Press, New York, NY. 359–70.

Frossard, J.L., L. Rubbia-Brandt, M.A. Wallig et al. 2003. Severe acute pancreatitis and reduced acinar cell apoptosis in the exocrine pancreas of mice deficient for the Cx32 gene. *Gastroenterology.* 124:481–93.

Giannoni, E., T. Sawa, L. Allen et al. 2006. Surfactant proteins A and D enhance pulmonary clearance of *Pseudomonas aeruginosa. Am J Respir Cell Mol Biol.* 34:704–10.

Giepmans, B.N. 2006. Role of connexin43-interacting proteins at gap junctions. *Adv Cardiol.* 42:41–56.

Goldberg, G.S., V. Valiunas, and P.R. Brink. 2004. Selective permeability of gap junction channels. *Biochim Biophys Acta.* 1662:96–101.

Guidot, D.M., and L.A. Brown. 2000. Mitochondrial glutathione replacement restores surfactant synthesis and secretion in alveolar epithelial cells of ethanol-fed rats. *Alcohol Clin Exp Res.* 24:1070–6.

Guidot, D.M., K. Modelska, M. Lois et al. 2000. Ethanol ingestion via glutathione depletion impairs alveolar epithelial barrier function in rats. *Am J Physiol Lung Cell Mol Physiol.* 279:L127–35.

Guo, Y., C. Martinez-Williams, C.E. Yellowley et al. 2001. Connexin expression by alveolar epithelial cells is regulated by extracellular matrix. *Am J Physiol Lung Cell Mol Physiol.* 280:L191–202.

Haddad, I.Y., G. Pataki, P. Hu et al. 1994. Quantitation of nitrotyrosine levels in lung sections of patients and animals with acute lung injury. *J Clin Invest.* 94:2407–13.

Hartshorn, K.L., E. Crouch, M.R. White et al. 1998. Pulmonary surfactant proteins A and D enhance neutrophil uptake of bacteria. *Am J Physiol.* 274:L958–69.

Hartshorn, K.L., M.R. White, D.R. Voelker et al. 2000. Mechanism of binding of surfactant protein D to influenza A viruses: Importance of binding to haemagglutinin to antiviral activity. *Biochem J.* 351 Pt 2:449–58.

Hawgood, S., and F.R. Poulain. 2001. The pulmonary collectins and surfactant metabolism. *Annu Rev Physiol.* 63:495–519.

Homolya, L., T.H. Steinberg, and R.C. Boucher. 2000. Cell to cell communication in response to mechanical stress via bilateral release of ATP and UTP in polarized epithelia. *J Cell Biol.* 150:1349–60.

Huang, S., T. Dudez, I. Scerri et al. 2003a. Defective activation of c-Src in cystic fibrosis airway epithelial cells results in loss of tumor necrosis factor-alpha-induced gap junction regulation. *J Biol Chem.* 278:8326–32.

Huang, S., L. Jornot, L. Wiszniewski et al. 2003b. Src signaling links mediators of inflammation to Cx43 gap junction channels in primary and transformed CFTR-expressing airway cells. *Cell Commun Adhes.* 10:279–85.

Ichimura, H., K. Parthasarathi, J. Lindert et al. 2006. Lung surfactant secretion by interalveolar Ca^{2+} signaling. *Am J Physiol Lung Cell Mol Physiol.* 291:L596–601.

Isakson, B.E., R.L. Lubman, G.J. Seedorf, and S. Boitano. 2001a. Modulation of pulmonary alveolar type II cell phenotype and communication by extracellular matrix and KGF. *Am J Physiol Cell Physiol.* 281:C1291–9.

Isakson, B.E., W.H. Evans, and S. Boitano. 2001b. Intercellular Ca^{2+} signaling in alveolar epithelial cells through gap junctions and by extracellular ATP. *Am J Physiol Lung Cell Mol Physiol.* 280:L221–8.

Isakson, B.E., C.E. Olsen, S. Boitano. 2006. Laminin-332 alters connexin profile, dye coupling and intercellular Ca^{2+} waves in ciliated tracheal epithelial cells. *Respir Res.* 2;7:105.

Isakson, B.E., G.J. Seedorf, R.L. Lubman et al. 2003. Cell-cell communication in heterocellular cultures of alveolar epithelial cells. *Am J Respir Cell Mol Biol.* 29:552–61.

Jakkula, M., T.D. Le Cras, S. Gebb et al. 2000. Inhibition of angiogenesis decreases alveolarization in the developing rat lung. *Am J Physiol Lung Cell Mol Physiol.* 279:L600–7.

Jean, J.C., Y. Liu, L.A. Brown et al. 2002. Gamma-glutamyl transferase deficiency results in lung oxidant stress in normoxia. *Am J Physiol Lung Cell Mol Physiol.* 283:L766–76.

Jiang, J.X., and S. Gu. 2005. Gap junction- and hemichannel-independent actions of connexins. *Biochim Biophys Acta.* 1711:208–14.

Johnson, L.N., and M. Koval. 2009. Cross-talk between pulmonary injury, oxidant stress, and gap junctional communication. *Antioxid Redox Signal.* 11:355–67.

Johnston, M.F., S.A. Simon, and F. Ramon. 1980. Interaction of anaesthetics with electrical synapses. *Nature.* 286:498–500.

Joshi, P.C., and D.M. Guidot. 2007. The alcoholic lung: Epidemiology, pathophysiology, and potential therapies. *Am J Physiol Lung Cell Mol Physiol.* 292:L813–23.

Kamata, T., M. Yamashita, M. Kimura et al. 2003. src homology 2 domain-containing tyrosine phosphatase SHP-1 controls the development of allergic airway inflammation. *J Clin Invest.* 111:109–19.

Kang, H.R., S.J. Cho, C.G. Lee et al. 2007. Transforming growth factor (TGF)-beta1 stimulates pulmonary fibrosis and inflammation via a Bax-dependent, bid-activated pathway that involves matrix metalloproteinase-12. *J Biol Chem.* 282:7723–32.

Kansui, Y., K. Fujii, K. Nakamura et al. 2004. Angiotensin II receptor blockade corrects altered expression of gap junctions in vascular endothelial cells from hypertensive rats. *Am J Physiol Heart Circ Physiol.* 287:H216–24.

Kasai, H., J.T. Allen, R.M. Mason et al. 2005. TGF-beta1 induces human alveolar epithelial to mesenchymal cell transition (EMT). *Respir Res.* 6:56.

Kasi, V.S., H.D. Xiao, L.L. Shang et al. 2007. Cardiac-restricted angiotensin-converting enzyme overexpression causes conduction defects and connexin dysregulation. *Am J Physiol Heart Circ Physiol.* 293:H182–92.

Kasper, M., O. Traub, T. Reimann et al. 1996. Upregulation of gap junction protein connexin43 in alveolar epithelial cells of rats with radiation-induced pulmonary fibrosis. *Histochem Cell Biol.* 106:419–24.

Kim, K.K., M.C. Kugler, P.J. Wolters et al. 2006. Alveolar epithelial cell mesenchymal transition develops *in vivo* during pulmonary fibrosis and is regulated by the extracellular matrix. *Proc Natl Acad Sci USA.* 103:13180–5.

King, T.J., K.E. Gurley, J. Prunty et al. 2005. Deficiency in the gap junction protein connexin32 alters p27Kip1 tumor suppression and MAPK activation in a tissue-specific manner. *Oncogene.* 24:1718–26.

King, T.J., and P.D. Lampe. 2004. The gap junction protein connexin32 is a mouse lung tumor suppressor. *Cancer Res.* 64:7191–6.

Knowles, M.R., and R.C. Boucher. 2002. Mucus clearance as a primary innate defense mechanism for mammalian airways. *J Clin Invest.* 109:571–7.

Kojima, T., Y. Kokai, H. Chiba et al. 2001. Cx32 but not Cx26 is associated with tight junctions in primary cultures of rat hepatocytes. *Exp Cell Res.* 263:193–201.

Kojima, T., M. Murata, M. Go et al. 2007. Connexins induce and maintain tight junctions in epithelial cells. *J Membr Biol.* 217:13–9.

Kojima, T., D.C. Spray, Y. Kokai et al. 2002. Cx32 formation and/or Cx32-mediated intercellular communication induces expression and function of tight junctions in hepatocytic cell line. *Exp Cell Res.* 276:40–51.

Kolliputi, N., R.S. Shaik, and A.B. Waxman. 2010. The inflammasome mediates hyperoxia-induced alveolar cell permeability. *J Immunol.* 184:5819–26.

Koval, M. 2002. Sharing signals: Connecting lung epithelial cells with gap junction channels. *Am J Physiol Lung Cell Mol Physiol.* 283:L875–93.

Koval, M. 2006. Claudins—Key pieces in the tight junction puzzle. *Cell Commun Adhes.* 13:127–38.

Kwak, B.R., F. Mulhaupt, N. Veillard et al. 2002. Altered pattern of vascular connexin expression in atherosclerotic plaques. *Arterioscler Thromb Vasc Biol.* 22:225–30.

Larson, D.M., K.H. Seul, V.M. Berthoud et al. 2000. Functional expression and biochemical characterization of an epitope- tagged connexin37. *Mol Cell Biol Res Commun.* 3:115–21.

Lee, Y.C., C.E. Yellowley, Z. Li, H.J. Donahue, and D.E. Rannels. Expression of functional gap junctions in cultured pulmonary alveolar epithelial cells. *Am J Physiol.* 272:L1105–14.

LeVine, A.M., J. Gwozdz, J. Stark et al. 1999. Surfactant protein-A enhances respiratory syncytial virus clearance in vivo. *J Clin Invest.* 103:1015–21.

LeVine, A.M., J.A. Whitsett, J.A. Gwozdz et al. 2000. Distinct effects of surfactant protein A or D deficiency during bacterial infection on the lung. *J Immunol.* 165:3934–40.

Liao, Y., K.H. Day, D.N. Damon et al. 2001. Endothelial cell-specific knockout of connexin 43 causes hypotension and bradycardia in mice. *Proc Natl Acad Sci USA.* 98:9989–94.

Lucattelli, M., S. Cicko, T. Muller et al. 2011. P2X7 receptor signaling in the pathogenesis of smoke-induced lung inflammation and emphysema. *Am J Respir Cell Mol Biol.* 44:423–9.

Martin, F.J., and A.S. Prince. 2008. TLR2 regulates gap junction intercellular communication in airway cells. *J Immunol.* 180:4986–93.

Matter, K., S. Aijaz, A. Tsapara et al. 2005. Mammalian tight junctions in the regulation of epithelial differentiation and proliferation. *Curr Opin Cell Biol.* 17:453–8.

Moreno, A.P., M. Chanson, S. Elenes et al. 2002. Role of the carboxyl terminal of connexin43 in transjunctional fast voltage gating. *Circ Res.* 90:450–7.

Moss, M., B. Bucher, F.A. Moore et al. 1996. The role of chronic alcohol abuse in the development of acute respiratory distress syndrome in adults. *JAMA.* 275:50–4.

Moss, M., D.M. Guidot, M. Wong-Lambertina et al. 2000. The effects of chronic alcohol abuse on pulmonary glutathione homeostasis. *Am J Respir Crit Care Med.* 161:414–9.

Moss, M., P.E. Parsons, K.P. Steinberg et al. 2003. Chronic alcohol abuse is associated with an increased incidence of acute respiratory distress syndrome and severity of multiple organ dysfunction in patients with septic shock. *Crit Care Med.* 31:869–77.

Muller, T., R. Paula Vieira, M. Grimm et al. 2011. A potential role for P2X7R in allergic airway inflammation in mice and humans. *Am J Respir Cell Mol Biol.* 44:456–64.

Nagasawa, K., H. Chiba, H. Fujita et al. 2006. Possible involvement of gap junctions in the barrier function of tight junctions of brain and lung endothelial cells. *J Cell Physiol.* 208h:123–32.

Nagata, K., K. Masumoto, G. Esumi et al. 2009. Connexin43 plays an important role in lung development. *J Pediatr Surg.* 44:2296–301.

Nakamura, K., T. Inai, K. Nakamura, and Y. Shibata. 1999. Distribution of gap junction protein connexin 37 in smooth muscle cells of the rat trachea and pulmonary artery. *Arch Histol Cytol.* 62:27–37.

Nusrat, A., J.A. Chen, C.S. Foley et al. 2000. The coiled-coil domain of occludin can act to organize structural and functional elements of the epithelial tight junction. *J Biol Chem.* 275:29816–22.

O'Riordan, D.M., J.E. Standing, K.Y. Kwon et al. 1995. Surfactant protein D interacts with *Pneumocystis carinii* and mediates organism adherence to alveolar macrophages. *J Clin Invest.* 95:2699–710.

Park, S.J., K.S. Lee, S.R. Kim et al. 2007. Change of connexin 37 in allergen-induced airway inflammation. *Exp Mol Med.* 39:629–40.

Parthasarathi, K., H. Ichimura, E. Monma et al. 2006. Connexin 43 mediates spread of Ca^{2+}-dependent proinflammatory responses in lung capillaries. *J Clin Invest.* 116:2193–200.

Patel, A.S., D. Reigada, C.H. Mitchell et al. 2005. Paracrine stimulation of surfactant secretion by extracellular ATP in response to mechanical deformation. *Am J Physiol Lung Cell Mol Physiol.* 289:L489–96.

Pelegrin, P., and A. Surprenant. 2006. Pannexin-1 mediates large pore formation and interleukin-1beta release by the ATP-gated P2X7 receptor. *EMBO J.* 25:5071–82.

Ransford, G.A., N. Fregien, F. Qiu et al. 2009. Pannexin 1 contributes to ATP release in airway epithelia. *Am J Respir Cell Mol Biol.* 41:525–34.

Reuben, A. 2007. Alcohol and the liver. *Curr Opin Gastroenterol.* 23:283–91.

Rignault, S., J.A. Haefliger, B. Waeber et al. 2007. Acute inflammation decreases the expression of connexin 40 in mouse lung. *Shock.* 28:78–85.

Riteau, N., P. Gasse, L. Fauconnier et al. 2010. Extracellular ATP is a danger signal activating P2X7 receptor in lung inflammation and fibrosis. *Am J Respir Crit Care Med.* 182:774–83.

Ruch, R.J., S. Porter, L.D. Koffler et al. 2001. Defective gap junctional intercellular communication in lung cancer: Loss of an important mediator of tissue homeostasis and phenotypic regulation. *Exp Lung Res.* 27:231–43.

Sano, H., Y. Kuroki, T. Honma et al. 1998. Analysis of chimeric proteins identifies the regions in the carbohydrate recognition domains of rat lung collectins that are essential for interactions with phospholipids, glycolipids, and alveolar type II cells. *J Biol Chem.* 273:4783–9.

Sarieddine, M.Z., K.E. Scheckenbach, B. Foglia et al. 2009. Connexin43 modulates neutrophil recruitment to the lung. *J Cell Mol Med.* 13:4560–70.

Scemes, E., S.O. Suadicani, G. Dahl et al. 2007. Connexin and pannexin mediated cell-cell communication. *Neuron Glia Biol.* 3:199–208.

Scerri, I., O. Tabary, T. Dudez et al. 2006. Gap junctional communication does not contribute to the interaction between neutrophils and airway epithelial cells. *Cell Commun Adhes.* 13:1–12.

Schneeberger, E.E., and R.D. Lynch. 2004. The tight junction: A multifunctional complex. *Am J Physiol Cell Physiol.* 286:C1213–28.

Seshiah, P.N., D.S. Weber, P. Rocic et al. 2002. Angiotensin II stimulation of NAD(P)H oxidase activity: Upstream mediators. *Circ Res.* 91:406–13.

Sharov, V.S., K. Briviba, and H. Sies. 1999. Peroxynitrite diminishes gap junctional communication: Protection by selenite supplementation. *IUBMB Life.* 48:379–84.

Sirianni, F.E., F.S. Chu, and D.C. Walker. 2003. Human alveolar wall fibroblasts directly link epithelial type 2 cells to capillary endothelium. *Am J Respir Crit Care Med.* 168:1532–7.

Sirianni, F.E., A. Milaninezhad, F.S. Chu et al. 2006. Alteration of fibroblast architecture and loss of basal lamina apertures in human emphysematous lung. *Am J Respir Crit Care Med.* 173:632–8.

Sittipunt, C., K.P. Steinberg, J.T. Ruzinski et al. 2001. Nitric oxide and nitrotyrosine in the lungs of patients with acute respiratory distress syndrome. *Am J Respir Crit Care Med.* 163:503–10.

Sohl, G., and K. Willecke. 2004. Gap junctions and the connexin protein family. *Cardiovasc Res.* 62:228–32.

Spray, D.C., Z.C. Ye, and B.R. Ransom. 2006. Functional connexin "hemichannels": A critical appraisal. *Glia.* 54:758–73.

Thompson, L.F., H.K. Eltzschig, J.C. Ibla et al. 2004. Crucial role for ecto-5'-nucleotidase (CD73) in vascular leakage during hypoxia. *J Exp Med.* 200:1395–405.

Travis, S.M., P.K. Singh, and M.J. Welsh. 2001. Antimicrobial peptides and proteins in the innate defense of the airway surface. *Curr Opin Immunol.* 13:89–95.

Traub, O., B. Hertlein, M. Kasper et al. 1998. Characterization of the gap junction protein connexin37 in murine endothelium, respiratory epithelium, and after transfection in human HeLa cells. *Eur J Cell Biol.* 77:313–22.

Trovato-Salinaro, A., E. Trovato-Salinaro, M. Failla et al. 2006. Altered intercellular communication in lung fibroblast cultures from patients with idiopathic pulmonary fibrosis. *Respir Res.* 27(7):122.

Turner, J.R. 2006. Molecular basis of epithelial barrier regulation: From basic mechanisms to clinical application. *Am J Pathol.* 169:1901–9.

Upham, B.L., K.S. Kang, H.Y. Cho et al. 1997. Hydrogen peroxide inhibits gap junctional intercellular communication in glutathione sufficient but not glutathione deficient cells. *Carcinogenesis.* 18:37–42.

Van Itallie, C.M., and J.M. Anderson. 2006. Claudins and epithelial paracellular transport. *Annu Rev Physiol.* 68:403–29.

Völmer, J.B., L.F. Thompson, and M.R. Blackburn. 2006. Ecto-5'-nucleotidase (CD73)-mediated adenosine production is tissue protective in a model of bleomycin-induced lung injury. *J Immunol.* 176:4449–58.

Wang, P.M., E. Fujita, and J. Bhattacharya. 2002. Vascular regulation of type II cell exocytosis. *Am J Physiol Lung Cell Mol Physiol.* 282:L912–6.

Ware, L.B., and M.A. Matthay. 2000. The acute respiratory distress syndrome. *N Engl J Med.* 342:1334–49.

Wentlandt, K., M. Kushnir, C.C. Naus et al. 2004. Ethanol inhibits gap-junctional coupling between P19 cells. *Alcohol Clin Exp Res.* 28:1284–90.

Wert, S.E., M. Yoshida, A.M. LeVine et al. 2000. Increased metalloproteinase activity, oxidant production, and emphysema in surfactant protein D gene-inactivated mice. *Proc Natl Acad Sci USA.* 97:5972–7.

Whitsett, J.A., and T.E. Weaver. 2002. Hydrophobic surfactant proteins in lung function and disease. *N Engl J Med.* 347:2141–8.

Willis, B.C., and Z. Borok. 2007. TGF-beta-induced EMT: Mechanisms and implications for fibrotic lung disease. *Am J Physiol Lung Cell Mol Physiol.* 293:L525–34.

Wiszniewski, L., L. Jornot, T. Dudez et al. 2006. Long-term cultures of polarized airway epithelial cells from patients with cystic fibrosis. *Am J Respir Cell Mol Biol.* 34:39–48.

Wiszniewski, L., J. Sanz, I. Scerri et al. 2007. Functional expression of connexin30 and connexin31 in the polarized human airway epithelium. *Differentiation.* 75:382–92.

Wong, C.W., T. Christen, I. Roth et al. 2006. Connexin37 protects against atherosclerosis by regulating monocyte adhesion. *Nat Med.* 12:950–4.

Wright, J.R. 2005. Immunoregulatory functions of surfactant proteins. *Nat Rev Immunol.* 5:58–68.

Yeh, H.I., S. Rothery, E. Dupont, S.R. Coppen, and N.J. Severs. 1998. Individual gap junction plaques contain multiple connexins in arterial endothelium. *Circ Res.* 83:1248–63.

Ying, X., Y. Minamiya, C. Fu et al. 1996. Ca^{2+} waves in lung capillary endothelium. *Circ Res.* 79:898–908.

Yoon, B.I., Y. Hirabayashi, Y. Kawasaki et al. 2004. Exacerbation of benzene pneumotoxicity in connexin 32 knockout mice: Enhanced proliferation of CYP2E1-immunoreactive alveolar epithelial cells. *Toxicology.* 195:19–29.

Zahler, S., A. Hoffmann, T. Gloe, and U. Pohl. 2003. Gap-junctional coupling between neutrophils and endothelial cells: A novel modulator of transendothelial migration. *J Leukoc
Biol.* 73:118–26.

Zhang, Z.Q., Y. Hu, B.J. Wang et al. 2004. Effective asymmetry in gap junctional intercellular
communication between populations of human normal lung fibroblasts and lung carcinoma cells. *Carcinogenesis* 25:473–482.

Zhang, J., W. Wang, J. Sun et al. 2010. Gap junction channel modulates pulmonary vascular
permeability through calcium in acute lung injury: An experimental study. *Respiration.*
80:236–45.

9 Effect of Oxidative Stress on Connexins in the Vasculature

Marie Billaud, Scott R. Johnstone, Katherine R. Heberlein, Adam C. Straub, and Brant E. Isakson

CONTENTS

9.1 INTRODUCTION

The key role of the vasculature is to provide oxygen for the metabolic demand of the tissues throughout the periphery of an organism. Although this seems like a straightforward task, oxygen itself is a highly reactive molecule that when combined with other substrates can have a multitude of effects, both deleterious and beneficial. For example, reactive oxygen species (ROS) can include hydroxyl radical and superoxide anion radical, which are both known for their deleterious effects on tissues. Conversely, nitric oxide (NO) is the most important ROS in the vasculature where it is well known for its beneficial effects. One of the defining features for each of these ROS molecules is the presence of a free radical, an unpaired electron in the outer orbital, which can cause the molecule to be highly reactive. Although most of the ROS originate from the superoxide anion radical that are mostly produced by mitochondria or NADPH oxidase (Demaurex and Scorrano, 2009), the endothelial cells (ECs) lining the inside of the blood vessels produce vast amount of NO via the enzyme endothelial NO synthase (eNOS, NOS3). In the vasculature, NO is quite beneficial by providing a relaxation signal to the vascular smooth muscle cells (VSMCs) to dilate, or provide a measure of inflammatory inhibition by preventing the binding of leukocytes to the endothelium. However, other ROS molecules, especially superoxide anion can be produced during pathological events in the vasculature (e.g., ischemic conditions) and have deleterious effects (for review, see Wolin, 2009). Although it has been hypothesized that ROS, and especially hydrophilic ROS such as superoxide anion and hydroxyl radical, could pass through gap junctions (Tang and Vanhoutte, 2008; Billaud et al., 2009), ROS molecules provide a different method of communication than gap junctions, which allow for direct intercellular transfer of molecules. This review focuses on the published observations and possible mechanisms that allow oxidative events by ROS to alter both connexin (Cx) expression and function, which by extension, would alter (or enhance) the status of the vasculature.

9.1.1 Vascular Structure

Mammalian vasculature is one of the most complex structures within the body, consisting of a network of arteries, arterioles, capillaries, and veins that if laid end to end would stretch more than 100,000 miles. While the basic structure of each of the different vessels are essentially the same, single layer of ECs separated from single or multiple layers of VSMCs by matrix layers called elastic lamina, the function of each of these systems varies greatly. Arteries, arterioles, and capillaries make up the oxygenation systems for the body and, depending on the vessel diameter, the cell composition and function, regulate tissue oxygenation and the blood pressure systems within the body. On the other hand, veins make up the low pressure, deoxygenation systems. Many vascular pathologies are associated to the arterial/oxygenation system such as atherosclerosis and hypertension. Many of these pathologies are directly mediated by ROS, which leads to a number of cellular changes. While several factors have been shown to mediate these processes, one of the most important may be regulation of gap junctions and their constitutive proteins the connexins (Cxs).

9.1.2 Vascular Cxs

To date, of the 20 Cxs identified, only Cx37, Cx40, Cx43, and Cx45 and, more recently, Cx32 have been demonstrated to be differentially expressed in vascular cells (Kwak et al., 2002; Okamoto et al., 2009; Johnstone et al., 2009a). In nondiseased large order vessels, for example, aortas and carotids, Cx32, Cx37, Cx40, Cx45, and Cx43 are expressed in vascular ECs (Van Kempen and Jongsma, 1999; Isakson et al., 2006b; Okamoto et al., 2009), although the latter Cx is thought to be poorly expressed and is even absent in the ECs of some vessels, such as intramural coronary arteries (Severs et al., 2001). It has been hypothesized that this weak EC expression could be explained by the fact that Cx43 is actively regulated by physical forces such as shear stress and mechanical load (Cowan et al., 1998; Gabriels and Paul, 1998; DePaola et al., 1999). In lower-order vessels such as cremaster and mesentery, Cx37, Cx40, and Cx43 are expressed in ECs (Hakim et al., 2008). The VSMCs of arteries are less coupled as compared to endothelium as revealed by dye diffusion experiments (Segal and Beny, 1992; Little et al., 1995; Sandow et al., 2003). Although Cx expression, particularly Cx43, is well characterized in the VSMC layers of large arteries, identification of VSMC Cxs in smaller muscular arteries has been more difficult (Hong and Hill, 1998; Looft-Wilson et al., 2004; Fanchaouy et al., 2005; Hakim et al., 2008). In addition to homocellular gap junctions within ECs in the intima and VSMCs, both cell types can communicate through a specialized structure, identified as myoendothelial junction (Heberlein et al., 2009a) (Figure 9.1a,c,e). This heterocellular junction is essentially found in resistance arteries and is absent of large arteries where the media and the intima are separated by a thick internal elastic lamina (Figure 9.1b,d,f). Among the five Cxs expressed in the vasculature, only Cx37, Cx40, and Cx43 have been identified at the myoendothelial junction (Mather et al., 2005; Haddock et al., 2006; Heberlein et al., 2009a; Straub et al., 2011).

In the vasculature, gap junctions play an essential role in the control of vascular tone by coordinating cellular responses within the arterial wall (for review, see Schmidt et al., 2008). The gap junctions, especially myoendothelial gap junctions, are thought to be involved in vasodilation mediated by the endothelium-derived hyperpolarizing factor (EDHF) (de Wit and Griffith, 2010). Several reports demonstrate that this vasodilation, usually initiated by stimuli such as shear stress, acetylcholine, or bradykinin, is altered by several gap junction inhibitors (Sandow et al., 2003, Coleman et al., 2001a,b). Furthermore, gap junctions have been shown to be involved in the propagation of calcium signaling and vasoconstriction. For example, gap junction inhibitors alter the myogenic tone, which is usually elicited by an increase in transmural pressure *in vivo* (Lagaud et al., 2002). Lastly, gap junctions control the propagation of vasoconstriction and vasodilation along the vessel length over several millimeters, effectively producing a coordinated change in vessel diameter (Segal and Duling, 1989; Segal et al., 1989; Xia and Duling, 1995).

9.1.3 Oxidative Stress and Reactive Oxygen Species

Oxygen is essential for cellular functions, but is also paradoxically a source of ROS that can lead to irreversible damages on cellular compounds such as DNA, lipids, and proteins. Oxidative stress is defined as a condition where cells and tissues are exposed to high level of ROS following an increase of their production or a decrease in their degradation. Oxidative stress is frequently related to different pathological conditions including cardiovascular diseases (CVDs). Superoxide anion and NO are at the origin of the formation of other ROS, including hydrogen peroxide, hydroxyl radical, and peroxynitrite.

The production of ROS is an essential component of vascular EC and VSMC functionality (Touyz et al., 2003) and ROS are generated by different sources within the cells. The main source of NO in the vasculature is eNOS, while superoxide anion can be produced by NADPH oxidase, the mitochondrial respiratory chain and

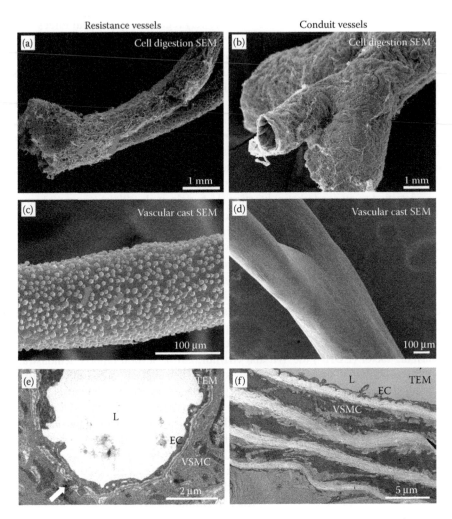

FIGURE 9.1   SEM and TEM images comparing the cellular organization of resistance and conduit vessels: SEM images of cellular digestion show elastic lamina in resistance vessel (cremaster, a) compared to conduit vessel (carotid, b). Sodium hydroxide (NaOH) removal of both smooth muscle and endothelial layers in each vessel type reveals the elastic lamina matrix between ECs and VSMCs to be a thin layer containing multiple holes corresponding to expected areas for myoendothelial junctions (MEJ, arrows a, c, e) in the resistance vessels, whereas in the large conduit vessels multiple thick layers of elastic lamina prevent direct communication between ECs and VSMCs (a and b). In c and d, PU4 vascular cast SEM images of VSMC denuded vessels reveals multiple, regular protrusions (arrows) representing the EC component of MEJs (mesentery, c), which are absent (as indicated by the "smoothness") from large conduit vessels (aorta, d). TEM of transverse sections of small resistance vessel (coronaries, e) and carotid artery (f) reveals singular layers of ECs and VSMCs in the resistance vessel with areas of cell–cell contact known as the MEJ (e, arrow) whereas carotid contains multiple layers of VSMCs separated from the ECs by a thick layer of IEL in the carotid (lumen is indicated by L).

several other oxidases such as xanthine oxidase, cyclo-oxygenase, lipo-oxygenase (for review, see Droge, 2002). Interestingly, eNOS can produce both NO and super-oxide anion: NO is produced in physiological condition where eNOS is homodi-meric, whereas superoxide anion will mostly be generated in pathological condition where eNOS is uncoupled (for review, see Forstermann and Munzel, 2006).

There is a plethora of molecules that act against oxidative stress in order to prevent cellular damages eventually caused by ROS. Among these antioxidant molecules are glutathione, ascorbic acid (vitamine C), α-tocopherol (vitamine E), catalase, and newly discovered enzymes such as peroxiredoxine (Wood et al., 2003; Forman et al., 2009; Wolin, 2009), which all act via different pathways to detoxify harmful com-ponents of the redox systems (Sihvo et al., 2003). Furthermore, NO and superox-ide dismutase (SOD) can be considered both as oxidant and antioxidant molecules. Indeed, superoxide anion reacts with NO faster than it reacts with SOD; therefore, the peroxynitrite produced only becomes a significant biological regulatory factor when superoxide levels are increased or when NO levels approach the concentra-tions of SOD (Wolin, 2009). Additionally, SOD can be considered as an antioxidant as it degrades the superoxide anion, but this degradation leads to the generation of hydrogen peroxide, which can consequently damage the cells if it is not degraded by the catalase.

Functionally, ROS are mostly associated to cellular damage; however, low levels of ROS are involved in several physiological pathways. Apart from NO, hydrogen peroxide is another vasodilator and several reports led to the hypothesis that hydro-gen peroxide could be the EDHF (Feletou and Vanhoutte, 2009). Furthermore, ROS play a role in the phenomenon of oxygen sensing in the tissues, which is essential to activate antioxidant defense (Droge, 2002). Lastly, ROS are involved in cellular apoptosis (Droge, 2002). When the levels of ROS are increased in cells, a series of cellular damages can occur, including DNA oxidation that can lead to irreversible mutation, peroxidation of lipids that leads to a disorganization of the plasma mem-brane proteins and lastly, ROS can nitrosylate proteins.

ROS play an important role in the etiology of cardiovascular pathologies, where both VSMCs and ECs produce increased ROS levels. The major sources of super-oxide anion in the VSMCs are the NADPH oxidase and the mitochondria (Lyle and Griendling, 2006), while several enzymatic systems such as eNOS, xanthine oxidase, and cyclooxygenase are at the origin of ROS production in ECs (Li and Shah, 2004). The superoxide anion has severe consequences on the arterial wall as it reacts with NO, thus decreasing NO bioavailability and increasing the formation of peroxynitrite. This phenomenon leads to an endothelial dysfunction characteris-tic of CVDs.

9.1.4 POSTTRANSLATIONAL MODIFICATIONS: PHOSPHORYLATION

Posttranslational modifications are critical for the regulation and control of protein function. Phosphorylation and S-nitrosylation are the main modifications of Cxs that regulate gap junction function. A number of oxidative stressors are known to act via phosphorylation and nitrosylation pathways, and therefore, may be associated to Cx posttranslational modifications in vascular diseases.

Cx phosphorylation has been well established to alter channel gating of conductive substrates as well as interactions between Cxs and other proteins (Maeda and Tsukihara, 2010; Solan and Lampe, 2005). Cx43 has been the prototypical Cx for studying phosphorylation and as a result, many studies have revealed that the C-terminus of Cx43 (as well as other Cxs) is a critical regulator of gap junction permeability (Lampe and Lau, 2000, 2004; Solan and Lampe, 2005). Cx43 is subjected to posttranslational modifications through phosphorylation of serine, threonine, and tyrosine residues in the C-terminus by a number of kinases, including c-SRC tyrosine phosphorylation, mitogen-activated protein kinase (MAPK), protein kinase C/A/G (PKC, PKA, PKG), P34^{cdc}, and casein kinases (CK1) serine phosphorylation (Table 9.1) (Solan and Lampe, 2005, 2007). Differential phosphorylations of sites within the C-terminus of Cx43 have been demonstrated to be important for either enhanced or reduced gap junction functionality. Phosphorylation by PKA and CK1 enhance gap junction permeability; while PKC, c-SRC, and MAPK phosphorylation reduce permeability of Cx43 gap junctions (Table 9.1) (Lampe et al., 2000; Solan

TABLE 9.1
Phosphorylation of Vascular Connexins

Connexin	Kinase	Phosphorylation Site	GJ Permeability	Reference
Cx37	GSK3	S321	−	Morel et al. (2010)
Cx40	PKA	Unknown	+/−	Bolon et al. (2008), van Rijen et al. (2000)
Cx43	v-Src	Y247	−	Kanemitsu et al. (1997)
Cx43	MAPK, CDC2, ERK1/2	S255	−	Moorby and Patel (2001), Warn-Cramer et al. (1996), Lampe et al. (1998), Kanemitsu et al. (1998), Solan and Lampe (2007)
Cx43	PKC, CDC2	S262	−	Doble et al. (2004), Solan and Lampe (2007), Lampe et al. (1998), Kanemitsu et al. (1998)
Cx43	v-Src	Y265	−	Kanemitsu et al. (1997)
Cx43	MAPK, PKC	S279/S282	−	Moorby and Patel (2001), Solan and Lampe (2007)
Cx43	CK1	S325/S328/ S330	+	Cooper and Lampe (2002), Solan and Lampe (2007)
Cx43	PKA	S364/S364	+	TenBroek et al. (2001), Yogo et al. (2002), Solan and Lampe (2007)
Cx43	PKC	S368	−/+	Solan et al. (2003)
Cx43	PKA	S369/S373	+	TenBroek et al. (2001), Yogo et al. (2002)
Cx45	PKC	Unknown	+	van Veen et al. (2000)

et al., 2003; Pahujaa et al., 2007). Moreover, there are species-specific differences in phosphorylation patterns for Cx43. For example, rat Cx43 is regulated by PKG, which makes Cx43 gap junctions less permeable, but human Cx43 does not, as the PKG phosphorylation site (Ser257) is absent from the C-terminus domain in humans (Moreno et al., 1994; Kwak et al., 1995). Lastly, there is increasing evidence suggesting that Cx phosphorylation, particularly Cx43 phosphorylation, can be altered in diseased states (Isakson et al., 2006c; Johnstone et al., 2009b).

The regulation of phosphorylation of other vascular Cxs is currently less clear. However, recent evidence has demonstrated that Cx37, which has a similar sequence homology to Cx43 C-terminus, can be phosphorylated by glycogen synthase kinase-3 (GSK-3) to reduce gap junction permeability (Morel et al., 2010). Cx40 has been shown to be phosphorylated on serine residues in response to hypoxia/reoxygenation through a PKA/PKC mechanism (Bolon et al., 2008). Lastly, Cx45 has also been demonstrated to be phosphorylated by PKA mechanism, however, specific sites of phosphorylation remain elusive (van Veen et al., 2000). Taken together, these data demonstrate that phosphorylation is a major regulator of Cx posttranslational modification that regulates gap junction communication. Table 9.1 demonstrates some of the current known effects of phosphorylation on gap junction function.

9.1.5 POSTTRANSLATIONAL MODIFICATIONS: NITROSYLATION

The covalent attachment of a NO group to a cysteine thiol side chain is called S-nitrosylation and serves as an important posttranslational modification that governs signal transduction throughout various aspects of cardiovascular physiology. Growing evidence suggests that dysregulation of S-nitrosylation lead to nitrosative stress, which is a critical mechanism in the initiation, and progression of a number of cardiovascular-related diseases (Lima et al., 2010). There is accumulating evidence suggesting that Cxs are regulated by S-nitrosylation in both physiological and pathophysiological states. Thus, there is an impetuous need for future studies aimed at understanding the molecular and cellular mechanisms used by cells to regulate protein S-nitrosylation mechanisms.

S-nitrosylation is regulated by NO produced by NOS, which can be expressed constitutively or induced in most cell types. The NOS enzyme family includes NOS1 (or neuronal (n)NOS), NOS2 (or inducible (i)NOS), and NOS3 (or endothelial (e)NOS) and are arguably the critical players for the initiation of S-nitrosylation (Jaffrey et al., 2001). A growing number of studies suggest that S-nitrosylation is spatially and temporally controlled and is tightly regulated by physiological stimuli such as receptor activation (e.g., α1D adrenergic receptor (Nozik-Grayck et al., 2006), NMDA receptor (Takahashi et al., 2007), estrogen receptor (Garban et al., 2005), and ryanodine receptor (Bellinger et al., 2009). Furthermore, S-nitrosylation is controlled by subcellular compartmentalization and direct interactions with substrate proteins (Stamler et al., 2001; Foster et al., 2003; Iwakiri et al., 2006; Lima et al., 2010; Qian et al., 2010; Straub et al., 2011). In addition, there is accumulating data showing that consensus protein motifs may also be critical for cysteine-specific S-nitrosylation and precise protein regulation (Stamler, 1994; Xue et al., 2010).

Modulating the extent of S-nitrosylation is critical for inhibiting excessive nitrosative stress and is achieved by a number of denitrosylases (for review, see Benhar et al., 2009). However, this process remains less studied. Once viewed as unregulated and spontaneous, the mechanisms of denitrosylation are becoming increasingly clear and a few studies have identified important denitrosylases critical for maintaining the S-nitrosylation/de-nitrosylation balance, such as S-nitrosoglutathione reductase (GSNOR) (Benhar et al., 2009; Liu et al., 2004). It should be noted that the process of denitrosylation should not be exclusively considered as an "off" switch, but rather, proteins can be constitutively S-nitrosylated and upon a stimulus can be de-nitrosylated, thereby eliciting signal transduction event. While the list of proteins regulated by S-nitrosylation/de-nitrosylation increases in the cardiovasculature, recent evidence has identified that Cxs are also regulated by S-nitrosylation (Retamal et al., 2006; Straub et al., 2011).

NO has been shown to have the ability to regulate gap junction permeability and electrical coupling in the vasculature. In microvascular ECs, Cx43 but not Cx40 is constitutively S-nitrosylated by eNOS on cysteine 271 (on the Cx43 C-terminus) at the myoendothelial junction to maintain gap junctions in a permeable state (Straub et al., 2011). Upon stimulation of VSMCs by phenylephrine, Cx43 gap junctions are denitrosylated by GSNOR, limiting gap junction channel conductance. This work correlated with other studies showing that Cx43 hemichannels in astrocytes also become S-nitrosylated in oxidative stress conditions (Retamal et al., 2006). These two studies indicate that Cx43 is indeed a target for S-nitrosylation and provides another mechanism for Cx43 gap junction permeability regulation.

There has been limited S-nitrosylation work performed on other Cxs, however, there is suggestive data showing that NO has the capacity to inhibit electrical current of Cx37 gap junctions in the microvasculature (Kameritsch et al., 2005). Although these studies infer Cx37 S-nitrosylation, future experiments directly testing if Cx37 is S-nitrosylated (i.e., via biotin switch assays and site-directed mutations) should identify critical cysteine targets of S-nitrosylation in Cx37 and assess whether it can alter channel permeability. Lastly, it is currently unknown whether Cx45 is S-nitrosylated. In conclusion, it is becoming increasingly clear that Cx proteins contain target sites for S-nitrosylation; however, future studies will have to be performed to identify whether Cx37 and Cx45 are regulated by S-nitrosylation. Table 9.2 demonstrates the known data on vascular Cx S-nitrosylation.

TABLE 9.2
S-Nitrosylation of Vascular Connexins

Connexin	S-Nitrosylation	Cysteine	Function	Reference
Cx37	Unknown	Unknown	NO decreases GJ permeability	Kameritsch et al. (2005)
Cx40	No	—	—	Straub et al. (2011)
Cx43	Yes	C271	Increases GJ permeability	Retamal et al. (2006), Straub et al. (2011)
Cx45	Unknown	Unknown	Unknown	—

9.2 METABOLIC SYNDROME

The metabolic syndrome is defined by an association of several metabolic disorders that increase the risk of CVDs. The metabolic syndrome has become a major health concern, especially in the developed countries such as North America where it affects approximately one-quarter of the population (Ford et al., 2002). Although pathophysiological causes leading to the metabolic syndrome remain unclear, several risk factors have been established such as central obesity, elevated triglycerides, diminished high-density lipoprotein (HDL), cholesterol, systemic hypertension, and elevated fasting glucose (Cleeman, 2001; Moebus et al., 2006). These risk factors lead to pathologies such as diabetes, obesity, or atherosclerosis where Cx expression and functions are altered.

9.2.1 ATHEROSCLEROSIS

Oxidative stress is a major component of the development of atherosclerosis (which is the underlying cause for the majority of CVDs) and has one of the highest mortality rates in the United States (Roger et al., 2011). Lifestyle choices including lack of exercise, cholesterol-rich diet, alcohol consumption, and smoking are all considered as significant risk factors in atherosclerotic disease progression and are all associated with lipid accumulation, production of oxidative by-products, and oxidative stress (Roberts et al., 2002; Ambrose and Barua, 2004; Mercado and Jaimes, 2007; Chen et al., 2011). Although our knowledge on atherosclerosis and its complications has increased, the mechanisms of initiation (atherogenesis) remain ill defined. Atherogenesis is typified by lipid deposition and oxidation, macrophage recruitment and VSMC proliferation within the wall of large and medium-sized arteries (Kwak et al., 2002). Concurrent with these stages of disease development is significant modulation of gap junctions and Cx proteins (Kwak et al., 2002, 2003, Solan et al., 2003).

Many studies have now detailed the roles of Cx37, Cx40, and Cx43 in atherosclerosis, yet less is known about the role of Cx45 (primarily due to poor reagents) (Kwak et al., 2002). In the earliest stages of atherogenesis, ECs become more permeable to inflammatory cells and accumulation of lipids and cytokines promote VSMCs to dedifferentiate from a quiescent phenotype to a highly motile, proliferative phenotype. Within the large order vessels, Cx37, Cx40, and Cx43 have been implicated in the progression of atherosclerotic disease state but through different mechanisms. Cx37 and Cx40 have been shown to reduce atherogenic potential in cells (Kwak et al., 2003; Chanson and Kwak, 2007; Chadjichristos et al., 2006a,b). A role for Cx37 in promoting atherosclerosis has been demonstrated with a genetic polymorphism of Cx37 in monocytes leading to dysfunctional hemichannels that inappropriately release antiadhesive molecules, thus enhancing the disease state (Boerma et al., 1999). Although Cx37 is known to become detectable in VSMCs in the advanced atherosclerotic disease state (Derouette et al., 2009; Wong et al., 2007) and a loss of Cx37 leads to increased plaque size in ApoE$^{-/-}$ Cx37$^{-/-}$ mice (Wong et al., 2006b), its expression is not found in VSMCs during early atherogenesis (Johnstone et al., 2009b). Cx43 on the other hand is reported to work through a different mechanism in atherosclerosis and may act, among other mechanisms, to promote VSMC proliferation directly. In atherogenesis, it is well recognized that VSMC proliferation is an event characterized by decreased expression of Cx43 (Chadjichristos et al., 2006b; Brisset et al., 2008) associated to increased VSMC content of atherosclerotic plaques (Kwak et al., 2003) and increasing neointimal formation (Liao et al., 2007). Indeed, following statin treatment, Cx43 expression returns to baseline conditions in VSMC and cell proliferation is reduced (Palmer et al., 1998; Kwak and Mach, 2001; Yeh et al., 2003a). Additionally, targeted deletion of Cx43 reduces VSMC response to platelet-derived growth factor (PDGF-BB) the receptors of which are targets of chemicals derived from cigarette smoke (Chadjichristos et al., 2008). How Cx43 modulates these responses in VSMC proliferation, especially as it relates to atherogenesis, is currently poorly defined.

Large-order vessels are particularly susceptible to accumulation of low-density lipoproteins (LDL) which, following oxidation, produce a number of by-products known to be integral to the progression of the disease state. An increase in blood LDL levels is often found in obesity and relates in particular to an increased body mass index (BMI) and increased risk of development of atherosclerosis (Codoner-Franch et al., 2010; Kilic, 2010; Buchner et al., 2011). High levels of circulating LDL build up in the intima and become oxidized, resulting in an inflammatory process. This promotes macrophage recruitment to remove oxidized LDL build up and results in the formation of the lipid core of the atherosclerotic plaque (Sigala et al., 2010). The recruitment of macrophages and subsequent inflammation has been shown to be mediated in part through release of signals from dysfunctional Cx37 (Wong et al., 2006a) and through inflammation by Cx32 (Okamoto et al., 2011).

However, a role must be considered for the pannexins (Panx), which are structural homologues of Cxs. It has been shown that VSMCs, ECs, and macrophages all express Panx1 (Locovei et al., 2006; Pelegrin and Surprenant, 2007; MacVicar and Thompson, 2010; Billaud et al., 2011b) and recent studies have identified the release of a "find me" signal through Panx1 that promotes the recruitment of macrophages to the site of cell apoptosis (Chekeni et al., 2011). These data give rise to the possibility that cellular damages mediated through the oxidative by-products of LDL leading to cell death may also result in the release of signals through Panx1 and thus recruitment of macrophages to the site of vessel injury.

Macrophages that are recruited to the site of atherosclerotic plaques engulf dying cells as well as lipids, that is, LDL. Lipid-rich macrophages break down the lipids via oxidation to produce several by-products referred to as oxidized phospholipids (OxPLs) (Boyle, 2005). Further recruitment of macrophages has been attributed to the build-up of OxPLs and signals released through Panx 1 channels (Chekeni et al., 2011), with Cx37 gap junction hemichannels expressed in ECs thought to provide a protective mechanism against recruitment of macrophages (Wong et al., 2006a). It has not been fully resolved whether these two subsets of proteins (Cxs and Panxs) can work to signal in parallel or whether they act in compensation of each during disease states.

While accumulation of lipids and macrophage recruitment make up an integral part of the atherosclerotic plaque it is thought that the production of OxPL may be the main contributor to the progression of the disease state. VSMCs underlying the forming plaque de-differentiate from a nonproliferative/quiescent to a proliferative/synthetic phenotype. These cells are prone to lipid uptake and display enhanced endoplasmic reticulum, which is believed to lead to mitochondrial production of

ROS and further cellular damages (Manderson et al., 1989; Bano and Nicotera, 2007; Demaurex and Scorrano, 2009). The change in phenotype is followed by VSMC migration across the internal elastic lamina, then proliferation and deposition of extracellular matrix (ECM) increases the size of the plaque (Manderson et al., 1989).

Oxidation of LDL gives rise to a series of OxPL species, for example, 1-palmitoyl-2-oxovaleroyl-*sn*-glycero-3-phosphorylcholine (POVPC), which accumulate in atherosclerotic lesions (Chatterjee et al., 2004; Chatterjee and Ghosh, 1996; Pidkovka et al., 2007). While only minor structural variations occur within these OxPLs it has been shown that some of them are relatively inert by-products such as 1-palmitoyl-2-glutaroyl-*sn*-glycero-3-phosphorylcholine (PGPC) whereas others such as POVPC can initiate a contractile phenotype in VSMCs and thus can lead to the progression of the disease state (Pidkovka et al., 2007; Johnstone et al., 2009b). Recently, it has been reported that OxPLs including POVPC promote DNA synthesis and induce VSMC proliferation (Johnstone et al., 2009b; Chatterjee et al., 2004; Heery et al., 1995). However, it is not clear how the specific OxPLs induce cell proliferation. A number of mechanisms have been shown to be important in this process, with activation of the general MAPK pathway appearing to be integral in this stage (Chatterjee et al., 2004). This pertains to Cxs (in particular Cx43) as acute exposure to POVPC reduces the expression of total Cx43 protein in VSMC but enhances its MAPK phosphorylated isoforms. Cx43 is now believed to be integral in the pathogenesis of this disease state (Johnstone et al., 2009b). Additionally, POVPC localizes primarily to the plasma membrane, which may be the site at which it interacts with and promotes phosphorylation of Cx43 MAPK-associated serines in VSMCs (Moumtzi et al., 2007; Johnstone et al., 2009b). The functional consequences of changes in Cx43 during atherosclerosis, coupled with the extensive evidence for POVPC involvement in the development of atherosclerosis, gives rise to the possibility that POVPC act as an atherogenic stimulus that promotes phosphorylation of Cx43.

In atherogenesis and other disease states such as carcinogenesis, it has also been hypothesized that phosphorylation of Cx43 is a key to cell cycle control and proliferation (Doble et al., 2004; Solan and Lampe, 2009; Johnstone et al., 2009b). Indeed, Cx43's cytoplasmically located C-terminus contains multiple sites for phosphorylation, which are integral in the modulation of the proteins function (Solan and Lampe, 2009). Differential site-specific phosphorylations of the Cx43 C-terminus have been observed throughout the stages of the cell cycle (Solan et al., 2003). Phosphorylation of serines S255/262/279/282 and tyrosine (Y247/265) residues increase in G1 (Solan and Lampe, 2008; Herrero-Gonzalez et al., 2010) and correlate with DNA synthesis stages of cell proliferation (Solan et al., 2003; Doble et al., 2004). In contrast, expression of phosphorylated Cx43-S368, a PKC-associated site, increases throughout the S > M (mitosis) phases of the cell cycle, but is decreased during G1 (Solan et al., 2003). In atherogenesis, Cx43 is phosphorylated at both MAPK (e.g., S255/262/279/282) and PKC (e.g., S368) associated sites, but cell proliferation only occurs following S279/282 phosphorylation (Johnstone et al., 2009b). It has still not been fully resolved whether Cx43 protein expression/phosphorylation and protein interactions or alterations in gap junctional communication promote cell proliferation in atherogenesis and other diseases (Lampe and Lau, 2004; Aasen et al., 2005; Norris et al., 2008; Solan and Lampe, 2009; Herrero-Gonzalez et al., 2010).

9.2.2 ANGIOPLASTY AND RESTENOSIS

In advanced stages of atherosclerosis, increased plaque size significantly reduces blood flow, increasing ischemia in the downstream tissues and potentially requiring medical intervention to restore blood flow. The principal medical plaque size reduction treatment is angioplasty through balloon-catheter insertion and inflation, which reduces plaque size via stretch of the vessel. Despite proven effectiveness in restoring initial blood flow, the most common postinterventional complication is restenosis, a renarrowing of the vessel resulting from EC and VSMC invasion of the space causing reocclusion of the vessel that occurs in a large number of patients. This complication means that a significant number of patients will require multiple interventions and stent implantation each of which increases the risk of complications.

Ischemia at the site of advanced atherosclerotic plaques leads to mitochondrial stress and release of ROS (Mayr et al., 2008). Ischemia followed by reoxygenation as occurs in advanced atherosclerosis followed by angioplasty (Mayr et al., 2008) promotes PKA activity and reduces cell coupling. Interestingly this does not occur in Cx40$^{-/-}$ cells suggesting that phosphorylation and communication via Cx40 in ECs following injury is important in the restenotic process (Bolon et al., 2008, Yu et al., 2010). However, in restenotic lesion in these Cx40$^{-/-}$ mice, all Cx isoforms are depleted at the wound edge indicating that Cx40 reductions in gap junctional communication are important in regrowth (Yu et al., 2010; Bolon et al., 2008).

Catheterization, balloon inflation, and stent placement can all cause significant damage to the vessel wall. Endothelial denudation requires endothelium regrowth, which in itself is a challenge given that normal EC turnover within the vasculature is typically slow. However, following denudation, EC proliferation and migration can take only a few weeks and relate to an upregulation in fibroblast growth factor expression in ECs (Lindner et al., 1989; Lindner and Reidy, 1993; Wempe et al., 1997). Endothelial denudation leaves lesions (exposed elastic lamina) which are prone to the accumulation of platelets, white blood cells, lipids and thrombus formation (Manderson et al., 1989). Mechanical injury through distension of the vessel leads to immediate cellular damage through increases in superoxide anion, release of damaging ROS from vascular cells and oxidative challenge identified through depletion of reduced glutathione pools (Souza et al., 2000). However, these initial changes are transient and levels of ROS are decreased 2–4 weeks post injury. Yet restenosis continues to be a problem in patients indicating that there are further issues following ROS damage that promote restenosis (Azevedo et al., 2000).

Following angioplasty, factors are released from surrounding cells to initiate EC regrowth and migration. The consequence of this release, however, can lead to further damage, cell growth, and renarrowing of the vessel known as restenosis. The release of growth factors and proinflammatory cytokines causes VSMC dedifferentiation (to a synthetic phenotype) promoting further thickening and remodeling within the vascular layers (Bundy et al., 2000, Brisset et al., 2009). Following angioplasty and stent placement, re-endothelialization of the vessel is required to repair post-balloon injury. In addition, release of NO can induce EC apoptosis, inducing a further loss of ECs from the vessel wall increasing denuded lesion size (Erdbruegger et al., 2010). Following injury, intracellular calcium levels at the wound edge increase in a process mediated by IP_3 release from intracellular stores; this phenomenon is believed to be facilitated through Cx hemichannel uptake of calcium at the site of injury and may further act to induce release of ROS (Thaulow and Jorgensen, 1997; Francesco, 2010).

Cxs have been demonstrated as integral in cell migration, differentiation, and proliferation in restenosis but may also act to promote the disease state. In other diseases such as chronic diabetes, high levels of Cx43 are associated to poor rates of wound repair (Wang et al., 2007a). However, in post-angioplasty and stent placement, vascular Cxs are significantly downregulated at the site of injury immediately post-injury (Yeh et al., 2000). Levels of Cx40 return to normal within 4 weeks, but both Cx37 and 43 become significantly increased at this time and may be critical in restenotic process (Chadjichristos et al., 2008; Yeh et al., 2000). Heterozygous deletion of Cx43 in mice lacking the LDL receptor ($Cx43^{+/-}LDLR^{-/-}$) display reduced intimal lesion formation and macrophage recruitment following balloon catheter vessel damage, implicating Cx43 as important in restenotic process (Chadjichristos et al., 2006b). Despite multiple common mechanisms that demonstrate reductions in Cx43 associated to enhanced proliferation, several studies have identified that enhanced Cx43 expression is associated with VSMC proliferation, for example, in the saphenous vein (Jia et al., 2008, 2011). While this may indicate differences between cellular systems (low pressure in vein versus high pressure in aorta) the authors suggest that phosphorylation pathways may be involved, and therefore, may reflect the need to assess Cx proteins other than by levels of expression alone, but more by levels of posttranslational modification.

9.2.3 Cx Genetics in Oxidative Stress

A hallmark for all vascular diseases in small vessels is the loss of regulation for vasomotor tone and blood flow. Cxs are a major regulator of cell–to-cell communication within the vasculature and as such, play a significant role in the maintenance of vasoreactivity. The disruption of Cx-mediated signaling events is often associated with vascular diseases, especially those that are induced by ROS (see above). Of the vascular Cxs, genetic abnormalities in Cx37 and Cx40 have been identified and are associated with various vascular diseases, whereas Cx43 and Cx45 currently have no known associated polymorphisms in oxidative stress.

Cx37 is known to have an important role in the maintenance of vasodilatory signals throughout the endothelium as well as facilitating heterocellular communication between ECs and VSMCs by forming gap junctions at myoendothelial junctions within the resistance vasculature (for review, see Heberlein et al., 2009b). In athero-prone ApoE knockout ($ApoE^{-/-}$) mice that do not contain the Cx37 gene, GJA4 ($GJA4^{-/-}$), there is >50% increase in both the development of atherosclerotic lesions and plaque area when the mice are exposed to a cholesterol-rich high-fat diet, indicating that the presence of Cx37 has a protective effect against the development of atherosclerosis (Wong et al., 2006a). A protective role for Cx37 may be inferred by its presence on monocytes, thereby regulating monocyte adhesion. Interestingly, changes in monocyte adhesion were also associated with a Cx37 C1019T polymorphism (a nonconserved amino acid substitution in the C-terminus), suggesting that its role in atherosclerosis development may be mediated, in part, through release of antiadhesive factors. The shift in amino acids may result in changes to the tertiary structure of Cx37, which leads to changes in signaling cascades potentially stemming from phosphorylation events (Boerma et al., 1999). Changes in the human Cx37 gene are currently linked to several CVDs such as hypertension, coronary artery disease, and myocardial infarction (Boerma et al., 1999; Yamada et al., 2002; Collings et al., 2007; Wong et al., 2007). Specifically, the Cx37 C1019T polymorphism is present in a large population of humans with increased carotid plaque formation, and the polymorphism appears to be an indicator of high atherosclerotic risk, especially in individuals with hypertensive-prone conditions (Boerma et al., 1999), highlighting the importance of screening for this genetic mutation to identify at-risk individuals. While it is not conclusively known how this genetic variation arises, individuals with the polymorphism displayed susceptibility to coronary artery disease and myocardial infarction independent of other common risk factors for CVD.

Cx40 is found predominantly in the endothelium and, similar to Cx37, has an associated genetic polymorphism which is linked to hypertension. In mice, Cx40 deficiency (Cx40$^{-/-}$) produces an increased blood pressure when compared to wild-type counterparts, and leads to irregular vasomotion (de Wit et al., 2003). In spontaneous hypertensive rats, increased levels in blood pressure are also associated with decreased levels of Cx40 expression in the endothelium (Rummery et al., 2002). In humans, substitutions in the Cx40 promoter region sites −44 (G◊A) and +71 (A◊G), are associated with CVDs (Firouzi et al., 2006). More specifically, the presence of the Cx40 haplotype −44A/+71G correlates with higher systolic blood pressures among healthy individuals and there is a significant association with these same allele changes and hypertension in men (Firouzi et al., 2006).

Together, these data suggest an important role for Cx40 in the regulation of vascular signaling in diseases such as hypertension. Furthermore, disruption of gap junction-mediated coordination between EC and VSMC as well as along the length of a vessel can exacerbate the pathological state. Similar to Cx37, these genetic polymorphisms show no association with lifestyle factors, that is, smoking or BMI which may induce oxidative stress.

9.2.4 TYPE 2 DIABETES AND OBESITY

Type 2 diabetes is characterized by high glucose levels in the blood as well as oxidative stress and chronic inflammation of several tissues (Pitocco et al., 2010). As of the year 2000, approximately 2.8% of the worldwide population suffered from type 2 diabetes, with this number expected to increase to 4.4% in 2030 (Wild et al., 2004). Similarly, obesity is a major concern in industrialized countries where the number of obese and overweight people has dramatically increased in the past decades (Wang et al., 2008). Although the consequences of diabetes and obesity on Cx expression and functions in the vasculature can be linked to observations as reported in atherosclerosis, several lines of evidence demonstrate direct consequences of diabetes and obesity on Cxs in the vasculature.

Several *in vitro* and *in vivo* studies have revealed a role of high glucose and high lipid content on Cx expression and function. In long-term hyperlipidemia *in vivo,* aortic endothelial gap junction function as well as Cx37 and Cx40 expression is downregulated (Yeh et al., 2003b). High glucose levels have been shown to affect gap junctional communication via a decrease of Cx43 expression due to decreased level of Cx43 mRNA or increased degradation through a proteasome-dependent mechanism in microvascular cells (Fernandes et al., 2004; Sato et al., 2002) or directly via decrease of permeability due to phosphorylation of Cx43 by PKC in cultured bovine aortic VSMCs (Kuroki et al., 1998). Interestingly, several lines of evidence lead to the hypothesis that Cx43 phosphorylation was induced by the oxidative stress generated by high level of glucose. Indeed, experiments in lens epithelial cells have shown that hydrogen peroxide directly activates the C1 domain of PKCγ, inducing a translocation of the PKCγ to Cx43-containing lipid rafts resulting in Cx43 phosphorylation on serine 368 (Lin and Takemoto, 2005). Moreover, this same study showed that activation of PKCγ by hydrogen peroxide decreases Cx43 gap junction plaques and gap junction activity (Lin and Takemoto, 2005). A recent study further demonstrated that a decrease in Cx43 expression in high glucose conditions and, consequently, inhibited gap junction function leads to a significant increase in apoptosis in microvascular ECs (Li and Roy, 2009).

Investigations on rat models of diabetes, induced by an enriched diet, have shown an increase in EDHF-dependent relaxation and in the number of gap junctions at the myoendothelial junction (MEJ) in saphenous artery (Chadha et al., 2010). It has been hypothesized that this increase in EDHF response would compensate for the altered NO-dependent response commonly observed in type 2 diabetes in animals, as well as in human (Triggle and Ding, 2010). Conversely, in the obese Zucker rat, a different rat model of diabetes, Cx40 expression as well as EDHF response are reduced in mesenteric arteries compared to control rats, whereas Cx43 expression is unchanged (Young et al., 2008). As Cx40 is essential for the conduction of vasodilation in arterioles, it has been hypothesized that this function could be altered in diabetes (de Wit et al., 2000; Triggle and Ding, 2010). In another study on spontaneously diabetic mice (db/db), the EDHF response is preserved while NO-dependent vasodilation is lost in mesenteric artery compared to control mice (Pannirselvam et al., 2002). Despite the contradiction between these studies, which could be explained by the difference in the animal model and/or the vascular bed, these studies bring evidence that Cx expression and functions are modified in diabetes.

As previously mentioned, endothelial function and especially NO production by NO synthase is altered in diabetes, and more generally in the metabolic syndrome. Zhang et al. investigated the potential role of NO on the changes in Cx expression in the renal circulation during diabetes (Zhang et al., 2006). In this study, Cx expression was compared between mice with overexpression or deletion of endothelial NO synthase (eNOS), both before and after induction of diabetes. Interestingly, in mice overexpressing eNOS, Cx40 and Cx43 expression pattern was similar to the Cx expression pattern in wild-type mice before and after diabetes induction (Zhang et al., 2006). Furthermore, while basal Cx40 and Cx43 expression in eNOS knockout mice do not differ from wild-type mice, the induction of diabetes produces no alteration of Cxs in the renal vasculature as opposed to wild-type animals. This suggests that NO and/or eNOS may contribute to the changes in Cx expression during diabetes.

9.2.5 INFLAMMATION

Oxidative stress, via the production of stressors such as ROS, or environmental factors such as exposure to cigarette smoke, can oftentimes lead to vascular injury that is typically accompanied by an inflammatory response. Vascular response to inflammation is multifaceted and following stimulation, pro-inflammatory cells interact with the endothelium, which is facilitated by communication between the two cell types. Equally as important in the response to vascular injury, is the coordination of vascular cells along the length of a vessel, in order to regulate vascular tone, blood flow, and the delivery of inflammatory mediators to the injured tissue. Several reports now indicate that in addition to paracrine or integrin signaling, homocellular and heterocellular gap junction channels may also facilitate crosstalk between pro-inflammatory cells and the endothelium (Jara et al., 1995; Oviedo-Orta et al., 2002). Additionally, the ability to regulate signals between vascular cells is essential for the control of vasomotor tone and subsequently, that of blood flow.

The formation of homocellular (i.e., EC-EC or VSMC-VSMC) or heterocellular (i.e., EC to VSMC) junctions at the myoendothelial junction allow for the propagation of signals throughout a vessel in order to maintain vasomotor tone and blood flow (for review, see Heberlein et al., 2009b). The coordination of vessel tone in response to vascular injury is important and allows for the effective delivery of inflammatory mediators to an injured tissue in a timely response. In a study that assessed homocellular communication by gap junctions in an EC monolayer, communication was determined by measuring intercellular resistance, where increased membrane resistance was inversely correlated to gap junction communication (Bolon et al., 2007). Using freshly isolated ECs from Cx null or mutated mice, inflammatory stimulation by bacterial lipopolysaccharide (LPS) lead to an increase in membrane resistance, which was shown to be a result of a decrease in Cx40. However, changes in Cx37 or Cx43 had no effect on LPS-mediated increases in membrane resistance (Bolon et al., 2007). A separate study looking at heterocellular communication of VSMCs and ECs using a coculture model showed that inflammatory stimulation by LPS, TNF-α, or interleukin-1B completely inhibited heterocellular communication through gap junctions (Hu and Cotgreave, 1997). Despite inhibition of gap junctional communication at the myoendothelial junction, there appeared to be no change in homocellular communication (Hu and Cotgreave, 1997). These data highlighted the distinct possibility that, in response to inflammation, cells can differentially regulate gap junction channels within subcellular compartments to promote or inhibit intercellular communication. Indeed, various studies now show that gap junction communication is largely regulated by posttranslational modifications and it is likely that highly localized posttranslational modification events such as phosphorylation or nitrosylation of Cxs (as described above) could lead to differential regulation of gap junctional communication (Heberlein et al., 2009a; Straub et al., 2011). It is important to note, that although the authors suggest that inflammation can selectively inhibit heterocellular communication (as indicated by dye transfer), future studies providing direct evidence for myoendothelial gap junctions, as well as looking at the effect of LPS on particular vascular Cxs and subsequent posttranslational modifications within the MEJ via TEM immunogold labeling *in vivo* and immunoblot analysis of MEJ fractions *in vitro*, would be important.

A hallmark of chronic inflammation is the migration of leukocytes from the blood to the point of tissue injury, in response to increases in pro-inflammatory molecules such as TNF-α or LPS (Hickey et al., 1997). In order for leukocyte migration to occur, the leukocytes first roll along, then adhere to the endothelium of the vessel wall via adhesion molecules. Inflammatory cells such as circulating leukocytes (Jara et al., 1995) and CD4+ Th0, Th1, and Th2 lymphocytes express Cx43 upon activation by LPS (Bermudez-Fajardo et al., 2007). Importantly, activation of Cx43 by LPS can lead to the formation of homocellular gap junction channels between leukocytes as well as functional heterocellular channels between leukocytes and ECs (Jara et al., 1995; Oviedo-Orta et al., 2002). However, early reports also show that there is a decrease in endothelial Cx40 and Cx37 expression in response to TNF-α, accompanied by decreased gap junction coupling (van Rijen et al., 1998; Zahler et al., 2003; Rignault et al., 2007). This would suggest that an uncoupling of cells rather than increased coupling is occurring during inflammation, emphasizing the current controversy surrounding the role for Cxs in inflammatory conditions. The discrepancy in the literature potentially might point to the ability for cells to differentially regulate Cxs within subcellular compartments to perform multiple functions within a single cell.

Once bound, adherence of leukocytes to the endothelium activates the EC and triggers a rise in intracellular EC calcium, which is essential for the continued migration of leukocytes through the vessel wall (Huang et al., 1993; Kielbassa-Schnepp et al., 2001). The presence of functional gap junctions would facilitate this process, allowing for the direct transfer of calcium signaling between the two cell types. Furthermore, in Cx43 null ECs there is a decrease in TNF-α-stimulated leukocyte adhesion, which was also seen when gap junctional communication was blocked *in vivo* (Veliz et al., 2008). An increase in leukocyte adhesion was seen following the elimination of Cx37 (Wong et al., 2006a). Interestingly, although TNF-α decreases Cx40 expression, these changes did not appear to have any effect on the trans-endothelial migration of neutrophils or other inflammatory cells (Rignault et al., 2007). However, endothelial Cx40 deficiency enhanced early neutrophil recruitment in another model of acute lung inflammation (Chadjichristos et al., 2010). This suggests that in addition to regulation of vasomotor tone, another role for Cxs in the inflammatory response may include the facilitation of leukocyte adhesion, but not necessarily the migration of inflammatory cells through the vessel wall.

It has also been demonstrated that Cx hemichannels at the EC membrane facilitate the release of intracellular ATP to promote leukocyte adhesion (Veliz et al., 2008). This theory was supported by experiments where blocking of Cx hemichannels at the luminal membrane causes a decrease in cellular ATP release and an inhibition of leukocyte adhesion following stimulation (Veliz et al., 2008). However it is important to note that, recent data now implicate pannexins in ATP release. Pannexin 1 (Panx1) can form a single channel in the cell membrane and release ATP (Chekeni et al., 2011). Like Cxs, Panx1 can also be reduced by gap junction inhibitors (Chekeni et al., 2011) and it is therefore possible that blocking pannexin channels on the EC membrane could explain the aforementioned results, thereby providing a new potential therapeutic target for vascular pathologies brought on by oxidative stress.

9.3 HYPERTENSION

9.3.1 Pulmonary Hypertension

Pulmonary hypertension (PH) is defined by a mean pulmonary arterial pressure superior to 25 mmHg. PH generally results from constriction of the small pulmonary arteries that supply blood to the lungs. Consequently, it becomes more difficult for the heart to pump blood through the lungs leading to hypertrophy of the right heart and eventually heart failure (Humbert et al., 2007). The narrowing of the small pulmonary arteries can be caused by different factors including sickle-cell disease, HIV infection, liver cirrhosis, autoimmune diseases, congenital heart diseases, and tuberculosis (Simonneau et al., 2009). Most importantly, hypoxia is a major risk factor of PH. Hypoxia is caused by insufficient oxygenation of the arterial blood as a result of diseases such as chronic obstructive pulmonary diseases (COPD), impaired control of breathing (such as sleep apnea), or residence at high altitude (above 2500 m) (Simonneau et al., 2009). Although PH is considered as a rare disease, the burden of PH is underestimated in regards to all these risk factors.

The pulmonary circulation is a low-pressure system that has the particularity to be highly sensitive to oxygen levels and therefore to oxidative stress. In contrast to the systemic circulation which produces dilation in response to hypoxemia, the pulmonary arteries constrict in response to alveolar hypoxia, a phenomenon known as hypoxic pulmonary vasoconstriction (HPV). Both adaptive responses are critical to the organ function: pulmonary circulation constricts in order to maintain the ventilation/perfusion ratio whereas the systemic circulation dilates to supply blood and oxygen to the stressed organs. The HPV is a sophisticated signalization pathway involving several ROS such as hydrogen peroxide and superoxide anion (for review, see Archer et al., 2008). In case of acute hypoxemia, the amount of hydrogen peroxide produced by VSMCs is decreased leading to an inactivation of voltage-dependent potassium channels and thus to a depolarization of the VSMCs, and successively an increase of intracellular calcium and a vasoconstriction (Post et al., 1992). As gap junctions are known to play a role in electrotonic diffusion of calcium in systemic vascular beds, one can hypothesize that they also play a role in HPV. Oxidative stress has been implicated in the pathogenesis of PH in several animal models as well as in human. A report on human tissue showed DNA damages, as well as protein damages and decreases in antioxidant enzymes in lung from patient with PH, demonstrating the presence of an oxidative stress (Bowers et al., 2004). These observations have also been reported in several animal models of PH (Redout et al., 2007; Farahmand et al., 2004; Fresquet et al., 2006). Interestingly, treatments using antioxidants, such as intra-tracheal administration of adenovirus containing the gene coding for SOD, or injection of ROS chelators, has been demonstrated to significantly attenuate the development of PH (Chen et al., 2001; Kamezaki et al., 2008).

There are few reports on the expression pattern of Cxs in the normal pulmonary circulation. Two studies on the main pulmonary artery of rat showed the presence of Cx37, Cx40, and Cx43 in the endothelium and Cx37 and Cx43 in smooth muscle (Ko et al., 1999; Nakamura et al., 1999). More recently, one report on the small pulmonary arteries showed a different expression pattern compared to the main pulmonary artery, especially in the VSMCs, where Cx37 and mostly Cx40 but not Cx43 were observed (Billaud et al., 2009). Functionally, the blocking of Cx37 and/or Cx43 using mimetic peptides leads to a decrease in contractile and calcium responses to serotonin in small pulmonary arteries, suggesting a role of gap junctions in pulmonary arterial reactivity (Billaud et al., 2009). Furthermore, blockade of gap junctions using carbenoxolone decreases the frequency of oscillations of the pulmonary arterial tone, shown to be due to variation in intracellular calcium levels. This finding suggests that gap junctions are involved in the pulmonary arterial vasomotion through coordination of intracellular calcium level between the cells of the pulmonary arterial wall (Burke et al., 2011).

PH is characterized by an extensive remodeling of the pulmonary arterial wall including hypertrophy and hyperplasia of VSMCs and/or ECs. A recent study on two rat models of PH showed that the pattern of expression of Cxs was modified in the small pulmonary arteries, especially in the VSMCs (Billaud et al., 2011a). Moreover, several studies reported a disorganization of Cx43 in the right ventricle in a rat model of PH. In contrast to controls that showed Cx43 localized to the intercalated disks, Cx43 was dispersed on the entire cell surface of rats with PH (Uzzaman et al., 2000). Moreover, the electrical conduction properties in the right ventricle were altered. This disorganization of Cx43 has been shown to be due to the phosphorylation of Cx43 leading to its internalization and degradation (Sasano et al., 2007). Interestingly, the disorganization of Cx43 as well as the impaired electrical conduction in the right ventricle are more pronounced with the progression of hypertrophy (Uzzaman et al., 2000).

Furthermore, PH is also characterized by a hyperreactivity to several agonists such as serotonin and endothelin (Humbert et al., 2004). As gap junctions have been shown to play a role in pulmonary arterial vasoreactivity in response to serotonin, it is possible that their functional role may be modified in PH. A recent report demonstrated that gap junction function in general was not altered in PH, but the role of Cxs in response to several vasoconstrictors was altered (Billaud et al., 2011a). Finally, inflammation processes play an important role in the development of PH as the level of several cytokines and chemokines is increased in patients with PH (for review, see Crosswhite and Sun, 2010). Although there is no evidence on a direct link between inflammation in PH and Cxs, it is more than likely that inflammation alters Cx expression patterns and function in PH.

9.3.2 RENAL HYPERTENSION

Renovascular hypertension is defined by an elevation of the blood pressure due to partial or complete occlusion of one or more renal arteries or their branches. This stenosis or occlusion causes a decrease in renal perfusion pressure, which activates the renin–angiotensin system leading to the release of renin and the production of angiotensin II. The activation of the renin–angiotensin system has direct effects on sodium excretion, sympathetic nerve activity, intrarenal prostaglandin concentrations, and NO production; leading ultimately to renal hypertension (Safian and Textor, 2001). Occlusion of a renal artery or renal arteriole occurs secondary to atherosclerosis in 90% of the patients, but can also be caused by fibromuscular dysplasia (Thatipelli and Misra, 2010). In the United States, the incidence of renal artery stenosis is estimated to be 1–5% of patients with hypertension (Derkx and Schalekamp, 1994; Ram, 1997).

As previously mentioned, renal arterial narrowing is mainly caused by atherosclerosis, which is known to be characterized by oxidative stress (see Section 9.2.1). ROS are known to have severe consequences on the renal microvasculature as superoxide anion causes vasoconstriction by increasing calcium in VSMCs and by scavenging the NO produced by ECs (Schnackenberg, 2002a,b; Lounsbury et al., 2000). Interestingly a study on mice deficient in SOD, an antioxidant enzyme, showed that these mice exhibit increased renal vascular resistance and oxidative stress (Briones and Touyz, 2010). Taken together, it is highly likely that oxidative stress is also present in renal hypertension in human.

Several studies have investigated the distribution of Cxs in the renal circulation and revealed that Cxs are not uniformly expressed in all blood vessels and vary between different species. However, in large renal arteries, ECs mostly express Cx37 and Cx40, whereas VSMCs express Cx43 and Cx45 (Wagner, 2008; Hanner et al., 2010). In renal microcirculation, Cx37 and Cx40 are found in ECs, whereas Cx37 and 43 are expressed in VSMCs. Interestingly, Cx37 but mostly Cx40 can be observed in the renin-producing cells of the juxtaglomerular apparatus (JGA), which are considered to be modified smooth muscle cells (Zhang and Hill, 2005; Wagner et al., 2007). Therefore, Cx40 is hypothesized to contribute to the regulation of glomerular function or renal endocrine function, such as renin production, and thus to the regulation of blood pressure.

The control of renal blood flow and blood pressure is also regulated by an autoregulatory mechanism, known as the Bayliss effect, consisting in a myogenic response activated by an increase in the perfusion of the renal circulation (Casellas et al., 1997). The myogenic response is initiated by a depolarization which opens voltage-dependent calcium channels and leads to an inward Ca^{2+} current (Davis and Hill, 1999). Therefore, it has been hypothesized that the electrotonic diffusion of Ca^{2+} may occur through the gap junctions connecting the smooth muscle cells. Although this hypothesis has not been verified in the renal circulation, a study in the cerebral circulation showed that the myogenic tone was attenuated by the gap junction inhibitor α-glycerrhetinic acid (Lagaud et al., 2002).

More generally, multiple studies in several vascular beds have shown that vascular gap junctions are responsible for the propagation of constriction or dilation along the length of vessels (Schmidt et al., 2008). These vascular conducted responses are believed to play an important role in the regulation of blood flow in microcirculation, and hypothetically in the regulation of the renal blood flow.

Gap junctions are also known to control renal blood flow through another autoregulatory mechanism, namely the tubuloglomerular feedback (TGF). The TGF is essential to the kidney function as it controls the glomerular filtration rate of the kidney.

In the structural unit of the kidney (the nephron), a specialized site of the JGA (the macula densa) is in close contact with its own afferent arteriole. Therefore, when the macula densa senses increasing sodium chloride concentrations in the tubule lumen, the macula densa sends vasoconstrictor signals to the afferent arteriole to decrease the renal blood flow and thereby decrease the glomerular filtration of the respective nephron (Moore et al., 1990). This sophisticated feedback loop requires communication from macula densa cells to the VSMCs of the afferent arteriole in order to transfer the vasoconstrictor signal. Cells of the juxtaglomerular apparatus (except those of the macula densa (Bell et al., 2003)) are coupled via gap junctions (Goligorsky et al., 1997) to establish a pathway for the transmission of the vasoconstrictor signal.

Several studies on the expression pattern of Cxs in renal hypertension reported modifications of expression level and/or distribution of Cxs. An animal model of renal hypertension induced by unilateral renal artery clipping showed an increase in Cx40 mRNA and protein in the clipped and nonclipped kidneys, whereas Cx43 was only increased in the nonclipped kidney (Haefliger et al., 2001). Moreover, as revealed by immunohistochemical labeling, the elevation of Cx40 protein was mostly observed in renin-producing cells in the wall of afferent arterioles (Haefliger et al., 2001). Further studies demonstrated that intrarenal infusion of Cx mimetic peptides which are known to decrease gap junction communication, led to vasoconstriction, increased blood pressure, and reduced renal blood flow (Lounsbury et al., 2002; Takenaka et al., 2008).

As previously mentioned, Cx40 is thought to play a major role in renin production; this hypothesis has been verified in Cx40 knockout mice. The deletion of Cx40 results in an increase of systemic blood pressure (de Wit et al., 2000, 2003). The consequences of Cx40 deletion on blood pressure was first hypothesized to be linked with increased peripheral resistance as small arterioles isolated from these mice exhibit segmental constriction and irregular vasomotion (de Wit et al., 2003, 2006). However, recent reports indicate that hypertension induced by the general deletion of Cx40 was due to an increase in renin secretion (Wagner et al., 2007; Krattinger et al., 2007). In wild-type mice, renin synthesis by renin-producing cells is downregulated by increase in blood pressure, but interestingly, cell-specific deletion of Cx40 in renin-producing cells leads to an inversion of this negative feedback (Wagner et al., 2010). This phenomenon could explain the increased hypertension observed in mice lacking Cx40, which has therefore been hypothesized to be the renal baroreceptor (Gomez and Sequeira Lopez, 2009).

Taken together, these results show that Cxs and gap junctional communication play an important role in the development of hypertension but their expression and function are also altered in renal hypertension.

9.4 ANGIOGENESIS IN CANCER

In addition to the classical biochemical signaling pathways, Cxs play a central role in cell growth, differentiation, and death (King et al., 2005). Decreased expression of key Cxs have been associated with the carcinogenesis process but the association has not been fully resolved (Mesnil et al., 2005). Early studies in HeLa cells reported that Cxs possessed tumor-suppressive properties (Mesnil et al., 1995). In addition to inhibiting tumor formation, Cxs have been implicated in their vascularization, possibly through responses to signals released under hypoxic conditions found within tumors. Many studies have suggested that Cx32 (King and Lampe, 2004a,b), Cx43 (Zhang et al., 1998; Kalvelyte et al., 2003; Avanzo et al., 2004), and Cx40 (Kwak et al., 2005) may play major roles in proliferative disorders such as the development of carcinoma, by promoting vascularization of tumors and capillary sprouting (Kwak et al., 2005; Frank et al., 2006; Haass et al., 2006; Kanczuga-Koda et al., 2006; Plante et al., 2010).

The formation of new blood vessels from preexisting vessels (angiogenesis) is essential in development, but can also promote several pathologies such as diabetic retinopathy and cancer. The process itself is driven largely by sprouting and migration of ECs, which are normally quiescent and nonmigratory. Activation of ECs involves several major signaling pathways, the most prominent being activation of the vascular endothelial growth factors (VEGFs) and their receptors (Martin, 1997). Activation of pro-angiogenic factors triggers a number of downstream effects that result in the degradation of ECM components, migration, and differentiation of ECs. Once activated, the migrating ECs incorporate a combination of three general mechanisms: chemotaxis, haptotaxis, and mechanotaxis (for review, see Lamalice et al., 2007). Chemotaxis, the directional migration toward chemoattractants, is the largest contributor to EC migration and is driven mainly by VEGF (Lamalice et al., 2006). Haplotaxis and mechanotaxis are also directional migrations, but are mediated by immobilized ligands such as integrins (Davis and Senger, 2005) and mechanical forces such as shear stress (Li et al., 2005), respectively. The result of these three mechanisms is the formation and extension of lamellipodia at the leading edge of the migrating ECs (Lamalice et al., 2007). These leading edges protrude onto and attach to the ECM components, which allow the ECs to contract the major cell body and move forward, culminating in the breakdown of adhesion proteins and release of the trailing end of the cell (Lamalice et al., 2007). The cells migrate to the targeted area, fusing together and remodeling into tubular structures with other migrating ECs (Lamalice et al., 2007) and eventually recruitment of pericytes and VSMCs leads to the formation of larger vessels (Iivanainen et al., 2003; vonTell et al., 2006).

The Cxs are integrally tied to the cellular migration, proliferation, and differentiation stages associated with angiogenesis and neo-vascularization of tumors. Enhanced Cx43 is associated with increased microvascular remodeling and angiogenesis in exercise (Bellafiore et al., 2007). However, several studies have identified that Cx43 overexpression reduces cell migration in a manner that is independent of gap junctional communication (Johnstone et al., 2010). It is not unexpected then that in the angiogenic response Cx43 gap junctional communication is reduced in tube formation (Ashton et al., 1999). Targeted knockdown of Cx43 has been shown to enhance VEGF signaling raising the possibility that Cx43 gap junctional signaling alters signaling involved in angiogenic responses (Shao et al., 2005). This was further demonstrated by studies showing that conditioned media from Cx43 overexpressing cells inhibits tube formation, all indicating that Cx43 in ECs is important in the migratory phase of angiogenesis (McLachlan et al., 2007). However, studies have demonstrated that in heterozygous deletions, a loss of Cx43 is associated with decreased angiogenesis in corneal injury models (Rodrigues et al., 2010).

Additionally (as previously discussed) many studies have identified that a loss of Cx43 in the VSMCs is key to VSMC phenotypic modulation and migration.

Knockdown of Cx43 in mice leads to an increase in the NO release possibly indicating increased eNOS levels (Liao et al., 2001). In ischemic/hypoxic models, acutely elevated levels of active eNOS are associated to a downregulation of Cx43 (but not Cx37 or Cx40) at the protein level. Despite this, application of L-NAME to ECs downregulates eNOS and inhibits endothelial migration but does not enhance expression of Cx43 or decrease associated proliferative events (Murohara et al., 1999). However, it should be noted in these studies that a Cx43 C-terminus antibody was used to stain (for flow cytometry) nonpermeabilized, nonfixed cells for Cx43 and all staining was negative, which may have altered the results as HUVECs should express Cx43 under normoxic conditions used (Murohara et al., 1999, Villars et al., 2002). Altogether, these studies illustrate that Cx43 expression can be modulated through eNOS associated activity or downstream signaling events.

9.5   FUTURE DIRECTIONS

An obstacle that remains in deciphering the exact role of Cxs in response to oxidative stress is the lack of a specific pharmacological inhibitor (reviewed in Rozental et al., 2001). For example, 18 β-glycyrrhetinic acid has been shown to alter levels of intracellular calcium in cells (Boitano and Evans, 2000). Indeed, although both carbenoxolone and octanol have been shown to potently inhibit gap junction function, they can have dramatically different effects when used in ex vivo preparations (e.g., LeBeau et al., 2002). In response to these issues, Cx mimetic peptides were developed. These peptides are approximately 8–12 amino acids in length and derived against the extracellular loops of certain vascular Cxs (reviewed in Evans and Boitano, 2001). The exact mechanism of action of these peptides, as with all of the reported gap junction inhibitors, is unknown. However, one hypothesis has been that the peptide acts by initially binding to exposed hemichannels before they dock in fixed gap junction conformations (Berthoud et al., 2000). The rational for this is the temporal delay associated with an uncoupling event which correlates with reported cycling of Cxs on the membrane of cells (reviewed in Martin and Evans, 2004). Evidence for this is demonstrated in Figure 9.2. In the figure, the Cx mimetic peptide ([37,43]gap27) is conjugated to a rhodamine dextran and appears to bind to undocked hemichannels on the apical location of the cell, which temporally moves to the lateral borders. This observation is in line with the rapid inhibition of the peptides on studies involving hemichannels (De Vuyst et al., 2006), but the more prolonged incubation time required for inhibition of gap junctional coupling (Boitano and Evans, 2000). The promise of these peptides is their built-in specificity and their capacity to be "washed out" for a return to function. Unfortunately, similar to carbenoxolone (Chekeni et al., 2011) it now appears that these peptides may also block pannexin channels (Wang et al., 2007b). For these reasons, the exact mechanism involved with the apparent gap junctional block associated with the Cx mimetics requires elucidation. Unfortunately, the use of global Cx knockout mice does not provide an alternative for pharmacological inhibition as many of these mice have been found to have extensive Cx compensatory expression (Simon and McWhorter, 2003; Isakson

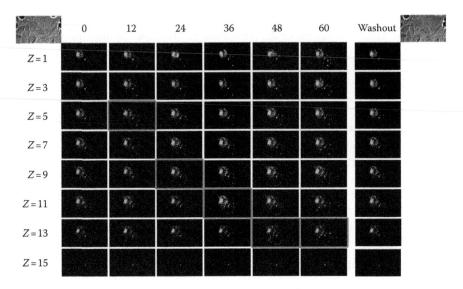

FIGURE 9.2   Spatial and temporal distribution of [37,43]gap27-rhodamine on HeLa cells transfected with Cx43-GFP. Top left and top right represent differential interphase contrast images of cells before application of the peptide and after washout of the peptide, respectively. The connexin mimetic [37,43]gap27-rhodamine was applied to HeLa cells transfected Cx43-GFP. The rows are confocal z-stacks going from the top of the cell ($Z = 1$) to the bottom of the cell ($Z = 15$) with 0.25 μm optical sections and the columns are images at each time point (in minutes). Images with gray borders indicate the farthest extent of the [37,43]gap27-rhodamine at each time point. Note that it takes at least 48 min for the borders of the transfected cells to show a positive rhodamine staining and that washout removes all trace of the peptide. Specificity of the peptide is indicated by the fact that it does not appear to bind to any other location in monolayer of cells.

et al., 2006a). Therefore, currently the conditional Cx knockout mice and the use of Cx siRNA appear to be the most specific mechanism for Cx null studies.

As demonstrated by the data above, several lines of evidence indicate that Cx expression and function is altered (or maintained) by physiological and pathological oxidative processes. However, there are several questions that remain, especially regarding the emerging role of pannexins. Indeed, the role of pannexins and their capacity to release ATP under pathological and homeostatic conditions as hexameric channels, has only recently been discovered and these proteins could play a major role in the pathologies described above. It will be interesting to observe how connexins and pannexins coordinate their functions in the vasculature in response to ROS. In summary, oxidative events and Cx function are intimately linked and their interplay can have dramatic effects on vascular function.

REFERENCES

Cleeman, J. I. 2001. Executive Summary of the Third Report of the National Cholesterol Education Program (NCEP) Expert Panel on Detection, Evaluation, and Treatment of High Blood Cholesterol in Adults (Adult Treatment Panel III). *JAMA*, 285, 2486–97.

Aasen, T., Graham, S. V., Edward, M., and Hodgins, M. B. 2005. Reduced expression of multiple gap junction proteins is a feature of cervical dysplasia. *Mol Cancer*, 4, 31.

Ambrose, J. A. and Barua, R. S. 2004. The pathophysiology of cigarette smoking and cardiovascular disease: An update. *J Am Coll Cardiol*, 43, 1731–7.

Archer, S. L., Gomberg-Maitland, M., Maitland, M. L., Rich, S., Garcia, J. G., and Weir, E. K. 2008. Mitochondrial metabolism, redox signaling, and fusion: A mitochondria-ROS-HIF-1alpha-Kv1.5 O2-sensing pathway at the intersection of pulmonary hypertension and cancer. *Am J Physiol Heart Circ Physiol*, 294, H570–8.

Ashton, A. W., Yokota, R., John, G., Zhao, S., Suadicani, S. O., Spray, D. C., and Ware, J. A. 1999. Inhibition of endothelial cell migration, intercellular communication, and vascular tube formation by thromboxane A(2). *J Biol Chem*, 274, 35562–70.

Avanzo, J. L., Mesnil, M., Hernandez-Blazquez, F. J., Mackowiak, II, Mori, C. M., Da Silva, T. C., Oloris, S. C. et al. 2004. Increased susceptibility to urethane-induced lung tumors in mice with decreased expression of connexin43. *Carcinogenesis*, 25, 1973–82.

Azevedo, L. C., Pedro, M. A., Souza, L. C., De Souza, H. P., Janiszewski, M., Da Luz, P. L., and Laurindo, F. R. 2000. Oxidative stress as a signaling mechanism of the vascular response to injury: The redox hypothesis of restenosis. *Cardiovasc Res*, 47, 436–45.

Bano, D. and Nicotera, P. 2007. Ca^{2+} signals and neuronal death in brain ischemia. *Stroke*, 38, 674–6.

Bell, P. D., Lapointe, J. Y., and Peti-Peterdi, J. 2003. Macula densa cell signaling. *Annu Rev Physiol*, 65, 481–500.

Bellafiore, M., Sivverini, G., Palumbo, D., Macaluso, F., Bianco, A., Palma, A., and Farina, F. 2007. Increased cx43 and angiogenesis in exercised mouse hearts. *Int J Sports Med*, 28, 749–55.

Bellinger, A. M., Reiken, S., Carlson, C., Mongillo, M., Liu, X., Rothman, L., Matecki, S., Lacampagne, A., and Marks, A. R. 2009. Hypernitrosylated ryanodine receptor calcium release channels are leaky in dystrophic muscle. *Nat Med*, 15, 325–30.

Benhar, M., Forrester, M. T., and Stamler, J. S. 2009. Protein denitrosylation: Enzymatic mechanisms and cellular functions. *Nat Rev Mol Cell Biol*, 10, 721–32.

Bermudez-Fajardo, A., Yliharsila, M., Evans, W. H., Newby, A. C., and Oviedo-Orta, E. 2007. CD4+ T lymphocyte subsets express connexin 43 and establish gap junction channel communication with macrophages in vitro. *J Leukoc Biol*, 82, 608–12.

Berthoud, V. M., Beyer, E. C., and Seul, K. H. 2000. Peptide inhibitors of intercellular communication. *Am J Physiol Lung Cell Mol Physiol*, 279, L619–22.

Billaud, M., Dahan, D., Marthan, R., Savineau, J. P., and Guibert, C. 2011a. Role of the gap junctions in the contractile response to agonists in pulmonary artery from two rat models of pulmonary hypertension. *Respir Res*, 12, 30.

Billaud, M., Lohman, A. W., Straub, A. C., Looft-Wilson, R., Johnstone, S. R., Araj, C. A., Best, A. K. et al. 2011b. Pannexin1 regulates alpha1-adrenergic receptor-mediated vasoconstriction. *Circ Res*, 109, 80–5.

Billaud, M., Marthan, R., Savineau, J. P., and Guibert, C. 2009. Vascular smooth muscle modulates endothelial control of vasoreactivity via reactive oxygen species production through myoendothelial communications. *PLoS One*, 4, e6432.

Boerma, M., Forsberg, L., Van Zeijl, L., Morgenstern, R., De Faire, U., Lemne, C., Erlinge, D., Thulin, T., Hong, Y., and Cotgreave, I. A. 1999. A genetic polymorphism in connexin 37 as a prognostic marker for atherosclerotic plaque development. *J Intern Med*, 246, 211–8.

Boitano, S. and Evans, W. H. 2000. Connexin mimetic peptides reversibly inhibit Ca(2+) signaling through gap junctions in airway cells. *Am J Physiol Lung Cell Mol Physiol*, 279, L623–30.

Bolon, M. L., Kidder, G. M., Simon, A. M., and Tyml, K. 2007. Lipopolysaccharide reduces electrical coupling in microvascular endothelial cells by targeting connexin40 in a tyrosine-, ERK1/2-, PKA-, and PKC-dependent manner. *J Cell Physiol*, 211, 159–66.

Bolon, M. L., Peng, T., Kidder, G. M., and Tyml, K. 2008. Lipopolysaccharide plus hypoxia and reoxygenation synergistically reduce electrical coupling between microvascular endothelial cells by dephosphorylating connexin40. *J Cell Physiol*, 217, 350–9.

Bowers, R., Cool, C., Murphy, R. C., Tuder, R. M., Hopken, M. W., Flores, S. C., and Voelkel, N. F. 2004. Oxidative stress in severe pulmonary hypertension. *Am J Respir Crit Care Med*, 169, 764–9.

Boyle, J. J. 2005. Macrophage activation in atherosclerosis: Pathogenesis and pharmacology of plaque rupture. *Curr Vasc Pharmacol*, 3, 63–8.

Briones, A. M. and Touyz, R. M. 2010. Oxidative stress and hypertension: Current concepts. *Curr Hypertens Rep*, 12, 135–42.

Brisset, A. C., Isakson, B. E., and Kwak, B. R. 2008. Connexins in vascular physiology and pathology. *Antioxid Redox Signal,* 11(2):267–85.

Brisset, A. C., Isakson, B. E., and Kwak, B. R. 2009. Connexins in vascular physiology and pathology. *Antioxid Redox Signal*, 11, 267–82.

Buchner, D. A., Yazbek, S. N., Solinas, P., Burrage, L. C., Morgan, M. G., Hoppel, C. L., and Nadeau, J. H. 2011. Increased mitochondrial oxidative phosphorylation in the liver is associated with obesity and insulin resistance. *Obesity (Silver Spring)*, 19(5):917–24.

Bundy, R. E., Marczin, N., Birks, E. F., Chester, A. H., and Yacoub, M. H. 2000. Transplant atherosclerosis: Role of phenotypic modulation of vascular smooth muscle by nitric oxide. *Gen Pharmacol*, 34, 73–84.

Burke, M. M., Bieger, D., and Tabrizchi, R. 2011. Agonist-induced periodic vasomotion in rat isolated pulmonary artery. *Fundam Clin Pharmacol*, 25(4):443–51.

Casellas, D., Bouriquet, N., and Moore, L. C. 1997. Branching patterns and autoregulatory responses of juxtamedullary afferent arterioles. *Am J Physiol*, 272, F416–21.

Chadha, P. S., Haddock, R. E., Howitt, L., Morris, M. J., Murphy, T. V., Grayson, T. H., and Sandow, S. L. 2010. Obesity up-regulates intermediate conductance calcium-activated potassium channels and myoendothelial gap junctions to maintain endothelial vasodilator function. *J Pharmacol Exp Ther*, 335, 284–93.

Chadjichristos, C. E., Derouette, J. P., and Kwak, B. R. 2006a. Connexins in atherosclerosis. *Adv Cardiol*, 42, 255–67.

Chadjichristos, C. E., Matter, C. M., Roth, I., Sutter, E., Pelli, G., Luscher, T. F., Chanson, M., and Kwak, B. R. 2006b. Reduced connexin43 expression limits neointima formation after balloon distension injury in hypercholesterolemic mice. *Circulation*, 113, 2835–43.

Chadjichristos, C. E., Morel, S., Derouette, J. P., Sutter, E., Roth, I., Brisset, A. C., Bochaton-Piallat, M. L., and Kwak, B. R. 2008. Targeting connexin 43 prevents platelet-derived growth factor-BB-induced phenotypic change in porcine coronary artery smooth muscle cells. *Circ Res*, 102, 653–60.

Chadjichristos, C. E., Scheckenbach, K. E., Van Veen, T. A., Richani Sarieddine, M. Z., de Wit, C., Yang, Z., Roth, I. et al. 2010. Endothelial-specific deletion of connexin40 promotes atherosclerosis by increasing CD73-dependent leukocyte adhesion. *Circulation*, 121, 123–31.

Chanson, M. and Kwak, B. R. 2007. Connexin37: A potential modifier gene of inflammatory disease. *J Mol Med*, 85, 787–95.

Chatterjee, S., Berliner, J. A., Subbanagounder, G. G., Bhunia, A. K., and Koh, S. 2004. Identification of a biologically active component in minimally oxidized low density lipoprotein (MM-LDL) responsible for aortic smooth muscle cell proliferation. *Glycoconj J*, 20, 331–338.

Chatterjee, S. and Ghosh, N. 1996. Oxidized low density lipoprotein stimulates aortic smooth muscle cell proliferation. *Glycobiology*, 6, 303–311.

Chekeni, F. B., Elliott, M. R., Sandilos, J. K., Walk, S. F., Kinchen, J. M., Lazarowski, E. R., Armstrong, A. J. et al. 2011. Pannexin 1 channels mediate "find-me" signal release and membrane permeability during apoptosis. *Nature*, 467, 863–7.

Chen, C. H., Pan, C. H., Chen, C. C., and Huang, M. C. 2011. Increased oxidative DNA damage in patients with alcohol dependence and its correlation with alcohol withdrawal severity. *Alcohol Clin Exp Res*, 35(2):338–44.

Chen, M. J., Chiang, L. Y., and Lai, Y. L. 2001. Reactive oxygen species and substance P in monocrotaline-induced pulmonary hypertension. *Toxicol Appl Pharmacol*, 171, 165–73.

Codoner-Franch, P., Boix-Garcia, L., Simo-Jorda, R., Del Castillo-Villaescusa, C., Maset-Maldonado, J., and Valls-Belles, V. 2010. Is obesity associated with oxidative stress in children? *Int J Pediatr Obes*, 5, 56–63.

Coleman, H. A., Tare, M., and Parkington, H. C. 2001a. EDHF is not K+ but may be due to spread of current from the endothelium in guinea pig arterioles. *Am J Physiol Heart Circ Physiol*, 280, H2478–83.

Coleman, H. A., Tare, M., and Parkington, H. C. 2001b. K+ currents underlying the action of endothelium-derived hyperpolarizing factor in guinea-pig, rat and human blood vessels. *J Physiol*, 531, 359–73.

Collings, A., Islam, M. S., Juonala, M., Rontu, R., Kahonen, M., Hutri-Kahonen, N., Laitinen, T., Marniemi, J., Viikari, J. S., Raitakari, O. T., and Lehtimaki, T. J. 2007. Associations between connexin37 gene polymorphism and markers of subclinical atherosclerosis: The Cardiovascular Risk in Young Finns study. *Atherosclerosis*, 195, 379–84.

Cooper, C. D. and Lampe, P. D. 2002. Casein kinase 1 regulates connexin-43 gap junction assembly. *J Biol Chem*, 277, 44962–8.

Cowan, D. B., Lye, S. J., and Langille, B. L. 1998. Regulation of vascular connexin43 gene expression by mechanical loads. *Circ Res*, 82, 786–93.

Crosswhite, P. and Sun, Z. 2010. Nitric oxide, oxidative stress and inflammation in pulmonary arterial hypertension. *J Hypertens*, 28, 201–12.

Davis, G. E. and Senger, D. R. 2005. Endothelial extracellular matrix: Biosynthesis, remodeling, and functions during vascular morphogenesis and neovessel stabilization. *Circ Res*, 97, 1093–107.

Davis, M. J. and Hill, M. A. 1999. Signaling mechanisms underlying the vascular myogenic response. *Physiol Rev*, 79, 387–423.

De Vuyst, E., Decrock, E., Cabooter, L., Dubyak, G. R., Naus, C. C., Evans, W. H., and Leybaert, L. 2006. Intracellular calcium changes trigger connexin 32 hemichannel opening. *EMBO J*, 25, 34–44.

de Wit, C. and Griffith, T. M. 2010. Connexins and gap junctions in the EDHF phenomenon and conducted vasomotor responses. *Pflugers Arch*, 459, 897–914.

de Wit, C., Hoepfl, B., and Wolfle, S. E. 2006. Endothelial mediators and communication through vascular gap junctions. *Biol Chem*, 387, 3–9.

de Wit, C., Roos, F., Bolz, S. S., Kirchhoff, S., Kruger, O., Willecke, K., and Pohl, U. 2000. Impaired conduction of vasodilation along arterioles in connexin40-deficient mice. *Circ Res*, 86, 649–55.

de Wit, C., Roos, F., Bolz, S. S., and Pohl, U. 2003. Lack of vascular connexin 40 is associated with hypertension and irregular arteriolar vasomotion. *Physiol Genomics*, 13, 169–177.

Demaurex, N. and Scorrano, L. 2009. Reactive oxygen species are NOXious for neurons. *Nat Neurosci*, 12, 819–20.

Depaola, N., Davies, P. F., Pritchard, W. F., Jr., Florez, L., Harbeck, N., and Polacek, D. C. 1999. Spatial and temporal regulation of gap junction connexin43 in vascular endothelial cells exposed to controlled disturbed flows in vitro. *Proc Natl Acad Sci USA*, 96, 3154–9.

Derkx, F. H. and Schalekamp, M. A. 1994. Renal artery stenosis and hypertension. *Lancet*, 344, 237–9.

Derouette, J. P., Wong, C., Burnier, L., Morel, S., Sutter, E., Galan, K., Brisset, A. C., Roth, I., Chadjichristos, C. E., and Kwak, B. R. 2009. Molecular role of Cx37 in advanced atherosclerosis: A micro-array study. *Atherosclerosis*, 206, 69–76.

Doble, B. W., Dang, X., Ping, P., Fandrich, R. R., Nickel, B. E., Jin, Y., Cattini, P. A., and Kardami, E. 2004. Phosphorylation of serine 262 in the gap junction protein connexin-43 regulates DNA synthesis in cell–cell contact forming cardiomyocytes. *J Cell Sci*, 117, 507–514.

Droge, W. 2002. Free radicals in the physiological control of cell function. *Physiol Rev*, 82, 47–95.

Erdbruegger, U., Dhaygude, A., Haubitz, M., and Woywodt, A. 2010. Circulating endothelial cells: Markers and mediators of vascular damage. *Curr Stem Cell Res Ther*, 5, 294–302.

Evans, W. H. and Boitano, S. 2001. Connexin mimetic peptides: Specific inhibitors of gap-junctional intercellular communication. *Biochem Soc Trans*, 29, 606–612.

Fanchaouy, M., Serir, K., Meister, J. J., Beny, J. L., and Bychkov, R. 2005. Intercellular communication: Role of gap junctions in establishing the pattern of ATP-elicited Ca^{2+} oscillations and Ca^{2+}-dependent currents in freshly isolated aortic smooth muscle cells. *Cell Calcium*, 37, 25–34.

Farahmand, F., Hill, M. F., and Singal, P. K. 2004. Antioxidant and oxidative stress changes in experimental cor pulmonale. *Mol Cell Biochem*, 260, 21–9.

Feletou, M. and Vanhoutte, P. M. 2009. EDHF: An update. *Clin Sci (Lond)*, 117, 139–55.

Fernandes, R., Girao, H., and Pereira, P. 2004. High glucose down-regulates intercellular communication in retinal endothelial cells by enhancing degradation of connexin 43 by a proteasome-dependent mechanism. *J Biol Chem*, 279, 27219–24.

Firouzi, M., Kok, B., Spiering, W., Busjahn, A., Bezzina, C. R., Ruijter, J. M., Koeleman, B. P. et al. 2006. Polymorphisms in human connexin40 gene promoter are associated with increased risk of hypertension in men. *J Hypertens*, 24, 325–30.

Ford, E. S., Giles, W. H., and Dietz, W. H. 2002. Prevalence of the metabolic syndrome among US adults: Findings from the third National Health and Nutrition Examination Survey. *JAMA*, 287, 356–9.

Forman, H. J., Zhang, H., and Rinna, A. 2009. Glutathione: Overview of its protective roles, measurement, and biosynthesis. *Mol Aspects Med*, 30, 1–12.

Forstermann, U. and Munzel, T. 2006. Endothelial nitric oxide synthase in vascular disease: From marvel to menace. *Circulation*, 113, 1708–14.

Foster, M. W., Mcmahon, T. J., and Stamler, J. S. 2003. S-nitrosylation in health and disease. *Trends Mol Med*, 9, 160–8.

Francesco, M. 2010. Ca2+ Signalling in damaged endothelium: Do connexin hemichannels aid in filling the gap? *Curr Drug Therapy*, 5, 277–287.

Frank, D. K., Szymkowiak, B., and Hughes, C. A. 2006. Connexin expression and gap junctional intercellular communication in human squamous cell carcinoma of the head and neck. *Otolaryngol Head Neck Surg*, 135, 736–43.

Fresquet, F., Pourageaud, F., Leblais, V., Brandes, R. P., Savineau, J. P., Marthan, R., and Muller, B. 2006. Role of reactive oxygen species and gp91phox in endothelial dysfunction of pulmonary arteries induced by chronic hypoxia. *Br J Pharmacol*, 148, 714–23.

Gabriels, J. E. and Paul, D. L. 1998. Connexin43 is highly localized to sites of disturbed flow in rat aortic endothelium but connexin37 and connexin40 are more uniformly distributed. *Circ Res*, 83, 636–43.

Garban, H. J., Marquez-Garban, D. C., Pietras, R. J., and Ignarro, L. J. 2005. Rapid nitric oxide-mediated S-nitrosylation of estrogen receptor: Regulation of estrogen-dependent gene transcription. *Proc Natl Acad Sci USA*, 102, 2632–6.

Goligorsky, M. S., Iijima, K., Krivenko, Y., Tsukahara, H., Hu, Y., and Moore, L. C. 1997. Role of mesangial cells in macula densa to afferent arteriole information transfer. *Clin Exp Pharmacol Physiol*, 24, 527–31.

Gomez, R. A. and Sequeira Lopez, M. L. 2009. Who and where is the renal baroreceptor?: The connexin hypothesis. *Kidney Int*, 75, 460–2.

Haass, N. K., Wladykowski, E., Kief, S., Moll, I., and Brandner, J. M. 2006. Differential induction of connexins 26 and 30 in skin tumors and their adjacent epidermis. *J Histochem Cytochem*, 54, 171–82.

Haddock, R. E., Grayson, T. H., Brackenbury, T. D., Meaney, K. R., Neylon, C. B., Sandow, S. L., and Hill, C. E. 2006. Endothelial coordination of cerebral vasomotion via myoendothelial gap junctions containing connexins 37 and 40. *Am J Physiol Heart Circ Physiol*, 291, H2047–56.

Haefliger, J. A., Demotz, S., Braissant, O., Suter, E., Waeber, B., Nicod, P., and Meda, P. 2001. Connexins 40 and 43 are differentially regulated within the kidneys of rats with renovascular hypertension. *Kidney Int*, 60, 190–201.

Hakim, C. H., Jackson, W. F., and Segal, S. S. 2008. Connexin isoform expression in smooth muscle cells and endothelial cells of hamster cheek pouch arterioles and retractor feed arteries. *Microcirculation*, 15, 503–14.

Hanner, F., Sorensen, C. M., Holstein-Rathlou, N. H., and Peti-Peterdi, J. 2010. Connexins and the kidney. *Am J Physiol Regul Integr Comp Physiol*, 298, R1143–55.

Heberlein, K. R., Straub, A. C., and Isakson, B. E. 2009a. The myoendothelial junction: Breaking through the matrix? *Microcirculation*, 16, 307–22.

Heberlein, K. R., Straub, A. C., and Isakson, B. E. 2009b. The myoendothelial junction: Breaking through the matrix? *Microcirculation*, 16(4):307–22.

Heery, J. M., Kozak, M., Stafforini, D. M., Jones, D. A., Zimmerman, G. A., Mcintyre, T. M., and Prescott, S. M. 1995. Oxidatively modified LDL contains phospholipids with platelet-activating factor-like activity and stimulates the growth of smooth muscle cells. *J Clin Invest*, 96, 2322–30.

Herrero-Gonzalez, S., Gangoso, E., Giaume, C., Naus, C. C., Medina, J. M., and Tabernero, A. 2010. Connexin43 inhibits the oncogenic activity of c-Src in C6 glioma cells. *Oncogene*, 29, 5712–23.

Hickey, M. J., Reinhardt, P. H., Ostrovsky, L., Jones, W. M., Jutila, M. A., Payne, D., Elliott, J., and Kubes, P. 1997. Tumor necrosis factor-alpha induces leukocyte recruitment by different mechanisms *in vivo* and *in vitro*. *J Immunol*, 158, 3391–400.

Hong, T. and Hill, C. E. 1998. Restricted expression of the gap junctional protein connexin 43 in the arterial system of the rat. *J Anat*, 192(Pt 4), 583–93.

Hu, J. and Cotgreave, I. A. 1997. Differential regulation of gap junctions by proinflammatory mediators in vitro. *J Clin Invest*, 99, 2312–6.

Huang, A. J., Manning, J. E., Bandak, T. M., Ratau, M. C., Hanser, K. R., and Silverstein, S. C. 1993. Endothelial cell cytosolic free calcium regulates neutrophil migration across monolayers of endothelial cells. *J Cell Biol*, 120, 1371–80.

Humbert, M., Khaltaev, N., Bousquet, J., and Souza, R. 2007. Pulmonary hypertension: From an orphan disease to a public health problem. *Chest*, 132, 365–7.

Humbert, M., Morrell, N. W., Archer, S. L., Stenmark, K. R., Maclean, M. R., Lang, I. M., Christman, B. W. et al. 2004. Cellular and molecular pathobiology of pulmonary arterial hypertension. *J Am Coll Cardiol*, 43, 13S–24S.

Iivanainen, E., Nelimarkka, L., Elenius, V., Heikkinen, S. M., Junttila, T. T., Sihombing, L., Sundvall, M. et al. 2003. Angiopoietin-regulated recruitment of vascular smooth muscle cells by endothelial-derived heparin binding EGF-like growth factor. *FASEB J*, 17, 1609–21.

Isakson, B. E., Damon, D. N., Day, K. H., Liao, Y., and Duling, B. R. 2006a. Connexin40 and connexin43 in mouse aortic endothelium: Evidence for coordinated regulation. *Am J Physiol Heart Circ Physiol*, 290, H1199–205.

Isakson, B. E., Damon, D. N., Day, K. H., Liao, Y., and Duling, B. R. 2006b. Connexin40 and connexin43 in mouse aortic endothelium: Evidence for coordinated regulation. *Am J Physiol Heart Circ Physiol*, 290, H1199–205.

Isakson, B. E., Kronke, G., Kadl, A., Leitinger, N., and Duling, B. R. 2006c. Oxidized phospholipids alter vascular connexin expression, phosphorylation, and heterocellular communication. *Arterioscler Thromb Vasc Biol*, 26, 2216–2221.

Iwakiri, Y., Satoh, A., Chatterjee, S., Toomre, D. K., Chalouni, C. M., Fulton, D., Groszmann, R. J., Shah, V. H., and Sessa, W. C. 2006. Nitric oxide synthase generates nitric oxide locally to regulate compartmentalized protein S-nitrosylation and protein trafficking. *Proc Natl Acad Sci USA*, 103, 19777–82.

Jaffrey, S. R., Erdjument-Bromage, H., Ferris, C. D., Tempst, P., and Snyder, S. H. 2001. Protein S-nitrosylation: A physiological signal for neuronal nitric oxide. *Nat Cell Biol*, 3, 193–7.

Jara, P. I., Boric, M. P., and Saez, J. C. 1995. Leukocytes express connexin 43 after activation with lipopolysaccharide and appear to form gap junctions with endothelial cells after ischemia–reperfusion. *Proc Natl Acad Sci USA*, 92, 7011–15.

Jia, G., Aggarwal, A., Yohannes, A., Gangahar, D. M., and Agrawal, D. K. 2011. Cross-talk between angiotensin II and IGF-1-induced connexin 43 expression in human saphenous vein smooth muscle cells. *J Cell Mol Med*, 15(8):1695–1702.

Jia, G., Cheng, G., Gangahar, D. M., and Agrawal, D. K. 2008. Involvement of connexin 43 in angiotensin II-induced migration and proliferation of saphenous vein smooth muscle cells via the MAPK-AP-1 signaling pathway. *J Mol Cell Cardiol*, 44(5), 882–90.

Johnstone, S., Isakson, B., and Locke, D. 2009a. Biological and biophysical properties of vascular connexin channels. *Int Rev Cell Mol Biol*, 278, 69–118.

Johnstone, S. R., Best, A. K., Wright, C. S., Isakson, B. E., Errington, R. J., and Martin, P. E. 2010. Enhanced connexin 43 expression delays intra-mitotic duration and cell cycle traverse independently of gap junction channel function. *J Cell Biochem*, 110, 772–82.

Johnstone, S. R., Ross, J., Rizzo, M. J., Straub, A. C., Lampe, P. D., Leitinger, N., and Isakson, B. E. 2009b. Oxidized phospholipid species promote *in vivo* differential cx43 phosphorylation and vascular smooth muscle cell proliferation. *Am J Pathol*, 175, 916–24.

Kalvelyte, A., Imbrasaite, A., Bukauskiene, A., Verselis, V. K., and Bukauskas, F. F. 2003. Connexins and apoptotic transformation. *Biochem Pharmacol*, 66, 1661–72.

Kameritsch, P., Khandoga, N., Nagel, W., Hundhausen, C., Lidington, D., and Pohl, U. 2005. Nitric oxide specifically reduces the permeability of Cx37-containing gap junctions to small molecules 19. *J Cell Physiol*, 203, 233–42.

Kamezaki, F., Tasaki, H., Yamashita, K., Tsutsui, M., Koide, S., Nakata, S., Tanimoto, A. et al. 2008. Gene transfer of extracellular superoxide dismutase ameliorates pulmonary hypertension in rats. *Am J Respir Crit Care Med*, 177, 219–26.

Kanczuga-Koda, L., Sulkowski, S., Lenczewski, A., Koda, M., Wincewicz, A., Baltaziak, M., and Sulkowska, M. 2006. Increased expression of connexins 26 and 43 in lymph node metastases of breast cancer. *J Clin Pathol*, 59, 429–33.

Kanemitsu, M. Y., Jiang, W., and Eckhart, W. 1998. Cdc2-mediated phosphorylation of the gap junction protein, connexin43, during mitosis. *Cell Growth Differ*, 9, 13–21.

Kanemitsu, M. Y., Loo, L. W., Simon, S., Lau, A. F., and Eckhart, W. 1997. Tyrosine phosphorylation of connexin 43 by v-Src is mediated by SH2 and SH3 domain interactions. *J Biol Chem*, 272, 22824–31.

Kielbassa-Schnepp, K., Strey, A., Janning, A., Missiaen, L., Nilius, B., and Gerke, V. 2001. Endothelial intracellular Ca^{2+} release following monocyte adhesion is required for the transendothelial migration of monocytes. *Cell Calcium*, 30, 29–40.

Kilic, T. 2010. [Mechanisms underlying obesity associated oxidative stress: The role of leptin and adiponectin]. *Anadolu Kardiyol Derg*, 10, 397–9.

King, T. J., Gurley, K. E., Prunty, J., Shin, J. L., Kemp, C. J., and Lampe, P. D. 2005. Deficiency in the gap junction protein connexin32 alters p27Kip1 tumor suppression and MAPK activation in a tissue-specific manner. *Oncogene*, 24, 1718–26.

King, T. J. and Lampe, P. D. 2004a. The gap junction protein connexin32 is a mouse lung tumor suppressor. *Cancer Res*, 64, 7191–6.

King, T. J. and Lampe, P. D. 2004b. Mice deficient for the gap junction protein Connexin32 exhibit increased radiation-induced tumorigenesis associated with elevated mitogen-activated protein kinase (p44/Erk1, p42/Erk2) activation. *Carcinogenesis*, 25, 669–80.

Ko, Y. S., Yeh, H. I., Rothery, S., Dupont, E., Coppen, S. R., and Severs, N. J. 1999. Connexin make-up of endothelial gap junctions in the rat pulmonary artery as revealed by immunoconfocal microscopy and triple-label immunogold electron microscopy. *J Histochem Cytochem*, 47, 683–92.

Krattinger, N., Capponi, A., Mazzolai, L., Aubert, J. F., Caille, D., Nicod, P., Waeber, G., Meda, P., and Haefliger, J. A. 2007. Connexin40 regulates renin production and blood pressure. *Kidney Int*, 72, 814–22.

Kuroki, T., Inoguchi, T., Umeda, F., Ueda, F., and Nawata, H. 1998. High glucose induces alteration of gap junction permeability and phosphorylation of connexin-43 in cultured aortic smooth muscle cells. *Diabetes*, 47, 931–6.

Kwak, B. R. and Mach, F. 2001. Statins inhibit leukocyte recruitment: New evidence for their anti-inflammatory properties. *Arterioscler Thromb Vasc Biol*, 21, 1256–1258.

Kwak, B. R., Mulhaupt, F., Veillard, N., Gros, D. B., and Mach, F. 2002. Altered pattern of vascular connexin expression in atherosclerotic plaques. *Arterioscler Thromb Vasc Biol*, 22, 225–30.

Kwak, B. R., Saez, J. C., Wilders, R., Chanson, M., Fishman, G. I., Hertzberg, E. L., Spray, D. C., and Jongsma, H. J. 1995. Effects of cGMP-dependent phosphorylation on rat and human connexin43 gap junction channels. *Pflugers Arch*, 430, 770–8.

Kwak, B. R., Silacci, P., Stergiopulos, N., Hayoz, D., and Meda, P. 2005. Shear stress and cyclic circumferential stretch, but not pressure, alter connexin43 expression in endothelial cells. *Cell Commun Adhes*, 12, 261–70.

Kwak, B. R., Veillard, N., Pelli, G., Mulhaupt, F., James, R. W., Chanson, M., and Mach, F. 2003. Reduced connexin43 expression inhibits atherosclerotic lesion formation in low-density lipoprotein receptor-deficient mice. *Circulation*, 107, 1033–1039.

Lagaud, G., Karicheti, V., Knot, H. J., Christ, G. J., and Laher, I. 2002. Inhibitors of gap junctions attenuate myogenic tone in cerebral arteries. *Am J Physiol Heart Circ Physiol*, 283, H2177–86.

Lamalice, L., Houle, F., and Huot, J. 2006. Phosphorylation of Tyr1214 within VEGFR-2 triggers the recruitment of Nck and activation of Fyn leading to SAPK2/p38 activation and endothelial cell migration in response to VEGF. *J Biol Chem*, 281, 34009–20.

Lamalice, L., Le Boeuf, F., and Huot, J. 2007. Endothelial cell migration during angiogenesis. *Circ Res*, 100, 782–94.

Lampe, P. D., Kurata, W. E., Warn-Cramer, B. J., and Lau, A. F. 1998. Formation of a distinct connexin43 phosphoisoform in mitotic cells is dependent upon p34cdc2 kinase. *J Cell Sci*, 111(Pt 6), 833–41.

Lampe, P. D. and Lau, A. F. 2000. Regulation of gap junctions by phosphorylation of connexins. *Arch Biochem Biophys*, 384, 205–215.

Lampe, P. D. and Lau, A. F. 2004. The effects of connexin phosphorylation on gap junctional communication. *Int J Biochem Cell Biol*, 36, 1171–86.

Lampe, P. D., Tenbroek, E. M., Burt, J. M., Kurata, W. E., Johnson, R. G., and Lau, A. F. 2000. Phosphorylation of connexin43 on serine368 by protein kinase C regulates gap junctional communication. *J Cell Biol*, 149, 1503–12.

Lebeau, F. E., Towers, S. K., Traub, R. D., Whittington, M. A., and Buhl, E. H. 2002. Fast network oscillations induced by potassium transients in the rat hippocampus in vitro. *J Physiol*, 542, 167–79.

Li, A. F. and Roy, S. 2009. High glucose-induced downregulation of connexin 43 expression promotes apoptosis in microvascular endothelial cells. *Invest Ophthalmol Vis Sci*, 50, 1400–7.

Li, J. M. and Shah, A. M. 2004. Endothelial cell superoxide generation: Regulation and relevance for cardiovascular pathophysiology. *Am J Physiol Regul Integr Comp Physiol*, 287, R1014–30.

Li, S., Huang, N. F., and Hsu, S. 2005. Mechanotransduction in endothelial cell migration. *J Cell Biochem*, 96, 1110–26.

Liao, Y., Day, K. H., Damon, D. N., and Duling, B. R. 2001. Endothelial cell-specific knockout of connexin 43 causes hypotension and bradycardia in mice. *Proc Natl Acad Sci USA*, 98, 9989–94.

Liao, Y., Regan, C. P., Manabe, I., Owens, G. K., Day, K. H., Damon, D. N., and Duling, B. R. 2007. Smooth muscle-targeted knockout of connexin43 enhances neointimal formation in response to vascular injury. *Arterioscler Thromb Vasc Biol*, 27, 1037–42.

Lima, B., Forrester, M. T., Hess, D. T., and Stamler, J. S. 2010. S-nitrosylation in cardiovascular signaling. *Circ Res*, 106, 633–46.

Lin, D. and Takemoto, D. J. 2005. Oxidative activation of protein kinase Cgamma through the C1 domain. Effects on gap junctions. *J Biol Chem*, 280, 13682–93.

Lindner, V. and Reidy, M. A. 1993. Expression of basic fibroblast growth factor and its receptor by smooth muscle cells and endothelium in injured rat arteries. An en face study. *Circ Res*, 73, 589–95.

Lindner, V., Reidy, M. A., and Fingerle, J. 1989. Regrowth of arterial endothelium. Denudation with minimal trauma leads to complete endothelial cell regrowth. *Lab Invest*, 61, 556–63.

Little, T. L., Xia, J., and Duling, B. R. 1995. Dye tracers define differential endothelial and smooth muscle coupling patterns within the arteriolar wall. *Circ Res*, 76, 498–504.

Liu, L., Yan, Y., Zeng, M., Zhang, J., Hanes, M. A., Ahearn, G., Mcmahon, T. J. et al. 2004. Essential roles of S-nitrosothiols in vascular homeostasis and endotoxic shock. *Cell*, 116, 617–28.

Locovei, S., Wang, J., and Dahl, G. 2006. Activation of pannexin 1 channels by ATP through P2Y receptors and by cytoplasmic calcium. *FEBS Lett*, 580, 239–44.

Looft-Wilson, R. C., Payne, G. W., and Segal, S. S. 2004. Connexin expression and conducted vasodilation along arteriolar endothelium in mouse skeletal muscle. *J Appl Physiol*, 97, 1152–8.

Lounsbury, K. M., Hu, Q., and Ziegelstein, R. C. 2000. Calcium signaling and oxidant stress in the vasculature. *Free Radic Biol Med*, 28, 1362–9.

Lounsbury, K. M., Stern, M., Taatjes, D., Jaken, S., and Mossman, B. T. 2002. Increased localization and substrate activation of protein kinase C delta in lung epithelial cells following exposure to asbestos. *Am J Pathol*, 160, 1991–2000.

Lyle, A. N. and Griendling, K. K. 2006. Modulation of vascular smooth muscle signaling by reactive oxygen species. *Physiology (Bethesda)*, 21, 269–80.

Macvicar, B. A. and Thompson, R. J. 2010. Non-junction functions of pannexin-1 channels. *Trends Neurosci*, 33, 93–102.

Maeda, S. and Tsukihara, T. 2010. Structure of the gap junction channel and its implications for its biological functions. *Cell Mol Life Sci*, 68(7):1115–29.

Manderson, J. A., Mosse, P. R., Safstrom, J. A., Young, S. B., and Campbell, G. R. 1989. Balloon catheter injury to rabbit carotid artery. I. Changes in smooth muscle phenotype. *Arteriosclerosis*, 9, 289–98.

Martin, P. 1997. Wound healing—Aiming for perfect skin regeneration. *Science*, 276, 75–81.

Martin, P. E. and Evans, W. H. 2004. Incorporation of connexins into plasma membranes and gap junctions. *Cardiovasc Res*, 62, 378–87.

Mather, S., Dora, K. A., Sandow, S. L., Winter, P., and Garland, C. J. 2005. Rapid endothelial cell-selective loading of connexin 40 antibody blocks endothelium-derived hyperpolarizing factor dilation in rat small mesenteric arteries. *Circ Res*, 97, 399–407.

Mayr, M., Sidibe, A., and Zampetaki, A. 2008. The paradox of hypoxic and oxidative stress in atherosclerosis. *J Am Coll Cardiol*, 51, 1266–7.

Mclachlan, E., Shao, Q., and Laird, D. W. 2007. Connexins and gap junctions in mammary gland development and breast cancer progression. *J Membr Biol*, 218, 107–21.

Mercado, C. and Jaimes, E. A. 2007. Cigarette smoking as a risk factor for atherosclerosis and renal disease: Novel pathogenic insights. *Curr Hypertens Rep*, 9, 66–72.

Mesnil, M., Crespin, S., Avanzo, J. L., and Zaidan-Dagli, M. L. 2005. Defective gap junctional intercellular communication in the carcinogenic process. *Biochim Biophys Acta*, 1719, 125–45.

Mesnil, M., Krutovskikh, V., Piccoli, C., Elfgang, C., Traub, O., Willecke, K., and Yamasaki, H. 1995. Negative growth control of HeLa cells by connexin genes: Connexin species specificity. *Cancer Res*, 55, 629–39.

Moebus, S., Hanisch, J. U., Neuhauser, M., Aidelsburger, P., Wasem, J., and Jockel, K. H. 2006. Assessing the prevalence of the Metabolic Syndrome according to Ncep ATP III in Germany: Feasibility and quality aspects of a two step approach in 1550 randomly selected primary health care practices. *Ger Med Sci*, 4, Doc07.

Moorby, C. and Patel, M. 2001. Dual functions for connexins: Cx43 regulates growth independently of gap junction formation. *Exp Cell Res*, 271, 238–48.

Moore, L. C., Casellas, D., Persson, A. E., Muller-Suur, R., and Morsing, P. 1990. Renal hemodynamic regulation by the renin-secreting segment of the afferent arteriole. *Kidney Int Suppl*, 30, S65–8.

Morel, S., Burnier, L., Roatti, A., Chassot, A., Roth, I., Sutter, E., Galan, K., Pfenniger, A., Chanson, M., and Kwak, B. R. 2010. Unexpected role for the human Cx37 C1019T polymorphism in tumour cell proliferation. *Carcinogenesis*, 31, 1922–31.

Moreno, A. P., Saez, J. C., Fishman, G. I., and Spray, D. C. 1994. Human connexin43 gap junction channels. Regulation of unitary conductances by phosphorylation. *Circ Res*, 74, 1050–7.

Moumtzi, A., Trenker, M., Flicker, K., Zenzmaier, E., Saf, R., and Hermetter, A. 2007. Import and fate of fluorescent analogs of oxidized phospholipids in vascular smooth muscle cells. *J Lipid Res.*, 48, 565–82.

Murohara, T., Witzenbichler, B., Spyridopoulos, I., Asahara, T., Ding, B., Sullivan, A., Losordo, D. W., and Isner, J. M. 1999. Role of endothelial nitric oxide synthase in endothelial cell migration. *Arterioscler Thromb Vasc Biol*, 19, 1156–61.

Nakamura, K., Inai, T., and Shibata, Y. 1999. Distribution of gap junction protein connexin 37 in smooth muscle cells of the rat trachea and pulmonary artery. *Arch Histol Cytol*, 62, 27–37.

Norris, R. P., Freudzon, M., Mehlmann, L. M., Cowan, A. E., Simon, A. M., Paul, D. L., Lampe, P. D., and Jaffe, L. A. 2008. Luteinizing hormone causes MAP kinase-dependent phosphorylation and closure of connexin 43 gap junctions in mouse ovarian follicles: One of two paths to meiotic resumption. *Development*, 135, 3229–3238.

Nozik-Grayck, E., Whalen, E. J., Stamler, J. S., Mcmahon, T. J., Chitano, P., and Piantadosi, C. A. 2006. S-nitrosoglutathione inhibits alpha1-adrenergic receptor-mediated vasoconstriction and ligand binding in pulmonary artery. *Am J Physiol Lung Cell Mol Physiol*, 290, L136–43.

Okamoto, T., Akiyama, M., Takeda, M., Akita, N., Yoshida, K., Hayashi, T., and Suzuki, K. 2011. Connexin32 protects against vascular inflammation by modulating inflammatory cytokine expression by endothelial cells. *Exp Cell Res*, 317, 348–55.

Okamoto, T., Akiyama, M., Takeda, M., Gabazza, E. C., Hayashi, T., and Suzuki, K. 2009. Connexin32 is expressed in vascular endothelial cells and participates in gap-junction intercellular communication. *Biochem Biophys Res Commun*, 382, 264–8.

Oviedo-Orta, E., Errington, R. J., and Evans, W. H. 2002. Gap junction intercellular communication during lymphocyte transendothelial migration. *Cell Biol Int*, 26, 253–63.

Pahujaa, M., Anikin, M., and Goldberg, G. S. 2007. Phosphorylation of connexin43 induced by Src: Regulation of gap junctional communication between transformed cells. *Exp Cell Res*, 313, 4083–90.

Palmer, A. M., Gopaul, N., Dhir, S., Thomas, C. R., Poston, L., and Tribe, R. M. 1998. Endothelial dysfunction in streptozotocin-diabetic rats is not reversed by dietary probucol or simvastatin supplementation. *Diabetologia*, 41, 157–64.

Pannirselvam, M., Verma, S., Anderson, T. J., and Triggle, C. R. 2002. Cellular basis of endothelial dysfunction in small mesenteric arteries from spontaneously diabetic (db/db –/–)
mice: Role of decreased tetrahydrobiopterin bioavailability. *Br J Pharmacol*, 136, 255–63.

Pelegrin, P. and Surprenant, A. 2007. Pannexin-1 couples to maitotoxin- and nigericin-induced
interleukin-1beta release through a dye uptake-independent pathway. *J Biol Chem*, 282,
2386–94.

Pidkovka, N. A., Cherepanova, O. A., Yoshida, T., Alexander, M. R., Deaton, R. A., Thomas,
J. A., Leitinger, N., and Owens, G. K. 2007. Oxidized phospholipids induce phenotypic
switching of vascular smooth muscle cells *in vivo* and *in vitro*. *Circ Res*, 101, 792–801.

Pitocco, D., Zaccardi, F., Di Stasio, E., Romitelli, F., Santini, S. A., Zuppi, C., and Ghirlanda,
G. 2010. Oxidative stress, nitric oxide, and diabetes. *Rev Diabet Stud*, 7, 15–25.

Plante, I., Stewart, M. K., Barr, K., Allan, A. L., and Laird, D. W. 2010. Cx43 suppresses mammary tumor metastasis to the lung in a Cx43 mutant mouse model of human disease.
Oncogene, 30(14):1681–92.

Post, J. M., Hume, J. R., Archer, S. L., and Weir, E. K. 1992. Direct role for potassium channel
inhibition in hypoxic pulmonary vasoconstriction. *Am J Physiol*, 262, C882–90.

Qian, J., Zhang, Q., Church, J. E., Stepp, D. W., Rudic, R. D., and Fulton, D. J. 2010. Role
of local production of endothelium-derived nitric oxide on cGMP signaling and
S-nitrosylation. *Am J Physiol Heart Circ Physiol*, 298, H112–8.

Ram, C. V. 1997. Renovascular hypertension. *Curr Opin Nephrol Hypertens*, 6, 575–9.

Redout, E. M., Wagner, M. J., Zuidwijk, M. J., Boer, C., Musters, R. J., Van Hardeveld, C.,
Paulus, W. J., and Simonides, W. S. 2007. Right-ventricular failure is associated with
increased mitochondrial complex II activity and production of reactive oxygen species.
Cardiovasc Res, 75, 770–81.

Retamal, M. A., Cortes, C. J., Reuss, L., Bennett, M. V., and Saez, J. C. 2006. S-nitrosylation
and permeation through connexin 43 hemichannels in astrocytes: Induction by oxidant
stress and reversal by reducing agents. *Proc Natl Acad Sci USA*, 103, 4475–4480.

Rignault, S., Haefliger, J. A., Waeber, B., Liaudet, L., and Feihl, F. 2007. Acute inflammation
decreases the expression of connexin 40 in mouse lung. *Shock*, 28, 78–85.

Roberts, C. K., Vaziri, N. D., and Barnard, R. J. 2002. Effect of diet and exercise intervention
on blood pressure, insulin, oxidative stress, and nitric oxide availability. *Circulation*,
106, 2530–2.

Rodrigues, L. C., Sinhorini, I. L., Avanzo, J. L., Oloris, S. C. S., Carneiro, C. S., Lima, C. E.,
Fukumasa, H., Costa-Pinto, F. A., and Dagli, M. L. 2010. Diminished angiogenesis in
the cornea of mice with heterologous deletion of Connexin 43 gene (GJA1). *Braz J Vet
Pathol*, 3, 24–30.

Roger, V. L., Go, A. S., Lloyd-Jones, D. M., Adams, R. J., Berry, J. D., Brown, T. M.,
Carnethon, M. R. et al. 2011. Heart disease and stroke statistics—2011 update: A report
from the American Heart Association. *Circulation*, 123, e18–e209.

Rozental, R., Srinivas, M., and Spray, D. C. 2001. How to close a gap junction channel.
Efficacies and potencies of uncoupling agents. *Methods Mol Biol*, 154, 447–476.

Rummery, N. M., Mckenzie, K. U., Whitworth, J. A., and Hill, C. E. 2002. Decreased endothelial size and connexin expression in rat caudal arteries during hypertension. *J Hypertens*,
20, 247–253.

Safian, R. D. and Textor, S. C. 2001. Renal-artery stenosis. *N Engl J Med*, 344, 431–42.

Sandow, S. L., Bramich, N. J., Bandi, H. P., Rummery, N. M., and Hill, C. E. 2003. Structure,
function, and endothelium-derived hyperpolarizing factor in the caudal artery of the
SHR and WKY rat. *Arterioscler Thromb Vasc Biol*, 23, 822–8.

Sasano, C., Honjo, H., Takagishi, Y., Uzzaman, M., Emdad, L., Shimizu, A., Murata, Y.,
Kamiya, K., and Kodama, I. 2007. Internalization and dephosphorylation of connexin43
in hypertrophied right ventricles of rats with pulmonary hypertension. *Circ J*, 71,
382–9.

Sato, T., Haimovici, R., Kao, R., Li, A. F., and Roy, S. 2002. Downregulation of connexin 43 expression by high glucose reduces gap junction activity in microvascular endothelial cells. *Diabetes*, 51, 1565–71.

Schmidt, V. J., Wolfle, S. E., Boettcher, M., and de Wit, C. 2008. Gap junctions synchronize vascular tone within the microcirculation. *Pharmacol Rep*, 60, 68–74.

Schnackenberg, C. G. 2002a. Oxygen radicals in cardiovascular-renal disease. *Curr Opin Pharmacol*, 2, 121–5.

Schnackenberg, C. G. 2002b. Physiological and pathophysiological roles of oxygen radicals in the renal microvasculature. *Am J Physiol Regul Integr Comp Physiol*, 282, R335–42.

Segal, S. S. and Beny, J. L. 1992. Intracellular recording and dye transfer in arterioles during blood flow control. *Am J Physiol*, 263, H1–7.

Segal, S. S., Damon, D. N., and Duling, B. R. 1989. Propagation of vasomotor responses coordinates arteriolar resistances. *Am J Physiol*, 256, H832–7.

Segal, S. S. and Duling, B. R. 1989. Conduction of vasomotor responses in arterioles: A role for cell-to-cell coupling? *Am J Physiol*, 256, H838–45.

Severs, N. J., Rothery, S., Dupont, E., Coppen, S. R., Yeh, H. I., Ko, Y. S., Matsushita, T., Kaba, R., and Halliday, D. 2001. Immunocytochemical analysis of connexin expression in the healthy and diseased cardiovascular system. *Microsc Res Tech*, 52, 301–22.

Shao, Q., Wang, H., Mclachlan, E., Veitch, G. I., and Laird, D. W. 2005. Down-regulation of Cx43 by retroviral delivery of small interfering RNA promotes an aggressive breast cancer cell phenotype. *Cancer Res*, 65, 2705–11.

Sigala, F., Kotsinas, A., Savari, P., Filis, K., Markantonis, S., Iliodromitis, E. K., Gorgoulis, V. G., and Andreadou, I. 2010. Oxidized LDL in human carotid plaques is related to symptomatic carotid disease and lesion instability. *J Vasc Surg*, 52, 704–13.

Sihvo, E. I., Ruohtula, T., Auvinen, M. I., Koivistoinen, A., Harjula, A. L., and Salo, J. A. 2003. Simultaneous progression of oxidative stress and angiogenesis in malignant transformation of Barrett esophagus. *J Thorac Cardiovasc Surg*, 126, 1952–7.

Simon, A. M. and Mcwhorter, A. R. 2003. Decreased intercellular dye-transfer and downregulation of non-ablated connexins in aortic endothelium deficient in connexin37 or connexin40. *J Cell Sci*, 116, 2223–36.

Simonneau, G., Robbins, I. M., Beghetti, M., Channick, R. N., Delcroix, M., Denton, C. P., Elliott, C. G., Gaine, S. P., Gladwin, M. T., Jing, Z. C., Krowka, M. J., Langleben, D., Nakanishi, N., and Souza, R. 2009. Updated clinical classification of pulmonary hypertension. *J Am Coll Cardiol*, 54, S43–54.

Solan, J. L., Fry, M. D., Tenbroek, E. M., and Lampe, P. D. 2003. Connexin43 phosphorylation at S368 is acute during S and G2/M and in response to protein kinase C activation. *J Cell Sci*, 116, 2203–11.

Solan, J. L. and Lampe, P. D. 2005. Connexin phosphorylation as a regulatory event linked to gap junction channel assembly. *Biochim Biophys Acta*, 1711, 154–63.

Solan, J. L. and Lampe, P. D. 2007. Key connexin 43 phosphorylation events regulate the gap junction life cycle. *J Membr Biol*, 217, 35–41.

Solan, J. L. and Lampe, P. D. 2008. Connexin 43 in LA-25 cells with active v-src is phosphorylated on Y247, Y265, S262, S279/282, and S368 via multiple signaling pathways. *Cell Commun Adhes*, 15, 75–84.

Solan, J. L. and Lampe, P. D. 2009. Connexin43 phosphorylation: Structural changes and biological effects. *Biochem J*, 419, 261–72.

Souza, H. P., Souza, L. C., Anastacio, V. M., Pereira, A. C., Junqueira, M. L., Krieger, J. E., Da Luz, P. L., Augusto, O., and Laurindo, F. R. 2000. Vascular oxidant stress early after balloon injury: Evidence for increased NAD(P)H oxidoreductase activity. *Free Radic Biol Med*, 28, 1232–42.

Stamler, J. S. 1994. Redox signaling: Nitrosylation and related target interactions of nitric oxide. *Cell*, 78, 931–6.

Stamler, J. S., Lamas, S., and Fang, F. C. 2001. Nitrosylation. the prototypic redox-based signaling mechanism. *Cell*, 106, 675–83.

Straub, A. C., Billaud, M., Johnstone, S. R., Best, A. K., Yemen, S., Dwyer, S. T., Looft-Wilson, R. et al. 2011. Compartmentalized connexin 43 S-nitrosylation/denitrosylation regulates heterocellular communication in the vessel wall. *Arterioscler Thromb Vasc Biol*, 31(2):399–407.

Takahashi, H., Shin, Y., Cho, S. J., Zago, W. M., Nakamura, T., Gu, Z., Ma, Y. et al. 2007. Hypoxia enhances S-nitrosylation-mediated NMDA receptor inhibition via a thiol oxygen sensor motif. *Neuron*, 53, 53–64.

Takenaka, T., Inoue, T., Kanno, Y., Okada, H., Meaney, K. R., Hill, C. E., and Suzuki, H. 2008. Expression and role of connexins in the rat renal vasculature. *Kidney Int*, 73, 415–22.

Tang, E. H. and Vanhoutte, P. M. 2008. Gap junction inhibitors reduce endothelium-dependent contractions in the aorta of spontaneously hypertensive rats. *J Pharmacol Exp Ther*, 327, 148–53.

Tenbroek, E. M., Lampe, P. D., Solan, J. L., Reynhout, J. K., and Johnson, R. G. 2001. Ser364 of connexin43 and the upregulation of gap junction assembly by cAMP. *J Cell Biol*, 155, 1307–18.

Thatipelli, M. and Misra, S. 2010. Endovascular intervention for renal artery stenosis. *Abdom Imaging*, 35, 612–21.

Thaulow, E. and Jorgensen, B. 1997. Clinical promise of calcium antagonists in the angioplasty patient. *Eur Heart J*, 18 Suppl B, B21–6.

Touyz, R. M., Tabet, F., and Schiffrin, E. L. 2003. Redox-dependent signalling by angiotensin II and vascular remodelling in hypertension. *Clin Exp Pharmacol Physiol*, 30, 860–6.

Triggle, C. R. and Ding, H. 2010. A review of endothelial dysfunction in diabetes: A focus on the contribution of a dysfunctional eNOS. *J Am Soc Hypertens*, 4, 102–15.

Uzzaman, M., Honjo, H., Takagishi, Y., Emdad, L., Magee, A. I., Severs, N. J., and Kodama, I. 2000. Remodeling of gap junctional coupling in hypertrophied right ventricles of rats with monocrotaline-induced pulmonary hypertension. *Circ Res*, 86, 871–8.

Van Kempen, M. J. and Jongsma, H. J. 1999. Distribution of connexin37, connexin40 and connexin43 in the aorta and coronary artery of several mammals. *Histochem Cell Biol*, 112, 479–86.

Van Rijen, H. V., Van Kempen, M. J., Postma, S., and Jongsma, H. J. 1998. Tumour necrosis factor alpha alters the expression of connexin43, connexin40, and connexin37 in human umbilical vein endothelial cells. *Cytokine*, 10, 258–264.

Van Rijen, H. V., Van Veen, T. A., Hermans, M. M., and Jongsma, H. J. 2000a. Human connexin40 gap junction channels are modulated by cAMP. *Cardiovasc Res*, 45, 941–51.

Van Veen, T. A., Van Rijen, H. V., and Jongsma, H. J. 2000b. Electrical conductance of mouse connexin45 gap junction channels is modulated by phosphorylation. *Cardiovasc Res*, 46, 496–510.

Veliz, L. P., Gonzalez, F. G., Duling, B. R., Saez, J. C., and Boric, M. P. 2008. Functional role of gap junctions in cytokine-induced leukocyte adhesion to endothelium in vivo. *Am J Physiol Heart Circ Physiol*, 295, H1056–66.

Villars, F., Guillotin, B., Amedee, T., Dutoya, S., Bordenave, L., Bareille, R., and Amedee, J. 2002. Effect of Huvec on human osteoprogenitor cell differentiation needs heterotypic gap junction communication. *Am J Physiol Cell Physiol*, 282, C775–85.

Von Tell, D., Armulik, A., and Betsholtz, C. 2006. Pericytes and vascular stability. *Exp Cell Res*, 312, 623–9.

Wagner, C. 2008. Function of connexins in the renal circulation. *Kidney Int*, 73, 547–55.

Wagner, C., de Wit, C., Kurtz, L., Grunberger, C., Kurtz, A., and Schweda, F. 2007. Connexin40 is essential for the pressure control of renin synthesis and secretion. *Circ Res*, 100, 556–63.

Wagner, C., Jobs, A., Schweda, F., Kurtz, L., Kurt, B., Lopez, M. L., Gomez, R. A., Van Veen, T. A., de Wit, C., and Kurtz, A. 2010. Selective deletion of Connexin 40 in renin-producing cells impairs renal baroreceptor function and is associated with arterial hypertension. *Kidney Int*, 78, 762–8.

Wang, C. M., Lincoln, J., Cook, J. E., and Becker, D. L. 2007a. Abnormal connexin expression underlies delayed wound healing in diabetic skin. *Diabetes*, 56, 2809–17.

Wang, J., Ma, M., Locovei, S., Keane, R. W., and Dahl, G. 2007b. Modulation of membrane channel currents by gap junction protein mimetic peptides: Size matters. *Am J Physiol Cell Physiol*, 293, C1112–9.

Wang, Y., Beydoun, M. A., Liang, L., Caballero, B., and Kumanyika, S. K. 2008. Will all Americans become overweight or obese? Estimating the progression and cost of the US obesity epidemic. *Obesity (Silver Spring)*, 16, 2323–30.

Warn-Cramer, B. J., Lampe, P. D., Kurata, W. E., Kanemitsu, M. Y., Loo, L. W., Eckhart, W., and Lau, A. F. 1996. Characterization of the mitogen-activated protein kinase phosphorylation sites on the connexin-43 gap junction protein. *J Biol Chem*, 271, 3779–86.

Wempe, F., Lindner, V., and Augustin, H. G. 1997. Basic fibroblast growth factor (bFGF) regulates the expression of the CC chemokine monocyte chemoattractant protein-1 (MCP-1) in autocrine-activated endothelial cells. *Arterioscler Thromb Vasc Biol*, 17, 2471–8.

Wild, S., Roglic, G., Green, A., Sicree, R., and King, H. 2004. Global prevalence of diabetes: Estimates for the year 2000 and projections for 2030. *Diabetes Care*, 27, 1047–53.

Wolin, M. S. 2009. Reactive oxygen species and the control of vascular function. *Am J Physiol Heart Circ Physiol*, 296, H539–49.

Wong, C. W., Christen, T., Pfenniger, A., James, R. W., and Kwak, B. R. 2007. Do allelic variants of the connexin37 1019 gene polymorphism differentially predict for coronary artery disease and myocardial infarction? *Atherosclerosis*, 191, 355–61.

Wong, C. W., Christen, T., Roth, I., Chadjichristos, C. E., Derouette, J. P., Foglia, B. F., Chanson, M., Goodenough, D. A., and Kwak, B. R. 2006a. Connexin37 protects against atherosclerosis by regulating monocyte adhesion. *Nat Med*, 12, 950–4.

Wong, R. C., Dottori, M., Koh, K. L., Nguyen, L. T., Pera, M. F., and Pebay, A. 2006b. Gap junctions modulate apoptosis and colony growth of human embryonic stem cells maintained in a serum-free system. *Biochem Biophys Res Commun*, 344, 181–8.

Wood, Z. A., Poole, L. B., and Karplus, P. A. 2003. Peroxiredoxin evolution and the regulation of hydrogen peroxide signaling. *Science*, 300, 650–3.

Xia, J. and Duling, B. R. 1995. Electromechanical coupling and the conducted vasomotor response. *Am J Physiol*, 269, H2022–30.

Xue, Y., Liu, Z., Gao, X., Jin, C., Wen, L., Yao, X., and Ren, J. 2010. GPS-SNO: Computational prediction of protein S-nitrosylation sites with a modified GPS algorithm. *PLoS One*, 5, e11290.

Yamada, Y., Izawa, H., Ichihara, S., Takatsu, F., Ishihara, H., Hirayama, H., Sone, T., Tanaka, M., and Yokota, M. 2002. Prediction of the risk of myocardial infarction from polymorphisms in candidate genes. *N Engl J Med*, 347, 1916–23.

Yeh, H. I., Lai, Y. J., Chang, H. M., Ko, Y. S., Severs, N. J., and Tsai, C. H. 2000. Multiple connexin expression in regenerating arterial endothelial gap junctions. *Arterioscler Thromb Vasc Biol*, 20, 1753–62.

Yeh, H. I., Lu, C. S., Wu, Y. J., Chen, C. C., Hong, R. C., Ko, Y. S., Shiao, M. S., Severs, N. J., and Tsai, C. H. 2003a. Reduced expression of endothelial connexin37 and connexin40 in hyperlipidemic mice: Recovery of connexin37 after 7-day simvastatin treatment. *Arterioscler Thromb Vasc Biol*, 23, 1391–1397.

Yeh, H. I., Lu, C. S., Wu, Y. J., Chen, C. C., Hong, R. C., Ko, Y. S., Shiao, M. S., Severs, N. J., and Tsai, C. H. 2003b. Reduced expression of endothelial connexin37 and connexin40 in hyperlipidemic mice: Recovery of connexin37 after 7-day simvastatin treatment. *Arterioscler Thromb Vasc Biol*, 23, 1391–7.

Yogo, K., Ogawa, T., Akiyama, M., Ishida, N., and Takeya, T. 2002. Identification and functional analysis of novel phosphorylation sites in Cx43 in rat primary granulosa cells. *FEBS Lett*, 531, 132–136.

Young, E. J., Hill, M. A., Wiehler, W. B., Triggle, C. R., and Reid, J. J. 2008. Reduced EDHF responses and connexin activity in mesenteric arteries from the insulin-resistant obese Zucker rat. *Diabetologia*, 51, 872–81.

Yu, G., Bolon, M., Laird, D. W., and Tyml, K. 2010. Hypoxia and reoxygenation-induced oxidant production increase in microvascular endothelial cells depends on connexin40. *Free Radic Biol Med*, 49, 1008–13.

Zahler, S., Hoffmann, A., Gloe, T., and Pohl, U. 2003. Gap-junctional coupling between neutrophils and endothelial cells: A novel modulator of transendothelial migration. *J Leukoc Biol*, 73, 118–26.

Zhang, J. and Hill, C. E. 2005. Differential connexin expression in preglomerular and postglomerular vasculature: Accentuation during diabetes. *Kidney Int*, 68, 1171–85.

Zhang, J. H., Kawashima, S., Yokoyama, M., Huang, P., and Hill, C. E. 2006. Increased eNOS accounts for changes in connexin expression in renal arterioles during diabetes. *Anat Rec A Discov Mol Cell Evol Biol*, 288, 1000–8.

Zhang, Z. Q., Zhang, W., Wang, N. Q., Bani-Yaghoub, M., Lin, Z. X., and Naus, C. C. 1998. Suppression of tumorigenicity of human lung carcinoma cells after transfection with connexin43. *Carcinogenesis*, 19, 1889–94.

10 Regulation of Gap Junctions and Cellular Coupling within the Microvasculature in Response to Acute Inflammation

Darcy Lidington and Karel Tyml

CONTENTS

10.1 INTRODUCTION

10.1.1 ROLE OF GAP JUNCTION INTERCELLULAR COUPLING IN THE CONTROL OF BLOOD FLOW

The primary role of microcirculation is to deliver nutrients and oxygen to tissues and to remove waste products. To meet the ever-changing microregional metabolic needs of various tissues in the body, feeding arterioles locally change their diameter (i.e., a vasomotor response) to alter blood flow specifically within perfusion modules (Berg et al. 1997; Delp and Laughlin 1998). Although neuronal control (e.g., sympathetic release of norepinephrine) of blood flow contributes to this regulation (Delp and Laughlin 1998), arterioles are also able to directly respond to cellular metabolites (e.g., adenosine, ATP, H$^+$) (Delp and Laughlin 1998). These signals may be highly localized and thus, to effectively alter blood flow, changes in arteriolar diameter need to occur in a coordinated manner over a relatively long length of the vessel (on the order of 1 mm).

Vascular cell communication, therefore, plays a key role in the coordination of microvascular blood flow control, by enhancing an arteriole's ability to mount a change in hemodynamic resistance (Tyml et al. 2001). This ability of arterioles to rapidly propagate arteriolar diameter changes has been termed the "arteriolar conducted response" (Gustafsson and Holstein–Rathlou 1999). Vascular communication also enable capillaries, which cannot alter their diameter (and are therefore unable to directly control blood flow), to participate in the regulation of organ perfusion: metabolite-initiated electrical signals originating in capillaries have the potential to travel along the capillary endothelium to the feeding arteriole (Berg et al. 1997; Mitchell et al. 1997; Song and Tyml 1993; Tyml et al. 1997; Yu et al. 2000).

There is general consensus that the microvascular endothelium (i) forms a continuous, electrically coupled layer between capillaries and arterioles (Beach et al. 1998; Yu et al. 2000), and (ii) is the primary cellular pathway for the longitudinal conduction of membrane hyper/depolarization within the arteriolar wall (Dora 2010; Emerson and Segal 2000; Haas and Duling 1997). Gap junction intercellular coupling between endothelial cells makes the conduction of electrical signals over large distances possible, by forming a low electrical resistance pathway (Haefliger et al. 2004; Johnstone et al. 2009). In addition to homocellular longitudinal coupling between endothelial cells (Tran and Welsh 2009), vascular gap junctions are also critically involved in radial intercellular coupling across the vascular wall (i.e., heterocellular coupling between endothelial and smooth muscle cells). Coupling between these two cell types is involved in the direct modulation vasomotor responses, including endothelial-derived hyperpolarizing factor (EDHF)-mediated arteriolar dilation (de Wit and Griffith 2010; Dora 2010).

10.1.2 CHAPTER SCOPE

Gap junction intercellular coupling plays a multifaceted role in the development of several inflammatory-based vascular pathologies (Brisset et al. 2009; Cronier et al. 2009; Hou et al. 2008; Johnson and Koval 2009; Morel et al. 2009). Although inflammation may represent the underlying basis of the disease process, it is important to note that inflammation can be chronic in nature (e.g., during the development of atherosclerotic plaques) or acute (e.g., rapid immune response to infection); therefore, different inflammatory processes may be mechanistically distinct in their effects.

This chapter will restrict its focus to the *acute* systemic inflammatory condition of sepsis, which develops over the time course of several hours and is characterized by a number of circulatory disorders, including decreased systemic vascular resistance, hypotension, and stoppage of blood flow in capillaries (Bone 1991; Lam et al. 1994; Nguyen et al. 2004). Specifically, this chapter will examine how key mediators and processes in septicemia alter the gap junction function/endothelial cell coupling to disrupt the normal homeostasis of microvascular blood flow. Because gap junctions play a key developmental role within the microcirculation (Dhein 2004; Saez et al. 2003), the models described here examine the physiological homeostasis of microvascular blood flow in mature (adult) animals to minimize vascular development as a confounding factor.

Although a considerable amount of research has been conducted in the field of vascular communication (de Wit and Griffith 2010; Tran and Welsh 2009), few studies have exploited *in vitro* and *in vivo* models simultaneously. This chapter, therefore, will focus on rat and mouse models of sepsis, where data from *in vitro* and *in vivo* experiments complement each other and provide mechanistic insight into the regulation of cellular communication during acute inflammatory responses. Although sepsis causes similar effects in the two species models, it should be noted that the two species may be mechanistically distinct at the gap junction level.

10.2 SEPSIS AND MODELS OF SEPSIS

Almost every type of microorganism has the potential to cause sepsis; however, Gram-negative bacteria appear to be the most common cause (Parrillo et al. 1990). The pathogenesis of sepsis is complex: sepsis progresses along a continuum from a localized nidus of infection, with local multiplication of the inciting organism, to the production and release of bacterial endotoxin (lipopolysaccharide; LPS) into the circulatory system (Parrillo 1993). LPS triggers a septic cascade causing (i) activation of immune cells and subsequent release of reactive oxygen species, proteinases, and inflammatory cytokines, including tumor necrosis factor alpha (TNFα), interleukin (IL)-1, and IL-6 (Ertel et al. 1991; Grisham et al. 1988; Jean-Baptiste 2007); (ii) activation of the coagulation and complement cascades (Fijnvandraat et al. 1994; Levi 2010); and (iii) increased production of the vasodilator, nitric oxide (NO) (Fortin et al. 2010; Kilbourn and Griffith 1992). The common end result of sepsis is the failure of the microcirculation to maintain adequate peripheral resistance (leading to hypotension) and organ perfusion. The progressive failure of microcirculatory homeostasis during septicemia causes microregional ischemia and reperfusion (I/R) (Armour et al. 2001; Bateman et al. 2001; Motterlini et al. 1998), a situation that can act as a "second hit" and further injure vascular cells by enhancing the innate immune response and stimulating the release of oxidant radicals (Khadaroo et al. 2003; Powers et al. 2006).

Several animal models have been established to investigate the mechanisms underlying vascular dysfunction in sepsis (Wichterman et al. 1980); the cecal ligation and perforation model is generally accepted as the most accurate representation of clinical sepsis, because it produces a peritoneal focus of infection, leading to sepsis characterized by a hyperdynamic state, hypotension, leukopenia, elevated cardiac output, and elevated plasma lactate concentrations (Lush and Kvietys 2000). This model, therefore, represents the *in vivo* model of sepsis highlighted in this chapter (Lidington et al. 2003; Tyml et al. 1998).

Because of the complexity of the *in vivo* environment and the interplay between signaling molecules and pathways, the determination of key mechanisms altering cellular function is difficult. Cell culture models (e.g., monolayers of endothelial cells) provide an environment that is substantially easier to manipulate and control. However, *in vitro* approaches represent a double-edged sword in terms of understanding the *in vivo* effects of sepsis because it is precisely the complexity and interplay that makes modeling sepsis *in vitro* challenging.

10.3   ASSESSING CELLULAR COMMUNICATION *IN VIVO* AND *IN VITRO*

Direct measurement of communicated signals *in situ* (e.g., conduction of membrane potential change) is technically difficult. Most often, cellular coupling in the microcirculation is assessed *in situ* by locally stimulating a capillary (Tyml et al. 1998) or arteriole (Gustafsson and Holstein–Rathlou 1999; Lidington et al. 2003) and measuring an upstream arteriolar diameter response as an index of conduction. The arterioles stimulated in these models lie very near the surface of a skeletal muscle tissue; it is possible, therefore, to treat these arterioles with signaling mediators (e.g., cytokines and LPS) and chemical inhibitors (usually in a superfusate; Lidington et al. 2003).

In vitro, cellular coupling is assessed most sensitively by monitoring the spread of electrical current between cells. In cell monolayers, this is modeled with a current-injection approach that relates the spatial decay of injected current to the degree of gap junction coupling between cells (more precisely, the electrical resistance of the intercellular path) (Larson et al. 1983; Lidington et al. 2000; Shiba 1971). The majority of *in vitro* data described in this chapter was obtained utilizing this approach.

Ouellette and coworkers (Ouellette et al. 2000) have developed a novel *in vitro* model to assess cellular coupling that takes advantage of an endothelial characteristic of forming capillary-like structures when grown on Matrigel. This model resembles the *in situ* approach for arterioles in the sense that the capillary-like structure is locally stimulated and an index of coupling is calculated based on local and conducted membrane potential responses (Ouellette et al. 2000). Compared to monolayers, this model better reflects the linear spread of electrical signals *in vivo*. In addition, the model evokes substantially larger electrical responses to locally applied agonists and is, therefore, particularly well suited to simultaneously discern alterations in local and communicated responses.

10.4 EFFECT OF INFLAMMATORY PROCESSES IN RAT MODEL OF SEPSIS

Although cellular communication within the vascular wall represents a key component of a vascular network's ability to sense, integrate, and respond to vasomotor signals, little is known about the effect of the disease in general, and sepsis in particular, on this component. Tyml and coworkers (Tyml et al. 1998) were the first to demonstrate that cell sensing (i.e., responsiveness of the cell to pharmacological stimuli) and/or communication along the rat skeletal muscle capillary is compromised by sepsis *in vivo*. Because direct measurement of electrical responses and communicated signals *in situ* is technically difficult, *in vitro* models using cultured endothelial cells have played a prominent role in advancing our mechanistic understanding of how acute inflammatory responses and mediators alter endothelial cell coupling.

Although rat microvascular endothelial cells (RMECs) express three connexins at the mRNA level (Cx37, Cx40, and Cx43), only one, Cx43, has demonstrated expression at the protein level (Lidington et al. 2000). This observation imparts the RMEC model with two primary advantages: (i) Cx43 is the best-characterized vascular connexin with respect to the structure and signaling motifs, and (ii) all the signaling effects of inflammatory processes that alter cellular coupling in RMECs may be routed through gap junctions composed of only one connexin isoform.

10.4.1 LIPOPOLYSACCHARIDE REDUCES RMEC COUPLING

Since separating the effect(s) of sepsis on capillary sensing (i.e., membrane potential response to stimulation) and/or cell-to-cell communication is not possible *in situ* (Tyml et al. 1998), an *in vitro* model of endothelial cells grown as capillary-like structures (Ouellette et al. 2000) was utilized to simultaneously assess local and communicated responses. LPS does not impact local membrane potential changes in response to ATP (receptor dependent) or KCl (receptor independent); however, LPS significantly attenuates their conduction along the length of the capillary-like structure (Lidington et al. 2002a). Attenuated conduction of electrical responses along the capillary-like structure's length could be due to a reduction in cellular coupling (i.e., a gap junction-dependent mechanism) or an increase in the "electrical leakiness" of the plasma membrane (i.e., a gap junction-independent mechanism of decreased transmembrane resistivity). The capillary-like structure model does not permit the separation of these variables; however, a current-injection approach in monolayers does. This approach demonstrated the attenuation of electronic spread via reduced cellular coupling (~50% increase in intercellular resistance), without alteration of transmembrane resistivity (Lidington et al. 2000).

The response to LPS is dose dependent, with the maximal response observed at an LPS concentration of 10 ng/mL, a concentration that is higher than what is observed clinically in the serum of septic patients (<500 pg/mL; Opal et al. 1999). The discrepancy may reflect several factors, including differences that arise as a result of maintaining cells in culture (typically, LPS concentrations of 1–10 ng/mL are used for cell culture studies, regardless of species) and the fact that humans are remarkably sensitive to endotoxin (~1000 times more sensitive than rodents) (Warren et al. 2010).

LPS requires 1 h to induce maximal reduction in RMEC coupling (Lidington et al. 2000). While this corresponds well with the previously reported values for monolayer permeability and phosphorylation changes in cultured bovine endothelial cells (Bannerman and Goldblum 1999), this time frame is rather slow for a typical gating response (which usually occurs within minutes, as exemplified by Hossain and coworkers (1998)). Since Cx43 possesses a half-life on the same order of magnitude as the LPS response (1.5 h; Laird 2006), it is intriguing to speculate that LPS mediates its effect by altering the life cycle of Cx43. An experimental approach of biotin-labeling cell surface Cx43, however, failed to detect any evidence of LPS-stimulated changes in cell surface Cx43 expression, internalization of Cx43 or Cx43 protein stability over a 1 h period (Figures 10.1 and 10.2).

Remarkably, the LPS-stimulated reduction in cellular coupling is fully reversed 1 h following the removal of LPS (i.e., a similar kinetic as the response to LPS; Lidington et al. 2000). Both the response to LPS (i.e., reduced cellular coupling) and its reversal following LPS removal are insensitive to cycloheximide-mediated protein synthesis inhibition (Lidington et al. 2000), which may be regarded as further evidence that alterations in Cx43 protein expression and/or life cycle are unlikely

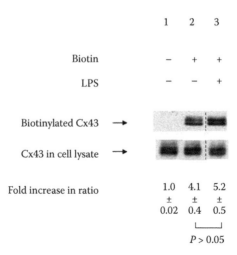

FIGURE 10.1 LPS does not alter cell surface Cx43. Rat microvascular endothelial cells were exposed to lipopolysaccharide (LPS) (10 μg/mL) or vehicle for 1 h, after which cell surface proteins were labeled with biotin. Biotinylated Cx43 was then captured with avidin-coated agarose beads and immunoblotted (Le et al. 1999). In the displayed representative experiment, the upper row shows captured, biotinylated Cx43, while the lower row displays a measure of total Cx43. Minimal levels of Cx43 were captured by the avidin-coated beads in the absence of biotinylation (very weak bands were detectable). Comparable levels of biotinylated Cx43 were detected in the vehicle-treated (lane 2) and LPS-treated (lane 3) cells. Also shown are the optical density ratios of biotinylated Cx43 blots to total Cx43 blots, normalized to control (untreated cells); $P > 0.05$, $n = 3$ independent experiments with cells from three rats. Lanes 2 and 3 originate from the same film but were not adjacently positioned on this film as shown in this figure. Assessment of total Cx43 protein (25 μg loaded for each sample) indicated no effect of LPS on total Cx43 expression.

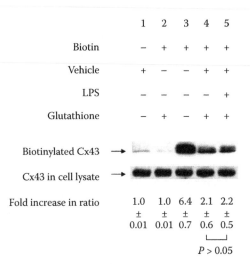

	1	2	3	4	5
Biotin	−	+	+	+	+
Vehicle	+	−	−	+	+
LPS	−	−	−	−	+
Glutathione	−	+	−	+	+

Biotinylated Cx43 →

Cx43 in cell lysate →

Fold increase in ratio   1.0   1.0   6.4   2.1   2.2
                          ±   ±   ±   ±   ±
                          0.01   0.01   0.7   0.6   0.5

                                        P > 0.05

FIGURE 10.2 LPS does not alter internalized Cx43. Rat microvascular endothelial cell surface proteins were labeled with biotin, treated with LPS (10 μg/mL) or vehicle for 1 h, after which the biotin labels were removed with glutathione. Biotinylated Cx43 was then captured with avidin-coated agarose beads and immunoblotted. In the displayed representative experiment, the upper row shows captured, biotinylated Cx43, while the lower row displays a measure of total Cx43. Minimal levels of Cx43 were captured by the avidin-coated beads in the absence of biotinylation (lane 1) or following prompt removal of the biotin label (lane 2); in both cases, very weak bands were detectable. Biotinylation, without glutathione treatment, leads to robust capture of Cx43 by the avidin-coated beads (lane 3). Comparable levels of internalized Cx43 (i.e., protected from glutathione-mediated biotin label removal) were detected in the vehicle-treated (lane 4) and LPS-treated (lane 5) cells. Also shown are the optical density ratios of biotinylated Cx43 blots to total Cx43 blots, normalized to control (lane 1), $P > 0.05$, $n = 3$ independent experiments with cells from three rats. Assessment of total Cx43 protein (25 μg loaded for each sample) indicated no effect of LPS on total Cx43 expression.

mechanisms. The reversible nature of the LPS response also argues against cytotoxicity and/or loss of viability as underlying reasons for the reduction in cellular coupling.

LPS is known to activate multiple, interacting kinase signaling pathways *in vitro*, including tyrosine and mitogen-activated protein (MAP) kinases (Arditi et al. 1995). Cx43 possesses several serine and tyrosine phosphorylation sites, which govern channel gating, protein–protein interactions, and Cx43 protein stability. In both RMEC monolayers and capillary-like structures, the LPS-stimulated reduction in cellular coupling is sensitive to tyrosine kinase inhibition (Lidington et al. 2000, 2002a). Tyrosine kinase inhibition also successfully reverses the effect of LPS on coupling (Lidington et al. 2000). Inhibition of p42/p44 MAP kinases, p38 MAP kinase, and protein kinase C (PKC) does not affect the LPS-stimulated reduction in cellular coupling (Lidington et al. 2000).

In agreement with these observations on cellular coupling, LPS stimulates Cx43 phosphorylation, with no apparent effect on serine phosphorylation (Lidington et al. 2002b). This correlation persists over several experimental interventions: both the

LPS-stimulated reduction in RMEC coupling and Cx43 tyrosine phosphorylation are (i) prevented by tyrosine kinase inhibition; (ii) prevented by ceramide treatment, a putative tyrosine phosphatase activator; and (iii) enhanced by tyrosine phosphatase inhibition (Lidington et al. 2002b). It is remarkable to note that the increase in Cx43 tyrosine phosphorylation is not associated with an electrophoretic mobility shift (Lidington et al. 2002b), as is often noted when Cx43 is serine phosphorylated (as exemplified by Hossain and coworkers (1998)). Indeed, visualization of Cx43 by Western blot reveals no obvious differences between Cx43 blotted from lysates prepared from control and LPS-treated RMECs: the phosphorylation is only evident when purified Cx43 is probed with phosphotyrosine antibodies or analyzed for phospho-amino acids (Lidington et al. 2002b).

Tyrosine phosphorylation of Cx43 can result in the activation of internalization (Thuringer 2004) or gating mechanisms (Cottrell et al. 2002): in the absence of an effect on Cx43 life cycle (Figures 10.1 through 10.3), the latter mechanism should be favored. It must be noted, however, that an LPS-stimulated gating mechanism has not been directly confirmed in RMECs. Typically, gating mechanisms are assessed using a dual whole-cell clamp technique; unfortunately, endothelial cells culture poorly at low confluencies and are generally very flat, making the successful achievement of reliable membrane seals problematic and technically challenging.

10.4.2 CYTOKINES AND NITRIC OXIDE DO NOT AFFECT RMEC COUPLING

Sepsis stimulates the production of several inflammatory cytokines and mediators, most notably IL-1β, IL-6, TNFα, interferon-γ (INF-γ), NO, and oxidants (Fortin et al. 2010; Jean-Baptiste 2007). There is a possibility that LPS acts via an indirect mechanism, whereby it stimulates the production of inflammatory mediators that then act in a paracrine manner to stimulate Cx43 phosphorylation. Should this not be the case, then there is the possibility that these factors could have effects on cellular coupling that are in addition to those of LPS.

Neither of these scenarios appears to occur in RMECs: (i) in monolayers, IL-1β, IL-6, and TNFα had no individual effect on cellular coupling, nor did they alter the LPS-stimulated reduction in cellular coupling (Tyml et al. 2001), and (ii) in capillary-like structures, neither NO nor NO synthase inhibition (with L-N^G-nitroarginine methyl ester, L-NAME) affected cellular coupling or the reduction in cellular coupling in response to LPS (Lidington et al. 2002a). In the latter case, two strategies were used to apply NO to the RMECs: "direct" addition via an NO donor (DETA NONOate; (Z)-1-[N-(2-aminoethyl)-N-(2-ammonioethyl)amino]diazen-1-ium-1,2-diolate) and cell-stimulated NO synthesis via the induction of inducible nitric oxide synthase (iNOS) expression in response to the combination of LPS and INF-γ. The free radicals peroxynitrite (applied via 3–morpholinosydnonimine (SIN-1), which simultaneously releases NO and superoxide) and hydrogen peroxide also have no effect on cellular coupling (D. Lidington, unpublished observations). Taken together, these observations support a direct signaling mechanism linking LPS to Cx43 tyrosine phosphorylation and ultimately, reduced cellular coupling; it is a striking observation that none of the other inflammatory mediators described above altered cellular coupling in RMECs.

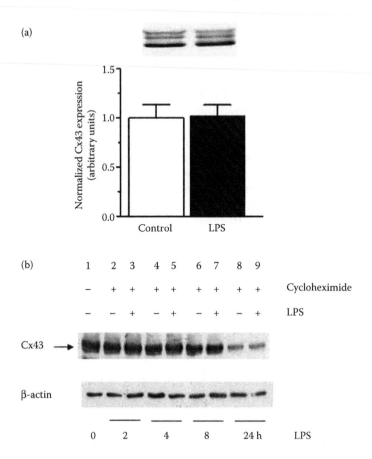

FIGURE 10.3 LPS does not alter Cx43 expression or degradation. (a) Shown is a representative Western blot and corresponding densitometry analysis indicating that LPS (10 µg/mL; 1 h) does alter Cx43 protein expression. In each experiment, 25 µg of protein samples were loaded (and visually confirmed by Coomassie staining before immunoblotting). All data points were normalized to the mean control value. (b) Protein synthesis inhibition (10 µg/mL cycloheximide) had a minimal effect on Cx43 expression over 8 h of treatment; after 24 h, however, Cx43 expression was significantly reduced. LPS co-treatment did not alter Cx43 expression at any of the four time points examined, compared to the respective control. β-Actin served as a loading control. Western blots shown are from two independent experiments (i.e., cells from two rats) in lanes 1, 2, 4, 6, and 8 (exp #1) and in lanes 3, 5, 7, and 9 (exp #2).

10.4.3 Hypoxia/Reoxygenation Reduces Coupling in RMECs

The sepsis-induced impairment of microvascular blood flow is accompanied by episodes of microregional ischemia and reperfusion (i.e., intermittent stoppage of capillary blood flow (Armour et al. 2001); I/R). Episodes of I/R (modeled *in vitro* as hypoxia/reoxygenation, H/R) can profoundly affect endothelial function, causing increased levels of oxidative stress, elevated intracellular calcium, and the activation of MAP kinases (Gourdin et al. 2009; Khan et al. 2004). H/R plays a key role in

I/R pathophysiology, which is complex and involves several biochemical processes (Bari et al. 1996; DeFily 1998). Thus, oxidative stress, which in RMECs does not itself affect coupling, can potentially enhance cellular responses to LPS, as has been demonstrated previously (Khadaroo et al. 2003; Powers et al. 2006).

In RMEC monolayers, H/R (1 h hypoxia at 0.1% O_2 followed by 5–15 min reoxygenation) stimulated a rapid (within 5 min), persistent (present for >60 min) reduction in cellular coupling (Rose et al. 2005). Mitochondrial inhibition (which mimics the energetic deficiency hypoxia causes) does not alter coupling (Rose et al. 2005), suggesting that the effect of H/R is due to the rapid reoxygenation, rather than the hypoxia itself. Like LPS, the H/R-mediated reduction in cellular coupling was dependent on tyrosine kinase activation; however, this is where the similarities between the two responses end. First, there is no evidence that Cx43 is tyrosine phosphorylated in response to H/R, and there is no electrophoretic mobility shift consistent with serine phosphorylation events (Rose et al. 2005). Additionally, the H/R-mediated reduction in cellular coupling is also dependent on p42/p44 MAP kinase. It is clear, therefore, that LPS and H/R stimulate distinct signaling mechanisms, both of which reduce cellular coupling in endothelial cells.

10.5 EFFECTS OF INFLAMMATORY PROCESSES IN MOUSE MODELS OF SEPSIS

The distribution of connexins has been shown to vary between individual vascular beds within one species, between the same vascular beds of different species, as well as during the progression of the disease (Hill et al. 2001; Severs et al. 2001; van Kempen and Jongsma 1999). It is not surprising, therefore, that mouse microvascular endothelial cells (MMECs) have been shown to possess a different connexin profile than RMECs, despite the fact that the cells originate from the same vascular bed within the *extensor digitorum longus* muscle. Specifically, MMECs express three vascular connexins at the protein level, Cx37, Cx40, and Cx43 (Bolon et al. 2007), which contrasts with the detection of only Cx43 protein in RMECs (Lidington et al. 2000).

The MMEC model may represent a better reflection in the *in vivo* situation, as it is clear that multiple connexins are expressed in endothelial cells of the rat and other species *in vivo* (Haddock et al. 2006; Heberlein et al. 2009; Looft-Wilson et al. 2004; Yeh et al. 1998). It is notable that Cx40 appears to be of prime importance, because (i) it displays a constitutive endothelial expression *in vivo* across species and vascular beds (van Kempen and Jongsma 1999) and (ii) it is a critical component in the conduction of electrically activated vasomotor responses along the endothelial layer (de Wit et al. 2000; Figueroa et al. 2003; Figueroa and Duling 2008; Schmidt et al. 2008). The demonstrated lack of Cx40 protein in the RMEC model (Lidington et al. 2000), therefore, is a distinct limitation that is not present in the MMEC model.

10.5.1 Lipopolysaccharide Reduces MMEC Coupling via Cx40

Consistent with observations in RMEC monolayers (Lidington et al. 2000), MMECs reduce their intercellular coupling in response to LPS (10 μg/mL, 1 h), without changes in the connexin protein expression (Bolon et al. 2007). The mechanism of action, however, appears to be different compared to the RMEC model. Using MMECs isolated from Cx37 knockout (Cx37$^{-/-}$), Cx40 knockout (Cx40$^{-/-}$), and nonfunctional Cx43 mutant (Cx43^{G60S}) mice, Bolon and coworkers (Bolon et al. 2007) convincingly demonstrate that the signaling mechanisms activated by LPS exclusively target Cx40. Because the expression of connexin proteins does not change in response to LPS (Bolon et al. 2007), a posttranslational mechanism targeting a connexin protein (e.g., connexin phosphorylation) is the likely mode of action.

In MMECs, LPS stimulation activates several mitogen-activated protein kinases (MAPK), including ERK1/2 (p42/p44 MAPK), p38 MAPK, and JNK1/2 (p46/p56 MAPK), as well as protein tyrosine kinases (PTKs; Bolon et al. 2007). Broad inhibition of PTKs (PP-2) and targeted inhibition of ERK1/2 (PD98059 and U0126) both inhibited the LPS-stimulated reduction in cellular coupling; p38 MAPK and JNK1/2 are apparently not involved in mediating the response to LPS (Bolon et al. 2007). In addition to MAPKs and PTKs, both protein kinase C (PKC) and protein kinase A (PKA) are involved in the response to LPS. Remarkably, the *activation* of either PKC or PKA blocks the effect of LPS on cellular coupling, while their inhibition reduces coupling in the absence of LPS (i.e., PKA and PKC inhibition mimics the effect of LPS in the sense that cellular coupling is reduced to a similar extent; Bolon et al. 2007).

The reduction in cellular coupling in MMECs correlates with *reduced Cx40 serine phosphorylation*, with convincing evidence presented by Bolon and coworkers (Bolon et al. 2008) that PKA is the kinase acting directly on Cx40. Like the tyrosine phosphorylation of Cx43 in RMECs, the PKA-dependent changes in Cx40 serine phosphorylation are not associated with an electrophoretic mobility shift (Bolon et al. 2008). While this lack of mobility shift contrasts the observations of van Rijen and coworkers (2000) (in SKHep1 cells), the fundamental mechanism of PKA-dependent Cx40 serine phosphorylation leading to enhanced cellular coupling is conserved between the two cell types.

Taken together, LPS stimulation modulates the activity of three distinct signaling inputs in MMECs, ERK1/2 (activates), PTK (activates), and PKC (inhibits), which act in concert to inhibit PKA activity. There is significant interplay between all of these signaling pathways, and it is not known at this time whether the connections between the signal mechanisms are serial, parallel, or a combination of both. However, given that PKC activation mimics the effect of LPS (i.e., decreases MMEC coupling), it is tempting to speculate that ERK1/2 and PTKs combine to inhibit PKC, which then leads to the serial inhibition of PKA to reduce coupling.

10.5.2 LIPOPOLYSACCHARIDE REDUCES CONDUCTED VASOCONSTRICTION IN MOUSE CREMASTER ARTERIOLES

As previously mentioned, most models that assess vascular communication *in situ* provide an opportunity to investigate the effect(s) of agents that are topically applied to the arteriole (e.g., LPS). When applied to mouse cremaster arterioles, LPS (10 µg/mL, 1 h) significantly reduces the conduction of electrically stimulated vasoconstriction (Tyml et al. 2001). Unfortunately, a comprehensive *in vitro/in situ* comparison of the underlying signaling mechanisms has not been undertaken. Several aspects of the LPS response in arterioles *in situ*, however, match results obtained from RMECs and MMECs *in vitro*: the LPS–stimulated reduction in conducted vasoconstriction *in situ* is (i) tyrosine kinase dependent (RMEC, MMEC), (ii) fully reversible following LPS washout (RMEC), and (iii) independent of NO production (RMEC; Lidington et al. 2000, 2002a, 2003; Tyml et al. 2001). Although caution must be observed when drawing mechanistic conclusions using results from two distinctly different *in vitro* models (RMEC and MMEC), it is clear that several core characteristics of the LPS responses display overlap *in vitro* and *in situ*.

10.5.3   TNFα Reduces MMEC Coupling via Cx40

The effects of cytokines have not been extensively explored in the MMEC model. TNFα is one cytokine that has been tested in the MMEC model; it stimulates a reduction in intercellular coupling in MMECs isolated from wild type, but not MMEC isolated from Cx40$^{-/-}$ mice (i.e., a Cx40-dependent mechanism; Figure 10.4). This is again in contrast with the RMEC model, where coupling was not affected by TNFα (Tyml et al. 2001).

10.5.4   Nitric Oxide Reduces MMEC Coupling via Cx37

Unlike RMECs, NO (via the NO donors DETA NONOate and *S*-nitroso-*N*-acetyl-l,l-penicillamine) elicits a rapid (within 10 min), dose-dependent, and rapidly reversible (within 10 min) reduction in MMEC intercellular coupling (McKinnon et al. 2009). These observations in MMECs are consistent with those of Kameritsch and coworkers (Kameritsch et al. 2005), who demonstrate that NO donors reduce

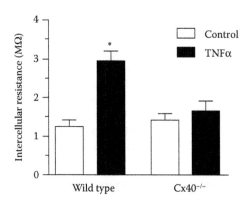

FIGURE 10.4   TNFα reduces interendothelial electrical coupling by a Cx40-dependent mechanism. Using an electrophysiological approach (Bolon et al. 2007), intracellular resistance (which is inversely related to electrical coupling) was measured in mouse microvascular endothelial cell (MMEC) monolayers. TNFα (1–10 ng/mL, applied for 3 h) increased electrical resistance in MMECs isolated from wild-type mice; preliminary data suggests that this response is lost in MMECs isolated from Cx40$^{-/-}$ mice. *$P < 0.05$, $n = 7$ for MMEC monolayers of wild-type cells, $n = 2$ MMEC monolayers of Cx40$^{-/-}$ cells.

cellular coupling in human umbilical vein endothelial cells (HUVECs), which also express Cx37, Cx40, and Cx43.

In contrast to the mechanism mediating the response to LPS, NO does not target Cx40; rather, the use of Cx37$^{-/-}$, Cx40$^{-/-}$, and Cx43^{G60S} mice convincingly demonstrate that Cx37 is the target (McKinnon et al. 2009). This is again consistent with the observations made by Kameritsch and coworkers (Kameritsch et al. 2005), who observed that NO donors reduce cellular coupling in immortalized HeLa cells transfected with Cx37 (HeLa cells have no constitutive connexin expression), but not those transfected with Cx43. Interestingly, recent evidence indicates that Cx37 possesses a binding motif for endothelial nitric oxide synthase (eNOS) (Pfenniger et al. 2010); this would position Cx37 in close proximity to the predominant source of endothelial nitric oxide under nonpathological conditions.

NO does not affect MMEC morphology, viability, connexin expression, or connexin distribution (McKinnon et al. 2009), suggestive of a posttranslational mechanism targeting Cx37. The two most prominent signaling actions of NO include (i) the activation of guanylyl cyclase, with subsequent production of cGMP and activation of cGMP-dependent effectors (e.g., protein kinase G; PKG), and (ii) the reaction with superoxide to form peroxynitrite, which can nitrosylate tyrosine residues and alter cell signaling processes (Pacher et al. 2007). Neither signaling mechanism appears to be involved, as the inhibition of guanylyl cyclase (1H-[1,2,4]oxadiazolo[4,3-a] quinoxalin-1-one; ODQ), scavenging of superoxide (to prevent the formation of peroxynitrite; MnTBAP), and scavenging of peroxynitrite (FeTPPS) all failed to prevent the NO-dependent reduction in MMEC coupling (McKinnon et al. 2009). The molecular mechanism(s) mediating the effect of NO on Cx37 permeability in MMECs remains unknown at present.

In the mouse cremaster muscle, conducted arteriolar vasoconstriction *in situ* is rapidly reduced following the application of NO donors (Lidington et al. 2003; McKinnon et al. 2007; Rodenwaldt et al. 2007), an effect that is rapidly reversible (McKinnon et al. 2007) and independent of Cx40 (Rodenwaldt et al. 2007). All these observations are consistent with results obtained from MMECs (McKinnon et al. 2009). Interestingly, conducted vasoconstriction is significantly reduced by the genetic deletion of Cx37 (Cx37$^{-/-}$) (McKinnon et al. 2006). Given that both Cx40 and Cx37 are highly expressed at endothelial cell junctions in cremaster arterioles (Looft-Wilson et al. 2004), this is not entirely unexpected.

There is one interesting discrepancy related to the use of the guanylyl cyclase inhibitor ODQ: Rodenwaldt and coworkers (Rodenwaldt et al. 2007) demonstrated that ODQ was *partially* effective in blocking the NO-stimulated reduction of conducted vasoconstriction in mouse cremaster arterioles *in situ*; McKinnon and coworkers observed no such ODQ effect in MMECs *in vitro* (McKinnon et al. 2009) or in mouse cremaster arterioles *in situ* (McKinnon et al. 2006). The only obvious difference in the experimental approach was that Rodenwaldt and coworkers (Rodenwaldt et al. 2007) additionally combined NO synthase inhibition (Nω-nitro-L-arginine; L-NNA) with ODQ to eliminate endogenous NO production; their investigation did not appear to examine the effects of L–NNA or ODQ in isolation. In separate investigations, ODQ (McKinnon et al. 2006) and a structurally distinct NO inhibitor (L–NAME) (Lidington et al. 2003) displayed no effect on conducted vasoconstriction. It appears, therefore, that L–NNA (or perhaps the combination of ODQ and L–NNA) has an unexpected effect that attenuates the effectiveness of NO in reducing vascular coupling.

10.5.5 HYPOXIA/REOXYGENATION REDUCES MMEC COUPLING VIA Cx40 AND SYNERGIZES WITH LPS

Consistent with the observations in RMECs, H/R rapidly and transiently decreases MMEC intercellular coupling, an effect that persists for at least 60 min following reoxygenation (Bolon et al. 2005). The underlying mechanism shares a striking similarity to the MMEC response to LPS: (i) it is Cx40 dependent, as the H/R-stimulated reduction in MMECs is absent in cells isolated from Cx40$^{-/-}$ mice, but not Cx37$^{-/-}$ or Cx43^{G60S} mice (Bolon et al. 2005, 2008); (ii) the H/R-stimulated reduction in coupling is absent in MMECs treated with inhibitors of ERK1/2 and PTKs (Bolon et al. 2008); (iii) the H/R response is PKA- and PKC-inhibition dependent and is lost if either PKA or PKC is chemically activated (Bolon et al. 2005, 2008); and (iv) the H/R-stimulated reduction in MMEC coupling correlates with reduced serine phosphorylation of Cx40 (Bolon et al. 2008).

The H/R-stimulated reduction of MMEC intercellular coupling and PKA activity requires abrupt reoxygenation; if oxygen levels are allowed to equilibrate slowly, these H/R effects are abolished (Bolon et al. 2005). The primary difference between rapid and slow reoxygenation scenarios is likely to lie in the generation of oxygen radicals. Two lines of evidence support this premise: (i) MMEC oxidative stress is increased following abrupt reoxygenation, but not slow reoxygenation and (ii) the H/R-stimulated reductions of intercellular coupling and PKA activity are abolished by pretreating MMECs with the antioxidant ascorbate (Bolon et al. 2005). Since both ERK1/2 and PTKs can be activated by oxygen radials (Paravicini and Touyz 2006; Yung et al. 2006), it tempting to speculate that the overall mechanism for LPS- and H/R-induced reductions in coupling is effectively identical (except that the common upstream signaling responses (ERK/PTK) are activated by a receptor in the case of LPS and by oxidative stress in H/R).

Despite the apparent commonality of the signaling mechanism, LPS and H/R combine to synergistically reduce cellular coupling (i.e., H/R + LPS reduces MMEC coupling to a greater extent than would be predicted by adding the individual effects; Bolon et al. 2008). It is especially important to note that subthreshold LPS concentrations gain significant importance when combined with H/R. As an illustration, 1 μg/mL LPS is required to reduce coupling in MMECs under normal conditions; when combined with H/R, LPS ranging from 10 ng/mL to 10 μg/mL display equivalent reductions in coupling that exceed 10 μg/mL LPS alone (Bolon et al. 2008). These results are in accordance with previous observations that oxidants can enhance inflammatory response (Khadaroo et al. 2003; Powers et al. 2006), and thus, care must be taken when attempting to translate relevant effective concentrations derived from *in vitro* systems to more complex *in vivo* systems. It is intriguing to note that the LPS + H/R synergy is a concurrent effect; "priming" the system with LPS before the H/R insult (a "two-hit" model; Powers et al. 2006) does not further enhance the response (Bolon et al. 2008).

The H/R model used for *in vitro* experiments is meant to model I/R *in vivo*, and although they are not directly comparable, it is clear that H/R plays a key role in I/R pathophysiology (Bari et al. 1996; DeFily 1998). In accordance with H/R results in MMECs, I/R transiently reduces conducted vasoconstriction in mouse cremaster arterioles *in situ*, by a PKA-inhibition-dependent mechanism (Bolon et al. 2005).

10.6 MECHANISMS MEDIATING REDUCED ARTERIOLAR CONDUCTION IN SEPSIS

Although sepsis causes numerous perturbations within the microcirculation (Pacher et al. 2007) and is therefore the subject of intense study, few investigations have addressed the effect of sepsis on vascular communication *in vivo*. On the basis of the *in vitro* (MMECs) and *in situ* (mouse cremaster arterioles) data presented here, we propose a conceptual scheme of the effect of sepsis on interendothelial electrical coupling (Figure 10.5). The scheme includes the documented reduction in cellular communication/arteriolar conduction in response to LPS, cytokines, NO, H/R, and I/R. Which of these mechanisms (and connexins) are critical in sepsis?

Remarkably, the sepsis-mediated reduction in conduction vasoconstriction can be fully reversed by superfusing the cremaster preparation with physiological saline (i.e., a washout), suggesting that a diffusible factor mediates the reduction in vascular communication (Lidington et al. 2003). In the 24-h cecal ligation and perforation model of sepsis, this factor is NO: (i) both NO synthase inhibition and NO scavenging

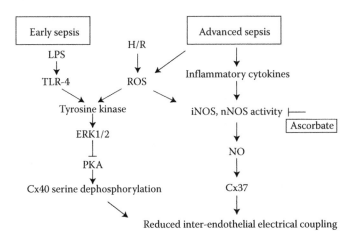

FIGURE 10.5 A schematic diagram of the effect of early and advanced sepsis on microvascular function involving interendothelial electrical coupling. This diagram is based on our work in mouse skeletal muscle *in vivo* and in mouse microvascular endothelial cells *in vitro*. LPS and its interaction with its receptor TLR-4 mimic early sepsis; inflammatory cytokines at 24 h of sepsis model mimic advanced sepsis. The arrows indicate activation and blunted-end lines indicate inhibition. (Adapted from Tyml, K. 2011. *Can J Physiol Pharmacol* 89: 1–12. With permission.)

reverse the sepsis-mediated reduction in conducted vasoconstriction (Lidington et al. 2003; McKinnon et al. 2006), and (ii) genetic deletion of neuronal nitric oxide synthase (nNOS knockout) prevents the conduction deficit (McKinnon et al. 2006). As predicted from experiments utilizing MMECs (McKinnon et al. 2009), inhibition of guanylyl cyclase (ODQ) has no effect on the sepsis-mediated conduction deficit (McKinnon et al. 2006).

In addition to its dependency on NO, the sepsis-mediated reduction in conducted vasoconstriction is also reversed by tyrosine kinase inhibition (Lidington et al. 2003) and by treatment with ascorbate (McKinnon et al. 2007). Since sepsis is associated with reactive oxygen species formation (Azevedo et al. 2006) and ascorbate is an antioxidant, it is logical to speculate that ascorbate acts by scavenging oxidants. However, the superoxide scavenger MnTBAP has no effect on the sepsis-stimulated reduction in conducted vasoconstriction. On the basis of NO synthase activity assays, it appears that ascorbate acts *upstream* of NO production, as it prevents the sepsis-mediated increase in nNOS activity (McKinnon et al. 2007). It is not known at this time whether the tyrosine kinase activity lies upstream of nNOS (i.e., required for activation of nNOS and subsequent NO production) or downstream (i.e., tyrosine kinases are activated by an NO-dependent mechanism).

10.7   SUMMARY, CHALLENGES, AND MODEL CONSIDERATIONS

This chapter has shown that although rat and mouse microvascular endothelial cells (RMECs and MMECs) are isolated from the same vascular bed, they appear to be different in terms of their connexin composition, responsiveness to inflammatory stimuli, and signaling mechanisms. In brief summary, RMECs have been shown to express only one connexin protein; they respond to LPS and H/R, but not cytokines or NO, and both LPS and H/R appear to reduce RMEC intercellular coupling by tyrosine phosphorylation of Cx43. In contrast, MMECs express three connexins, respond to LPS, H/R, cytokines, and NO, and target two connexins (Cx40 and Cx37) via distinct mechanisms to reduce MMEC coupling. It is intriguing to note that although Cx43 is present in both RMECs and MMECs, it appears to be a signaling target only in RMECs. It is, therefore, clear that species variation in connexin composition and signaling pathways represent significant challenges for cross-comparison of experimental models and ultimately the translation of experimental data to clinically relevant practice. Any comparison across species must be made with due caution.

Since the distribution of connexin proteins varies between individual vascular beds within a species (Hill et al. 2001), there was significant potential that the MMEC model (isolated from the *extensor digitorum longus* muscle) would not accurately predict the effects in cremaster arterioles. Nevertheless, a remarkable parallel between the mechanisms operating in MMECs *in vitro* and mouse cremaster arterioles *in situ* was observed. While this indicates that MMECs are a relatively good experimental system for mechanistic study in *skeletal muscle* microvascular beds, the results from MMECs may not be broadly applicable across distinctly different vascular beds, such as the cerebral, pulmonary, and renal microcirculation. Thus, due caution must again be applied when comparing experimental results from different models, even when the same species is used.

On the basis of the result that the sepsis, induced by cecal ligation and perforation, reduced conducted arteriolar vasoconstriction solely by an NO (and putatively Cx37)-dependent mechanism, there is an inclination to dismiss the other factors and processes (i.e., LPS, cytokines, and H/R) as nonfactors. This would be a poor assumption. Sepsis must be viewed as a progression, with distinct mediators and signaling processes gaining and losing prominence at different points along the progression (Hinshaw 1996; Lange et al. 2011). Confounding issues include the fact that plasma cytokines rise and fall at different time points, LPS concentrations may be higher in vascular beds closer to the infectious focus, and H/R may be more relevant in smaller arterioles or capillary beds further downstream, since blood flow does not stop in the 50 μm branch order used for communication assessment. Systemic effects, including altered sympathetic/parasympathetic activity, blood pressure, and shear forces (as a result of altered flow and blood viscosity) can all potentially contribute to or modulate the effects of sepsis on vascular communication.

In addition to the progressive nature of sepsis, it must be highlighted that not all models of sepsis are directly comparable (Dyson and Singer 2009; Freise et al. 2001). Some models (e.g., LPS injection) can be highly controlled and standardized, reproduce several characteristics of sepsis, and yet fail to create an infectious focus or the protracted immune reaction that characterizes sepsis (Freise et al. 2001). Although the cecal ligation and perforation model of sepsis is decisively closer to clinical reality (Freise et al. 2001; Lush and Kvietys 2000), the inability to reliably control the bacterial challenge often leads to poor interlaboratory standardization (Freise et al. 2001).

Adding to the complexity, the progression and response to sepsis displays some species specificity. For example, in small rodents, plasma NO concentrations can reach molar concentrations (Harbrecht et al. 1992; Liaudet et al. 1998); however, NO overproduction in humans rarely increases more than 50% above the background, despite major microcirculatory failure (Cauwels 2007; Feihl et al. 2001). Taken together, comparisons of vascular function during inflammatory responses will vary depending on the species, vascular bed, inflammatory model used, and the time point of assessment. Such comparison considerations are applicable to all vascular functions (e.g., vasodilator and constrictor responses) and would not be restricted to cellular communication.

A final level of complexity resides at the level of connexins. Connexins display a broad overlap in function, which is why the genetic deletion of one vascular connexin does not generally result in a profound vascular phenotype (notably, deletion of Cx40 is associated with hypertension (de Wit et al. 2003). The overlap in function, coupled, perhaps, with compensatory expression changes in the other connexin proteins, may explain why cellular coupling remained similar between wild-type and Cx37[-/-], Cx40[-/-], and Cx43[G60S] MMECs (Bolon et al. 2007). While not all connexin functions are universally compensated (Wolfle et al. 2007), the use of genetic deletion models may not always identify contributing mechanisms. There is an additional point to be made regarding deletion models: gap junctions are composed of 12 connexin subunits (six connexin subunits contributed from each cell); if two connexins are allowed to coalesce freely, up to 196 unique gap junction channels are formed (Cottrell and Burt 2001). Genetic deletion of a single connexin protein, therefore, eliminates 195 unique gap junction channels. Thus, connexin regulation under native conditions is likely a "blend" of regulatory mechanisms acting on different connexins within a single gap junction channel. This potentially allows a single connexin protein iso-form to dominate the regulation of all channels (Burt and Steele 2003; Cottrell et al. 2002; Kumar and Gilula 1996).

10.8 CONCLUSION

The regulation of endothelial connexins under inflammatory conditions is complex. In endothelial cells, three vascular connexins, Cx43, Cx40, and Cx37, are the targets of inflammatory mediators/processes, each acted up by a unique mechanism of regu-lation. Although there is a remarkable mechanistic comparability within the mouse models presented (comparing *in vitro* and *in vivo* results), important interspecies dif-ferences between the rat and mouse are evident. These species differences, coupled with the dynamic nature of inflammatory responses and the inherent complexity of gap junction physiology, contribute to the significant challenges in understanding the mechanistic basis of how vascular communication is impaired in sepsis.

ACKNOWLEDGMENTS

We thank Drs. Y. Ouellette, M. Bolon, R. McKinnon, K. Rose, F. Li, J. Wilson, H. Dietrich, J. Kidder, D. Laird, A. Simon, F. Wu, G. Yu, X. Wang and T. Peng, and Mr S. Swarbreck for their discussion, work, and technical assistance. The financial sup-port from the Heart and Stroke Foundation of Ontario is also acknowledged (grant T6325).

REFERENCES

Arditi, M., Zhou, J., Torres, M. et al. 1995. Lipopolysaccharide stimulates the tyrosine phosphorylation of mitogen-activated protein kinases p44, p42, and p41 in vascular endothelial cells in a soluble CD14-dependent manner. Role of protein tyrosine phos-phorylation in lipopolysaccharide-induced stimulation of endothelial cells. *J Immunol* 155: 3994–4003.

Armour, J., Tyml, K., Lidington, D., and Wilson, J. X. 2001. Ascorbate prevents microvascular dysfunction in the skeletal muscle of the septic rat. *J Appl Physiol* 90: 795–803.

Azevedo, L. C., Janiszewski, M., Soriano, F. G., and Laurindo, F. R. 2006. Redox mechanisms of vascular cell dysfunction in sepsis. *Endocr Metab Immune Disord Drug Targets* 6: 159–64.

Bannerman, D. D., Goldblum, S. E. 1999. Direct effects of endotoxin on the endothelium: Barrier function and injury. *Lab Invest* 79: 1181–99.

Bari, F., Errico, R. A., Louis, T. M., and Busija, D. W. 1996. Differential effects of short-term hypoxia and hypercapnia on N-methyl–D-aspartate-induced cerebral vasodilatation in piglets. *Stroke* 27: 1634–9.

Bateman, R. M., Jagger, J. E., Sharpe, M. D. et al. 2001. Erythrocyte deformability is a nitric oxide-mediated factor in decreased capillary density during sepsis. *Am J Physiol Heart Circ Physiol* 280: H2848–56.

Beach, J. M., McGahren, E. D., and Duling, B. R. 1998. Capillaries and arterioles are electri-cally coupled in hamster cheek pouch. *Am J Physiol* 275: H1489–96.

Berg, B. R., Cohen, K. D., and Sarelius, I. H. 1997. Direct coupling between blood flow and metabolism at the capillary level in striated muscle. *Am J Physiol* 272: H2693–700.

Bolon, M. L., Kidder, G. M., Simon, A. M., and Tyml, K. 2007. Lipopolysaccharide reduces electrical coupling in microvascular endothelial cells by targeting connexin40 in a tyrosine-, ERK1/2-, PKA-, and PKC-dependent manner. *J Cell Physiol* 211: 159–66.

Bolon, M. L., Ouellette, Y., Li, F., and Tyml, K. 2005. Abrupt reoxygenation following hypoxia reduces electrical coupling between endothelial cells of wild-type but not connexin40 null mice in oxidant- and PKA-dependent manner. *FASEB J* 19: 1725–7.

Bolon, M. L., Peng, T., Kidder, G. M., and Tyml, K. 2008. Lipopolysaccharide plus hypoxia and reoxygenation synergistically reduce electrical coupling between microvascular endothelial cells by dephosphorylating connexin40. *J Cell Physiol* 217: 350–9.

Bone, R. C. 1991. Gram-negative sepsis. Background, clinical features, and intervention. *Chest* 100: 802–8.

Brisset, A. C., Isakson, B. E., and Kwak, B. R. 2009. Connexins in vascular physiology and pathology. *Antioxid Redox Signal* 11: 267–82.

Burt, J. M., Steele, T. D. 2003. Selective effect of PDGF on connexin 43 versus connexin40 comprised gap junction channels. *Cell Commun Adhes* 10: 287–91.

Cauwels, A. 2007. Nitric oxide in shock. *Kidney Int* 72: 557–65.

Cottrell, G. T., Burt, J. M. 2001. Heterotypic gap junction channel formation between heteromeric and homomeric Cx40 and Cx43 connexons. *Am J Physiol Cell Physiol* 281: C1559–67.

Cottrell, G. T., Wu, Y., and Burt, J. M. 2002. Cx40 and Cx43 expression ratio influences heteromeric/heterotypic gap junction channel properties. *Am J Physiol Cell Physiol* 282: C1469–82.

Cronier, L., Crespin, S., Strale, P. O., Defamie, N., and Mesnil, M. 2009. Gap junctions and cancer: New functions for an old story. *Antioxid Redox Signal* 11: 323–38.

de Wit, C., Griffith, T. M. 2010. Connexins and gap junctions in the EDHF phenomenon and conducted vasomotor responses. *Pflugers Arch* 459: 897–914.

de Wit, C., Roos, F., Bolz, S. S. et al. 2000. Impaired conduction of vasodilation along arterioles in connexin40-deficient mice. *Circ Res* 86: 649–55.

de Wit, C., Roos, F., Bolz, S. S., and Pohl, U. 2003. Lack of vascular connexin 40 is associated with hypertension and irregular arteriolar vasomotion. *Physiol Genomics* 13: 169–77.

DeFily, D. V. 1998. Control of microvascular resistance in physiological conditions and reperfusion. *J Mol Cell Cardiol* 30: 2547–54.

Delp, M. D., Laughlin, M. H. 1998. Regulation of skeletal muscle perfusion during exercise. *Acta Physiol Scand* 162: 411–9.

Dhein, S. 2004. Pharmacology of gap junctions in the cardiovascular system. *Cardiovasc Res* 62: 287–98.

Dora, K. A. 2010. Coordination of vasomotor responses by the endothelium. *Circ J* 74: 226–32.

Dyson, A., Singer, M. 2009. Animal models of sepsis: Why does preclinical efficacy fail to translate to the clinical setting? *Crit Care Med* 37: S30–7.

Emerson, G. G., Segal, S. S. 2000. Endothelial cell pathway for conduction of hyperpolarization and vasodilation along hamster feed artery. *Circ Res* 86: 94–100.

Ertel, W., Morrison, M. H., Wang, P. et al. 1991. The complex pattern of cytokines in sepsis. Association between prostaglandins, cachectin, and interleukins. *Ann Surg* 214: 141–8.

Feihl, F., Waeber, B., and Liaudet, L. 2001. Is nitric oxide overproduction the target of choice for the management of septic shock? *Pharmacol Ther* 91: 179–213.

Figueroa, X. F., Duling, B. R. 2008. Dissection of two Cx37-independent conducted vasodilator mechanisms by deletion of Cx40: Electronic versus regenerative conduction. *Am J Physiol Heart Circ Physiol* 295: H2001–7.

Figueroa, X. F., Paul, D. L., Simon, A. M. et al. 2003. Central role of connexin40 in the propagation of electrically activated vasodilation in mouse cremasteric arterioles in vivo. *Circ Res* 92: 793–800.

Fijnvandraat, K., Peters, M., Derkx, B., van, D. S., and ten Cate, J. W. 1994. Endotoxin induced coagulation activation and protein C reduction in meningococcal septic shock. *Prog Clin Biol Res* 388: 247–54.

Fortin, C. F., McDonald, P. P., Fulop, T., and Lesur, O. 2010. Sepsis, leukocytes, and nitric oxide (NO): An intricate affair. *Shock* 33: 344–52.

Freise, H., Bruckner, U. B., and Spiegel, H. U. 2001. Animal models of sepsis. *J Invest Surg* 14: 195–212.

Gourdin, M. J., Bree, B., and De, K. M. 2009. The impact of ischaemia–reperfusion on the blood vessel. *Eur J Anaesthesiol* 26: 537–47.

Grisham, M. B., Everse, J., and Janssen, H. F. 1988. Endotoxemia and neutrophil activation *in vivo*. *Am J Physiol* 254: H1017–22.

Gustafsson, F., Holstein-Rathlou, N. 1999. Conducted vasomotor responses in arterioles: Characteristics, mechanisms, and physiological significance. *Acta Physiol Scand* 167: 11–21.

Haas, T. L., Duling, B. R. 1997. Morphology favors an endothelial cell pathway for longitudinal conduction within arterioles. *Microvasc Res* 53: 113–20.

Haddock, R. E., Grayson, T. H., Brackenbury, T. D. et al. 2006. Endothelial coordination of cerebral vasomotion via myoendothelial gap junctions containing connexins 37 and 40. *Am J Physiol Heart Circ Physiol* 291: H2047–56.

Haefliger, J. A., Nicod, P., and Meda, P. 2004. Contribution of connexins to the function of the vascular wall. *Cardiovasc Res* 62: 345–56.

Harbrecht, B. G., Billiar, T. R., Stadler, J. et al. 1992. Inhibition of nitric oxide synthesis during endotoxemia promotes intrahepatic thrombosis and an oxygen radical-mediated hepatic injury. *J Leukoc Biol* 52: 390–4.

Heberlein, K. R., Straub, A. C., and Isakson, B. E. 2009. The myoendothelial junction: Breaking through the matrix? *Microcirculation* 16: 307–22.

Hill, C. E., Phillips, J. K., and Sandow, S. L. 2001. Heterogeneous control of blood flow amongst different vascular beds. *Med Res Rev* 21: 1–60.

Hinshaw, L. B. 1996. Sepsis/septic shock: Participation of the microcirculation: An abbreviated review. *Crit Care Med* 24: 1072–8.

Hossain, M. Z., Ao, P., and Boynton, A. L. 1998. Rapid disruption of gap junctional communication and phosphorylation of connexin 43 by platelet-derived growth factor in T51B rat liver epithelial cells expressing platelet-derived growth factor receptor. *J Cell Physiol* 174: 66–77.

Hou, C. J., Tsai, C. H., and Yeh, H. I. 2008. Endothelial connexins are down-regulated by atherogenic factors. *Front Biosci* 13: 3549–57.

Jean-Baptiste, E. 2007. Cellular mechanisms in sepsis. *J Intensive Care Med* 22: 63–72.

Johnson, L. N., Koval, M. 2009. Cross-talk between pulmonary injury, oxidant stress, and gap junctional communication. *Antioxid Redox Signal* 11: 355–67.

Johnstone, S., Isakson, B., and Locke, D. 2009. Biological and biophysical properties of vascular connexin channels. *Int Rev Cell Mol Biol* 278: 69–118.

Kameritsch, P., Khandoga, N., Nagel, W. et al. 2005. Nitric oxide specifically reduces the permeability of Cx37-containing gap junctions to small molecules. *J Cell Physiol* 203: 233–42.

Khadaroo, R. G., Kapus, A., Powers, K. A. et al. 2003. Oxidative stress reprograms lipopolysaccharide signaling via Src kinase-dependent pathway in RAW 264.7 macrophage cell line. *J Biol Chem* 278: 47834–41.

Khan, T. A., Bianchi, C., Ruel, M., Voisine, P., and Sellke, F. W. 2004. Mitogen-activated protein kinase pathways and cardiac surgery. *J Thorac Cardiovasc Surg* 127: 806–11.

Kilbourn, R. G., Griffith, O. W. 1992. Overproduction of nitric oxide in cytokine-mediated and septic shock. *J Natl Cancer Inst* 84: 827–31.

Kumar, N. M., Gilula, N. B. 1996. The gap junction communication channel. *Cell* 84: 381–8.

Laird, D. W. 2006. Life cycle of connexins in health and disease. *Biochem J* 394: 527–43.

Lam, C., Tyml, K., Martin, C., and Sibbald, W. 1994. Microvascular perfusion is impaired in a rat model of normotensive sepsis. *J Clin Invest* 94: 2077–83.

Lange, M., Hamahata, A., Traber, D. L. et al. 2011. Specific inhibition of nitric oxide synthases at different time points in a murine model of pulmonary sepsis. *Biochem Biophys Res Commun* 404: 877–81.

Larson, D. M., Kam, E. Y., and Sheridan, J. D. 1983. Junctional transfer in cultured vascular endothelium: I. Electrical coupling. *J Membr Biol* 74: 103–13.

Le, T. L., Yap, A. S., and Stow, J. L. 1999. Recycling of E-cadherin: A potential mechanism for regulating cadherin dynamics. *J Cell Biol* 146: 219–32.

Levi, M. 2010. The coagulant response in sepsis and inflammation. *Hamostaseologie* 30: 10–6.

Liaudet, L., Rosselet, A., Schaller, M. D. et al. 1998. Nonselective versus selective inhibition of inducible nitric oxide synthase in experimental endotoxic shock. *J Infect Dis* 177: 127–32.

Lidington, D., Ouellette, Y., Li, F., and Tyml, K. 2003. Conducted vasoconstriction is reduced in a mouse model of sepsis. *J Vasc Res* 40: 149–58.

Lidington, D., Ouellette, Y., and Tyml, K. 2000. Endotoxin increases intercellular resistance in microvascular endothelial cells by a tyrosine kinase pathway. *J Cell Physiol* 185: 117–25.

Lidington, D., Ouellette, Y., and Tyml, K. 2002a. Communication of agonist-induced electrical responses along "capillaries" *in vitro* can be modulated by lipopolysaccharide, but not nitric oxide. *J Vasc Res* 39: 405–13.

Lidington, D., Tyml, K. and Ouellette, Y. 2002b. Lipopolysaccharide-induced reductions in cellular coupling correlate with tyrosine phosphorylation of connexin 43. *J Cell Physiol* 193: 373–9.

Looft-Wilson, R. C., Payne, G. W., and Segal, S. S. 2004. Connexin expression and conducted vasodilation along arteriolar endothelium in mouse skeletal muscle. *J Appl Physiol* 97: 1152–8.

Lush, C. W., Kvietys, P. R. 2000. Microvascular dysfunction in sepsis. *Microcirculation* 7: 83–101.

McKinnon, R. L., Bolon, M. L., Wang, H. X. et al. 2009. Reduction of electrical coupling between microvascular endothelial cells by NO depends on connexin 37. *Am J Physiol Heart Circ Physiol* 297: H93–101.

McKinnon, R. L., Lidington, D., Bolon, M. et al. 2006. Reduced arteriolar conducted vasoconstriction in septic mouse cremaster muscle is mediated by nNOS-derived NO. *Cardiovasc Res* 69: 236–44.

McKinnon, R. L., Lidington, D., and Tyml, K. 2007. Ascorbate inhibits reduced arteriolar conducted vasoconstriction in septic mouse cremaster muscle. *Microcirculation* 14: 697–707.

Mitchell, D., Yu, J., and Tyml, K. 1997. Comparable effects of arteriolar and capillary stimuli on blood flow in rat skeletal muscle. *Microvasc Res* 53: 22–32.

Morel, S., Burnier, L., and Kwak, B. R. 2009. Connexins participate in the initiation and progression of atherosclerosis. *Semin Immunopathol* 31: 49–61.

Motterlini, R., Kerger, H., Green, C. J., Winslow, R. M., and Intaglietta, M. 1998. Depression of endothelial and smooth muscle cell oxygen consumption by endotoxin. *Am J Physiol* 275: H776–82.

Nguyen, H. B., Rivers, E. P., Knoblich, B. P. et al. 2004. Early lactate clearance is associated with improved outcome in severe sepsis and septic shock. *Crit Care Med* 32: 1637–42.

Opal, S. M., Scannon, P. J., Vincent, J. L. et al. 1999. Relationship between plasma levels of lipopolysaccharide (LPS) and LPS-binding protein in patients with severe sepsis and septic shock. *J Infect Dis* 180: 1584–9.

Ouellette, Y., Lidington, D., Naus, C. G., and Tyml, K. 2000. A new *in vitro* model for agonist-induced communication between microvascular endothelial cells. *Microvasc Res* 60: 222–31.

Pacher, P., Beckman, J. S., and Liaudet, L. 2007. Nitric oxide and peroxynitrite in health and disease. *Physiol Rev* 87: 315–424.

Paravicini, T. M., Touyz, R. M. 2006. Redox signaling in hypertension. *Cardiovasc Res* 71: 247–58.

Parrillo, J. E. 1993. Pathogenetic mechanisms of septic shock. *N Engl J Med* 328: 1471–7.

Parrillo, J. E., Parker, M. M., Natanson, C. et al. 1990. Septic shock in humans. Advances in the understanding of pathogenesis, cardiovascular dysfunction, and therapy. *Ann Intern Med* 113: 227–42.

Pfenniger, A., Derouette, J. P., Verma, V. et al. 2010. Gap junction protein Cx37 interacts with endothelial nitric oxide synthase in endothelial cells. *Arterioscler Thromb Vasc Biol* 30: 827–34.

Powers, K. A., Szaszi, K., Khadaroo, R. G. et al. 2006. Oxidative stress generated by hemorrhagic shock recruits Toll-like receptor 4 to the plasma membrane in macrophages. *J Exp Med* 203: 1951–61.

Rodenwaldt, B., Pohl, U., and de Wit, C. 2007. Endogenous and exogenous NO attenuates conduction of vasoconstrictions along arterioles in the microcirculation. *Am J Physiol Heart Circ Physiol* 292: H2341–8.

Rose, K., Ouellette, Y., Bolon, M., and Tyml, K. 2005. Hypoxia/reoxygenation reduces microvascular endothelial cell coupling by a tyrosine and MAP kinase dependent pathway. *J Cell Physiol* 204: 131–8.

Saez, J. C., Berthoud, V. M., Branes, M. C., Martinez, A. D., and Beyer, E. C. 2003. Plasma membrane channels formed by connexins: Their regulation and functions. *Physiol Rev* 83: 1359–400.

Schmidt, V. J., Wolfle, S. E., Boettcher, M., and de Wit, C. 2008. Gap junctions synchronize vascular tone within the microcirculation. *Pharmacol Rep* 60: 68–74.

Severs, N. J., Rothery, S., Dupont, E. et al. 2001. Immunocytochemical analysis of connexin expression in the healthy and diseased cardiovascular system. *Microsc Res Tech* 52: 301–22.

Shiba, H. 1971. Heaviside's "Bessel cable" as an electric model for flat simple epithelial cells with low resistive junctional membranes. *J Theor Biol* 30: 59–68.

Song, H., Tyml, K. 1993. Evidence for sensing and integration of biological signals by the capillary network. *Am J Physiol* 265: H1235–42.

Thuringer, D. 2004. The vascular endothelial growth factor-induced disruption of gap junctions is relayed by an autocrine communication via ATP release in coronary capillary endothelium. *Ann N Y Acad Sci* 1030: 14–27.

Tran, C. H., Welsh, D. G. 2009. Current perspective on differential communication in small resistance arteries. *Can J Physiol Pharmacol* 87: 21–8.

Tyml, K. 2011. Role of connexins in microvascular dysfunction during inflammation. *Can J Physiol Pharmacol* 89: 1–12.

Tyml, K., Song, H., Munoz, P., and Ouellette, Y. 1997. Evidence for K+ channels involvement in capillary sensing and for bidirectionality in capillary communication. *Microvasc Res* 53: 245–53.

Tyml, K., Wang, X., Lidington, D., and Ouellette, Y. 2001. Lipopolysaccharide reduces intercellular coupling *in vitro* and arteriolar conducted response *in vivo*. *Am J Physiol Heart Circ Physiol* 281: H1397–406.

Tyml, K., Yu, J., and McCormack, D. G. 1998. Capillary and arteriolar responses to local vasodilators are impaired in a rat model of sepsis. *J Appl Physiol* 84: 837–44.

van Kempen, M. J., Jongsma, H. J. 1999. Distribution of connexin 37, connexin40 and connexin 43 in the aorta and coronary artery of several mammals. *Histochem Cell Biol* 112: 479–86.

van Rijen, H. V., van Veen, T. A., Hermans, M. M. and Jongsma, H. J. 2000. Human connexin40 gap junction channels are modulated by cAMP. *Cardiovasc Res* 45: 941–51.

Warren, H. S., Fitting, C., Hoff, E. et al. 2010. Resilience to bacterial infection: Difference between species could be due to proteins in serum. *J Infect Dis* 201: 223–32.

Wichterman, K. A., Baue, A. E. and Chaudry, I. H. 1980. Sepsis and septic shock—A review of laboratory models and a proposal. *J Surg Res* 29: 189–201.

Wolfle, S. E., Schmidt, V. J., Hoepfl, B. et al. 2007. Connexin 45 cannot replace the function of connexin40 in conducting endothelium-dependent dilations along arterioles. *Circ Res* 101: 1292–9.

Yeh, H. I., Rothery, S., Dupont, E., Coppen, S. R. and Severs, N. J. 1998. Individual gap junction plaques contain multiple connexins in arterial endothelium. *Circ Res* 83: 1248–63.

Yu, J., Bihari, A., Lidington, D. and Tyml, K. 2000. Gap junction uncouplers attenuate arteriolar response to distal capillary stimuli. *Microvasc Res* 59: 162–8.

Yung, L. M., Leung, F. P., Yao, X., Chen, Z. Y. and Huang, Y. 2006. Reactive oxygen species in vascular wall. *Cardiovasc Hematol Disord Drug Targets* 6: 1–19.

11 Impact of Microglial Activation on Astroglial Connexin Expression and Function in Brain Inflammation

Christian Giaume, Nicolas Froger, Juan Andrés Orellana, Mauricio Retamal, and Juan Carlos Sáez

CONTENTS

11.1 INTRODUCTION

The central nervous system (CNS) is capable of dynamic immune and inflammatory responses mediated by the activation of microglia, the brain resident macrophage population, and astrocyte reactivity. A growing body of evidence shows that the innate immune response exerts a dichotomous role in the brain. Under physiological conditions, microglia exhibit a resting phenotype that is associated with the production of anti-inflammatory and neurotrophic factors. Microglia shift to an activated phenotype in response to a wide range of insults that activate specific signaling pathways, thereby promoting an inflammatory response necessary to further engage the immune system. The sustained inflammation resulting in brain injury and pathology implies either persistence of an inflammatory stimulus or failure in normal resolution mechanisms. A constant stimulus may result from the formation of endogenous factors, that is, protein aggregates as in the case of neurodegenerative diseases. Under these conditions, rather than serving as protective, neuroinflammation may turn into a damaging process leading to further recruitment of other cell types involved in innate immune response, which worsen the disease progression, characterized by neuronal dysfunction and even cell death as for instance in progressive neurodegenerative pathologies.

In the CNS, astrocytes constitute a rather unique example of multifunctional cells that establish tight morphological and functional interactions with neurons and the vasculature (Giaume et al., 2010). They represent the brain cell population that expresses the largest amount of connexins (Cxs) and exhibits a high degree of intercellular communication mediated by gap junction channels (GJCs) (see Giaume and McCarthy, 1996; Giaume et al., 2010). Moreover, under defined conditions that are mainly related to pathological situations, neuronal and glial Cxs can also work as hemichannels (HCs) allowing exchange of ions and small molecules between the intracellular and extracellular media (see Orellana et al., 2009). However, the expression of Cxs in astrocytes is not exclusive, and microglia as well as neurons also express these proteins. In addition, these cell types express another family of membrane proteins, the pannexins (Panxs) that also form HCs. Finally, the expression of astroglial Cxs is modified after brain injury and other pathologies, including neurodegenerative diseases (see Orellana et al., 2009; Giaume et al., 2010).

In this chapter, we summarize and discuss recent data indicating that, in an inflammatory context, Cx expression level and function in astrocytes depend on the state of microglial activation. A consequence of these properties is neuronal fate that, in certain conditions, depends on the balance of the two channel functions of astroglial Cxs, that is, GJCs and HCs. Finally, we will discuss recent data suggesting that the anti-inflammatory properties of cannabinoids (CBs) and the stimulation of cannabinoid receptors in glia may represent a way to prevent the deleterious effects of astroglial Cx channels on neuronal survival.

11.2 CONNEXIN AND PANNEXIN EXPRESSION IN MICROGLIA AND ASTROCYTES

11.2.1 MICROGLIA

In vivo, low levels of Cx43 are detected in only 5% of resting microglia, whereas at stab wounds, a recruitment of Cx43 positive microglia is observed (Eugenín et al., 2003). *In vitro*, Cx43 is not detected in nonactivated cultures of microglia (Rouach et al., 2002a; Même et al., 2006). However, upon activation with certain proinflammatory agents, microglia show the expression of Cx43 and establish intercellular communication via GJCs as evaluated by dye-coupling experiments. Indeed, treatments with TNF-α plus IFN-γ or LPS plus IFN-γ upregulate GJC formation in a Cx43-dependent manner, since microglia cultured from Cx43 knockout (KO) mice do not exhibit such upregulation when treated with the same mixtures of proinflammatory agents (Eugenín et al., 2003). Similarly, *Staphylococcus aureus*-derived peptidoglycan induces Cx43 expression in microglia that is associated with a functional gap junctional communication (Garg et al., 2005). Interestingly, treatment with TNF-α alone increases the surface levels of Cx32 in microglia (Takeuchi et al., 2006). In contrast, cultured human or mouse microglia treated with LPS, granulocyte-macrophage colony-stimulating factor, INF-γ, or TNF-α do not show changes in Cx expression (Même et al., 2006; Dobrenis et al., 2005). The intracellular transduction pathway involved in the regulation of microglial GJC formation requires elevation of intracellular free Ca^{2+} concentration (Martínez and Sáez, 1999), while enhanced intracellular cAMP or cGMP levels or activation of protein kinase C (PKC) do not induce dye coupling in cultured microglia (Martínez and Sáez, 1999). These observations are consistent with the hypothesis that activated but not resting microglia might express functional GJCs, depending on the treatment leading to their activation status.

Activated microglia release glutamate through HCs (Takeuchi et al., 2006) and their neurotoxic effect is inhibited by glutaminase or by the GJC and HC blocker carbenoxolone (D'Hondt et al., 2009). Over time, the microglia-mediated neurotoxicity is progressive and could contribute to the progressive nature of diverse neurodegenerative diseases (Block and Hong, 2005). Moreover, Panx1 expression has been detected in microglia (Shestopalov and Panchin, 2008), although evidence of functional Panx1 HCs is still missing. Nevertheless, ATP and LPS are able to induce uptake of impermeant DNA intercalatants (for instance, YOPRO, a carbocyanine monomer, or ethidium bromide) in microglia via $P2X_7$ receptor activation (Monif et al., 2009), an effect that might be explained by activation of Panx1 HC.

11.2.2 ASTROCYTES

In adult brain, Cx43 and Cx30 are the main Cxs detected in astrocytes (Dermietzel et al., 1991; Giaume et al., 1991; Kunzelmann et al., 1999; Nagy et al., 1999). Their relative levels vary according to the developmental stage and the region studied (Nagy and Rash, 2000). Cx43 is expressed early in development, around the 12th day of gestation in rat radial glial processes. As development proceeds, Cx43 expression increases and is found as immunoreactive puncta throughout the brain. Cx30 is expressed in astrocytes later in juvenile rodents during the third postnatal week showing a punctuate pattern of staining (Kunzelmann et al., 1999; Nagy and Rash, 2000). Although Cx30 and Cx43 are clearly coexpressed at gap junctional plaques of mature gray matter astrocytes, the regional distribution of Cx30 is more heterogeneous than that of Cx43 (Nagy et al., 1999), the major difference being the lack of Cx30 expression in astrocytes of white matter tracts. Although single-cell RT–PCR performed in hippocampal astrocytes has detected the presence of other Cx mRNAs (Blomstrand et al., 2004), no intercellular communication, assessed by dye-coupling experiments, was observed in astrocytes recorded from Cx43/Cx30 double knockout mice (Wallraff et al., 2006; Rouach et al., 2008), strengthening the notion that Cx43 and Cx30 are the major astroglial Cxs. Besides their role in cell-to-cell communication mediated either by GJCs or by HCs, Cxs expressed in astroglial cells also appeared to be involved in other functions, such as adhesion (Elias et al., 2007) and modulation of purinergic receptors (Scemes, 2008). *In vitro*, astrocytes only express Cx43 (Dermietzel et al., 1991; Giaume et al., 1991), unless cultures are maintained for a long period of time, that is, more than 10 weeks (Kunzelmann et al., 1999). In addition, astrocyte/neuron cocultures demonstrate that neurons upregulate the expression of Cx43 and induce that of Cx30 in subsets of astrocytes preferentially located in close proximity to neuronal soma (Rouach et al., 2000; Koulakoff et al., 2008). Finally, Panxs can also form HCs at the cell surface of diverse brain mammalian cells, including astrocytes (Thompson et al., 2008; Iglesias et al., 2009). However, although detected in these cells under resting conditions by immunofluorescence, Western blot analysis, and electrophysiology (Iglesias et al., 2009), these glycoproteins have not been found at the surface of astrocytes by other groups (Huang et al., 2007; see Orellana et al., 2009). Such discrepancy could be originated from several possible sources of variations, including culture media, brain region, and age of the animals as well as desired cell density before the experiments.

11.3 ACTIVATED MICROGLIA REGULATE CX43 CHANNELS IN ASTROCYTES

Microglia are the principal immune cells in the CNS and under normal conditions their principal role is to degrade cellular debris. After brain injury, however, they become active and release cytokines and reactive oxygen species (ROS) such as nitric oxide (NO) and H_2O_2, which could decrease neuronal survival. Thus, microglia play a fundamental role in brain pathologies associated with inflammation such as human immunodeficiency virus (HIV), stroke, Down syndrome, neurodegenerative processes, and multiple sclerosis. As mentioned above, brain inflammation significantly affects Cx expression and function in astrocytes that in turn affects the homeostasis of the whole CNS. In pure cultured microglia, treatment with LPS induces production and release of cytokines such as IL-1β, IL-6, IL-8, IL-10, TNF-α, and INFγ, among others. While astrocytes under controlled conditions are highly coupled through GJCs, it has been reported that the addition of resting microglial cells to primary cultures of astrocytes decreases their Cx43 expression and GJC coupling (Rouach et al., 2002a). Moreover, LPS treatment in astrocyte–microglia cocultures induces a drastic decrease of Cx43 expression and GJC function in astrocytes (Même et al., 2006). This regulation can be mimicked using conditioned medium from pure microglia cultures pretreated with LPS or IL-1β plus TNF-α (Même et al., 2006). While IL-1β apparently has an inhibitory effect per se on GJCs (John et al., 1999), at lower ineffective concentration, it potentiates the inhibitory effect of ATP and endothelin-1 on Cx43-mediated gap junctional communication in astrocytes (Même et al., 2004). Additionally, cytokines released by microglia also have an effect on HC activity. Indeed, exposure of astrocytes to LPS or IL-1β plus TNF-α increases HC activity detected by unitary conductance measurements as well as by uptake of fluorescent dyes and glucose analogs. As illustrated in Figure 11.1, this phenomenon was observed in parallel to the inhibition of gap junctional communication, demonstrating the occurrence of an opposite regulation of the two Cx43 functions (Retamal et al., 2007). The decrease in GJC permeability and the increase in HC activity induced by LPS or IL-1β plus TNF-α is blocked by inhibition of p38 MAP kinase, suggesting that this intracellular signaling pathway is downstream implicated in this dual effect. On the other hand, dithiothreitol (DTT) or an nitric

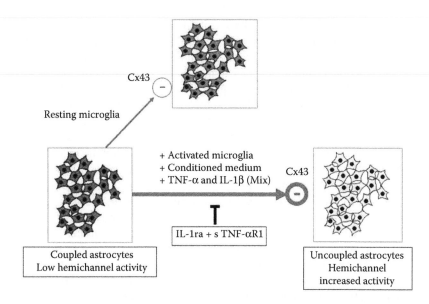

FIGURE 11.1 Microglia regulates Cx43 expression and channel functions in astrocytes. When astrocytes are cocultured with resting microglia, the expression and function of Cx are weakly inhibited; these effects become strong when microglial cells are specifically activated by LPS. Alternatively, similar strong inhibitions are also observed when cultured astrocytes are treated either by conditioned medium harvest from LPS-activated microglia or by a mixture of two proinflammatory cytokines. These inhibitory effects are prevented in the presence of an IL-1β antagonist and soluble TNF-α.

oxide synthase (NOS) inhibitor (L-name) inhibits only the increase of HC activity, indicating that NO is specifically involved in this regulation (Retamal et al., 2007). In agreement with this property, NO induces S-nitrosylation of Cx43 HCs in astrocytes under metabolic inhibition, which is directly correlated with HC opening (Retamal et al., 2006). Thus, it is proposed that NO produced by activated microglia is an important mediator of the increase of HC opening in astrocytes.

Interestingly, endogenous and synthetic CBs that are known to have anti-inflammatory properties prevent the release of cytokines from microglia after LPS activation and also prevent HC opening and loss of GJC function (Froger et al., 2009). In astrocytes, the opening of HCs has been correlated with ATP and glutamate release (Cotrina et al., 1998; Ye et al., 2003). Most of the brain neurons express receptors for several neurotransmitters (i.e., P2X, P2Y, and N-methyl-D-aspartate (NMDA) receptors). An excessive release of these neurotransmitters in the extracellular space provokes a massive and constant activation of these receptors that could lead to neuronal death.

11.4 ASTROGLIAL CX43 HEMICHANNELS CONTRIBUTE TO NEURONAL DEATH

During the past 20 years, the protective role of astrocytes against neuronal death has been well described in different models of neuronal toxicity (Rosenberg and

Aizenman, 1989; Desagher et al., 1996). Considering the high level of Cx expression in astrocytes, the involvement of Cxs in neuronal death regulation has been a frequently addressed question. Numerous studies have investigated whether the level of gap junctional communication can modify the neuroprotective role of astrocytes. They first investigated the consequences of the blockade of Cx43 GJCs on the modulation of neurotoxicity. Indeed, some authors reported a decrease of neuroprotection exerted by astrocytes using GJC uncoupling drugs (Blanc et al., 1998; Ozog et al., 2002) in neuron–astrocytes cocultures. In contrast, a similar drug treatment was found to reinforce the neuroprotection either in a trauma model *in vitro* (Frantseva et al., 2002b) or in models of ischemia (Frantseva et al., 2002a; de Pina-Benabou et al., 2005). An important improvement for such studies consisted in the use of mutant mice whose Cx43 gene was genetically abolished (Cx43$^{-/-}$ mice). In *in vitro* trauma model performed with brain slices from Cx43$^{-/-}$ mice, Frantseva et al. (2002b) also described a protective effect due to Cx43 absence, similar to the effect observed with the pharmacological inhibition of GJCs. But, *in vivo* experiments performed in Cx43$^{+/-}$ mice (in which only one allele of the gene was invalidated because Cx43$^{-/-}$ animals do not survive after birth) have revealed that the reduction of Cx43 expression level in these mice increases the infarct size and cell death following cerebral ischemia (Nakase et al., 2003; Nakase and Naus, 2004). In summary, studies investigating the role of astroglial Cxs have provided results either in favor or against a neuroprotective action of Cx43 in astrocytes. But it is important to note that all these studies never differentiated between the two channel functions of Cx43, that is, GJCs versus HCs. As mentioned above, the mixture of IL-1β plus TNF-α activates Cx43 HCs and closes GJCs in astrocyte cultures (Même et al., 2006; Retamal et al., 2007). In these conditions, the dual regulation of Cx43 functions provides a model to assess the impact of Cx43 GJCs versus HCs on neuronal death. For this purpose, a model of NMDA-induced excitotoxicity in neuron–astrocyte cocultures was used to discriminate the consequence of the dual regulation of astroglial Cx43 channel functions induced by proinflammatory treatment (IL-1β plus TNF-α). Interestingly, the NMDA-induced neuronal death in astrocyte–neuron cocultures is reinforced by this proinflammatory treatment, while such a potentiating effect is not observed in cultures of neurons alone. This observation suggests that the protective role of astrocytes against excitotoxicity is impaired under proinflammatory conditions when Cx channel functions are affected. Using neuron–astrocyte cocultures with astrocytes lacking Cx43 expression (from Cx43$^{-/-}$ mice) and wild-type neurons revealed that the NMDA excitotoxicity is not reinforced by the proinflammatory treatment applied on these cocultures with Cx43-deficient astrocytes (Froger et al., 2010). This result suggests that astroglial Cx43 is involved in the potentiated neurotoxicity induced by proinflammatory treatments. To discriminate between the implication of GJCs and HCs, since mimetic peptides specific for Cx43 HC blockade are available (Gap26 and Gap27) and when used for short-term treatment (<30 min) prevent only HCs in astrocytes (Retamal et al., 2007). Interestingly, the application of these peptides prevents the potentiated neurotoxicity response induced by proinflammatory cytokines (Froger et al., 2010). These observations further demonstrate the crucial role of astroglial Cx43 HCs in the potentiation of excitotoxicity that occurs under an inflammatory context and strengthens the hypothesis of synergic interactions between inflammatory toxicity and glutamate toxicity, in which Cx43 HCs contribute to accentuate the neuronal susceptibility to adverse conditions.

11.5 ALTERATION OF ASTROGLIAL CONNEXIN EXPRESSION IN BRAIN PATHOLOGIES

There is accumulating evidence indicating that the expression of Cxs in astrocytes is modified during reactive gliosis associated with most, if not all, acute or chronic brain injuries and pathologies. Most studies in this field have examined the level of Cx43 mRNA and/or protein (see Orellana et al., 2009; Giaume et al., 2010). In most neurodegenerative diseases, and after ischemic and excitotoxic injuries, an increased expression of Cx43 was detected. However, in the latter case, the increase in Cx43 and Cx30 was restricted to reactive astrocytes located in a ring surrounding the damaged area while a strong downregulation occurred in the reactive astrocytes located within the core of the lesion (Koulakoff et al., 2008). The consequences of these modifications on intercellular communication via either GJCs or HCs have started to be examined *in vivo*. Unfortunately, these analyses included in most cases only gap junctional communication assessed by dye coupling. In this way, increased Cx expression has been correlated with increased dye coupling in astrocytes. However, the reduction in gap junctional communication induced by ischemia has been proposed to be mediated by dephosphorylation of Cx43 (Li and Nagy, 2000). The consequence of changes in the expression of astroglial Cxs on neuronal function and/or survival is not yet clear, since both beneficial and deleterious effects have been reported (see above). The working hypothesis is that astrocytes can redistribute potentially detrimental ions and metabolites over long distances within their communicating networks, hence contributing to the protection of neurons from an excess of toxic substances. In support of this hypothesis, the infarct size following ischemia is enlarged in Cx43 KO mice compared to wild-type mice (Siushansian et al., 2001; Nakase et al., 2003). In contrast, the blockade of Cx43 functions either with pharmacological agents or with mimetic peptides was reported to limit the extent of damages generated by brain or spinal cord injury (Frantseva et al., 2002b; Cronin et al., 2008; O'Carroll et al., 2008). The relative contribution of the two Cx channel functions to these effects has not yet been fully dissected but certainly will give some clue to the apparent contradictory literature in this field. Hence, the role played by astroglial Cxs in pathological situations is likely diverse and certainly depends on several factors, such as time, cell type, location, and the type of brain injury or pathology.

11.6 CANNABINOID SYSTEM AS A POSSIBLE THERAPEUTIC STRATEGY AGAINST CONNEXIN CHANNEL-MEDIATED NEURODEGENERATION

CBs are known to exert anti-inflammatory properties (Cabral and Griffin-Thomas, 2008) and their receptors are expressed in glial cells, in particular in astrocytes (Stella, 2004, 2010). As astrocytes are well-known targets for proinflammatory agents, the effect of both synthetic and endogenous CBs has been tested on Cx43 channels in cells treated with proinflammatory agents. As mentioned above, Cx43 functions, GJCs versus HCs, are oppositely regulated by proinflammatory treatment (Retamal et al., 2007). Under these conditions, the addition of CBs, either synthetic (such as WIN-55,212-2) or endogenous (such as methanandamine), into the culture medium prevents this dual regulation of Cx43 channels; thus, astrocytes maintain a low level of HC activity and a high level of gap junctional communication (Froger et al., 2009). Pharmacological investigations reveal that the regulation of Cx43 channels exerted by CBs is preferentially mediated through CB1 receptors, although the implication of non-CB1/CB2 receptor pathways cannot be excluded (Froger et al., 2009; see also Venance et al., 1995).

As the opening of Cx43 HCs occurring in inflammatory conditions reinforces the NMDA-induced excitotoxicity, any pharmacological action that consists in reducing or blocking Cx43 HC activity in astrocytes may represent a new strategy to protect neurons against excitotoxic damages. In this context, the effect of CBs agonists on NMDA-induced excitotoxicity in inflammatory conditions, that is, TNF-α and IL-1β, have been tested. As shown in Figure 11.2, these experiments demonstrated that both synthetic and endogenous CBs (WIN-55,212-2 and methamandamide, respectively) abolish the enhanced excitotoxicity induced by proinflammatory cytokines (Froger

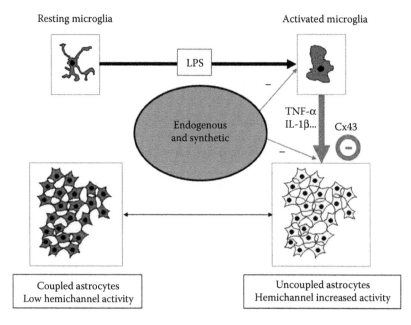

FIGURE 11.2 Cannabinoids (CBs) prevent the downregulation of Cx43 expression and function in astrocytes. In coculture, the stimulation of microglia by LPS induces the release of proinflammatory cytokines that reduce the expression of Cx43 and generates an opposite regulation of Cx43 channels, an inhibition of gap junctional communication and an activation of hemichannels. These effects are prevented by treatment with synthetic and endogenous CBs that act at two levels: in microglia on the production of TNF-α and IL-1β and in astrocytes on Cx43 channels themselves.

et al., 2010). Accordingly, as a working hypothesis, it can be proposed that glial CB receptor stimulation may constitute a new target as a neuroprotection strategy, to prevent excitotoxicity-induced neuronal damages in an inflammatory context, like that observed in neurodegenerative diseases.

11.7 CONCLUSIONS

Brain inflammation is manifest by infiltration of circulating immune cells, microglial activation and proliferation, and reactive astrocytosis. These responses impair brain energy metabolism (Steele and Robinson, 2010), modify blood–brain barrier integrity, produce microvascular hyalinosis, and affect mitochondrial function and neuronal activity as well as cognitive functions (Di Filippo et al., 2010; Giaume et al., 2010; Ownby, 2010; Allaman et al., 2011). Numerous inflammatory factors are known to contribute to these different dysfunctions. Here, we have gathered data indicating that the functional state of astroglial Cx channels is also affected. They are in favor of an increase in HC activity that under resting condition is very low possibly to maintain the cell integrity. Thus, inflammatory conditions enhance inside–outside exchanges of ions and small molecules exacerbating neuronal death. Putting the light on the contribution of astroglial HCs will certainly help to reconsider and clarify the controversy on the good or bad roles of astroglial Cx channels in neuronal death and survival. Indeed, the role Cx-based channels are rather alternating in the literature between a protective and a deleterious role to neuronal susceptibility to injuries (Rouach et al., 2002b; Perez Velazquez et al., 2003; Nakase and Naus, 2004; Farahani et al., 2005). This controversy might be explained because up to now only GJCs were considered as the only Cx43-based channel involved in neuroprotection.

To go further and improve our knowledge about the contribution of Cx channels in neuronal death and survival, new tools to study their function are required. Indeed, so far, pharmacological and genetic approaches used to study glial Cxs have shown their limitations (Giaume and Theis, 2010). What we need in the future are tools that allow discriminating between the contribution of microglial and astroglial Cxs (and neuronal), between GJC and HC functions, and also between Panx and Cx HCs. Also, it becomes more and more relevant to identify what signaling molecules are passing through these channels in normal and inflammatory conditions. The achievement of these goals will certainly help to design new therapeutic strategies that will target glia instead of neurons to prevent or at least minimize neuronal damages in brain injury and pathologies. Hereof, the observations made with synthetic and endogenous CBs might indicate new directions that could be taken in the coming years.

ACKNOWLEDGMENTS

CG and NF were supported by the CRPCEN and France Alzheimer. JCS and JA were supported by Fondecyt (1070591) anillo ACT71 grants. MAR was supported by Fondecyt de Iniciacion (11080061). Additional support was provided by an ECOS-CONICYT grant to CG and JCS.

REFERENCES

Allaman I, Belanger M, Magistretti PJ. 2011. Astrocyte–neuron metabolic relationships: For better and for worse. *Trends Neurosci* 34:76–87.

Blanc EM, Bruce-Keller AJ, Mattson MP. 1998. Astrocytic gap junctional communication decreases neuronal vulnerability to oxidative stress-induced disruption of Ca^{2+} homeostasis and cell death. *J Neurochem* 70:958–970.

Block ML, Hong JS. 2005. Microglia and inflammation-mediated neurodegeneration: Multiple triggers with a common mechanism. *Prog Neurobiol* 76:77–98.

Blomstrand F, Venance L, Siren AL, Ezan P, Hanse E, Glowinski J, Ehrenreich H, Giaume C. 2004. Endothelins regulate astrocyte gap junctions in rat hippocampal slices. *Eur J Neurosci* 19:1005–1015.

Cabral GA, Griffin-Thomas L. 2008. Cannabinoids as therapeutic agents for ablating neuroinflammatory disease. *Endocr Metab Immune Disord Drug Targets* 8:159–172.

Cotrina ML, Lin JH, Alves-Rodrigues A, Liu S, Li J, Azmi-Ghadimi H, Kang J, Naus CC, Nedergaard M. 1998. Connexins regulate calcium signaling by controlling ATP release. *Proc Natl Acad Sci USA* 95:15735–15740.

Cronin M, Anderson PN, Cook JE, Green CR, Becker DL. 2008. Blocking connexin 43 expression reduces inflammation and improves functional recovery after spinal cord injury. *Mol Cell Neurosci* 39:152–160.

D'Hondt C, Ponsaerts R, De Smedt H, Bultynck G, Himpens B. 2009. Pannexins, distant relatives of the connexin family with specific cellular functions? *Bioessays* 31:953–974.

de Pina-Benabou MH, Szostak V, Kyrozis A, Rempe D, Uziel D, Urban-Maldonado M, Benabou S, Spray DC, Federoff HJ, Stanton PK, Rozental R. 2005. Blockade of gap junctions *in vivo* provides neuroprotection after perinatal global ischemia. *Stroke* 36:2232–2237.

Dermietzel R, Hertberg EL, Kessler JA, Spray DC. 1991. Gap junctions between cultured astrocytes: Immunocytochemical, molecular, and electrophysiological analysis. *J Neurosci* 11:1421–1432.

Desagher S, Glowinski J, Premont J. 1996. Astrocytes protect neurons from hydrogen peroxide toxicity. *J Neurosci* 16:2553–2562.

Di Filippo M, Chiasserini D, Tozzi A, Picconi B, Calabresi P. 2010. Mitochondria and the link between neuroinflammation and neurodegeneration. *J Alzheimers Dis* 20(Suppl 2): S369–379.

Dobrenis K, Chang HY, Pina-Benabou MH, Woodroffe A, Lee SC, Rozental R, Spray DC, Scemes E. 2005. Human and mouse microglia express connexin 36, and functional gap junctions are formed between rodent microglia and neurons. *J Neurosci Res* 82:306–315.

Elias LA, Wang DD, Kriegstein AR. 2007. Gap junction adhesion is necessary for radial migration in the neocortex. *Nature* 448:901–907.

Eugenín EA, Brañes MC, Berman JW, Saez JC. 2003. TNF-alpha plus IFN-gamma induce connexin 43 expression and formation of gap junctions between human monocytes/macrophages that enhance physiological responses. *J Immunol* 170:1320–1328.

Farahani R, Pina-Benabou MH, Kyrozis A, Siddiq A, Barradas PC, Chiu FC, Cavalcante LA, Lai JC, Stanton PK, Rozental R. 2005. Alterations in metabolism and gap junction expression may determine the role of astrocytes as "good samaritans" or executioners. *Glia* 50:351–361.

Frantseva MV, Kokarovtseva L, Naus CC, Carlen PL, MacFabe D, Perez Velazquez JL. 2002b. Specific gap junctions enhance the neuronal vulnerability to brain traumatic injury. *J Neurosci* 22:644–653.

Frantseva MV, Kokarovtseva L, Perez Velazquez JL. 2002a. Ischemia-induced brain damage depends on specific gap-junctional coupling. *J Cereb Blood Flow Metab* 22:453–462.

Froger N, Orellana JA, Calvo CF, Amigou E, Kozoriz MG, Naus CC, Sáez JC, Giaume C. 2010. Inhibition of cytokine-induced connexin 43 hemichannel activity in astrocytes is neuroprotective. *Mol Cell Neurosci* 45:37–46.

Froger N, Orellana JA, Cohen-Salmon M, Ezan P, Amigou E, Sáez JC, Giaume C. 2009. Cannabinoids prevent the opposite regulation of astroglial connexin43 hemichannels and gap junction channels induced by pro-inflammatory treatments. *J Neurochem* 111:1383–1397.

Garg S, Md Syed M, Kielian T. 2005. *Staphylococcus aureus*-derived peptidoglycan induces Cx43 expression and functional gap junction intercellular communication in microglia. *J Neurochem* 95:475–483.

Giaume C, Fromaget C, el Aoumari A, Cordier J, Glowinski J, Gros D. 1991. Gap junctions in cultured astrocytes: Single-channel currents and characterization of channel-forming protein. *Neuron* 6:133–143.

Giaume C, Koulakoff A, Roux L, Holcman D, Rouach N. 2010. Astroglial networks: A step further in neuroglial and gliovascular interactions. *Nat Rev Neurosci* 11:87–99.

Giaume C, McCarthy KD. 1996. Control of gap-junctional communication in astrocytic networks. *Trends Neurosci* 19:319–325.

Giaume C, Theis M. 2010. Pharmacological and genetic approaches to study connexin-mediated channels in glial cells of the central nervous system. *Brain Res Rev* 63:160–176.

Huang Y, Grinspan JB, Abrams CK, Scherer SS. 2007. Pannexin1 is expressed by neurons and glia but does not form functional gap junctions. *Glia* 55:46–56.

Iglesias R, Dahl G, Qiu F, Spray DC, Scemes E. 2009. Pannexin 1: The molecular substrate of astrocyte "hemichannels". *J Neurosci* 29:7092–7097.

John GR, Scemes E, Suadicani SO, Liu JS, Charles PC, Lee SC, Spray DC, Brosnan CF. 1999. IL-1β differentially regulates calcium wave propagation between primary human fetal astrocytes via pathways involving P2 receptors and gap junction channels. *Proc Natl Acad Sci USA* 96:11613–8.

Koulakoff A, Ezan P, Giaume C. 2008. Neurons control the expression of connexin 30 and connexin 43 in mouse cortical astrocytes. *Glia* 56:1299–1311.

Kunzelmann P, Schroder W, Traub O, Steinhauser C, Dermietzel R, Willecke K. 1999. Late onset and increasing expression of the gap junction protein connexin 30 in adult murine brain and long-term cultured astrocytes. *Glia* 25:111–119.

Li WE and Nagy JI. 2000. Connexin 43 phosphorylation state and intercellular communication in cultured astrocytes following hypoxia and protein phosphatase inhibition. *Eur J Neurosci* 12:2644–2650.

Martínez AD, Sáez JC. 1999. Arachidonic acid-induced dye uncoupling in rat cortical astrocytes is mediated by arachidonic acid byproducts. *Brain Res* 816:411–423.

Même W, Calvo CF, Froger N, Ezan P, Amigou E, Koulakoff A, Giaume C. 2006. Proinflammatory cytokines released from microglia inhibit gap junctions in astrocytes: Potentiation by beta-amyloid. *FASEB J* 20:494–496.

Même W, Ezan P, Venance L, Glowinski J, Giaume C. 2004. ATP-induced inhibition of gap junctional communication is enhanced by interleukin-1 beta treatment in cultured astrocytes. *Neuroscience* 126:95–104.

Monif M, Reid CA, Powell KL, Smart ML, Williams DA. 2009. The P2X7 receptor drives microglial activation and proliferation: A trophic role for P2X7R pore. *J Neurosci* 29:3781–3791.

Nagy JI, Patel D, Ochalski PA, Stelmack GL. 1999. Connexin 30 in rodent, cat and human brain: Selective expression in gray matter astrocytes, co-localization with connexin 43 at gap junctions and late developmental appearance. *Neuroscience* 88:447–468.

Nagy JI, Rash JE. 2000. Connexins and gap junctions of astrocytes and oligodendrocytes in the CNS. *Brain Res Brain Res Rev* 32:29–44.

Nakase T, Fushiki S, Naus CC. 2003. Astrocytic gap junctions composed of connexin 43 reduce apoptotic neuronal damage in cerebral ischemia. *Stroke* 34:1987–1993.

Nakase T, Naus CC. 2004. Gap junctions and neurological disorders of the central nervous system. *Biochim Biophys Acta* 1662:149–158.

O'Carroll SJ, Alkadhi M, Nicholson LF, Green CR. 2008. Connexin 43 mimetic peptides reduce swelling, astrogliosis, and neuronal cell death after spinal cord injury. *Cell Commun Adhes* 15:27–42.

Orellana JA, Sáez PJ, Shoji KF, Schalper KA, Palacios-Prado N, Velarde V, Giaume C, Bennett MV, Sáez JC. 2009. Modulation of brain hemichannels and gap junction channels by pro-inflammatory agents and their possible role in neurodegeneration. *Antioxid Redox Signal* 11(2):369–399.

Ownby RL. 2010. Neuroinflammation and cognitive aging. *Curr Psychiatry Rep* 12:39–45.

Ozog MA, Siushansian R, Naus CC. 2002. Blocked gap junctional coupling increases glutamate-induced neurotoxicity in neuron–astrocyte co-cultures. *J Neuropathol Exp Neurol* 61:132–141.

Perez Velazquez JL, Frantseva MV, Naus CC. 2003. Gap junctions and neuronal injury: Protectants or executioners? *Neuroscientist* 9:5–9.

Retamal MA, Cortés CJ, Reuss L, Bennett MV, Sáez JC. 2006. S-nitrosylation and permeation through connexin 43 hemichannels in astrocytes: Induction by oxidant stress and reversal by reducing agents. *Proc Natl Acad Sci USA* 103:4475–4480.

Retamal MA, Froger N, Palacios-Prado N, Ezan P, Sáez PJ, Sáez JC, Giaume C. 2007. Cx43 hemichannels and gap junction channels in astrocytes are regulated oppositely by pro-inflammatory cytokines released from activated microglia. *J Neurosci* 27:13781–13792.

Rosenberg PA, Aizenman E. 1989. Hundred-fold increase in neuronal vulnerability to glutamate toxicity in astrocyte-poor cultures of rat cerebral cortex. *Neurosci Lett* 103:162–168.

Rouach N, Avignone E, Meme W, Koulakoff A, Venance L, Blomstrand F, Giaume C. 2002b. Gap junctions and connexin expression in the normal and pathological central nervous system. *Biol Cell* 94:457–475.

Rouach N, Calvo CF, Glowinski J, Giaume C. 2002a. Brain macrophages inhibit gap junctional communication and downregulate connexin 43 expression in cultured astrocytes. *Eur J Neurosci* 15:403–407.

Rouach N, Glowinski J, Giaume C. 2000. Activity-dependent neuronal control of gap-junctional communication in astrocytes. *J Cell Biol* 149:1513–26.

Rouach N, Koulakoff A, Abudara V, Willecke K, Giaume C. 2008. Astroglial metabolic networks sustain hippocampal synaptic transmission. *Science* 322:1551–1555.

Scemes E. 2008. Modulation of astrocyte P2Y1 receptors by the carboxyl terminal domain of the gap junction protein Cx43. *Glia* 56:145–153.

Shestopalov VI, Panchin Y. 2008. Pannexins and gap junction protein diversity. *Cell Mol Life Sci* 65:376–394.

Siushansian R, Bechberger JF, Cechetto DF, Hachinski VC, Naus CC. 2001. Connexin 43 null mutation increases infarct size after stroke. *J Comp Neurol* 440:387–394.

Steele ML, Robinson SR. 2010. Reactive astrocytes give neurons less support: Implications for Alzheimer's disease. *Neurobiol Aging* 33(2):423.e1–13.

Stella N. 2004. Cannabinoid signaling in glial cells. *Glia* 48:267–277.

Stella N. 2010. Cannabinoid and cannabinoid-like receptors in microglia, astrocytes, and astrocytomas. *Glia* 58:1017–1030.

Takeuchi H, Jin S, Wang J, Zhang G, Kawanokuchi J, Kuno R, Sonobe Y, Mizuno T, Suzumura A. 2006. Tumor necrosis factor-alpha induces neurotoxicity via glutamate release from hemichannels of activated microglia in an autocrine manner. *J Biol Chem* 281:21362–21368.

Thompson RJ, Jackson MF, Olah ME, Rungta RL, Hines DJ, Beazely MA, MacDonald JF, MacVicar BA. 2008. Activation of pannexin-1 hemichannels augments aberrant bursting in the hippocampus. *Science* 322:1555–1559.

Venance L, Piomelli D, Glowinski J, Giaume C. 1995. Inhibition by anandamide of gap junctions and intercellular calcium signalling in striatal astrocytes. *Nature* 376:590–594.

Wallraff A, Kohling R, Heinemann U, Theis M, Willecke K, Steinhauser C. 2006. The impact of astrocytic gap junctional coupling on potassium buffering in the hippocampus. *J Neurosci* 26:5438–5447.

Ye ZC, Wyeth MS, Baltan-Tekkok S, Ransom BR. 2003. Functional hemichannels in astrocytes: A novel mechanism of glutamate release. *J Neurosci* 23:3588–3596.

12 A Role for Connexins in Inflammatory Disorders of the Epidermis

Steven Donnelly, Catherine S. Wright,
Maurice A. M. van Steensel,
Malcolm B. Hodgins, and Patricia E. Martin

CONTENTS

12.1 STRUCTURE OF SKIN AND EPIDERMAL BARRIER

The human skin is a complex organ that provides a protective barrier against fluid loss from within, and against assault from the external environment by chemicals (irritants), physical agents (mechanical forces, UV radiation), and biological pathogens (Proksch et al., 2008). Moreover, skin contributes to thermal regulation by sweating. The skin consists of three layers: the hypodermis, loose connective tissue that contains the subcutaneous fat; the dermis, a connective tissue (rich in collagens and glycosaminoglycans synthesized by dermal fibroblasts) containing blood supply and nerves, which supports the epidermis and the epidermal appendages (hair follicles sebaceous glands, and sweat glands); and the epidermis, the outer, stratified squamous epithelium of keratinocytes from which the appendages are derived and which forms the skin barrier (Figure 12.1). Other resident cells of the epidermis are melanocytes, responsible for skin pigmentation, Langerhans cells (antigen-presenting dendritic cells), Merkel cells and other sensory receptors and nerve endings. In normal

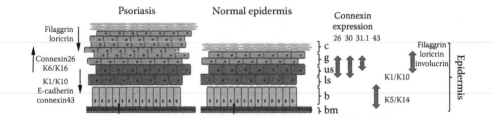

FIGURE 12.1 Key epidermal characteristics in normal and psoriatic epidermis. Representative image of normal and psoriatic epidermis, illustrating hyperproliferation events and changes in the expression of key epidermal proteins in psoriasis. In normal epidermis, expression of Cx26 and Cx30 is found in the upper granular (g) and upper spinous layers, with Cx31.1 only noted in the granular. Cx43 expression is predominantly found in the basal (b) layer of the epidermis and dermal fibroblast cells in the dermis. Filaggrin, loricrin, and involucrin are key markers for the upper epidermal layers and the differentiation status can be determined by the expression of Keratin partners K5/K14 in the basal layers and K1/K10 in the upper layers. During psoriasis, an example of an epidermal inflammatory disorder, hyperproliferation is evident with K6/K16 replacing K1/K10 expression, a decrease in E-cadherin and Cx43 expression, while Cx26 expression is markedly enhanced. Both filaggrin and loricrin are also downregulated in psoriatic epidermis. These events likely contribute to altered inflammatory-mediated consequences in the epidermis in which Cx26 is emerging to play a pivotal role. See text for details. c, cutaneous; g, granular; us, upper spinous; ls, lower spinous; b, basal; bm, basement membrane. Not pictured: HF (hair follicle) regions, ORS/IRS (outer/inner root sheath), and MD (medulla).

epidermis, a steady state exists between production of new cells from keratinocytes resident in the basal layer and the loss of terminally differentiated cells from the skin surface (Blanpain and Fuchs, 2009). Histologically, four layers of keratinocyte differentiation can be recognized within the epidermis: the basal, spinous, granular, and stratum corneum layers. Each represents a stage in progression through the cell terminal differentiation program (Figure 12.1). Epidermal physical barrier function against water loss and penetration by chemical, physical, and biological agents resides mainly within the stratum corneum, although recent work suggests that a network of tight junctions within the granular layer may also be important in some circumstances (Kirschner et al., 2009). The specific immunological barrier function of the epidermis is largely associated with the Langerhans cells (Romani et al., 2006, 2010), although keratinocytes also seem to play a role, particularly in the innate immune response (Lai et al., 2010; Lai and Gallo, 2008).

Central to the organization of the epidermis is a complex network of cell-to-cell adhesion and communication junctions, which can be termed the epidermal junctional nexus. These include adherens junctions, desmosomes, tight junctions, and gap junctions; their constituent proteins are differentially expressed throughout the epidermal layers, possibly reflecting a gradual evolution of junctional properties as keratinocytes differentiate (Proksch et al., 2008). The carboxyl terminal domains of each class of protein have multiple interaction sites for cytoskeletal elements. The gap junction component, connexin43 (Cx43), adherens junction component E-cadherin, and components of the tight junction interact with cytoskeletal linker proteins such as ZO-1 and β-catenin that link with the polarized actin cytoskeletal network (Laird, 2006; Shaw et al., 2007). Connexins also interact with the microtubule network, most likely via kinesin motor proteins, involved in trafficking of connexins to the plasma membrane (Fort et al., 2012; Martin et al., 2001; Shaw et al., 2007). Desmosomal cadherins interact with keratin intermediate filaments that constitute the major element of the keratinocyte cytoskeleton, possibly providing mechanical support for microfilaments and microtubules (Garrod and Chidgey, 2008).

Keratins were among the first proteins whose expression could be shown to change during epidermal keratinocyte differentiation (Figure 12.1). Keratins are a diverse family of proteins that are expressed in pairs from two main groups, type 1 (acidic or subfamily A) and type 2 (basic or subfamily B). The K5/14 keratin pair is the characteristic of the basal keratinocytes, giving way to the K1/K10 pair as basal cells commit to terminal differentiation. Changes in expression profiles of keratins were also among the first molecular changes to be identified in epidermal diseases, for example, in psoriatic epidermis, expression of the K1/10 pair is lost, being replaced by the K6/K16 pair that is usually associated with rapidly turning over, thick epidermis (Ekanayake-Mudiyanselage et al., 1998). Not surprisingly, the striking morphological changes that mark the passage of keratinocytes through the basal, spinous, granular, and stratum corneum layers are marked by changes in most of the cells' major molecular constituents—proteins, lipids, and nucleic acids (reviewed by Proksch et al., 2008). In this chapter, we concentrate on the connexins. In addition to forming gap junction intercellular communication channels, connexins, like their homologs, the pannexins, which do not seem to form gap junctions (D'Hondt et al., 2009), are now known to form "hemichannels" that can open to release adenosine-5′-triphosphate (ATP) and other active metabolites from the cell, resulting in activation of purinergic receptor pathways. It also appears that the adhesive properties of connexins in gap junctions may have important functions, independent of channel activity. We will review recent work that suggests that these alternative roles of connexins may be important in the epidermis.

12.2   CONNEXIN EXPRESSION AND FUNCTION IN EPIDERMIS

The epidermis is an avascular tissue. It is thought that intercellular junctions, in addition to providing cell-to-cell adhesion, allow cells to communicate within and between the layers that mark stages in differentiation. This may help to maintain homeostasis and to coordinate responses to various stimuli, for example, damage to the stratum corneum results in a burst of basal cell division. Changing expression of junctional proteins might restrict communication as keratinocytes differentiate. The human epidermis appears to express as many as 10 different connexin genes differentially throughout its layers, although not all have been detected as proteins (Di et al., 2001). Antibody staining has shown connexin abundance to be lowest in the basal and highest in the spinous layers (similar to desmosomal proteins), correlating with abundance of gap junctions seen by electron microscopy. Connexins Cx26, Cx30, Cx30.3, Cx31, Cx31.1, Cx40, Cx43, and Cx45 have all been detected by antibody staining. However, connexins have not reliably been detected in the stratum corneum, although this could simply reflect the difficulty of conducting protein analysis in this highly cross-linked tissue. As with keratins, abnormalities of connexin expression are associated with disease states, for example, Cx26, which is normally restricted mainly to human palmoplantar epidermis (and even there in only low abundance), is greatly upregulated in psoriatic epidermis (Labarthe et al., 1998), in human papilloma virus-infected epidermis (Lucke et al., 1999), and in other states of pathological hyperproliferation (Hivnor et al., 2004; Lucke et al., 1999). However, in spite of the differing permeability properties of gap junctions composed of different connexins and of the differential expression of connexins within epidermal layers, studies of human and mouse epidermis have identified unrestricted gap junctional transfer of small molecule tracers of varying size and charge (e.g., Lucifer yellow, neurobiotin) between keratinocytes from the basal, up to the granular layer (but not to the cornified layers) (Choudhry et al., 1997). This poses the question: what changes in function do the changing connexin profiles represent?

The fact that mutations in genes encoding connexins (including Cx26, 30, 30.3, 31) are causal in several inherited skin disorders that have distinct histopathological features strongly suggests that specificity of connexin expression does play an important role in the correct functioning of the epidermis (de Zwart-Storm et al., 2009). Recent evidence suggests that hemichannel activity and possibly cell-to-cell adhesion may have significant roles related to the changing Cx expression profiles during keratinocyte differentiation, particularly in relation to the maintenance of the epidermal barrier.

12.3   CONNEXIN EXPRESSION AND EPIDERMAL BARRIER

Defects of the epidermal barrier give rise to many different medical problems often initiating cell signaling cascades promoting an inflammatory response and/or wound repair. A variety of these events result in altered connexin expression and signaling, suggesting that connexin expression or function may be involved in maintaining barrier integrity (Djalilian et al., 2006). For example, the pathogenesis of psoriasis, a chronic inflammatory condition characterized by gross epidermal thickening (plaques) and a dysfunctional barrier, involves both the activation of T-lymphocytes and aberrant proliferation and differentiation of keratinocytes (Wagner et al., 2010). Accumulating evidence suggests that susceptibility to psoriasis is linked to the epidermal differentiation complex and that the epidermal junctional nexus is altered, possibly leading to the hyperproliferative phenotype (Chung et al., 2005; Hampton et al., 2007; Kirschner et al., 2009; Li et al., 2008; Tschachler, 2007). Indeed, Connexin26 is strongly upregulated while Cx43 and E-cadherin are downregulated in psoriatic plaques (Chung et al., 2005; Hivnor et al., 2004; Labarthe et al., 1998; Li et al., 2008). Such information suggests that impaired cell-to-cell adhesion and interaction with cytoskeletal networks play a role in barrier recovery and may contribute to disease progression.

Studies in mouse models have determined that the transcription factor *kruppel like factor 4* (KLF4) is required to establish and maintain the murine epidermal barrier (Djalilian et al., 2006), protecting against water loss. The ability of the epidermis to function as a permeability barrier occurs in anticipation of birth, usually at day 18 of a 20-day gestation period in mice. During fetal development, Cx26 is expressed in the epidermis and KLF4 expression is low. Activation of KLF4 expression, usually day 18 of a 20-day gestation period, is associated with a decrease in the expression of Cx26 and induction of correct barrier formation. KLF4 is expressed in the spinous layer as the cells lose adhesion with the basement membrane and enter their terminal differentiation program. An elegant study showed that in KLF4$^{-/-}$ mice, Cx26 was not downregulated before birth. These animals died perinatally due to water loss across the skin surface (Djalilian et al., 2006). Furthermore, ectopic expression of Cx26 in suprabasal keratinocytes (from an involucrin (inv) promoter) impaired barrier function and resulted in an inflammatory response that was mediated via increased Cx26 hemichannel activity, stimulating purinergic signaling events (see Section 3.3). These results suggest that correct KLF4 expression and subsequently correct regulation of Cx26 is necessary for epidermal terminal differentiation, further supporting the idea that high levels of Cx26 in suprabasal keratinocytes promote proliferation and impair barrier function (Djalilian et al., 2006). The association between Cx26 and hyperproliferation is complicated by the fact that, in psoriatic epidermis, most Cx26 gap junctions are found in differentiating nonproliferative keratinocytes. However, in related stratified squamous epithelia (oral, vaginal) that have naturally high mitotic rates and relatively poor water-barrier function, Cx26 is also abundantly expressed (Lucke et al., 1999).

The link between impaired barrier function and epidermal inflammatory/immune responses is well illustrated by the discovery that common atopic dermatitis is associated with mutations in the gene encoding filaggrin, the same gene that is mutated in the inherited scaling disease ichthyosis vulgaris and in the FlakyTail mouse, both of which have an impaired epidermal barrier (Moniaga et al., 2010; Smith et al., 2006). Filaggrin is a late-differentiation protein of epidermal keratinocytes, important for barrier formation. Atopic dermatitis has long been associated with altered skin microflora and dysfunction of skin immune responses. The identification of filaggrin being involved in maintaining epidermal barrier integrity and skin immune responses provides strong support for the concept of the barrier as a modulator of "outside-in" triggered skin immune responses (Cork et al., 2009; Sandilands et al., 2009; Smith et al., 2006; Weidinger et al., 2008).

12.4 EPIDERMAL BARRIER, SKIN MICROBIOME AND INFLAMMATION

The skin surface is an ecosystem with microbial communities that live in varied pathophysiological and topographical niches. Using 16s rRNA phylotyping, Grice et al. (2009) characterized the diversity of the human skin microbiota from different skin niches from 10 healthy subjects and showed that the diversity of our skin flora is much more varied than that established by traditional methods.

The difference between harmless flora and pathogenic organisms is often determined by the skin's ability to resist infection rather than by the microbe itself, for example, *Staphylococcus epidermidis* is usually harmless but is a frequent cause of nosocomial infections, usually in immune-compromised patients. On the other hand, *S. epidermidis* and *Pseudomonas aeruginosa* produce inhibitors active against other organisms such as *Staphylococcus aureus*, streptococci, and some fungal pathogens.

This suggests that *S. epidermidis* is tolerated on the skin surface as a defense against other more pathogenic microbes. Commensal and pathogenic microbes are not normally able to penetrate deeper into the epidermis than the stratum corneum, yet their products elicit defense responses within the living keratinocyte layers beneath the epidermal barrier (Cogen et al., 2008). Peptidoglycan (PGN), a key component of the Gram-positive cell wall, plays a key role in eliciting an innate immune response in diverse cell types, including keratinocytes, resulting in the release of inflammatory cytokines, including interleukin 6 (IL-6) and tumor necrosis factor α (TNF-α). A basal innate immune response to commensal organisms such as *S. epidermidis* may therefore prime keratinocytes, through Toll-like receptor 2 (TLR2) signaling, so that they respond more effectively to pathogenic challenge, and suggests a fine balance between different organisms of the skin microflora and the integrity of the epidermal barrier (Holland et al., 2009; Lai et al., 2010).

Disruption to this may lead to opportunistic infection even from skin commensals (Buchau and Gallo, 2007), underlying the fact that a fine interplay between keratinocytes and immune cells plays a critical role in the pathogenesis of diseases such as psoriasis and atopic dermatitis (Tonel and Conrad, 2009). Studies of inherited connexin disorders are now providing evidence that the connexins may also play an important role in the interaction between keratinocytes and skin surface microflora.

12.5 PROINFLAMMATORY MEDIATORS MODIFY CONNEXIN EXPRESSION AND FUNCTION

During injury or infection, release of cytokines and chemokines is an early response stage of the innate immune system and activates the migration of inflammatory cells to the affected area. These cytokines can induce changes in connexin expression profiles in diverse tissue types. For example, in mammary epithelial cells, transforming growth factor β (TGF-β) induced Cx43 expression (Tacheau et al., 2008a) and interleukin 1β (IL-1β) induced Cx43 expression in chondrocytes (Tonon and D'Andrea, 2000). TNF-α has tissue-specific effects on connexin expression. In endothelial cells, exposure to TNF-α reduced Cx37 and Cx40 expression but had limited effect on Cx43 (van Rijen et al., 1998). Further, in glioma cells, application of TNF-α impaired gap junctional intercellular communication (GJIC) mediated via Cx43 (Haghikia et al., 2008b) but following injury of skeletal muscle cells, Cx43 was upregulated. Keratinocytes are a major source of inflammatory cytokines and have been shown to produce IL-1, IL-6, and TNF-α (Proksch et al., 2008; Wang et al., 2004). Following exposure of mouse epidermal dendritic cells to keratinocyte preconditioned media, GJIC was enhanced (Corvalan et al., 2007). This was associated with TNF-α and IL-1β expression as antibodies targeted to these two cytokines reduced the incidence of dye coupling between the dendritic cells. However, the TNF-α/IL-1β-induced response was attenuated by coexposure to IL-6, suggesting a complex signaling network regulating GJIC and expression (Corvalan et al., 2007). Other tissue-specific effects have been observed in corneal fibroblasts where TNF-α inhibited GJIC, while in Schwann cells, application of TNF-α reduced Cx46 expression, GJIC, and cell proliferation (Chandross et al., 1996). A recent study in HaCat cells, an immortalized human keratinocyte cell line, determined that TNF-α decreased Cx43 expression and Cx43-mediated gap junction coupling (Tacheau et al., 2008b). Interestingly, TNF-α is overexpressed in the epidermis of psoriasis skin (Kristensen et al., 1993) and TNF-α antagonists are effective therapeutic agents in many patients (Richardson and Gelfand, 2008). A recent study has shown that skin isolated from psoriatic plaques is deficient in filaggrin and loricrin expression and that TNF-α inhibits the expression of these critical barrier proteins in primary human keratinocytes (Kim et al., 2011).

Bacterial pathogens and their toxins are major proinflammatory mediators, eliciting their responses via Toll-like receptor signaling mechanisms ultimately resulting in cytokine release (Kumar et al., 2009; Lai and Gallo, 2008). Bacterial toxins and cell wall components appear to have diverse and tissue-specific effects on connexin expression and signaling properties that may be related to disease progression. Examples are widespread and include some of the following. In primary astroglial and microglial cells, lipopolysaccharide (LPS) from the Gram-negative bacteria *Escherichia coli* decreased GJIC and Cx43 expression (Haghikia et al., 2008a; Hinkerohe et al., 2010). Other studies have shown that in microglial cells, LPS in combination with interferon gamma enhanced GJIC and increased Cx43 expression (Eugenin et al., 2001). Interestingly, a similar effect was observed in mouse epidermal and bone marrow-derived dendritic cells (Matsue et al., 2006). Within the neuronal network, *S. aureus* infection is a major causative agent of deep brain abscesses and is associated with decreased Cx43 expression and increased Cx26 expression in astrocytes; a decrease in gap junctional coupling also occurred (Esen et al., 2007). In contrast, in microglial cells, PGN, isolated from *S. aureus*, increased GJIC and Cx43 mRNA expression (Garg et al., 2005). Similar observations were found in endothelial cells challenged with PGN isolated from *S. epidermidis* or *S. aureus*, causative agents of infective endocarditis (Donnelly et al., 2012; Robertson et al., 2010). In contrast, in keratinocytes, PGN from *S. epidermidis* had no effect on IL-6 expression but that isolated from *S. aureus* induced IL-6 and Cx26 expression but had no impact on Cx43 expression (Donnelly et al., 2012).

In addition to altering connexin expression profiles, both LPS and PGN have been shown to elicit hemichannel signaling events. Challenge of cocultures of astrocytes and microglial cells (Retamal et al., 2007a), HeLa cells expressing Cx43 or Cx26 (Robertson et al., 2010), and human mesenchymal cells (Fruscione et al., 2010) with LPS has been reported to induce hemichannel signaling events. Exposure of endothelial cells to PGN acutely activated hemichannel signaling (within 15 min of exposure) that was linked with induction of IL-6, TNF-α, and TLR2 expression (Robertson et al., 2010). Blocking the acute hemichannel response or removing extracellular ATP by treatment with apyrase attenuated the PGN-induced cytokine and TLR2 responses strongly, indicating a very acute role of hemichannel signaling in the early innate immune response (Robertson et al., 2010).

12.6 EPITHELIAL BARRIER INTEGRITY AND CONNEXIN SIGNALING

In epithelial cells, barrier integrity is increasingly associated with appropriate hemichannel signaling events. This has been highlighted in a number of recent studies on enteric pathogens. In the mouse, the attaching and effacing (A/E) patho-gen *Citrobacter rodentium* caused an increase in Cx43 expression with associated increase in hemichannel activity in the apical membrane of infected colonocytes, suggesting a role for Cx43 in diarrheal disease (Guttman et al., 2009). Other studies revealed that *Shigella flexineri*, the causative agent of bacilli dysentery, induced the opening of Cx26 hemichannels by an actin and phospholipase-C-dependent pathway that ultimately led to an increase in further bacterial invasion and dissemination in epithelial cells (Tran Van Nhieu et al., 2003). *Shigella* also induced hemichannel signaling in human intestinal epithelial cells (Clair et al., 2008) and in HeLa cells expressing wild-type Cx26 but not in cells transfected to express the Cx26 muta-tion R143W, one of the mutations associated with recessive nonsyndromic hearing loss (Man et al., 2007). R143W is prevalent in African countries such as Ghana where carriers present with a thicker epidermis than noncarriers (Brobby et al., 1998; Hamelmann et al., 2001). Indeed, recessive mutations in Cx26 are the most com-mon cause of nonsyndromic hearing loss where defective channel function is a key cause of deafness (Martinez et al., 2009). Recent population studies of carriers of Cx26 recessive-deafness mutations in Africa, Italy, and the United States suggest that the epidermis of carriers was thicker than normal controls (D'Adamo et al., 2009; Guastalla et al., 2009). These observations led to the suggestion that the high carrier frequency of Cx26 recessive mutations could be a result of heterozygous advantage to barrier dysfunction both in epithelial and epidermal tissue (D'Adamo et al., 2009; Guastalla et al., 2009; Meyer et al., 2002). However, other explanations have also been advanced for the prevalence of recessive Cx26 mutations in various human populations.

As discussed in the following sections the situation is very different in a range of dominantly acting Cx26 mutations that are associated with hearing impairment and epidermal dysplasias and in disorders where epidermal Cx26 is overexpressed, such as in psoriasis.

The functional studies of GJIC have not provided any evidence for specific spa-tial restriction of communication associated with expression of different connexins throughout the multiple layers of the epidermis. This takes us back to the question of whether or not the apparently specific roles of connexins (especially Cx26) in skin inflammation could involve some other function(s) of the proteins. Some clues are emerging from the study of ectopic Cx26 expression and connexin mutants that cause hereditary skin disorders, and suggest a link between connexin hemichannels, the epidermal barrier response to microorganisms, and activation of keratinocyte innate immune responses.

12.7 CONNEXIN 26 MUTATIONS AND ECTODERMAL DYSPLASIA

There are many mutations in Cx26 present in the human population. Recessive mutations throughout the Cx26 coding region or deletions of the gene cause deaf-ness without any obvious skin pathology (although epidermal thickening in carriers has been reported) (D'Adamo et al., 2009; Guastalla et al., 2009), suggesting that the human skin does not require functional Cx26; it remains possible that closely related Cx30 may substitute, at least partly, for loss of Cx26. However, there are also many autosomal dominant mutations that, in addition to causing deafness, result in a spectrum of skin disorders (de Zwart-Storm, 2009). In the case of Cx26, multiple syndromes with unique pathologies are associated with a plethora of mutations in Cx26. In some cases, the abnormal Cx26 arrests within the ER–Golgi intermediate compartment/Golgi regions of the cell (Bakirtzis et al., 2003; de Zwart-Storm et al., 2008a); in others, the mutant protein traffics to the cell membrane (de Zwart-Storm et al., 2011; Donnelly et al., 2012; Easton et al., 2012; Schutz et al., 2010). Interestingly, this latter type of mutant is found in the most severe skin disorder keratitis–ichthyosis–deafness (KID) syndrome, in which inflammation, possibly involving skin microbiota, is usually present.

KID syndrome (OMIM#148210), first described in 1915 by Burns and named in 1981 by Skinner et al. after a study of 18 patients (Skinner et al., 1981), is a rare ectodermal dysplasia characterized by sensorineural hearing loss, hyperkeratosis of the palms and soles (palmoplantar keratoderma, PPK), erytherokeratoderma on the extremities and face, follicular hyperkeratosis, photophobia, and corneal vascularization that can lead to progressively poor visual acuity and ultimately blindness. To date, there have been around 100 reported cases of KID syndrome with these patients reportedly more prone to Gram-positive bacterial and fungal skin infections. Some patients have also been reported to have a predisposition to squamous cell carcinoma. Examples of case reports include Djalilian et al. (2010), Haruna et al. (2010), Kelly et al. (2008), Koppelhus et al. (2011), Mazereeuw-Hautier et al. (2007), Nyquist et al. (2007), Richard et al. (2002), and van Steensel et al. (2002). The Cx26 mutations causing KID syndrome cluster in the N-terminus (G12R, N14K, S17F) and the proximal extracellular loop 1 (E1) domain (A40V, G45E, D50N, D50Y) of the Cx26 protein with the exception of a recently discovered mutation A88V, which is located in the second transmembrane domain. In contrast, other Cx26 mutations linked with "milder" syndromes are predominantly found on the distal domain of E1 and on E2 (reviewed by de Zwart-Storm, 2009). There are subtle differences between the phenotypes associated with KID mutations, G45E having been described as the lethal form, while D50N is the most common mutation, perhaps because it generates a less severe form of the disease (Janecke et al., 2005).

Spatial localization studies of mutant Cx26 proteins show differences between Wt Cx26, KID mutants, and non-KID mutants. Wt Cx26 shows trafficking to the plasma membrane (PM) with the formation of gap junction clusters between adjacent cells. Many non-KID mutants such as D66H (associated with Vohwinklel's syndrome) and H73R (associated with Palmoplantar keratoderma) are trapped within the cell, in particular around the nucleus and they have limited or no junctional activity (Bakirtzis et al., 2003; de Zwart-Storm et al., 2008a,b). In contrast, while KID syndrome mutations show some cytoplasmic aggregation, they also retain the ability to reach the plasma membrane where they are found at discrete punctate locations rather than incorporated into large gap junction plaque-like areas (de Zwart-Storm et al., 2011; Donnelly et al., 2012; Easton et al., 2012). Although gap junction formation appears low, the fact that these mutations can be detected in the plasma membrane suggests a preference for hemichannel formation.

Three-dimensional reconstructions of the crystal structure of Cx26 with a 3.5 Å resolution has determined that the channel pore is narrowed at the entrance to the funnel by the six amino-terminal helices lining the wall of the channel, which dictates the molecular size restriction at the channel entrance (Maeda et al., 2009; Oshima et al., 2011). It is possible that mutations in the flexible Cx26 N-terminus may alter the permeability of the hemichannel, suggesting that the localization of KID mutations are in critical areas involved in gating and docking. Thus the pathogenesis of KID syndrome could be due, at least in part, to altered hemichannel signaling. Indeed, Lee et al. (2009) assessed hemichannel functionality in *Xenopus* oocytes by directly injecting the mRNA of Wt Cx26 and Cx26 KID syndrome mutations G12R, N14K, S17F, and D50N by monitoring membrane current. An increase in membrane current was observed in G12R, N14K, and D50N; this was concluded to be due to reduced resistance, indicating open hemichannels. Other studies have shown that KID mutations G45E and A40V also have increased hemichannel activity (Janecke et al., 2005). Thus, increased hemichannel activity in the *Xenopus* oocyte system appears to be a common feature among KID-causing mutations. However, the mutation S17F showed a complete lack of connexin function (Richard et al., 2002). One possible explanation for this exception would be that interaction of S17F with Wt connexins resulted in the formation of "leaky" hemichannels. This mutation resulted in severe developmental defects in a transgenic mouse model when the mutation was under the control of the endogenous Cx26 promoter (Schutz et al., 2010). Neonates had grossly impaired barrier integrity and only heterozygous animals were viable, displaying many KID-like symptoms (Schutz et al., 2010).

In mammalian cells, hemichannel activity can be triggered by low-calcium conditions, and observed by monitoring ATP release in the presence and absence of hemichannel blockers such as carbenoxolone (De Vuyst et al., 2007; Donnelly et al., 2012; Easton et al., 2012; Robertson et al., 2010; Wright et al., 2012). Transfection of HeLa cells with KID mutants, followed by exposure to a brief challenge in zero-calcium conditions 24 h posttransfection resulted in a rapid release of ATP that was not observed in non-KID mutations, again emphasizing the ability of KID mutations to retain hemichannel activity (Easton et al., 2012). Interestingly, transfection of KID mutations into HaCat cells cultured in low-calcium conditions results in cell death, while cell viability can be extended by maintaining the cells in higher extracellular calcium conditions. It is possible that KID-hemichannels are more sensitive to environmental triggers that might somehow disrupt epidermal barrier function against the skin flora, triggering inflammatory/immune responses (Donnelly et al., 2012). This could also include the sensitivity of Cx26 hemichannels to temperature (Steffens et al., 2008) or a range of bacterial proinflammatory mediators (Donnelly et al., 2012).

12.8   CONSEQUENCES OF ENHANCED HEMICHANNEL ACTIVITY FOR EPIDERMAL INFLAMMATION

The recent research has shown that connexins, in addition to forming gap junction channels, can form hemichannels that upon appropriate stimulation release secondary messenger mediators such as ATP and NADP into the extracellular space, where they can bind to and activate purinergic receptors (Evans et al., 2006). The impact of such events in triggering inflammatory responses is becoming increasingly clear in numerous tissues, with highly tissue-specific connexin subtypes involved. Within the epidermis, enhanced Cx26 expression and hemichannel activity are emerging to play a key role in hyperproliferative inflammatory disorders, including psoriasis and KID syndrome. The precise mechanisms triggering these events are subject to further research; however, clear links between exacerbated hemichannel activity and alterations in keratinocyte physiology and inflammation are emerging.

Hemichannels behave like large membrane conductance channels that are normally closed but can act as channels for secondary messengers such as ATP, nicotinamide adenine dinucleotide (NAD^+), glutamate, and prostaglandins (Braet et al., 2003; De Vuyst et al., 2007). Connexin hemichannels can be activated to release ATP by many stimuli, including mechanical cell stimulation (Bao et al., 2004; Cherian et al., 2005), elevation of intracellular inositol triphosphate (InsP3) (Braet et al., 2003), and exposure to low or zero extracellular calcium (as previously described in this chapter) (De Vuyst et al., 2007), and by challenge with a range of bacterial cell wall components (including PGN isolated from *Staphylococcus* spp.) (Braet et al., 2003; Robertson et al., 2010) and membrane depolarization or metabolic inhibition (Retamal et al., 2007b). As ATP is a highly charged molecule and cannot pass across the membrane by simple diffusion, connexin and pannexin hemichannels provide a passive route for its transport (D'Hondt et al., 2009). The release of ATP has downstream signaling consequences on purinergic receptor signaling cascades, a potent signaling pathway that plays a critical role in differentiation and proliferation of the epidermis as well as regulation of innate immune responses (Burnstock, 2006). Ultimately, the purinergic signaling cascade induces cytokine release and influences intracellular calcium dynamics within the epidermis. Alterations in the balance of these events can have profound consequences on the keratinocyte terminal differentiation program (Dixon et al., 1999).

A role for Cx26 hemichannel signaling in such epidermal events was first suggested by Djalilian et al. (2006), in relation to the increased abundance of Cx26 seen in psoriatic plaques. The ectopic expression of Cx26 in the inv-Cx26 mouse is associated with gross barrier deficiency and promoted a psoriasiform response. This was associated with enhanced ATP release from keratinocytes overexpressing the Cx26 protein that was returned to basal levels upon blocking the response with carbenoxolone (Djalilian et al., 2006). However, the situation is complicated by the fact that epidermal overexpression of Cx26 D66H that does not traffic efficiently to the plasma membrane had previously been shown to produce a similar histological phenotype in mouse skin (Bakirtzis et al., 2003). Furthermore, in psoriasis, the Cx26 is localized into gap junction plaques (Labarthe et al., 1998).

Keratinocytes continually release a basal level of ATP irrespective of skin damage and respond to this by a regulated ATP-induced intracellular calcium flux that can be blocked by purinergic receptor agonists (Dixon et al., 1999). The terminal differentiation program of keratinocytes is highly sensitive to subtle changes in intracellular calcium. Under normal circumstances, elevation of intracellular calcium promotes differentiation; however, excessive ATP stimulation can alter these calcium dynamics and inhibit terminal differentiation, leaving the cells in a "hyperproliferative state" (Denda et al., 2002; Pillai and Bikle, 1992). Release of ATP from keratinocytes also activates, via the puringeric signaling pathway, the release of cytokines, including IL-6, and can activate $CD39^+$ Langerhans cells that infiltrate and modulate the skin inflammatory responses (Granstein et al., 2005; Mizumoto et al., 2002) (see Chapter 15). Persistent leakage of ATP from keratinocytes, caused by excessive Cx26 expression (in psoriasis) or dysfunctional hemichannels (in KID syndrome), may influence these events and contribute to the pathogenic state. As Langerhans cells need to be primed, not just with presentation of major histocompatibility complex but also cytokine levels, it is conceivable that the downstream effects of excessive hemichannel signaling may stimulate the Langerhans cells more easily than normal. Despite this suggestion of a link between Cx26 hemichannel activity and activation of Langerhans cells, there is no evidence that Cx-mediated communication via the Langerhans cells is involved in epidermal inflammatory events (Zimmerli et al., 2008). Although cytokine levels were not recorded in the studies by Djalilian et al. (2006), the psorisoform lesions studied in the inv-Cx26 mice displayed areas of lymphocyte infiltrate of $CD3^+$, $CD4^+$, and $CD8^+$ cells, indicating an exacerbated inflammatory response. Whether such events are seen in mouse models or human skin biopsies of KID syndrome remains to be resolved.

12.9 CONCLUSIONS AND FUTURE PERSPECTIVES

An intriguing feature of the clinical reports on KID syndrome is the increased sensitivity of patients to Gram-positive bacterial and fungal skin infections (van Steensel, personal communication, and in reports such as those by Haruna et al., 2010; Mazereeuw-Hautier et al., 2007). It remains a future challenge to elucidate the molecular mechanisms by which disturbances in the balance of Cx26 expression and hemichannel activity might lead to a wide variety of hyperproliferative disorders. For example, it was recently reported that PGN isolated from *S. epidermidis* elicited in endothelial cells acute hemichannel opening that triggered IL-6 and TLR2 expression, key events in the innate immune response (Robertson et al., 2010). In contrast, KID mutant hemichannels expressed in keratinocytes were more sensitive to PGN isolated from *S. aureus*, an opportunistic pathogen than *S. epidermidis*, a skin commensal, suggesting that environmental challenges presented by skin flora play a role in disease progression (Donnelly et al., 2012).

Nevertheless, aberrant Cx26 hemichannel signaling is indicative of a range of epidermal inflammatory disorders. Connexins are increasingly identified as potent therapeutic targets either by use of antisense technologies or highly specific peptides that can inhibit or enhance channel behavior (Evans and Leybaert, 2007; Gourdie et al., 2006; Kjolbye et al., 2007; Qiu et al., 2003). Although not addressed in this review, dynamic changes in Cx43 occur during skin wound healing events and both Cx43 and Cx26 expression levels are enhanced in chronic nonhealing wounds (Brandner et al., 2004). Using antisense oligodeoxynucleotides (ODNs) targeted to Cx43, Becker and colleagues have elegantly demonstrated that reduction in Cx43 gene expression can improve wound healing rates in normal and diabetic animal models with an associated reduction in wound inflammation and scarring (Mori et al., 2006; Qiu et al., 2003; Wang et al., 2007). In normal mice, neutrophil number was reduced in the edges of antisense ODN-treated wounds, which in turn reduced the release of proinflammatory cytokines. This dampening of the early inflammatory response following wounding corresponded with a decrease in the amount of granulation tissue present in wounds and scars, improving reepithelialization and scar formation (Qiu et al., 2003). Macrophage numbers at wound edges were also seen to be reduced with antisense ODN treatment (Mori et al., 2006) (see Prequel). As macrophages are associated with the late inflammatory response and reepithelialization, it can be seen that Cx43 plays an essential role in coordinating these events. Antisense ODNs were also able to increase wound healing rates in diabetic rats with a similar reduction in granulation tissue and increase in reepithelialization (Wang et al., 2007), indicating an abrogated inflammatory response. Other studies have shown that connexin mimetic peptides targeted specifically to the extracellular loops of Cx43 can improve wound closure rates in *in vitro* and *ex vivo* mouse, porcine, and human organotypic skin models (Kandyba et al., 2008; Pollok et al., 2011; Wright et al., 2009, 2012). The mechanism of peptide action in these events remains unresolved, but recent evidence suggests that they block hemichannel activity and influence cell adhesion and extracellular matrix deposition thereby enhancing cell migration rates (Wright et al., 2012), and are also effective in *in vitro* conditions mimicking raised glucose and insulin levels seen in type II diabetes. Such peptides are reversible and highly connexin specific (Martin et al., 2005; Wright et al., 2009). Ultimately, such technologies could be tailored to additionally block Cx26-mediated hemichannel activity with topical application to localized skin lesions, which would enhance the integrity of the epidermal barrier and provide relief for epidermal inflammatory disorders.

ACKNOWLEDGMENTS

We would like to thank Dr. Scott Johnstone for contributing Figure 12.1. This work was supported by a GCU PhD studentship to SD; grants from The Cunningham Trust (PEM and CW) and Tenovus Scotland (PEM, S09/5). MvS is supported by GROW, the Dutch Cancer Society KWF (KWF UM2009-4352) and the Association for International Cancer Research AICR (11-0678).

REFERENCES

Bakirtzis G, Choudhry R, Aasen T, Shore L, Brown K, Bryson S, Forrow S et al. 2003. Targeted epidermal expression of mutant Connexin 26(D66H) mimics true Vohwinkel syndrome and provides a model for the pathogenesis of dominant connexin disorders. *Hum Mol Genet* 12(14):1737–1744.

Bao L, Sachs F, Dahl G. 2004. Connexins are mechanosensitive. *Am J Physiol Cell Physiol* 287(5):C1389–1395.

Blanpain C, Fuchs E. 2009. Epidermal homeostasis: A balancing act of stem cells in the skin. *Nat Rev Mol Cell Biol* 10(3):207–217.

Braet K, Vandamme W, Martin PE, Evans WH, Leybaert L. 2003. Photoliberating inositol-1,4,5-trisphosphate triggers ATP release that is blocked by the connexin mimetic peptide gap 26. *Cell Calcium* 33(1):37–48.

Brandner JM, Houdek P, Husing B, Kaiser C, Moll I. 2004. Connexins 26, 30, and 43: Differences among spontaneous, chronic, and accelerated human wound healing. *J Invest Dermatol* 122(5):1310–1320.

Brobby GW, Muller-Myhsok B, Horstmann RD. 1998. Connexin 26 R143W mutation associated with recessive nonsyndromic sensorineural deafness in Africa. *N Engl J Med* 338(8):548–550.

Buchau AS, Gallo RL. 2007. Innate immunity and antimicrobial defense systems in psoriasis. *Clin Dermatol* 25(6):616–624.

Burnstock G. 2006. Purinergic signalling. *Br J Pharmacol* 147(Suppl 1):S172–181.

Chandross KJ, Spray DC, Cohen RI, Kumar NM, Kremer M, Dermietzel R, Kessler JA. 1996. TNF alpha inhibits Schwann cell proliferation, connexin46 expression, and gap junctional communication. *Mol Cell Neurosci* 7(6):479–500.

Cherian PP, Siller-Jackson AJ, Gu S, Wang X, Bonewald LF, Sprague E, Jiang JX. 2005. Mechanical strain opens connexin 43 hemichannels in osteocytes: A novel mechanism for the release of prostaglandin. *Mol Biol Cell* 16(7):3100–3106.

Choudhry R, Pitts JD, Hodgins MB. 1997. Changing patterns of gap junctional intercellular communication and connexin distribution in mouse epidermis and hair follicles during embryonic development. *Dev Dyn* 210(4):417–430.

Chung E, Cook PW, Parkos CA, Park YK, Pittelkow MR, Coffey RJ. 2005. Amphiregulin causes functional downregulation of adherens junctions in psoriasis. *J Invest Dermatol* 124(6):1134–1140.

Clair C, Combettes L, Pierre F, Sansonetti P, Tran Van Nhieu G. 2008. Extracellular-loop peptide antibodies reveal a predominant hemichannel organization of connexins in polarized intestinal cells. *Exp Cell Res* 314(6):1250–1265.

Cogen AL, Nizet V, Gallo RL. 2008. Skin microbiota: A source of disease or defence? *Br J Dermatol* 158(3):442–455.

Cork MJ, Danby SG, Vasilopoulos Y, Hadgraft J, Lane ME, Moustafa M, Guy RH, Macgowan AL, Tazi-Ahnini R, Ward SJ. 2009. Epidermal barrier dysfunction in atopic dermatitis. *J Invest Dermatol* 129(8):1892–1908.

Corvalan LA, Araya R, Branes MC, Saez PJ, Kalergis AM, Tobar JA, Theis M, Willecke K, Saez JC. 2007. Injury of skeletal muscle and specific cytokines induce the expression of gap junction channels in mouse dendritic cells. *J Cell Physiol* 211(3):649–660.

D'Adamo P, Guerci VI, Fabretto A, Faletra F, Grasso DL, Ronfani L, Montico M, Morgutti M, Guastalla P, Gasparini P. 2009. Does epidermal thickening explain GJB2 high carrier frequency and heterozygote advantage? *Eur J Hum Genet* 17(3):284–286.

D'Hondt C, Ponsaerts R, De Smedt H, Bultynck G, Himpens B. 2009. Pannexins, distant relatives of the connexin family with specific cellular functions? *Bioessays* 31(9):953–974.

De Vuyst E, Decrock E, De Bock M, Yamasaki H, Naus CC, Evans WH, Leybaert L. 2007. Connexin hemichannels and gap junction channels are differentially influenced by lipopolysaccharide and basic fibroblast growth factor. *Mol Biol Cell* 18(1):34–46.

de Zwart-Storm EA, Hamm H, Stoevesandt J, Steijlen PM, Martin PE, van Geel M, van Steensel MA. 2008a. A novel missense mutation in GJB2 disturbs gap junction protein transport and causes focal palmoplantar keratoderma with deafness. *J Med Genet* 45(3):161–166.

de Zwart-Storm EA, Martin, P.E., and van Steensel. M.A.M. 2009 Gap junction diseases of the skin—Novel insights from new mutations. *Expert Rev Dermatol* 4:455–468.

de Zwart-Storm EA, Rosa RF, Martin PE, Foelster-Holst R, Frank J, Bau AE, Zen PR et al. 2011. Molecular analysis of connexin26 asparagine14 mutations associated with syndromic skin phenotypes. *Exp Dermatol* 20(5):408–412.

de Zwart-Storm EA, van Geel M, van Neer PA, Steijlen PM, Martin PE, van Steensel MA. 2008b. A novel missense mutation in the second extracellular domain of GJB2, p.Ser183Phe, causes a syndrome of focal palmoplantar keratoderma with deafness. *Am J Pathol* 173(4):1113–1119.

Denda M, Inoue K, Fuziwara S, Denda S. 2002. P2X purinergic receptor antagonist accelerates skin barrier repair and prevents epidermal hyperplasia induced by skin barrier disruption. *J Invest Dermatol* 119(5):1034–1040.

Di WL, Rugg EL, Leigh IM, Kelsell DP. 2001. Multiple epidermal connexins are expressed in different keratinocyte subpopulations including connexin 31. *J Invest Dermatol* 117(4):958–964.

Dixon CJ, Bowler WB, Littlewood-Evans A, Dillon JP, Bilbe G, Sharpe GR, Gallagher JA. 1999. Regulation of epidermal homeostasis through P2Y2 receptors. *Br J Pharmacol* 127(7):1680–1686.

Djalilian AR, Kim JY, Saeed HN, Holland EJ, Chan CC. 2010. Histopathology and treatment of corneal disease in keratitis, ichthyosis, and deafness (KID) syndrome. *Eye (Lond)* 24(4):738–740.

Djalilian AR, McGaughey D, Patel S, Seo EY, Yang C, Cheng J, Tomic M, Sinha S, Ishida-Yamamoto A, Segre JA. 2006. Connexin 26 regulates epidermal barrier and wound remodeling and promotes psoriasiform response. *J Clin Invest* 116(5):1243–1253.

Donnelly S, English, G, Lang S, de Zwart-Storm E, van Steensel MAM, and Martin PE. 2012. Differential susceptibility of Cx26 mutations associated with epidermal dysplasia's to Peptidoglycan derived from *S. aureus* and *S. epidermidis*. *Exp Dermatol* 8: 592–598.

Easton JA, Donnelly S, Kamps MA, Steijlen PM, Martin PE, Tadini G, Janssens R, Happle R, van Geel M, van Steensel MA. 2012. Porokeratotic eccrine nevus may be caused by somatic connexin26 mutations. *J Invest Dermatol* 132(9): 2181–2191.

Ekanayake-Mudiyanselage S, Aschauer H, Schmook FP, Jensen JM, Meingassner JG, Proksch E. 1998. Expression of epidermal keratins and the cornified envelope protein involucrin is influenced by permeability barrier disruption. *J Invest Dermatol* 111(3):517–523.

Esen N, Shuffield D, Syed MM, Kielian T. 2007. Modulation of connexin expression and gap junction communication in astrocytes by the Gram-positive bacterium *S. aureus*. *Glia* 55(1):104–117.

Eugenin EA, Eckardt D, Theis M, Willecke K, Bennett MV, Saez JC. 2001. Microglia at brain stab wounds express connexin 43 and *in vitro* form functional gap junctions after treatment with interferon-gamma and tumor necrosis factor-alpha. *Proc Natl Acad Sci USA* 98(7):4190–4195.

Evans WH, De Vuyst E, Leybaert L. 2006. The gap junction cellular internet: Connexin hemichannels enter the signalling limelight. *Biochem J* 397(1):1–14.

Evans WH, Leybaert L. 2007. Mimetic peptides as blockers of connexin channel-facilitated intercellular communication. *Cell Commun Adhes* 14(6):265–273.

Fort AG, Murray, J.W., Dandachi, N., Davidson, M.W., Dermietzel, R., Wolkoff, A.W., Spray, D.C. 2012. *In vitro* motility of liver connexin vesicles along microtubules utilizes kinesin motors. *J Biol Chem* 286(26): 22875–22885.

Fruscione F, Scarfi S, Ferraris C, Bruzzone S, Benvenuto F, Guida L, Uccelli A et al. 2010. Regulation of human mesenchymal stem cell functions by an autocrine loop involving NAD(+) release and P2Y11-mediated signaling. *Stem Cells Dev* 20(7): 1183–1198.

Garg S, Md Syed M, Kielian T. 2005. *Staphylococcus aureus*-derived peptidoglycan induces Cx43 expression and functional gap junction intercellular communication in microglia. *J Neurochem* 95(2):475–483.

Garrod D, Chidgey M. 2008. Desmosome structure, composition and function. *Biochim Biophys Acta* 1778(3):572–587.

Gourdie RG, Ghatnekar GS, O'Quinn M, Rhett MJ, Barker RJ, Zhu C, Jourdan J, Hunter AW. 2006. The unstoppable connexin43 carboxyl-terminus: New roles in gap junction organization and wound healing. *Ann N Y Acad Sci* 1080:49–62.

Granstein RD, Ding W, Huang J, Holzer A, Gallo RL, Di Nardo A, Wagner JA. 2005. Augmentation of cutaneous immune responses by ATP gamma S: Purinergic agonists define a novel class of immunologic adjuvants. *J Immunol* 174(12):7725–7731.

Grice EA, Kong HH, Conlan S, Deming CB, Davis J, Young AC, Bouffard GG et al. 2009. Topographical and temporal diversity of the human skin microbiome. *Science* 324(5931):1190–1192.

Guastalla P, Guerci VI, Fabretto A, Faletra F, Grasso DL, Zocconi E, Stefanidou D et al. 2009. Detection of epidermal thickening in GJB2 carriers with epidermal US. *Radiology* 251(1):280–286.

Guttman JA, Lin AE, Li Y, Bechberger J, Naus CC, Vogl AW, Finlay BB. 2009. Gap junction hemichannels contribute to the generation of diarrhea during infectious enteric disease. *Gut* 59:218–226.

Haghikia A, Ladage K, Hinkerohe D, Vollmar P, Heupel K, Dermietzel R, Faustmann PM. 2008a. Implications of antiinflammatory properties of the anticonvulsant drug leveti-racetam in astrocytes. *J Neurosci Res* 86(8):1781–1788.

Haghikia A, Ladage K, Lafenetre P, Hinkerohe D, Smikalla D, Haase CG, Dermietzel R, Faustmann PM. 2008b. Intracellular application of TNF-alpha impairs cell to cell com-munication via gap junctions in glioma cells. *J Neurooncol* 86(2):143–152.

Hamelmann C, Amedofu GK, Albrecht K, Muntau B, Gelhaus A, Brobby GW, Horstmann RD. 2001. Pattern of connexin 26 (GJB2) mutations causing sensorineural hearing impairment in Ghana. *Hum Mutat* 18(1):84–85.

Hampton PJ, Ross OK, Reynolds NJ. 2007. Increased nuclear beta-catenin in suprabasal involved psoriatic epidermis. *Br J Dermatol* 157(6):1168–1177.

Haruna K, Suga Y, Oizumi A, Mizuno Y, Endo H, Shimizu T, Hasegawa T, Ikeda S. 2010. Severe form of keratitis-ichthyosis-deafness (KID) syndrome associated with septic complications. *J Dermatol* 37(7):680–682.

Hinkerohe D, Smikalla D, Schoebel A, Haghikia A, Zoidl G, Haase CG, Schlegel U, Faustmann PM. 2010. Dexamethasone prevents LPS-induced microglial activation and astroglial impairment in an experimental bacterial meningitis co-culture model. *Brain Res* 1329:45–54.

Hivnor C, Williams N, Singh F, VanVoorhees A, Dzubow L, Baldwin D, Seykora J. 2004. Gene expression profiling of porokeratosis demonstrates similarities with psoriasis. *J Cutan Pathol* 31(10):657–664.

Holland DB, Bojar RA, Farrar MD, Holland KT. 2009. Differential innate immune responses of a living skin equivalent model colonized by *Staphylococcus epidermidis* or *Staphylococcus aureus*. *FEMS Microbiol Lett* 290(2):149–155.

Janecke AR, Hennies HC, Gunther B, Gansl G, Smolle J, Messmer EM, Utermann G, Rittinger O. 2005. GJB2 mutations in keratitis-ichthyosis-deafness syndrome including its fatal form. *Am J Med Genet A* 133A(2):128–131.

Kandyba EE, Hodgins MB, Martin PE. 2008. A murine living skin equivalent amenable to live-cell imaging: Analysis of the roles of connexins in the epidermis. *J Invest Dermatol* 128(4):1039–1049.

Kelly B, Lozano A, Altenberg G, Makishima T. 2008. Connexin 26 mutation in keratitis-ich-thyosis-deafness (KID) syndrome in mother and daughter with combined conductive and sensorineural hearing loss. *Int J Dermatol* 47(5):443–447.

Kim BE, Howell MD, Guttman E, Gilleaudeau PM, Cardinale IR, Boguniewicz M, Krueger JG, Leung DY. 2011. TNF-alpha downregulates filaggrin and loricrin through c-Jun N-terminal kinase: Role for TNF-alpha antagonists to improve skin barrier. *J Invest Dermatol* 131(6):1272–1279.

Kirschner N, Poetzl C, von den Driesch P, Wladykowski E, Moll I, Behne MJ, Brandner JM. 2009. Alteration of tight junction proteins is an early event in psoriasis: Putative involve-ment of proinflammatory cytokines. *Am J Pathol* 175(3):1095–1106.

Kjolbye AL, Haugan K, Hennan JK, Petersen JS. 2007. Pharmacological modulation of gap junction function with the novel compound rotigaptide: A promising new principle for prevention of arrhythmias. *Basic Clin Pharmacol Toxicol* 101(4):215–230.

Koppelhus U, Tranebjaerg L, Esberg G, Ramsing M, Lodahl M, Rendtorff ND, Olesen HV, Sommerlund M. 2011. A novel mutation in the connexin 26 gene (GJB2) in a child with clinical and histological features of keratitis-ichthyosis-deafness (KID) syndrome. *Clin Exp Dermatol* 36(2):142–148.

Kristensen M, Chu CQ, Eedy DJ, Feldmann M, Brennan FM, Breathnach SM. 1993. Localization of tumour necrosis factor-alpha (TNF-alpha) and its receptors in normal and psoriatic skin: Epidermal cells express the 55-kD but not the 75-kD TNF receptor. *Clin Exp Immunol* 94(2):354–362.

Kumar H, Kawai T, Akira S. 2009. Toll-like receptors and innate immunity. *Biochem Biophys Res Commun* 388(4):621–625.

Labarthe MP, Bosco D, Saurat JH, Meda P, Salomon D. 1998. Upregulation of connexin 26 between keratinocytes of psoriatic lesions. *J Invest Dermatol* 111(1):72–76.

Lai Y, Cogen AL, Radek KA, Park HJ, Macleod DT, Leichtle A, Ryan AF, Di Nardo A, Gallo RL. 2010. Activation of TLR2 by a small molecule produced by *Staphylococcus epidermidis* increases antimicrobial defense against bacterial skin infections. *J Invest Dermatol* 130(9):2211–2221.

Lai Y, Gallo RL. 2008. Toll-like receptors in skin infections and inflammatory diseases. *Infect Disord Drug Targets* 8(3):144–155.

Laird DW. 2006. Life cycle of connexins in health and disease. *Biochem J* 394(Pt 3):527–543.

Lee JR, Derosa AM, White TW. 2009. Connexin mutations causing skin disease and deafness increase hemichannel activity and cell death when expressed in *Xenopus* oocytes. *J Invest Dermatol* 129(4):870–878.

Li Z, Peng Z, Wang Y, Geng S, Ji F. 2008. Decreased expression of E-cadherin and beta-catenin in the lesional skin of patients with active psoriasis. *Int J Dermatol* 47(2):207–209.

Lucke T, Choudhry R, Thom R, Selmer IS, Burden AD, Hodgins MB. 1999. Upregulation of connexin 26 is a feature of keratinocyte differentiation in hyperproliferative epidermis, vaginal epithelium, and buccal epithelium. *J Invest Dermatol* 112(3):354–361.

Maeda S, Nakagawa S, Suga M, Yamashita E, Oshima A, Fujiyoshi Y, Tsukihara T. 2009. Structure of the connexin 26 gap junction channel at 3.5 A resolution. *Nature* 458(7238):597–602.

Man YK, Trolove C, Tattersall D, Thomas AC, Papakonstantinopoulou A, Patel D, Scott C et al. 2007. A deafness-associated mutant human connexin 26 improves the epithelial barrier in vitro. *J Membr Biol* 218(1–3):29–37.

Martin PE, Blundell G, Ahmad S, Errington RJ, Evans WH. 2001. Multiple pathways in the trafficking and assembly of connexin 26, 32 and 43 into gap junction intercellular communication channels. *J Cell Sci* 114(Pt 21):3845–3855.

Martin PE, Wall C, Griffith TM. 2005. Effects of connexin-mimetic peptides on gap junction functionality and connexin expression in cultured vascular cells. *Br J Pharmacol* 144(5):617–627.

Martinez AD, Acuna R, Figueroa V, Maripillan J, Nicholson B. 2009. Gap-junction channels dysfunction in deafness and hearing loss. *Antioxid Redox Signal* 11(2):309–322.

Matsue H, Yao J, Matsue K, Nagasaka A, Sugiyama H, Aoki R, Kitamura M, Shimada S. 2006. Gap junction-mediated intercellular communication between dendritic cells (DCs) is required for effective activation of DCs. *J Immunol* 176(1):181–190.

Mazereeuw-Hautier J, Bitoun E, Chevrant-Breton J, Man SY, Bodemer C, Prins C, Antille C et al. 2007. Keratitis-ichthyosis-deafness syndrome: Disease expression and spectrum of connexin 26 (GJB2) mutations in 14 patients. *Br J Dermatol* 156(5):1015–1019.

Meyer CG, Amedofu GK, Brandner JM, Pohland D, Timmann C, Horstmann RD. 2002. Selection for deafness? *Nat Med* 8(12):1332–1333.

Mizumoto N, Kumamoto T, Robson SC, Sevigny J, Matsue H, Enjyoji K, Takashima A. 2002. CD39 is the dominant Langerhans cell-associated ecto-NTPDase: Modulatory roles in inflammation and immune responsiveness. *Nat Med* 8(4):358–365.

Moniaga CS, Egawa G, Kawasaki H, Hara-Chikuma M, Honda T, Tanizaki H, Nakajima S et al. 2010. Flaky tail mouse denotes human atopic dermatitis in the steady state and by topical application with *Dermatophagoides pteronyssinus* extract. *Am J Pathol* 176(5):2385–2393.

Mori R, Power KT, Wang CM, Martin P, Becker DL. 2006. Acute downregulation of connexin43 at wound sites leads to a reduced inflammatory response, enhanced keratinocyte proliferation and wound fibroblast migration. *J Cell Sci* 119(Pt 24):5193–5203.

Nyquist GG, Mumm C, Grau R, Crowson AN, Shurman DL, Benedetto P, Allen P et al. 2007. Malignant proliferating pilar tumors arising in KID syndrome: A report of two patients. *Am J Med Genet A* 143(7):734–741.

Oshima A, Tani K, Toloue MM, Hiroaki Y, Smock A, Inukai S, Cone A, Nicholson BJ, Sosinsky GE, Fujiyoshi Y. 2011. Asymmetric configurations and N-terminal rearrangements in connexin26 gap junction channels. *J Mol Biol* 405(3):724–735.

Pillai S, Bikle DD. 1992. Adenosine triphosphate stimulates phosphoinositide metabolism, mobilizes intracellular calcium, and inhibits terminal differentiation of human epidermal keratinocytes. *J Clin Invest* 90(1):42–51.

Pollok S, Pfeiffer AC, Lobmann R, Wright CS, Moll I, Martin PE, Brandner JM. 2011. Connexin 43 mimetic peptide Gap27 reveals potential differences in the role of Cx43 in wound repair between diabetic and non-diabetic cells. *J Cell Mol Med* 15(4): 861–873.

Proksch E, Brandner JM, Jensen JM. 2008. The skin: An indispensable barrier. *Exp Dermatol* 17(12):1063–1072.

Qiu C, Coutinho P, Frank S, Franke S, Law LY, Martin P, Green CR, Becker DL. 2003. Targeting connexin43 expression accelerates the rate of wound repair. *Curr Biol* 13(19):1697–1703.

Retamal MA, Froger N, Palacios-Prado N, Ezan P, Saez PJ, Saez JC, Giaume C. 2007a. Cx43 hemichannels and gap junction channels in astrocytes are regulated oppositely by proinflammatory cytokines released from activated microglia. *J Neurosci* 27(50):13781–13792.

Retamal MA, Schalper KA, Shoji KF, Bennett MV, Saez JC. 2007b. Opening of connexin 43 hemichannels is increased by lowering intracellular redox potential. *Proc Natl Acad Sci USA* 104(20):8322–8327.

Richard G, Rouan F, Willoughby CE, Brown N, Chung P, Ryynanen M, Jabs EW et al. 2002. Missense mutations in GJB2 encoding connexin-26 cause the ectodermal dysplasia keratitis-ichthyosis-deafness syndrome. *Am J Hum Genet* 70(5):1341–1348.

Richardson SK, Gelfand JM. 2008. Update on the natural history and systemic treatment of psoriasis. *Adv Dermatol* 24:171–196.

Robertson J, Lang S, Lambert PA, Martin PE. 2010. Peptidoglycan derived from *Staphylococcus epidermidis* induces Connexin43 hemichannel activity with consequences on the innate immune response in endothelial cells. *Biochem J* 432(1):133–143.

Romani N, Clausen BE, Stoitzner P. 2010. Langerhans cells and more: Langerin-expressing dendritic cell subsets in the skin. *Immunol Rev* 234(1):120–141.

Romani N, Ebner S, Tripp CH, Flacher V, Koch F, Stoitzner P. 2006. Epidermal Langerhans cells—Changing views on their function *in vivo*. *Immunol Lett* 106(2):119–125.

Sandilands A, Sutherland C, Irvine AD, McLean WH. 2009. Filaggrin in the frontline: Role in skin barrier function and disease. *J Cell Sci* 122(Pt 9):1285–1294.

Schutz M, Auth T, Gehrt A, Bosen F, Korber I, Strenzke N, Moser T, Willecke K. 2010. The connexin26 S17F mouse mutant represents a model for the human hereditary keratitis-ichthyosis-deafness syndrome. *Hum Mol Genet* 20(1):28–39.

Shaw RM, Fay AJ, Puthenveedu MA, von Zastrow M, Jan YN, Jan LY. 2007. Microtubule plus-end-tracking proteins target gap junctions directly from the cell interior to adherens junctions. *Cell* 128(3):547–560.

Skinner BA, Greist MC, Norins AL. 1981. The keratitis, ichthyosis, and deafness (KID) syndrome. *Arch Dermatol* 117(5):285–289.

Smith FJ, Irvine AD, Terron-Kwiatkowski A, Sandilands A, Campbell LE, Zhao Y, Liao H et al. 2006. Loss-of-function mutations in the gene encoding filaggrin cause ichthyosis vulgaris. *Nat Genet* 38(3):337–342.

Steffens M, Gopel F, Ngezahayo A, Zeilinger C, Ernst A, Kolb HA. 2008. Regulation of connexons composed of human connexin26 (hCx26) by temperature. *Biochim Biophys Acta* 1778(5):1206–1212.

Tacheau C, Fontaine J, Loy J, Mauviel A, Verrecchia F. 2008a. TGF-beta induces connexin43 gene expression in normal murine mammary gland epithelial cells via activation of p38 and PI3K/AKT signaling pathways. *J Cell Physiol* 217(3):759–768.

Tacheau C, Laboureau J, Mauviel A, Verrecchia F. 2008b. TNF-alpha represses connexin43 expression in HaCat keratinocytes via activation of JNK signaling. *J Cell Physiol* 216(2):438–444.

Tonel G, Conrad C. 2009. Interplay between keratinocytes and immune cells—Recent insights into psoriasis pathogenesis. *Int J Biochem Cell Biol* 41(5):963–968.

Tonon R, D'Andrea P. 2000. Interleukin-1beta increases the functional expression of connexin 43 in articular chondrocytes: Evidence for a Ca^{2+}-dependent mechanism. *J Bone Miner Res* 15(9):1669–1677.

Tran Van Nhieu G, Clair C, Bruzzone R, Mesnil M, Sansonetti P, Combettes L. 2003. Connexin-dependent inter-cellular communication increases invasion and dissemination of *Shigella* in epithelial cells. *Nat Cell Biol* 5(8):720–726.

Tschachler E. 2007. Psoriasis: The epidermal component. *Clin Dermatol* 25(6):589–595.

van Rijen HV, van Kempen MJ, Postma S, Jongsma HJ. 1998. Tumour necrosis factor alpha alters the expression of connexin43, connexin40, and connexin37 in human umbilical vein endothelial cells. *Cytokine* 10(4):258–264.

van Steensel MA, van Geel M, Nahuys M, Smitt JH, Steijlen PM. 2002. A novel connexin 26 mutation in a patient diagnosed with keratitis-ichthyosis-deafness syndrome. *J Invest Dermatol* 118(4):724–727.

Wagner EF, Schonthaler HB, Guinea-Viniegra J, Tschachler E. 2010. Psoriasis: What we have learned from mouse models. *Nat Rev Rheumatol* 6(12):704–714.

Wang CM, Lincoln J, Cook JE, Becker DL. 2007. Abnormal connexin expression underlies delayed wound healing in diabetic skin. *Diabetes* 56(11):2809–2817.

Wang XP, Schunck M, Kallen KJ, Neumann C, Trautwein C, Rose-John S, Proksch E. 2004. The interleukin-6 cytokine system regulates epidermal permeability barrier homeostasis. *J Invest Dermatol* 123(1):124–131.

Weidinger S, O'Sullivan M, Illig T, Baurecht H, Depner M, Rodriguez E, Ruether A et al. 2008. Filaggrin mutations, atopic eczema, hay fever, and asthma in children. *J Allergy Clin Immunol* 121(5):1203–1209 e1201.

Wright CS, Pollok S, Flint DJ, Brandner JM, Martin PE. 2012. The connexin mimetic peptide Gap27 increases human dermal fibroblast migration in hyperglycemic and hyperinsulinemic conditions in vitro. *J Cell Physiol* 227(1):77–87.

Wright CS, van Steensel MA, Hodgins MB, Martin PE. 2009. Connexin mimetic peptides improve cell migration rates of human epidermal keratinocytes and dermal fibroblasts in vitro. *Wound Repair Regen* 17(2):240–249.

Zimmerli SC, Kerl K, Hadj-Rabia S, Hohl D, Hauser C. 2008. Human epidermal Langerhans cells express the tight junction protein claudin-1 and are present in human genetic claudin-1 deficiency (NISCH syndrome). *Exp Dermatol* 17(1):20–23.

Section 3

Connexin-Based Therapeutical Approaches in Inflammatory Diseases

13 Translating Basic Research on Cx43 Gap Junctions into Therapies for Reducing Scarring and Cardiac Arrhythmia

Robert G. Gourdie, J. Matthew Rhett,
Emily L. Ongstad, Joseph A. Palatinus,
and Michael P. O'Quinn

CONTENTS

13.1   INTRODUCTION

The gap junction (GJ) has been frustratingly refractory to the tools of classical pharmacology. However, in recent years, a number of approaches for targeting the connexin subunits of GJs in therapeutic modulation of the immune and wound-healing responses have been described.

Researchers working in this area are grappling with the mechanism of how such interventions work. The most easy-to-understand connection to mode-of-action comes from the canonical assignments of the GJ in intercellular transfer of small molecules between cells. In one example, GJs appear to participate in cell-to-cell spread of injury signals (Lin et al., 1998). However, this may be only part of the story.

There are also strong indications that connexins have functions that are independent of their role in cell-to-cell communication. Support for nonjunctional roles include data indicating that connexin 43 (Cx43) regulates TGF-β1 signaling (Dai et al., 2007), as well as functions in the plasma membrane for connexin hemichannels that are not docked within GJ aggregates (Saez et al., 2003; Elias et al., 2007; Cotrina et al., 2008; Bao et al., 2011).

This chapter focuses on the research of our laboratory of the last few years on the Cx43 GJ protein and in particular the carboxyl terminus of this protein. Of course, ours is one of the number of groups working in this fast-moving area—and many of these other contributions are covered in the Prequel and Chapter 3 of this book.

The work of the lab has involved cell and molecular biological approaches, as well as experiments translating lessons from this basic research into skin and cardiac injury models. One aspect of our tack has been to use the tractability of skin wounding models to better understand scar amelioration (Ghatnekar et al., 2009) and inhibition of arrhythmia in the heart (Palatinus et al., 2010; O'Quinn et al., 2011).

As with other laboratories in this field, we are struggling to understand the specific mechanisms. Nonetheless, piece-by-piece the puzzle is beginning to be assembled.

13.2   THE STARTING POINT PROVIDED BY CX43 IN THE HEART

The path taken by the lab had its beginnings in the work on connexins in cardiac conduction. In the mammalian heart, GJs are central to the intercellular propagation of electrical signals that trigger contraction of cardiac muscle (Rohr, 2004). The main GJ protein expressed in the ventricular myocardium of mammals is Cx43 (Beyer et al., 1987; Gourdie et al., 1991; Becker et al., 1998). One important feature of Cx43 GJs in the mammalian ventricle is that these structures show highly ordered patterns of spatial organization at specialized zones of electromechanical coupling between myocytes called intercalated disks (Figure 13.1).

Studies of humans with ischemic heart disease and animal models of this pathology have identified the occurrence of a striking phenomenon in which GJs are remodeled from intercalated disks to myocyte lateral surfaces (Luke and Saffitz, 1991; Smith et al., 1991; Peters et al., 1997; Peters and Wit, 2000; Severs et al., 2008). This lateralized remodeling of GJs is the most evident in a narrow zone of muscle bordering ischemic injuries and is thought to contribute directly to the reentrant arrhythmias that accompany myocardial infarction (Smith et al., 1991; Peters et al., 1997; Akar and Rosenbaum, 2003; Yao et al., 2003; Cabo et al., 2006; Duffy, 2008).

The spatial order of GJ organization in the heart appears to have significant implications for cardiac electrical behavior in both health and disease (Hesketh et al., 2009). Important questions thus are how is this order generated and maintained in the healthy heart and how does it become disrupted in disease?

13.3   ZONULA OCCLUDENS-1 AND CX43 GJ REMODELING

One candidate for regulating GJ spatial organization emerged in the late 1990s. The actin-binding protein Zonula Occludens (ZO)-1 was shown by Japanese and Dutch

FIGURE 13.1 (**See color insert.**) Highly ordered patterns of Cx43 GJs (green) at interca-lated disks between cardiomyoctes in the healthy rat ventricle. Following myocardial infarc-tion, Cx43 redistributes to myocyte lateral surfaces at the injury border zone—that is, the cell edges marked by wheat germ agglutinin (WGA) labeling (red) in the image shown. Nuclei are marked by blue DAPI labeling.

scientists to interact with Cx43 (Giepmans and Moolenaar, 1998; Toyofuku et al., 1998). Originally discovered in association with tight junctions (Stevenson et al., 1986), ZO-1 is a member of the membrane-associated guanylate kinase (MAGUK) family of proteins that function in protein targeting, signal transduction, and deter-mination of cell polarity (Anderson, 1996; Hartsock and Nelson, 2008).

MAGUKs, such as ZO-1, synaptic protein PSD95 and the Drosophila tumor suppressor discs large (dlg) are characterized by amino-terminal (NT) protein–protein binding motifs, including one or more PSD95/dlg/ZO-1 (PDZ) domains. Initial immunoprecipitation and yeast 2-hybrid studies showed that a 4 amino acid PDZ-binding ligand at the carboxyl-terminus of Cx43 (Figure 13.2a) was neces-sary for interaction with the second of three PDZ domains on ZO-1 (Giepmans and Moolenaar, 1998; Toyofuku et al., 1998).

Ongoing research has indicated that the ZO-1-binding domain comprises up to the carboxyl terminal-most 19 amino acids of Cx43 and this sequence likely interacts with a ZO-1 dimer (Jin et al., 2000; Sorgen et al., 2004). It is now known that numer-ous connexins interact with ZO-1 via a carboxyl terminal PDZ-binding sequence similar to that of Cx43 (Kausalya et al., 2001; Laing et al., 2001, 2005; Nielsen et al., 2002, 2003; Fanning et al., 2007; Bouvier et al., 2008; Flores et al., 2008). A second MAGUK, ZO-2, has been shown to interact with Cx43 in a cell-cycle-dependent manner (Singh et al., 2005).

Initially, ZO-1 function at the GJ was assumed to be analogous to its pre-sumed role at tight junctions. Namely, ZO-1 was hypothesized to be a scaffolding

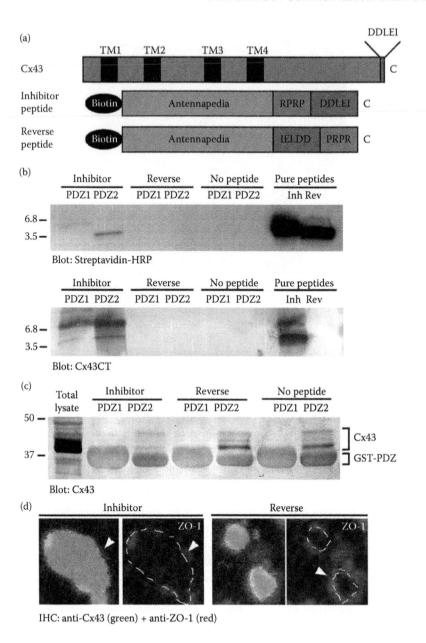

FIGURE 13.2 **(See color insert.)** αCT1 binds to the PDZ2 domain of ZO-1 and blocks interaction with Cx43 carboxyl terminus. (a) Schemetic of domain structure of Cx43 and αCT1 and reverse control (REV) peptides. (b) Blots αCT1 and REV (~3.5 kDa) detected by streptavidin-HRP (top) and Ab recognizing Ct of Cx43 (bottom). αCT1 was pulled down by GST-PDZ2, but not GST-PDZ1. (c) Endogenous Cx43 pulled from HeLa lysates by GST-PDZ2 reduced in presence of αCT1. (d) αCT1 increases GJ size (green) and decreases ZO-1 (red) localization at the GJ edge compared to REV control peptide. (From Hunter, A. W. et al. 2006. *Mol Biol Cell* 16: 5686–5698.)

molecule that stabilized GJs by anchoring the carboxyl terminal of connexins to the actin cytoskeleton (Toyofuku et al., 1998). However, subsequent reports have indicated that interactions between connexins and ZO-1 might include other more dynamic functions.

Our group provided one of the first reports that Cx43/ZO-1 interaction was involved in the remodeling of GJs in myocytes and other cell types including in the internalization of GJs from the membrane (Barker et al., 2002, 2008; Segretain et al., 2004). Support for this was provided by other independent studies (Barker et al., 2001, 2002; Defamie et al., 2001; Toyofuku et al., 2001; Hunter et al., 2003; Duffy et al., 2004; Segretain et al., 2004; Zhu et al., 2005; Gourdie et al., 2006; Hunter et al., 2006; Gilleron et al., 2008; Hunter and Gourdie, 2008).

In subsequent work, we demonstrated that ZO-1 localizes to the periphery of Cx43 GJs (Zhu et al., 2005), and that blockade of this PDZ2-mediated interaction, either via genetic- or competitive inhibition-based approaches decreased this co-localization (Figure 13.2b–d) (Hunter et al., 2005; Gourdie et al., 2006). Inhibition of the Cx43/ZO-1 interaction also gave rise to a GJ remodeling defect—a loss of plaque size control—which led to aberrantly large GJs (Hunter et al., 2005).

The concept that ZO-1 participates in GJ size regulation has been corroborated in studies of human patients with ischemic cardiomyopathy and congestive heart failure (Bruce et al., 2008). This work demonstrated that increased levels of Cx43/ZO-1 interaction in diseased hearts from transplant patients correlated with decreases in the abundance and size of GJs at intercalated disks in the ventricular myocardium.

Further support comes from studies of transgenic mice, where genetically manipulated, heart-specific loss of ZO-1 binding to Cx43 was found to result in the generation of abnormally large GJs mislocalized to the margins of intercalated disks (Maass et al., 2007).

In 2011, we extended our work on ZO-1 regulation of the extent of GJ aggregates by using *in situ* protein–protein interaction assays. This approach brought to light a previously unappreciated domain around plaques where Cx43/ZO-1 interaction is concentrated—a sector immediately proximal to the GJ that we have termed Perinexus. It is hypothesized that one aspect of ZO-1 remodeling of GJs involves governing the flow of Cx43 connexon hemichannels from this perinexal reservoir. Inhibition of Cx43/ZO-1 interaction results in rapid size growth of plaques as connexons switch localization from the perinexus to aggregation into GJs (Rhett et al., 2011).

13.4   THE CX43 CARBOXYL TERMINUS AND SKIN WOUND HEALING

The project took an unexpected turn when one of the tools that had been developed for experiments on the cell biology of GJs turned out to have effects in a "scratch wound" assay of cultured cells.

To disrupt ZO-1 interaction with Cx43, a synthetic peptide (αCT1) incorporating the carboxyl terminal ZO-1-binding domain of Cx43 was synthesized and characterized (Figure 13.2a). This ligand for the ZO-1 PDZ2 domain was linked at its NT to an antennapedia penetration sequence to enable transport from the outside of cells into the cytoplasm (Drin et al., 2003).

In "scratch wound" assays on 2D cultures of NIH-3T3 fibroblasts it was determined that αCT1 had effects on the rate of cell migration. To determine whether the Cx43 mimetic peptide influenced fibroblast motility *in vivo*, the peptide was applied to small skin wounds on mice (at 60 μM) in a 20% F127 pluronic gel. αCT1 was provided acutely at the time of injury and subsequently 24 h later (Ghatnekar et al., 2009).

Pluronic gel has also been used by Green, Becker, and colleagues in the delivery of Cx43 antisense to skin wounds (see Chapters 14 and 15, respectively). This vehicle has the convenient property of being liquid at temperatures near freezing, aiding in application to wounds, but gelling at body temperature. Peptide is thus released from the pluronic thermogel in a controlled manner over several hours following application. The gel also has mild surfactant properties that may aid in uniform dispersal of αCT1 within skin wounds.

The first notable effect of αCT1 observed in skin wounds was on neutrophil inflammatory cells, not dermal fibroblasts. Similar to what has been reported in response to Cx43 antisense (see Chapters 14 and 15, respectively), there was a 40–50% reduction in the numerical density of neutrophils over the first 24–48 h following wounding (Figure 13.3a–f). There was also evidence of neutrophil retention in vascular structures local to the injury, suggestive of the peptide prompting a failure of leukocytic cells to undergo transmigration from the blood (Figure 13.3d). Following the acute inflammatory phase, the rate of closure was increased in αCT1-treated wounds relative to controls (Ghatnekar et al., 2006, 2009).

In fully healed wounds treated with αCT1, the area of scar progenitor (granulation) tissue was decreased in histological sections compared to controls. Mechanical testing at 3 months indicated that the breaking strain and stress of healed peptide-treated wounds extended to failure were also significantly increased compared to controls. Interestingly, these mechanical properties were not significantly different at 1 month. This suggested that downstream effects on scar remodeling continued to emerge up to a quarter of a year after treatment with αCT1.

Similar to the histological results seen in mice, peptide treatment of skin wounds on pigs also resulted in a reduction in the area of granulation tissue. This result may be of pertinence to translation, as pigs are said to have skin that resembles humans better than rodents in a number of aspects (Forbes, 1969; Hall and Giaccia, 2005).

13.5 MODULATION OF IMPLANT TOLERANCE BY THE CX43 CARBOXYL TERMINUS

As αCT1 reduced inflammation and scarring of cutaneous wounds, we were curious to explore the effect of the mimetic peptide on other models of wound healing. One injury scenario of interest to surgeons is the differentiation of contraction capsule around implants placed in the body (Caffee et al., 1988). Implants undergo what is termed the foreign body reaction, which typically results in the differentiation of a fibrous capsule of scar tissue surrounding the implant.

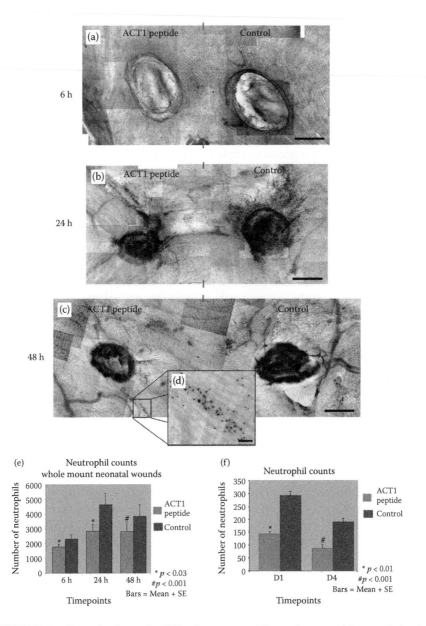

FIGURE 13.3 **(See color insert.)** αCT1 reduces neutrophil recruitment to skin wounds (a–c). Neutrophil recruitment in paired excisional wounds on the back of neonatal mice over a 6, 24, and 48 h time course. The density of neutrophils at αCT1-treated wounds was visibly lower than the matching control wound on the same animal—particularly at 24 h (b). Inset (d) shows higher magnification of individually labeled neutrophils associated with a vascular structure adjacent to a peptide-treated wound. (e and f) Bar graphs showing significant reduction in the density of neutrophils at αCT1-peptide treated wounds (blue bars) relative to the paired control wounds (purple bars) in neonates (e) and adults (f). Scale a–c = 250 μm, inset = 50 μm. (From Ghatnekar, G. S. et al. 2009. *Regen Med* 4(2): 205–223.)

Myofibroblasts and other mechanically active cells are found within the capsule, and these cells participate in active processes that can lead to deformation, impairment of function or even extrusion of the implant from the body.

Mike Yost and his colleagues at the University of South Carolina (USC) have developed an experimental model of device implantation that involved the surgical placement of a small silicone disk in subdermal muscle in rats. We worked with Yost and his team to develop a method for coating silicone implants with pluronic gel containing αCT1.

As was observed in wounded skin, significantly reduced levels of neutrophils were measured acutely adjacent to implants coated with αCT1 compared to vehicle controls (Soder et al., 2009). Additionally, collagen and fibrous protein in the contraction capsule were reduced at 4 weeks after implantation.

Meaningful to potential usefulness in therapy, αCT1 reduced numbers of myofibroblasts present in the contraction capsule. Associated with this change in cellular composition, piezoelectric sensors placed within the implant to measure mechanical properties *in situ* provided recordings that were consistent with reduced capsular compliance. Further studies are required to determine whether the lower numbers of myofibroblasts and reduced force exertion by the scar capsule resulting from the αCT1 drug coating results in longer-term maintenance of implants.

13.6   HOW DOES THE CX43 CARBOXYL TERMINUS INFLUENCE WOUND HEALING?

The essential question that emerges is how does αCT1 mediate its impacts on the injury response? An aspect of mode-of-action suggested by results from the skin wound and silicone disk implantation models is via effects during the inflammatory phase of wound healing (e.g., Figure 13.3).

In seminal investigations, earlier workers noted that the propensity of embryos to heal without scarring correlated inversely with the maturation of the cellular immune response (Whitby and Ferguson, 1991; Martin, 1997; Rhett et al., 2008). Subsequently, the concept that inflammation might be counter-productive to healing found support from experiments in transgenic mouse models (Ashcroft et al., 1999; Martin et al., 2003), studies in athymic mice (Gawronska-Kozak et al., 2006), antisense knockdown *in vivo* (Qiu et al., 2003; Mori et al., 2006), and via direct neutralization of neutrophils in peripheral blood prior to skin wounding (Dovi et al., 2003).

Inflammatory cells provide signals that actively promote granulation and fibrosis (Clark, 1988; Werner and Grose, 2003; Martin and Leibovich, 2005). These signals include reactive oxygen species, chemokines, and cytokines that induce the chemotaxis and collagen deposition of fibroblasts. By inhibiting inflammatory cell transmigration and/or differentiation in wounds, αCT1 may indirectly alter the signaling cascades that are necessary for normal granulation tissue differentiation and scar remodeling.

Alternately, the Cx43 mimetic peptide may have direct effects on cells contributing materially to repaired tissues. There is mounting evidence that bone marrow derivatives other than inflammatory cells contribute to wound repair. Granulation tissue might in part be derived from fibroblast-like progenitors (fibrocytes) emigrating from the peripheral circulation (Stramer et al., 2007; Wang et al., 2007). Putting this another way, αCT1 may inhibit scarring by decreasing recruitment of fibrocytes to injuries via mechanisms similar to those accounting for reduced neutrophil recruitment in treated wounds (e.g., Figure 13.3).

A second clue to understanding the mechanism may come from reports of remodeled Cx43 expression observed in cells immediately proximal to injured tissues. Disruption to Cx43 expression is a common feature of epidermal cells at the edge of skin wounds and in cardiomyocytes in the injury border zone of the myocardial infarction (Smith et al., 1991; Goliger and Paul, 1995; Richards et al., 2004). The wound border in skin and heart also appears to be characterized by altered Cx43 phosphorylation—particularly at the serine at amino acid position 368 of the Cx43 sequence (Cx43-pS368).

As discussed earlier, GJ size growth results from targeting ZO-1 interaction with Cx43 using αCT1 or Cx43 mutants incompetent to interact with ZO-1. Related to these observations *in vitro*, application of αCT1 to mouse skin wounds was found to cause increases in the extent of Cx43 GJ in epidermal cells proximal to the injuries *in vivo*. Our data from cultured cells provide evidence that increased aggregation of connexons into GJs occurs at the expense of functional hemichannels in the plasma membrane (Hunter et al., 2005; Rhett et al., 2011).

Changes in connexin abundance levels, phosphorylation and switching between hemichannel and GJ channel function in tissues adjacent to injuries are likely part of broader responses of tissues to the stress. GJs may enhance the intercellular propagation of injury signals. But looking at this same phenomenon from the other side, adaptive mechanisms working through connexins may act in the compartmentalization of wounded tissues—preventing excessive injury spread and focusing repair processes on the most vulnerable cells proximal to injuries. This important theme of Cx43 possibly having a role in "quarantining" damaged cells will be returned to in the concluding discussion of this chapter.

13.7   THE CX43 CARBOXYL TERMINUS, CARDIAC INJURY, AND ARRHYTHMIA

The wound-healing response of the heart following myocardial infarction is one of the major pre-occupations of modern medicine. Myocardial infarction is a common end point of ischemic heart disease, a pathology that is one of the biggest killers of humans in developed countries including those in Europe, Asia, and the Americas.

As introduced earlier, our work on skin injuries had its origins in studies of Cx43 and its role of conduction of cardiac electrical activation. In the last few years, we have returned to the heart. This work has led to some new and unexpected directions.

A wound to the heart self-evidently exerts a more exacting toll than most cutaneous injuries. The scar tissue generated by healing of a myocardial infarction is problematic for cardiac mechanical function, but a key source of further trouble to the patient stems from the narrow zone of surviving muscle bordering the scar—the infarct border zone (Smith et al., 1991; Peters et al., 1997; Cabo and Boyden, 2006).

The structure of the infarct border zone tends to be highly nonuniform, with regions of fibrotic tissue interspersed with loose strands of surviving myocardium. Also manifest in the heterogeneity of these tissues is significant remodeling of normal patterns of Cx43 organization between border zone myocytes. This heterogeneous organization of the infarct border is thought to give rise to slowed conduction, reentrant arrhythmias, and ultimately an increased chance of undergoing sudden cardiac death following a myocardial infarction (Kleber and Rudy, 2004; Spach et al., 2004).

As αCT1 inhibited remodeling of Cx43 *in vitro* and reduced scar tissue following wounding of skin *in vivo*, we hypothesized that αCT1 might inhibit tissue and GJ remodeling in the arrhythmia-prone tissues bordering infarcts. The approaches taken in heart borrowed methodologically from our experience with cutaneous wounds.

13.8 TECHNICAL INNOVATIONS IN THE CARDIAC INJURY STUDIES

Two technical innovations were developed as part of our study of Cx43 carboxyl terminal peptide effects on cardiac injury (O'Quinn et al., 2011). The first of these was a standardized model of ventricular injury and the second was a method for peptide delivery to injured hearts *in vivo*.

Commonly used coronary artery ligation models in rodents for simulating a myocardial infarction typically result in injuries with high variability in size and shape. We developed a cryo-injury technique that produced a wound of consistent geometry on the mouse left ventricle, similar in repeatability to those that we were familiar with in work on skin. The cryo-injury protocol was based on a method described by van den Bos and coworkers (van den Bos et al., 2005), except that we modified their protocol to obtain nontransmural injuries of the left ventricle.

The rationale for generating a nontransmural injury was to obtain extended regions of injury border zone (cryo-IBZ) adjacent to the freeze wound. As with the infarct border of ischemic injuries, collagen- and periostin-rich fibrous tissue interspersed with myocytes in the cryo-IBZ. Significant remodeling of Cx43 distribution was also characterized in the narrow zone of muscle adjacent to the cryo-scar (O'Quinn et al., 2008). In novel observations, increased levels of immunolabeling for ventricular- and atrial-myosin-light-chain-2 proteins in myocardial cells were additionally noted to occur within the cryo-IBZ (O'Quinn et al., 2011).

The second technical innovation became necessary when it was found that the pluronic thermogel used for skin wounds would not remain localized following application to cryo-lesions on the heart. To achieve focal delivery of peptide, αCT1 was reformulated into a "dry" methylcellulose patch that adhered stably when placed on the ventricular wound immediately following cryo-injury.

Western blotting revealed that αCT1 was maintained in and around cryo-injured tissues for up to 48 h following placement of the methylcellulose patch. Correlated with acute, focal delivery of αCT1, ZO-1 colocalization with Cx43 was reduced in the cryo-IBZ 24 hours after treatment—consistent with an inhibition of the interaction between these two proteins. No αCT1 was detected in ventricular muscle distal from the injury nor did there appear to be effects on ZO-1 association with Cx43 in remote tissues, as assessed by double immunolabeling.

13.9   HOW DOES THE CX43 CARBOXYL TERMINUS INFLUENCE CARDIAC ARRHYTHMIA?

The cryo-IBZ of ventricles from control mice showed profound downregulation and lateralized redistribution of Cx43 from intercalated disks 1 week following cryo-injury. However, in line with the hypothesis, αCT1-treated cryo-IBZ tissues demonstrated much lower levels of Cx43 remodeling at the 1 week subacute time point. There was a trend of increasing GJ size, but the most notable effect was a reduction of lateralization of Cx43 in myocardial tissues proximal to the cryo-scar.

One week after ventricular injury, treatment with peptide also significantly improved systolic contractility of the left ventricle, increased efficiency of ventricular activation, decreased propensity of ventricles to develop arrhythmias in response to programmed stimulation, and reduced arrhythmia severity (Figure 13.4).

The observation that αCT1 promoted Cx43 maintenance in intercalated disks (i.e., a decrease in lateralization) following cryo-injury proved to be a turning point. In 2006, Janis Burt and coworkers had reported that a specific phospho-isoform of Cx43, Cx43-pS368 remained stabilized at intercalated disks following induction of transient ischemia, even as total Cx43 underwent remodeling from disks to lateral myocyte surfaces (Ek-Vitorin et al., 2006).

Inspired by these studies, we found that αCT1-treatment following cryo-injury was associated with upregulation of Cx43 phosphorylation at serine 368 (S368) in myocardium bordering the cryo-injury (O'Quinn et al., 2011). Perhaps most interesting of all, this elevation in Cx43-pS368 was sustained over the first week, consistent with the time course for peptide effects on the cellular distribution of Cx43 and the propensity of treated hearts to resist provoked arrhythmia during the sub-acute period following cryo-injury.

Again, it is emphasized this was a localized change. Increased phosphorylation only occurred in those myocardial tissues proximal to the injury (i.e., in the cryo-IBZ). Ventricular myocardial tissues remote from the cryo-injury showed no significant change in Cx43-pS368 in response to treatment.

13.10   αCT1 PROMOTES PKC-EPSILON PHOSPHORYLATION OF THE CX43 CARBOXYL TERMINUS

The serine at amino acid position 368 of Cx43 is a consensus site for phosphorylation by the kinase PKC-ε. In assays of S368 phosphorylation by this kinase using a GST-Cx43 carboxyl terminal substrate, we found that αCT1 caused a dose-dependent increase in levels of Cx43-pS368, relative to baseline and control reactions (Figure 13.5).

Further assays indicated that PDZ2 domain of ZO-1 alone had no direct effect on PKC-ε-mediated phosphorylation of Cx43. Thus, in addition to inhibiting ZO-1 interaction with Cx43, these results indicated that αCT1 may independently target PKC-ε. This being said, combination of PDZ2 with αCT1 in PKC-ε reactions caused strong inhibition of Cx43-pS368 generation.

The PKC-ε consensus domain on Cx43 that includes S368 also incorporates sequences at the carboxyl terminus of Cx43 involved in ZO-1 binding (Sorgen et al., 2004; Fanning et al., 2007). Physical interaction between Cx43 and PKC-ε

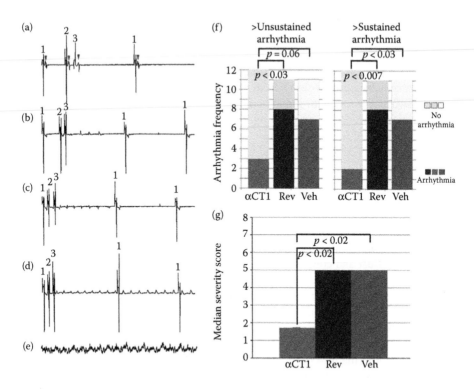

FIGURE 13.4 **(See color insert.)** αCT1 reduces Cx43 GJ remodeling and arrhythmic propensity. Tracings from pacing protocols on isolated perfused ventricles illustrate no arrhythmia (a), three spontaneous premature beats (b), resolving tachycardia (c), sustained tachycardia (d), and fibrillation (e). The green numbers in figures a–d label the s1, s2, and s3 stimuli. The blue arrows in (a) show the stimulated ventricular action potentials. (f) Numbers of hearts displaying arrhythmias (dark red and blue colors) that were unsustained (left-hand bar graph) or sustained (right-hand bar graph) in αCT1, Rev, and Veh groups following pacing. Lighter colors within bars indicate hearts within groups in which arrhythmia was not induced by pacing. (g) Graphical representation of the median severity of arrhythmia for the three treatment groups ($p < 0.02$ αCT1/Rev, $p < 0.02$ αCT1/Veh). $n \geq 11$ (mice/group). (From O'Quinn, M. P. et al. 2011. *Circ Res* 108(6): 704–715.)

has been confirmed in co-immunoprecipitations from myocardial tissues and this interaction is increased by pharmacophores activating PKC-ε (Miura et al., 2007). Taken together with our own data on the combinatorial effects of PDZ2 and αCT1 on PKC-ε activity, this literature suggests the potential for regulatory interactions between PKC-ε and ZO-1 at the carboxyl-terminus of Cx43.

αCT1 appears to pharmacologically induce a cardioprotective state in injured myocardial tissues resembling that occurring in response to transient ischemia. The full parallel is yet to be enumerated, but reduced injury severity prompted by ischemic preconditioning also occurs via PKC-ε activation (Srisakuldee et al., 2009), is dependent on Cx43 gene dosage (Schwanke et al., 2002) and is associated with increase in the phosphorylation of Cx43 at S368 (Srisakuldee et al., 2009).

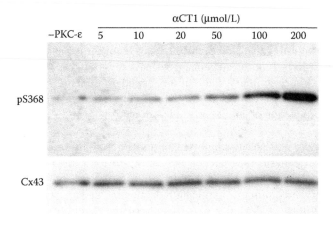

FIGURE 13.5 αCT1 prompts a dose-dependent increase in Cx43 phosphorylation at serine 368. PKC-ε catalytic competence for phosphorylating GST-Cx43 carboxyl terminus at S368 was assessed in biochemical assays *in vitro*. Blots of Cx43-pS368 (upper panel) and total Cx43 (lower panel) in mixtures (from left to right) including a no-kinase control that includes substrate (GST-Cx43-CT aas 255-382), but no PKC-ε enzyme (−PKC-ε), and then blot lanes of reaction solutions containing both substrate and enzyme, as well as increasing concentrations of αCT1 (i.e., at 5, 10, 20, 50, 100, and 200 μmol/L). (From O'Quinn, M. P. et al. 2011. *Circ Res* 108(6): 704–715.)

An unknown in this novel aspect of mode of action is whether increase in Cx43-pS368 result from direct activation of PKC-ε enzyme by αCT1 or whether the peptide interacts with the Cx43 substrate in a manner that increases efficiency of phosphorylation. Ongoing work in the laboratory is focused on this question.

13.11 CONCLUDING REMARKS

At the molecular level, the picture that emerges of αCT1 is one of the multimodal effector. On the one hand, the data indicate that αCT1 is a potent inhibitor of ZO-1 interaction with Cx43 *in vitro* and *in vivo*. Studies in HeLa cells indicate that this inhibition occurs without recourse to changes in Cx43 phosphorylation at serine 368 (Rhett et al., 2011).

On the other hand, both in *in vitro* phosphorylation assays and in the context of cardiac injury *in vivo*, the Cx43 mimetic peptide appears to enhance Cx43-S368 phosphorylation, independent of ZO-1. It should be added that Belgium workers have recently reported that a synthetic peptide similar to αCT1 interrupts intramolecular interactions between the cytoplasmic loop and carboxyl terminal domains of Cx43 (Ponsaerts et al., 2010).

This complex mode of action is fully consistent with the emerging role of the Cx43 carboxyl terminus as a regulatory hub, involved in a dynamic set of interactions with cytoskeletal, signal transduction, membrane channel, and other junctional molecules. What comes as something of a surprise is that so much information is imparted in only the last nine amino acids of Cx43.

A further interesting question is whether endogenous fragments of Cx43 are generated *in vivo*. Merry Lindsey and colleagues have reported that MMP-7 is activated in response to ischemic injury and that this peptidase directly interacts with Cx43 (Lindsey et al., 2006). Most pertinently, studies of proteolytic cleavage of Cx43 *in vitro* suggest production of native carboxyl terminal fragments by MMP-7 very similar to αCT1.

At the broader cell and tissue level, the most important prospect is that Cx43 phosphorylation patterns may encode the facility to form distinct tissue compartments in response to injury. Ek-Vitorin and colleagues have shown that changes in S368 phosphorylation prompt significant reduction in conductance by unitary Cx43 channels (Ek-Vitorin et al., 2006). Compartmentalized change in Cx43-pS368 levels is characteristic in cells proximal to both cutaneous and cardiac injuries (Richards et al., 2004; O'Quinn et al., 2011). The ability to rapidly quarantine wounded cells and their neighbors, from otherwise healthy surrounding tissues, may be a key adaptive response. Confirmation of such a mechanism could set the stage for future therapies that enhance regenerative healing following injury, disease, or surgery.

REFERENCES

Akar, F. G. and D. S. Rosenbaum. 2003. Transmural electrophysiological heterogeneities underlying arrhythmogenesis in heart failure. *Circ Res* 93(7): 638–645.

Anderson, J. M. 1996. Cell signalling: MAGUK magic. *Curr Biol* 6(4): 382–384.

Ashcroft, G. S., X. Yang, A. B. Glick et al. 1999. Mice lacking Smad3 show accelerated wound healing and an impaired local inflammatory response. *Nat Cell Biol* 1(5): 260–266.

Baker, S. M., N. Kim, A. M. Gumpert et al. 2008. Acute internalization of gap junctions in vascular endothelial cells in response to inflammatory mediator-induced G-protein coupled receptor activation. *FEBS Lett* 582(29): 4039–4046.

Barker, R. J., R. L. Price, and R. G. Gourdie. 2001. Increased co-localization of connexin43 and ZO-1 in dissociated adult myocytes. *Cell Commun Adhes* 8(4–6): 205–208.

Barker, R. J., R. L. Price, and R. G. Gourdie. 2002. Increased association of ZO-1 with connexin43 during remodeling of cardiac gap junctions. *Circ Res* 90(3): 317–324.

Bao, B., J. Jiang, T. Yanase et al. 2011. Connexon-mediated cell adhesion drives microtissue self-assembly. *FASEB J* 25(1): 255–264.

Becker, D. L., J. E. Cook, C. S. Davies et al. 1998. Expression of major gap junction connexin types in the working myocardium of eight chordates. *Cell Biol Int* 22(7–8): 527–543.

Beyer, E. C., D. L. Paul, and D. A. Goodenough. 1987. Connexin43: A protein from rat heart homologous to a gap junction protein from liver. *J Cell Biol* 105(6 Pt 1): 2621–2629.

Bouvier, D., F. Kieken, A. Kellezi et al. 2008. Structural changes in the carboxyl terminus of the gap junction protein connexin 40 caused by the interaction with c-Src and zonula occludens-1. *Cell Commun Adhes* 15(1): 107–118.

Bruce, A. F., S. Rothery, E. Dupont et al. 2008. Gap junction remodelling in human heart failure is associated with increased interaction of connexin43 with ZO-1. *Cardiovasc Res* 77(4): 757–765.

Cabo, C. and P. A. Boyden. 2006. Heterogeneous gap junction remodeling stabilizes reentrant circuits in the epicardial border zone of the healing canine infarct: A computational study. *Am J Physiol Heart Circ Physiol* 291(6): H2606–2616.

Cabo, C., J. Yao, P. A. Boyden et al. 2006. Heterogeneous gap junction remodeling in reentrant circuits in the epicardial border zone of the healing canine infarct. *Cardiovasc Res* 72(2): 241–249.

Caffee, H. H., N. P. Mendenhall, W. M. Mendenhall et al. 1988. Postoperative radiation and implant capsule contraction. *Ann Plast Surg* 20(1): 35–38.

Clark, R. A. F. 1988. General considerations and overview of wound repair. In: *The Molecular and Cellular Biology of Wound Repair*, Clark, R. A. F. and Henson, P. M., eds. Plenum Press, Inc., New York, pp. 3–33.

Cotrina, M. L., J. H. Lin, and M. Nedergaard. 2008. Adhesive properties of connexin hemichannels. *Glia* 56(16): 1791–1798.

Dai, P., T. Nakagami, H. Tanaka et al. 2007. Cx43 mediates TGF-beta signaling through competitive Smads binding to microtubules. *Mol Biol Cell* 18(6): 2264–2273.

Defamie, N., B. Mograbi, C. Roger et al. 2001. Disruption of gap junctional intercellular communication by lindane is associated with aberrant localization of connexin43 and zonula occludens-1 in 42GPA9 Sertoli cells. *Carcinogenesis* 22(9): 1537–1542.

Dovi, J. V., L. K. He, and L. A. DiPietro. 2003. Accelerated wound closure in neutrophil-depleted mice. *J Leukoc Biol* 73(4): 448–455.

Drin, G., S. Cottin, E. Blanc et al. 2003. Studies on the internalization mechanism of cationic cell-penetrating peptides. *J Biol Chem* 278(33): 31192–31201.

Duffy, H. S. 2008. Cardiac connections—The antiarrhythmic solution? *N Engl J Med* 358(13): 1397–1398.

Duffy, H. S., A. W. Ashton, P. O'Donnell et al. 2004. Regulation of connexin43 protein complexes by intracellular acidification. *Circ Res* 94(2): 215–222.

Ek-Vitorin, J. F., T. J. King, N. S. Heyman et al. 2006. Selectivity of connexin 43 channels is regulated through protein kinase C-dependent phosphorylation. *Circ Res* 98(12): 1498–1505.

Elias, L. A., D. D. Wang, and A. R. Kriegstein. 2007. Gap junction adhesion is necessary for radial migration in the neocortex. *Nature* 448(7156): 901–907.

Fanning, A. S., M. F. Lye, J. M. Anderson et al. 2007. Domain swapping within PDZ2 is responsible for dimerization of ZO proteins. *J Biol Chem* 282(52): 37710–37716.

Flores, C. E., X. Li, M. V. Bennett et al. 2008. Interaction between connexin35 and zonula occludens-1 and its potential role in the regulation of electrical synapses. *Proc Natl Acad Sci USA* 105(34): 12545–12550.

Forbes, P. D. 1969. Vascular supply of the skin and hair in swine. *Advances in Biology of Skin*. W. Montagna and R. L. Dobson. Oxford, Pergamon Press. IX: 419–432.

Gawronska-Kozak, B., M. Bogacki, J. S. Rim et al. 2006. Scarless skin repair in immunodeficient mice. *Wound Repair Regen* 14(3): 265–276.

Ghatnekar, G. S., J. Jourdan, and R. G. Gourdie. 2006. Novel connexin-based peptides accelerate closure and reduce inflammation and scarring in skin wounds. *Wound Repair Regen* 14(62).

Ghatnekar, G. S., M. P. O'Quinn, L. J. Jourdan et al. 2009. Connexin43 carboxyl-terminal peptides reduce scar progenitor and promote regenerative healing following skin wounding. *Regen Med* 4(2): 205–223.

Giepmans, B. N. and W. H. Moolenaar. 1998. The gap junction protein connexin43 interacts with the second PDZ domain of the zona occludens-1 protein. *Curr Biol* 8(16): 931–934.

Gilleron, J., C. Fiorini, D. Carette et al. 2008. Molecular reorganization of Cx43, Zo-1 and Src complexes during the endocytosis of gap junction plaques in response to a non-genomic carcinogen. *J Cell Sci* 121(Pt 24): 4069–4078.

Goliger, J. A. and D. L. Paul. 1995. Wounding alters epidermal connexin expression and gap junction-mediated intercellular communication. *Mol Biol Cell* 6(11): 1491–1501.

Gourdie, R. G., G. S. Ghatnekar, M. O'Quinn et al. 2006. The unstoppable connexin43 car-
    boxyl-terminus: New roles in gap junction organization and wound healing. *Ann N Y
    Acad Sci* 1080:49–62.
Gourdie, R. G., C. R. Green, and N. J. Severs. 1991. Gap junction distribution in adult mam-
    malian myocardium revealed by an anti-peptide antibody and laser scanning confocal
    microscopy. *J Cell Sci* 99 (Pt 1): 41–55.
Hall, E. J. and A. J. Giaccia. 2005. *Radiobiology for the Radiologist*, 6th edition. Lippincott
    Williams & Wilkins, Philadelphia.
Hartsock, A. and W. J. Nelson. 2008. Adherens and tight junctions: Structure, function and
    connections to the actin cytoskeleton. *Biochim Biophys Acta* 1778(3): 660–669.
Hesketh, G. G., J. E. Van Eyk, and G. F. Tomaselli. 2009. Mechanisms of gap junction traffic
    in health and disease. *J Cardiovasc Pharmacol* 54(4): 263–272.
Hunter, A. W., R. J. Barker, C. Zhou et al. 2006. Interaction between connexin43 and ZO1
    regulates gap junction size. *Mol Biol Cell* 16: 5686–5698.
Hunter, A. W., R. J. Barker, C. Zhu et al. 2005. Zonula occludens-1 alters connexin43 gap junction
    size and organization by influencing channel accretion. *Mol Biol Cell* 16(12): 5686–5698.
Hunter, A. W. and R. G. Gourdie. 2008. Microtubules and actin differentially influence remod-
    eling of connexin43 gap junctions. *FASEB J* 22(800.4).
Hunter, A. W., J. Jourdan, and R. G. Gourdie. 2003. Fusion of GFP to the carboxyl terminus
    of connexin43 increases gap junction size in HeLa cells. *Cell Commun Adhes* 10(4–6):
    211–214.
Jin, C., A. F. Lau, and K. D. Martyn. 2000. Identification of connexin-interacting proteins:
    Application of the yeast two-hybrid screen. *Methods* 20(2): 219–231.
Kausalya, P. J., M. Reichert, and W. Hunziker. 2001. Connexin45 directly binds to ZO-1 and
    localizes to the tight junction region in epithelial MDCK cells. *FEBS Lett* 505(1): 92–96.
Kleber, A. G. and Y. Rudy. 2004. Basic mechanisms of cardiac impulse propagation and asso-
    ciated arrhythmias. *Physiol Rev* 84(2): 431–488.
Laing, J. G., B. C. Chou, and T. H. Steinberg. 2005. ZO-1 alters the plasma membrane local-
    ization and function of Cx43 in osteoblastic cells. *J Cell Sci* 118(Pt 10): 2167–2176.
Laing, J. G., R. N. Manley-Markowski, M. Koval et al. 2001. Connexin45 interacts with zonula
    occludens-1 and connexin43 in osteoblastic cells. *J Biol Chem* 276(25): 23051–23055.
Lin, J. H., H. Weigel, M. L. Cotrina et al. 1998. Gap-junction-mediated propagation and
    amplification of cell injury. *Nat Neurosci* 1(6): 494–500.
Lindsey, M. L., G. P. Escobar, R. Mukherjee et al. 2006. Matrix metalloproteinase-7 affects
    connexin-43 levels, electrical conduction, and survival after myocardial infarction.
    Circulation 113(25): 2919–2928.
Luke, R. A. and J. E. Saffitz. 1991. Remodeling of ventricular conduction pathways in healed
    canine infarct border zones. *J Clin Invest* 87(5): 1594–1602.
Maass, K., J. Shibayama, S. E. Chase et al. 2007. C-terminal truncation of connexin43 changes
    number, size, and localization of cardiac gap junction plaques. *Circ Res* 101(12): 1283–1291.
Martin, P. 1997. Wound healing—Aiming for perfect skin regeneration. *Science* 276(5309):
    75–81.
Martin, P. and S. J. Leibovich. 2005. Inflammatory cells during wound repair: The good, the
    bad and the ugly. *Trends Cell Biol* 15(11): 599–607.
Martin, P., D. D'Souza, J. Martin et al. 2003. Wound healing in the PU.1 null mouse—Tissue
    repair is not dependent on inflammatory cells. *Curr Biol* 13(13): 1122–1128.
Miura, T., T. Yano, K. Naitoh et al. 2007. Delta-opioid receptor activation before ischemia
    reduces gap junction permeability in ischemic myocardium by PKC-epsilon-mediated
    phosphorylation of connexin 43. *Am J Physiol Heart Circ Physiol* 293(3): H1425–1431.
Mori, R., K. T. Power, C. M. Wang et al. 2006. Acute downregulation of connexin43 at wound
    sites leads to a reduced inflammatory response, enhanced keratinocyte proliferation and
    wound fibroblast migration. *J Cell Sci* 119(Pt 24): 5193–5203.

Nielsen, P. A., A. Baruch, V. I. Shestopalov et al. 2003. Lens connexins alpha3Cx46 and alpha8Cx50 interact with zonula occludens protein-1 (ZO-1). *Mol Biol Cell* 14(6): 2470–2481.

Nielsen, P. A., D. L. Beahm, B. N. Giepmans et al. 2002. Molecular cloning, functional expression, and tissue distribution of a novel human gap junction-forming protein, connexin-31.9. Interaction with zona occludens protein-1. *J Biol Chem* 277(41): 38272–38283.

O'Quinn, M. P., B. S. Harris, K. W. Hewett et al. 2008. A peptide containing the carboxy-terminal domain of connexin43 reduces arrhythmias and improves cardiac function after myocardial injury. *Circulation* 118(18S2): S495.

O'Quinn, M. P., J. A. Palatinus, B. S. Harris, and R. G. Gourdie. 2011. A peptide mimetic of the connexin43 carboxyl terminus reduces gap junction remodeling and induced arrhythmia following ventricular injury. *Circ Res* 108(6): 704–715.

Palatinus, J. A., J. M. Rhett, and R. G. Gourdie. 2010. Translational lessons from scarless healing of cutaneous wounds and regenerative repair of the myocardium. *J Mol Cell Cardiol* 48(3): 550–557.

Peters, N. S., J. Coromilas, N. J. Severs et al. 1997. Disturbed connexin43 gap junction distribution correlates with the location of reentrant circuits in the epicardial border zone of healing canine infarcts that cause ventricular tachycardia. *Circulation* 95(4): 988–996.

Peters, N. S. and A. L. Wit 2000. Gap junction remodeling in infarction: Does it play a role in arrhythmogenesis? *J Cardiovasc Electrophysiol* 11(4): 488–490.

Ponsaerts, R., E. De Vuyst, M. Retamal et al. 2010. Intramolecular loop/tail interactions are essential for connexin 43-hemichannel activity. *FASEB J* 24(11): 4378–4395.

Qiu, C., P. Coutinho, S. Frank et al. 2003. Targeting connexin43 expression accelerates the rate of wound repair. *Curr Biol* 13(19): 1697–1703.

Rhett, J. M., G. S. Ghatnekar, J. A. Palatinus et al. 2008. Novel therapies for scar reduction and regenerative healing of skin wounds. *Trends Biotechnol* 26(4): 173–180.

Rhett, J. M., L. J. Jourdan, and R. G. Gourdie. 2011. Connexin43 connexon to gap junction transition is regulated by zonula occludens-1 (ZO-1). *Mol Biol Cell* 22: 1516–1528.

Richards, T. S., C. A. Dunn, W. G. Carter et al. 2004. Protein kinase C spatially and temporally regulates gap junctional communication during human wound repair via phosphorylation of connexin43 on serine368. *J Cell Biol* 167(3): 555–562.

Rohr, S. 2004. Role of gap junctions in the propagation of the cardiac action potential. *Cardiovasc Res* 62(2): 309–322.

Saez, J. C., V. M. Berthoud, M. C. Branes et al. 2003. Plasma membrane channels formed by connexins: Their regulation and functions. *Physiol Rev* 83(4): 1359–1400.

Schwanke, U., I. Konietzka, A. Duschin et al. 2002. No ischemic preconditioning in heterozygous connexin43-deficient mice. *Am J Physiol Heart Circ Physiol* 283(4): H1740–1742.

Segretain, D., C. Fiorini, X. Decrouy et al. 2004. A proposed role for ZO-1 in targeting connexin 43 gap junctions to the endocytic pathway. *Biochimie* 86(4–5): 241–244.

Severs, N. J., A. F. Bruce, E. Dupont et al. 2008. Remodelling of gap junctions and connexin expression in diseased myocardium. *Cardiovasc Res* 80(1): 9–19.

Singh, D., J. L. Solan, S. M. Taffet et al. 2005. Connexin 43 interacts with zona occludens-1 and -2 proteins in a cell cycle stage-specific manner. *J Biol Chem* 280(34): 30416–30421.

Smith, J. H., C. R. Green, N. S. Peters et al. 1991. Altered patterns of gap junction distribution in ischemic heart disease. An immunohistochemical study of human myocardium using laser scanning confocal microscopy. *Am J Pathol* 139(4): 801–821.

Soder, B. L., J. T. Propst, T. M. Brooks et al. 2009. The connexin43 carboxyl-terminal peptide ACT1 modulates the biological response to silicone implants. *Plast Reconstr Surg* 123(5): 1440–1451.

Sorgen, P. L., H. S. Duffy, P. Sahoo et al. 2004. Structural changes in the carboxyl terminus of the gap junction protein connexin43 indicates signaling between binding domains for c-Src and zonula occludens-1. *J Biol Chem* 279(52): 54695–54701.

Spach, M. S., J. F. Heidlage, R. C. Barr et al. 2004. Cell size and communication: Role in structural and electrical development and remodeling of the heart. *Heart Rhythm* 1(4): 500–515.

Srisakuldee, W., M. M. Jeyaraman, B. E. Nickel et al. 2009. Phosphorylation of connexin-43 at serine 262 promotes a cardiac injury-resistant state. *Cardiovasc Res* 83(4): 672–681.

Stevenson, B. R., J. D. Siliciano, M. S. Mooseker et al. 1986. Identification of ZO-1: A high molecular weight polypeptide associated with the tight junction (zonula occludens) in a variety of epithelia. *J Cell Biol* 103(3): 755–766.

Stramer, B. M., R. Mori, and P. Martin. 2007. The inflammation-fibrosis link? A Jekyll and Hyde role for blood cells during wound repair. *J Invest Dermatol* 127(5): 1009–1017.

Toyofuku, T., Y. Akamatsu, H. Zhang et al. 2001. c-Src regulates the interaction between connexin-43 and ZO-1 in cardiac myocytes. *J Biol Chem* 276(3): 1780–1788.

Toyofuku, T., M. Yabuki, K. Otsu et al. 1998. Direct association of the gap junction protein connexin-43 with ZO-1 in cardiac myocytes. *J Biol Chem* 273(21): 12725–12731.

van den Bos, E. J., B. M. Mees, M. C. de Waard et al. 2005. A novel model of cryoinjury-induced myocardial infarction in the mouse: A comparison with coronary artery ligation. *Am J Physiol Heart Circ Physiol* 289(3): H1291–1300.

Wang, J., H. Jiao, T. L. Stewart et al. 2007. Accelerated wound healing in leukocyte-specific, protein 1-deficient mouse is associated with increased infiltration of leukocytes and fibrocytes. *J Leukoc Biol* 82(6): 1554–1563.

Werner, S. and R. Grose. 2003. Regulation of wound healing by growth factors and cytokines. *Physiol Rev* 83(3): 835–870.

Whitby, D. J. and M. W. Ferguson. 1991. The extracellular matrix of lip wounds in fetal, neonatal and adult mice. *Development* 112(2): 651–668.

Yao, J. A., W. Hussain, P. Patel et al. 2003. Remodeling of gap junctional channel function in epicardial border zone of healing canine infarcts. *Circ Res* 92(4): 437–443.

Zhu, C., R. J. Barker, A. W. Hunter et al. 2005. Quantitative analysis of ZO-1 colocalization with Cx43 gap junction plaques in cultures of rat neonatal cardiomyocytes. *Microsc Microanal* 11(3): 244–248.

14 Connexin-Based Therapeutic Approaches to Inflammation in the Central Nervous System

Jie Zhang, Simon J. O'Carroll, Helen V. Danesh-Meyer, Henri C. Van der Heyde, David L. Becker, Louise F. B. Nicholson, and Colin R. Green

CONTENTS

14.1 INTRODUCTION

There are multiple studies reporting changes to connexin expression associated with central nervous system (CNS) injury mechanisms and inflammatory processes (for review, see Kielian 2008; Chew et al. 2010). Areas of maximal tissue damage may show lower connexin levels (generally reflecting cell death), but the most commonly reported response is connexin43 (Cx43) upregulation, either in the wounded area or in tissue immediately surrounding the damaged sector. This upregulation correlates closely with a number of inflammatory events, in particular secondary damage or lesion spread, white blood cell infiltration or glial cell activation, and endothelial cell activation leading to a breakdown in the blood–brain barrier. Gap junction modulation has been identified as a potential neuroprotective target in events such as traumatic brain injury, optic nerve and retinal ischemia, spinal cord injury or repair, and for chronic neurodegenerative diseases. Section 14.2 briefly introduces the key inflammatory events involving gap junction communication and hemichannel roles relevant to this chapter (for more specific details refer to the Prequel and Chapters 1 and 3 of this book). Examples of both *ex vivo* and *in vivo* models are then presented in which communication levels have been modulated following CNS injury, at the level of a gene knockout or through transient modulation, primarily with connexin-specific antisense oligonucleotides or functional connexin peptidomimetics.

14.2 CNS INFLAMMATORY RESPONSES TO INJURY AND DISEASE

14.2.1 Connexin-Related Changes with CNS Injury

In addition to the possibility of neutrophil and macrophage invasion with CNS injury, both microglia and astrocytes respond by proliferation and migration and these changes are associated with increased Cx43 expression. Gap junctions have also been implicated in various CNS disorders, such as neurodegenerative disease (Vis et al. 1998), brain tumors (Naus et al. 1991), epilepsy (Fonseca et al. 2002), and brain ischemia (Budd and Lipton 1998; Oguro et al. 2001; Contreras et al. 2004). Following ischemic injury to the brain, for example, Cx43 expression is increased (Rami et al. 2001; Hossain et al. 1994), with inhibition of astrocytic gap junction permeability by octanol restricting the flow of undesirable neurotoxins that would otherwise exacerbate neuronal damage (Rami et al. 2001). Following focal cerebral ischemia induced by photothrombosis, reactive astrocytes envelope the lesion and there is an upregulation of Cx43 mRNA and protein expression (Haupt et al. 2007). In a postmortem human brain study following ischemia, the examination of 236 samples showed enhanced Cx43 immunoreactivity in the penumbral areas of both acute and chronic cortical infarcts (Nakase et al. 2006). Increased Cx43 immunoreactivity has also been reported surrounding kainic-acid-induced lesions in the rat brain (Sawchuk et al. 1995) and similar patterns have been observed following spinal cord injury (Cronin et al. 2008).

There has been conflicting data as to whether gap junctions are protective or detrimental after brain injury (Contreras et al. 2004; Farahani et al. 2005). A number of studies indicate that Cx43 expression may be detrimental (Frantseva et al. 2002a, 2002b) with others, conversely, reporting that a lack of Cx43 leads to an increase in the size of the injury (Nakase et al. 2004; Siushansian et al. 2001). Studies that report a loss of connexin as detrimental have primarily used models in which connexin expression was prevented by gene knockout. Knockout, though, also prevents the subsequent recovery of gap junction communication that is essential for spatial buffering, neuronal survival, and normal tissue coupling function. Studies where connexin expression proved to be detrimental have used transient gap junction channel down regulation or transient channel block to bring about benefits. Such treatments prevent connexin signaling during the early stages (first 24–48 h) of injury (or as a treatment for a chronic condition), but allow subsequent recovery of connexin expression and gap junction communication essential for normal function.

Death signals can be propagated via Cx43 gap junctions; Cx43 expression in injury-resistant cells leads to decreased survival once gap junctions are formed (Lin et al. 1998). Excessive intracellular Ca^{2+} and water accumulation is a direct trigger for the initial necrotic cell burst but astrocytic gap junctions then contribute to propagation of a Ca^{2+} wave both directly and indirectly (Rouach et al. 2002). Evidence for direct gap junction involvement includes the following: gap junctions are permeable to Ca^{2+} (Rouach et al. 2002); Ca^{2+} wave spread is not directed by speed and direction of any Ca^{2+} gradient (Finkbeiner 1992); gap junction uncoupling agents block Ca^{2+} propagation (Finkbeiner 1992; Blomstrand et al. 1999); cells devoid of gap junctions only spread Ca^{2+} waves once Cx43 is expressed (Charles et al. 1992); and the intracellular messenger of the Ca^{2+} cascade, IP_3, diffuses through Cx43 gap junctions (Frame and de Feijter 1997; Suadicani et al. 2004). Given that Cx43 is the main constituent of astrocytic gap junctions and is capable of coordinating synchronous activity over large populations of glial cells, its upregulation and subsequent enhanced gap junction communication may play a role in propagating death signals and Ca^{2+} waves extensively and rapidly in the glial network during the acute phase (Lin et al. 1998).

There are several mechanisms proposed by which hemichannels can contribute to cell death. The opening of astrocyte hemichannels could lead to propagation of Ca^{2+} waves by releasing ATP, NAD^+, and glutamate from astrocytes, and spread of IP3 between neighboring astrocytes (Rouach et al. 2002; Evans and Martin 2002; Rodriguez-Sinovas et al. 2007; Stout et al. 2002; Ye et al. 2003). ATP acting on P_2Y purinergic receptors on adjacent cells raises intracellular Ca^{2+} leading to propagation of a Ca^{2+} wave from cell to cell (Goodenough and Paul 2003) and spreading of a cell death signal from one cell to the next (Krutovskikh et al. 2002). Glutamate is known to "kill" susceptible neurons when released in an uncontrolled manner (Zipfel et al. 2000). Astrocytes have an as yet unidentified mechanism for sensing low extracellular Ca^{2+} concentrations and respond by releasing Ca^{2+} from intracellular stores, increasing intracellular Ca^{2+} concentration and enhancing hemichannel opening (Saez et al. 2003).

14.2.2 Edema

Direct necrosis resulting from injury is not amenable to connexin modulation, but edema in cells not directly damaged, but within the penumbra, is prevalent and a target for protective treatment. Hemichannels have been implicated in cell swelling, which is a hallmark of ischemic cell death (Rodriguez-Sinovas et al. 2007). Overexpression of connexin46 in *Xenopus* oocytes or Cx43 in single cells, for example, results in cell swelling and lysis when the extracellular Ca^{2+} concentration is lowered, a condition promoting hemichannel opening (Paul et al. 1991; Quist et al. 2000). Cytotoxic edema in astrocytes is likely due to the presence of a number of molecules in the extracellular space shortly after injury, including glutamate, lactate, K^+, nitric oxide, arachidonate, reactive oxygen species, and ammonia (Kimelberg 1992), many of which can pass through or regulate hemichannels. In astrocytes, the abnormal release of excitatory amino acids can occur via Cx43 hemichannels (Ye et al. 2003) that have a higher probability of opening under conditions of membrane depolarization and low extracellular Ca^{2+} (Saez et al. 2003). Release of excitatory amino acids leads to overstimulation of glutamate receptors on neurons and oligodendrocytes and excessive accumulation of intracellular Ca^{2+} and water (Zipfel et al. 2000; Tekkok and Goldberg 2001; Wilke et al. 2004). Membrane deformation and depolarization can also lead to activation of voltage-sensitive ion channels that allow ions to move down their electrochemical gradients, increasing extracellular K^+ and decreasing extracellular Na^+ and Ca^{2+} concentrations (Nilsson et al. 1993; Marmarou 2007). Chloride ions and water then follow Na^+ and Ca^{2+} passively into the cells (Marmarou 2007). By down regulating Cx43 expression after injury, the number of Cx43 hemichannels in the membrane is decreased, leading to a significant reduction in swelling (Cronin et al. 2008; O'Carroll et al. 2008; Zhang et al. 2010).

14.2.3 Vessel Hemorrhage and Leak

Inflammation follows injury, and chronic inflammation is a common feature of neurodegenerative diseases, with inflammatory cells then able to migrate across the normally impermeable blood–brain barrier. The mechanism by which this occurs has been thought to be mechanistic, involving changes in the expression of multiple genes in the endothelial cells (Velardo et al. 2004) that lead to increased permeability and dysfunction facilitating movement across the blood vessel wall (DiStasi and Ley 2009). TNF-α has been reported to play a part in causing such changes leading to an increase in vascular permeability, although other molecules such as vascular endothelial growth factor may also have a role. Leukocytes are then able to migrate across the compromised wall and become involved in upregulation of adhesion molecules such as intercellular adhesion molecule-1 (ICAM-1), in addition to the upregulation of lymphocyte function-associated antigen-1 (LFA-1). LFA-1 on leukocytes and ICAM-1 on the endothelial cells interact to form a "ring-like structure" around the migrating cell, which then mediates its passage and remains present for the entire migratory process (DiStasi and Ley 2009).

Interactions between leukocytes and endothelial cells though are early events in acute and chronic inflammation and wound defense and repair (McIntyre et al. 2003). These correlate with Cx43 expression in both the endothelial and the inflammatory cells and, in general, are associated with macrophage infiltration and endothelial cell activation (Oviedo-Orta and Howard Evans 2004; see also Section 1.1 of Chapter 1 for more details). The gap junction blockers octanol and 18α-glycyrrhetinic acid have been reported to induce a profound reduction in transendothelial monocyte/macrophage migration in an *in vitro* blood–brain barrier model (Eugenin et al. 2003), and Cx43 is upregulated independently of neutrophils in inflamed mouse lungs (Sarieddine et al. 2009), but not detected in mice genetically protected from inflammation. In heterozygous Cx43$^{-/+}$ mice, airway neutrophil count reduced by over half, and conversely, neutrophil numbers increased in challenged mice expressing a modified Cx43 with higher channel conductivity. Blocking peptides decreased adhesion of neutrophils to endothelial cells and reduced neutrophil transmigration in the inflamed lungs by two-thirds. Down regulation of Cx43 using antisense oligodeoxynucleotides (AsODNs) also attenuates recruitment of both neutrophils and macrophages at skin wound sites (Coutinho et al. 2005; Mori et al. 2006; Qiu et al. 2003) and following spinal cord injury (Cronin et al. 2008).

However, evidence is emerging suggesting that the entry of inflammatory cells might not always be so well regulated. Indeed, it is possible that gap junctions or hemichannels might play a role in mediating the physical destruction of the endothelial cells, creating holes in the vasculature through which cells can migrate (Cronin et al. 2008). Postmortem analysis of human brain tissue from patients with Alzheimer's and Parkinson's disease (both chronic neuroinflammatory diseases) has found that there are morphological changes to the microvasculature indicating capillary dysfunction (Guan et al. 2012; Farkas and Luiten 2001). It has also been reported that none of the proangiogenic growth factors (VEGF, FGF, PDGF) added exogenously are able to prevent vascular regression in wounds (Gosain et al. 2006). These authors concluded that antiangiogenic signals that mediate vessel regression in wounds are strongly dominant over proangiogenic factors during the later stages of wound healing where Cx43 expression may be a dominant factor. In a rat bacteremic sepsis lung model, for example, there was a sevenfold increase in plasma levels entering the airspace (Conhaim et al. 2008). Latex particles perfused into the inflamed lungs were found clustered, not unlike the sequestration of neutrophils during pulmonary sepsis, indicating a previously unreported lung event at the micro level. This serum leak and inanimate latex bead infiltration is consistent with capillary leak reported as a result of connexin upregulation (Cronin et al. 2008). In the Cronin study of traumatic spinal cord injury, a dramatic elevation of Cx43 expression was observed in the walls of small blood vessels within the white matter as early as 6 h after injury (Cronin et al. 2008). This upregulation was accompanied by vascular leakage of fluorescently labeled albumin and accumulation of blood-borne neutrophils. Suppression of Cx43 upregulation using AsODNs reduced vascular leakage and neutrophil recruitment, suggesting a role for Cx43 in the pathogenesis of endothelial cell dysfunction (Cronin et al. 2008). Following retinal ischemia, vessel leak is also one of the very first signs of inflammation (Figure 14.1).

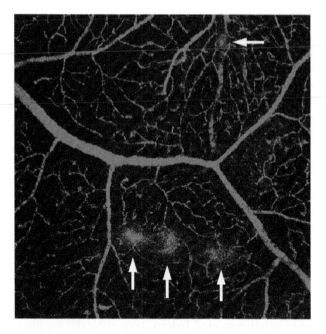

FIGURE 14.1   Rat retinal blood vessels highlighted an hour after ischemia using Evans Blue dye delivered by intraperitoneal injection. Ischemia was induced by raising intraocular pressure for 1 h followed by reperfusion when pressure was reduced to normal. Dye is seen leaking from the compromised capillary bed in a number of places indicating that vessel leak is one of the primary responses to injury (arrows).

14.3   TOOLS

14.3.1   NONSPECIFIC CHANNEL BLOCKERS

Connexin gap junction and hemichannel communication can be inhibited by a number of compounds, including halothane, octanol, heptanol, 18α-glycyrrhetinic acid, carbenoxolone, and flufenamic acid. For further review, see Salameh and Dhein (2005). The first three induce radical changes in membrane fluidity and the remainder have secondary effects inside the cell that in turn alter connexin communication (Evans and Boitano 2001). Quinoline compounds also affect gap junction activity (see, e.g., Das et al. 2008). None of these compounds are considered to be very selective as they can modify other membrane channels, as in the case of heptanol and octanol, or act on gap junctions indirectly through a number of signaling pathways, as in the case of 18α-glycyrrhetinic acid (Evans and Boitano 2001). They have however provided useful information. In cerebral ischemia, gap junctions remain open and may allow the passage of apoptotic and necrotic signals between dying cells and potentially viable cells, increasing cell death (Thompson et al. 2006; Cotrina et al. 1998). Global gap junction blockade using carbenoxolone, flufenamic acid, heptanol, or octanol reduced the spreading wave of astrocytic apoptosis in rat hippocampal cultures exposed to ischemic conditions (Nodin et al. 2005), and in a rodent model of stroke, pretreatment with octanol significantly decreased mean infarction volume following middle cerebral artery occlusion (Rawanduzy et al. 1997). Octanol pretreatment also inhibited the development of experimentally induced waves of spreading depression, and reduced neuronal cell death in the pyramidal subfields of the hippocampus following transient forebrain ischemia induced by bilateral carotid artery occlusion (Rami et al. 2001). Similarly, global gap junction blockade with carbenoxolone has been shown to reduce neuronal death following intrauterine hypoxia–ischemia (de Pina-Benabou et al. 2005). In one attempt to identify the mechanisms by which gap junction blockade reduces neuronal cell death in the hippocampus, pretreatment with the gap junction blockers, carbenoxolone, glycyrrhetinic acid, and endothelin, all resulted in decreased numbers of TUNEL-positive neurons compared to the contralateral side (Velazquez et al. 2006).

Two specific gap junction modulators have also been investigated—connexin-specific AsODNs and connexin peptidomimetics.

14.3.2 ANTISENSE OLIGODEOXYNUCLEOTIDES

To counter the detrimental effects of connexin upregulation seen after injury, it is possible to specifically and effectively modulate connexin protein expression using AsODN (see, e.g., Becker et al. 1999; Green et al. 2001; Law et al. 2006; Frantseva et al. 2002a, 2002b). The most commonly used Cx43-specific AsODN is a single-strand DNA of 30 deoxynucleotides with an unmodified backbone that binds specifically to complementary sequences on an accessible region of the rat Cx43 mRNA, blocking protein translation (Law et al. 2006). Unmodified AsODNs typically have a cytoplasmic half-life of about 20 min, and a sustained delivery vehicle, Pluronic F-127 gel, has been employed (Becker et al. 1999; Green et al. 2001). Pluronic gel is liquid at cold temperatures (4°C) and sets into a soft gel at physiological temperatures. The gel is also a mild surfactant aiding AsODN penetration into cells. The sustained delivery of the AsODN to tissue avoids the need to use high doses or modified backbones that have lower affinity, lower efficiency at entering cells and reduced specificity (Green et al. 2001). It is also possible to deliver antisense with osmotic minipumps or to use modifications such as short interfering RNA (siRNA). The use of siRNA, which often requires transfection reagents or viral delivery to achieve efficacy, has so far had limited application *in vivo* for connexin studies. Connexins typically have a half-life of around 1.5–2 h (Gaietta et al. 2002; Leithe et al. 2006) and antisense is blocking new protein translation. Antisense has no effect on existing protein levels and cell uncoupling will in part be dependent upon protein turnover rate. Cronin et al. (2006), for example, showed different knockdown and protein recovery rates in spinal cord dorsal, ventral, white, or gray matter, which may reflect dose delivered to the different tissues, cell metabolic rates, and protein turnover variance.

The advantages of an antisense approach include ease of use, controllable dose, low cost, transient protein knockdown, site directable (especially with gel delivery systems), no compensation by other members of the connexin family noted to date, and clinically relevant approach. The importance of transient knockdown has been highlighted in Section 14.2.1; in the CNS where normal astrocyte coupling is essential for neuronal function, this is especially relevant. Rapid penetration of the Cx43 AsODN is seen in all areas of spinal cord, for example, with knockdown apparent within 2 h and recovery of Cx43 levels over 48–72 h (Cronin et al. 2006). This was achieved with just 250 µL of 1 µM Cx43 AsODN applied to intact rat spinal cords with intact *dura mater*. The design of antisense is a science in itself and multiple methods have been developed. The approach used by Law et al. (2006) was to expose *in vitro* transcribed connexin mRNA to deoxyribozymes complementing the sense coding sequence. Deoxyribozymes have a small catalytic domain of about 15 nucleotides flanked by two mRNA recognition and binding domains, each 7–9 nucleotides long (Santoro and Joyce 1997). They have intrinsic catalytic activity and those that cleaved the connexin mRNA were used as the basis for design of AsODN, thus taking into account tertiary mRNA configurations rather than relying upon computed predictions.

The Cx43 AsODN 30mer described above has been used in several experimental disease models, including brain slice models (Yoon et al. 2010b), optic nerve ischemia (Danesh-Meyer et al. 2008), skin wound healing (Mori et al. 2006; Qiu et al. 2003; Wang et al. 2007), skin burns (Coutinho et al. 2005), and spinal cord injury (Cronin et al. 2008; O'Carroll et al. 2008; Zhang et al. 2010), demonstrating anti-inflammatory and wound-healing-promoting effects. Other studies have also shown connexin 32 and 26 or Cx43 antisense effects in brain slice cultures after trauma (Frantseva et al. 2002a) and ischemia (Frantseva et al. 2002b).

14.3.3 PEPTIDOMIMETICS

An alternative approach uses connexin mimetic peptides. In most cases, these are small peptide sequences designed to match the extracellular regions of the connexin molecule that is normally involved in the docking of two connexons to form a gap junction channel. Mimetic peptides may impair the interactions of the extracellular loops by binding to recognition sites on the connexon (Berthoud et al. 2000) and have been reported to inhibit both gap junction and hemichannel signaling in *in vitro* models (Boitano and Evans 2000; Braet et al. 2003; De Vuyst et al. 2007; Kwak and Jongsma 1999; O'Carroll et al. 2008; Martin et al. 2005). Connexin mimetic peptides are also said to regulate both hemichannels and gap junctions independently of each other, with short incubation times (or low concentrations) preventing hemichannel opening without affecting gap junction communication, while longer incubations (or higher concentrations) can interfere with gap junction communication (Leybaert et al. 2003; O'Carroll et al. 2008). Owing to sequence conservation in the connexin extracellular loops, mimetic peptides are less likely to be connexin isoform specific (in contrast to AsODNs, which can be designed to be totally isoform specific).

In recent years, a number of these peptides have been investigated for connexin activity. In one experimental model, a series of peptides covering most of the two extracellular loops of connexin32 were analyzed for their ability to inhibit gap junction-mediated contractions in aggregates of chick myocytes (Warner et al. 1995). More recently, an analysis of overlapping peptides that covered the entire extracellular loops of Cx43 was completed (O'Carroll et al. 2008). Other peptides in wide use for Cx43 studies are Gap26 (Chaytor et al. 1997) and Gap27 (Boitano and Evans

2000). For more detailed summaries, a number of reviews are available (Evans et al. 2006; Berthoud et al. 2000; Spray et al. 2002; Evans and Leybaert 2007; Evans and Boitano 2001; Dahl 2007).

Most reports indicate that hemichannels can release ATP, prostaglandins, glutamate, and glutathione and that this process can be interrupted by mimetic peptides (Evans et al. 2006). Much of the evidence comes from cell culture models and then by inference to observations in intact tissues. The primacy of the hemichannel target for mimetic peptides is demonstrated by the rapid action of peptides causing lower conductivity in isolated hemichannels (Romanov et al. 2007) and rapid effects on voltage conduction in other tissues within as short a time as 10 min (reported in Evans and Leybaert 2007; Evans et al. 2006). The uptake of dyes such as propidium iodide or ethidium bromide through low calcium-induced opening of hemichannels also provides a direct measure of rapid mimetic peptide hemichannel function. Cx43 mimetic peptide added simultaneously with propidium iodide in lowered extracellular calcium conditions blocks dye uptake, indicating immediate hemichannel closure (O'Carroll et al. 2008). More recently, Gap26 peptide added 12 min after the start of dye uptake in a hypoxia/high-glucose model caused an immediate attenuation of the dye uptake curve, again indicating rapid hemichannel closure by mimetic peptides (Orellana et al. 2010). The specificity of mimetic peptides for hemichannels has been demonstrated in astrocytes where both Gap26 and La^{3+} blocked hemichannel activity (Orellana et al. 2010). La^{3+} was also used as a control by O'Carroll et al. (2008) in their screening for mimetic peptide activity. La^{3+} is said to not close pannexin channels. Furthermore, Cx43 null astrocytes have no hemichannel activity (supporting the idea that there is no confounding conduction role for pannexins present in this model), and neither of the two pannexin mimetic peptides (panx1 and Elb) nor probenecid (all presumptive pannexin blockers) had any effect on channel activity in the hypoxia–high-glucose reoxygenation astrocyte model of Orellana et al. (2010).

Mimetic peptides may act in a number of ways to affect gap junctional communication (i.e., cell to cell as opposed to hemichannels effects). It was originally thought that mimetic peptides may directly interfere with the ability of connexons to dock (Berthoud et al. 2000). Evidence for this is countered by studies showing that peptides blocking gap junction communication do not appear to alter rates of connexin synthesis, alter hemichannel trafficking processes to the membrane or disrupt the formation and extent of existing gap junction plaques (Martin et al. 2005; Berman et al. 2002). Peptides may also diffuse into a gap junction plaque to compete with docking sites or to alter the conformation of the docked hemichannels to cause them to undock. Again, evidence that peptides block gap junction communication but do not change plaque characteristics argues against this mode of action (Berman et al. 2002). One possibility is that peptides bind to and lock connexons in a closed state prior to docking, with the closed connexons subsequently incorporated into the junction plaques. Thus cell–cell communication may reduce over time, dependent upon protein turnover rates and the removal of previously functional channels from plaques and their replacement by locked closed connexons (Evans and Leybaert 2007). It is more likely that peptides perfuse into plaque regions to close communication channels directly (Evans and Boitano 2001). Indirect support for this comes from the cell settlement (parachute) assay used by O'Carroll et al. (2008). Calcein AM dye-loaded cells were parachuted onto a recipient monolayer of cells without dye and allowed to settle for 30 min to enable gap junctions to start forming between the hemichannels already present in the complementary membranes (a process that can occur in minutes). As these started to form (indicated by initiation of dye transfer) Cx43 mimetic peptide was added and further dye spread was prevented over the next 60 min. This blocking of cell–cell communication by peptides over a relatively short time frame argues against a major altered connexon turnover effect but rather for a direct hemichannel closure mechanism. Peptide blockade effect on gap junctions is thought to be reversible.

14.4   PRECLINICAL MODELS

The role of gap junctions and the potential of therapeutic approaches to inflammation in the CNS has been demonstrated in a number of *ex vivo* and *in vivo* animal models. The results from some of these are outlined in the following two sections, starting with *ex vivo* models.

14.4.1   *Ex Vivo* Models

14.4.1.1   Brain Slice Cultures

The air–liquid interface organotypic brain slice culture method allows the culture of thin slices of immature brain tissue that, for prolonged periods, retain much of the morphological and physiological characteristics of the intact brain (Stoppini et al. 1991; Bahr 1995). In mature brain tissue that lacks the pluripotency of immature tissue, the damage incurred by "slicing" of the brain is not easily repaired. Tissue response to CNS injury is clearly dependent on the age of the animal (Towfighi et al. 1997; Putnam et al. 2005) and while immature tissue is more resilient to organotypic slice culture preparation, adult brain tissue better reflects the pathology of the human condition. In the first week of a culture period, cells are regularly lost and slices thin down from 300–500 μm to 100–150 μm. Slices become ensheathed within a thick astrocytic scar, leaving a thin layer of live neurons in the middle of the slice (Stoppini et al. 1991). Accompanying this reduction in neuronal cell number is a visible swelling and edema within the slice in the first days of culture. Based on the premise that toxic signals released by damaged cells at the cut surface of a slice are propagated via intercellular gap junctions and hemichannels (Orellana et al. 2010; Decrock et al. 2009), blockade of intercellular gap junction communication and thereby reduced gap junction-mediated bystander cell death should lead to improved viability of brain slices obtained from mature rats. Indeed, a recent study has shown that application of 2 μM concentration of unmodified Cx43-specific AsODN in Pluronic F-127 gel to the upper surface of a freshly cut whole brain slices leads to a significant improvement in the morphology within a whole brain slice taken from 14–40-day-old rats (Yoon et al. 2010b).

Cx43 protein was transiently down regulated in brain slices processed for Cx43 immunohistochemistry 7 h after AsODN treatment with the average number of plaques being just 17.2% that of the no-treatment control slices (Yoon et al. 2010b).

Most of the Cx43 label present in control and AsODN-treated slices was colocalized with glial fibrillary acidic protein (GFAP), indicating that these are predominantly astrocytic gap junctions. The 7-h time point was chosen because an almost complete turnover of preexisting Cx43 gap junctions was expected to be achieved by then (Laird 2006; Gaietta et al. 2002). At the cellular level there was a marked difference between the morphology of the untreated control and Cx43 AsODN-treated slices after 14 days in culture with the latter appearing remarkably similar to freshly cut slices. The majority of the hippocampal neurons survived and were viable as indicated by the large number of cells labeled with 5-chloromethylfluorescein diacetate (CMFDA) and Neuronal-N, with preservation of the intrinsic neuronal organization within the hippocampus. Thus, short-term knockdown of Cx43 during the period when glial cells and Cx43 are known to be upregulated as a result of "slicing damage" (Cronin et al. 2008; Danesh-Meyer et al. 2008), leads to reduced astrogliosis, a marked reduction in edema and inflammation, and an optimal retention of cellular morphology (Yoon et al. 2010b). Cx43 AsODN treatment also allows the culture of 150-μm-thick whole brain slices (since die back from the cut edges is reduced, a thicker starting slice is not required) facilitating imaging and electrophysiological recording, and providing a simple but effective procedure to maintain adult CNS tissue *in vitro.*

Others have also shown that slice cultures afford a good model to investigate the role of gap junctions in the spread of injury following traumatic brain injury and ischemia (Frantseva et al. 2002a, 2002b). Frantseva and colleagues have shown that in organotypic slices from Cx43 knockout mice, as well as following knockout by incubation with AsODN, a significant reduction in cell death is achieved. Gap junction blockers carbenoxolone and octanol alleviated the trauma-induced impairment of synaptic function as measured by electrophysiological field potential recordings. Similarly, the gap junction blocker carbenoxolone significantly decreased the spread of cell death, as measured by propidium iodide staining in an *in vitro* model of hypoxia–ischemia where organotypic hippocampal slices were submerged in glucose-free deoxygenated medium. The knockdown of two neuronal connexins, connexins 32 and 26, resulted in significant neuroprotection 48 h after the hypoxic-hypoglycemic episode indicating that gap junctional communication contributes to the propagation of hypoxic injury and that specific gap junctions could afford a novel target to reduce brain damage.

14.4.1.2   Brain Slice Models of Epilepsy

It is becoming increasingly evident that gap junction and hemichannel-mediated purinergic signaling is involved in an amplification cascade of cytotoxicity in pathological conditions (Parpura et al. 1994; Braet et al. 2004; Bruzzone et al. 2001; Charles et al. 1992; Cornell-Bell and Finkbeiner 1991). Increased cytosolic Ca^{2+} and membrane depolarization during epileptiform events may induce hemichannel opening, allowing release of ATP into the extracellular space (Albowitz et al. 1997; van der Linden et al. 1993). ATP then activates G-protein-coupled P_2Y receptors on neighboring cells, eventually leading to an elevation of cytosolic Ca^{2+} and ATP release. This type of Ca^{2+} wave can be stopped by gap junction blockers (Enkvist and McCarthy 1992).

Organotypic slice culture models have been used extensively to investigate mechanisms of epileptiform activity and electrophysiological studies have demonstrated that pharmacological blockade of gap junctions can reduce or block seizure-like discharges in a range of *in vitro* models of epilepsy (Li et al. 2001; de Curtis et al. 1998; Margineanu and Klitgaard 2001), while agents that promote opening of gap junctions enhance neuronal bursting (Köhling et al. 2001). Gap junction blockers have been shown to prevent neuronal network synchronization in neocortical slices obtained from patients with mesial temporal lobe epilepsy (Gigout et al. 2006) as well as in a rodent *in vivo* model where epilepsy is induced by 4-aminopyridine (Szente et al. 2002). In Cx43 and Cx30 knockout mice, the threshold for generation of epileptiform discharge is reduced (Wallraff et al. 2006). These studies have focused on the role of Cx43 gap junctions in the generation of seizure-like discharges and neuronal synchronization, but gap junction communication may also be involved in epileptic lesion spread. There is formation of an extensive astrocytic syncytium coupled with marked colocalization of Cx43 in hippocampal tissue surgically removed as a treatment for intractable mesial temporal lobe epilepsy (Fonseca et al. 2002) and astrocytic Cx43 gap junction coupling may contribute to both lesion spread and the severity of cell death following epileptiform activity.

Yoon et al. (2010a) used Cx43 mimetic peptides to regulate hemichannel and gap junction-mediated intercellular communication in a hippocampal slice culture model of epileptiform injury. Bicuculline methyl chloride (BMC) was used to establish epileptiform activity; the causal relationship between epileptiform events and BMC lesion has been established elsewhere (Muller et al. 1993). Epileptiform injury in hippocampal slice cultures (that had been maintained for 12–14 days prior to commencement of treatment) was induced by 48 h of exposure to 100 μM BMC. During the 24 h recovery period following BMC, lesion spread was observed in the CA1 region of the hippocampus. A Cx43 mimetic peptide (peptide 5, O'Carroll et al. 2008), applied during either the BMC treatment or the recovery period, produced concentration- and exposure time-dependent neuroprotection as measured by propidium iodide labeling of dead cells at the end of the recovery period. It is essential, however, to differentiate between the BMC insult period and the recovery period where distinct mechanisms appear to be operating (Yoon et al. 2010a).

During the BMC insult period, peptide concentrations between 5 and 50 μM (sufficient to block hemichannels) had a protective effect while a substantial gap junction blockade with 500 μM peptide exacerbated the lesion. Although peptide was applied for 48 h, which was expected to result in continuous blockade of gap junction and hemichannels, the dose-dependent effects seen during this period argue for some distinction between hemichannel and gap junction uncoupling effects. Gap junction communication appears to be essential for tissue survival at the time of insult when the slices are undergoing excessive neuronal firing and epileptiform events are occurring. Blockade of astrocytic gap junctions may, therefore, disturb neuronal ionic homeostasis and reduce the coordinated response to injury (Blanc and Bruce-Keller 1998; Nakase et al. 2003; Ozog et al. 2002; Siushansian et al. 2001). Functional astrocytic communication will effectively buffer extracellular ions and neurotoxic substances released as a result of epileptic discharge. In addition, termination of an epileptiform activity is thought to occur as the presynaptic vesicles become slowly depleted and voltage-dependent electrogenic ion pumps are activated in an attempt to restore normal ionic balance (McCormick and Contreras 2001).

There is also evidence that termination of epileptiform events requires spatial K^+ buffering by the glial syncytium (Bragin et al. 1997).

In contrast, all doses (both high and low) applied during the postinsult recovery period protected the CA1 region from further damage (Yoon et al. 2010b). Once the BMC insult was over and the medium changed to BMC-free and serum-containing medium (recovery period), toxic signals from the damaged neurons may spread via Cx43 gap junctions amplifying the damage. Although hemichannels may also be implicated (the lower doses were still effective), the trend in the postrecovery period was the reverse of that seen in the insult period, with increased neuroprotection at higher doses. Propidium iodide label was used as an indicator of cell loss but it may also indicate open hemichannels. MAP2 immunolabeling of neurons strengthened the claim that the former was the case (Yoon et al. 2010a).

The application of Cx43 mimetic peptides *in vitro* has shown that they are able to block intercellular Ca^{2+} propagation and Ca^{2+}-induced ATP release/dye uptake, and suppress electrical coupling (Chaytor et al. 1999; Dora et al. 1999; Evans and Boitano 2001; Leybaert et al. 2003). Hemichannels may be opening (or be upregulated) during the convulsant period in response to depolarization and low extracellular Ca^{2+} (Evans et al. 2006) and hemichannel-mediated purinergic signaling may contribute to the lesion spread. Although the effect of gap junction blockade in neuronal synchronization has been examined in both acute slices and *in vivo* animal experiments, and astrocytes coupled via gap junctions may recruit larger masses of neurons into the epileptic foci (Bennett and Zukin 2004; Köhling et al. 2001; Li et al. 2001; Margineanu and Klitgaard 2001; Ross et al. 2000; Carlen et al. 2000; de Curtis et al. 1998; Gigout et al. 2006; Jahromi et al. 2002; Perez-Velazquez et al. 1994; Samoilova et al. 2008; Szente et al. 2002; Wallraff et al. 2006), the data from Yoon et al. (2010a) would suggest that complete uncoupling of gap junctions during the insult period itself may be detrimental.

14.4.1.3 Spinal Cord (and Peripheral Nerve Grafting)

The spinal cord is a part of the CNS and as with brain slices, spinal cord segments placed into organotypic culture and treated with gap junction channel modulators show reduced edema and inflammation, and improved neuronal survival. When placed into culture, nerve segments swell despite the absence of blood flow, with material squeezed from the ends of the cut dura. Both Cx43-specific AsODN (Zhang et al. 2010) and connexin mimetic peptides (O'Carroll et al. 2008) significantly reduced the swelling in a dose-dependent manner. Furthermore, treatment with AsODN and peptides within the first hours after removal from the animal ("injury") and placing into organotypic culture reduced Cx43 levels, astrocytosis (GFAP levels), and loss of Neuronal-N and SMI-32 positive neurons in a concentration- and time-dependent manner (O'Carroll et al. 2008).

The ability to culture spinal cord segments for up to 5 days afforded the opportunity for studies investigating the effects of Cx43 AsODN in promoting or enabling axon regeneration following peripheral nerve grafting (Zhang et al. 2010). In Cx43 AsODN-treated segments, good graft to spinal cord integration was observed with large aggregates of SMI32-positive axon fibers at the interface, some of which crossed the border and entered the peripheral nerve graft. In Pluronic gel, vehicle-treated or medium-only control segments integration was poor, axon aggregates at the border were rarely seen, and there was a decrease in the number of axon extensions into the graft. The amount of axon outgrowth into the grafted peripheral nerve was quantified by measuring the total length of outgrowth and the number of axons crossing the graft interface. On average, the total length of axon outgrowth was over 10 times more with Cx43 AsODN treatment than in controls (Brown-Forsythe analysis, $p = 0.024$) and the average number of axons crossing the graft interface after Cx43 AsODN treatment was more than fivefold greater than controls ($p = 0.01$). Thus, the environment generated by Cx43 AsODN treatment promoted axon regeneration from spinal cords into peripheral nerve grafts. Edema was reduced, but the "hostile" astrocyte environment had also been inhibited with astrocytes maintained in a growth-supporting phenotype allowing axon regeneration to occur in the presence of a stimulus such as a peripheral nerve graft (Zhang et al. 2010).

14.4.1.4 Optic Nerve Ischemia

The effect of oxygen–glucose deprivation (OGD) has been investigated on isolated rat optic nerve segments cultured *ex vivo* and the effect of Cx43 AsODN treatments evaluated (Danesh-Meyer et al. 2008). In this study, optic nerves were removed from 21–25-day-old Wistar rats, and the optimal concentration of Cx43-specific AsODN determined using two parameters: percentage swelling and cell death as identified using propidium iodide nuclear staining. To create OGD conditions, the isolated optic nerve segments were maintained for 2 h in a closed tube containing neurobasal medium without glucose and glutamine, and which had been previously bubbled with a 95% nitrogen:5% carbon dioxide gas mixture. The segments were then removed and placed into organotypic culture similar to that used for the brain slice cultures described above. Pluronic F-127 gel with or without Cx43 AsODN was layered over the segments. Upregulation of Cx43 was seen within 2 h following exposure of the optic nerve to OGD and peaked at day 3. Cx43 AsODN treatment reversed this upregulation and significantly reduced swelling, which occurred from the cut ends of the meninges surrounding nerve segments by over 50% ($p < 0.05$). More dead cells were found in the center of control optic nerve segments than treated nerve segments ($p < 0.05$) and controls showed evidence of capillary breakdown. Increased numbers of astrocytes and activated (macrophage phenotype) microglia were observed in controls compared to Cx43 AsODN-treated nerves ($p < 0.01$).

This study again indicates that application of Cx43-specific AsODN may modulate several facets of the CNS inflammatory process: it reduced tissue edema, retained greater vascular integrity (despite the absence of blood flow), and reduced differentiation of inflammatory cells. In addition to an overall dampening of the inflammatory response, there was less lesion spread with Cx43 AsODN treatment restricting most damage to the cut ends of the optic nerve section, with reduced cell death centrally resulting after the OGD. The swelling evident at 6 h is comparable to the onset of swelling in the brain following ischemia (Roelfsema et al. 2004), but in optic nerve, which is essentially white matter track with no neural cell bodies present, swelling must be predominantly within astrocytes or oligodendrocytes. CNS axon injury has been shown previously to rapidly activate microglial and astroglial cells close to axotomized neurons, with Cx43 upregulation seen within hours (Aldskogius and

Kozlova 1998). The Danesh-Meyer et al. (2008) study demonstrated similar changes with activation of microglial and astroglial cells in control specimens extending to the middle of the optic nerve sections, but with Cx43 AsODN limiting both astrocyte activation and the differentiation of microglial cells to the cut edges.

14.4.2 BRAIN LESIONS AND INFLAMMATION

Under normal conditions, coupled astrocytes regulate the neural environment, maintain potassium homeostasis, and provide a link between neurons and blood vessels (Blanc and Bruce-Keller 1998; Lee et al. 1994). Following physical brain damage, Cx43 expression is upregulated in the region of damaged tissue (Hossain et al. 1994), paralleled by astrogliosis (Hozumi et al. 1990; Vijayan et al. 1990; Cancilla et al. 1992; Hossain et al. 1994) with, subsequently, undamaged neighboring cells destroyed in a spreading penumbra. Nonspecific gap junction channel blockers (octanol and halothane) can reduce final infarct volumes *in vivo* by 30–80% in the brain after focal ischemic stroke (Rawanduzy et al. 1997). In another *in vivo* study, unmodified AsODN (at 2 µM final concentration) delivered in Pluronic F-127 gel was used to specifically knockdown Cx43 expression following unilateral stereotaxic mechanical lesioning of rat cerebral cortex. Quantitative analysis of Cx43 immunolabeling showed that protein levels were strongly reduced along the edge of Cx43 AsODN-treated lesions, particularly at the basal and medial edges of the lesion compared with controls, and Cx43 AsODN-treated lesions remained discrete, with neuronal survival evident right to the edge of the needle puncture. In control lesions (gel-only or gel-plus-sense ODNs), neuronal loss (shown by Neuronal-N labeling) and gliosis (GFAP immunohistochemistry) occurred within 24 h and continued for up to 48 h. In Cx43 AsODN-treated animals, secondary lesion spread was reduced compared to control groups, especially at deeper cortical levels (Figure 14.2). The pooled 24 and 48 h mean lesion size (central section of the lesion volume in a coronal plane) for Cx43 AsODN-treated lesions was 1.45 mm^2 (±0.55), significantly smaller than control lesions, which were 2.1 mm^2 (±0.6) (one-tailed Mann-Whitney U test P value 0.030). Maximum lesion size was reached within the first 24 h in treated lesions whereas control lesions continued to spread for up to 48 h. Other connexin levels (connexins 26 and 32) were not altered by the Cx43 AsODN treatments. Specifically knocking down Cx43 expression within the first few hours after injury, as with octanol- or halothane-induced gap junction channel closure, is effective in attenuating lesion spread following damage to the CNS. As noted earlier, the knockdown here was transient, and quite distinct to the increased lesion spread observed with long-term channel reduction in knockout mice (see, e.g., Nakase et al. 2004; Siushansian et al. 2001).

14.4.3 SPINAL CORD INJURY

Within 1–2 h of spinal cord injury, there is an acute phase that involves hemorrhage, inflammation, and edema. Cytotoxic edema (intracellular swelling) occurs in astrocytes and is likely due to the increase of a number of molecules in the extracellular space shortly after injury, including glutamate, lactate, K$^+$, nitric oxide, arachidonate,

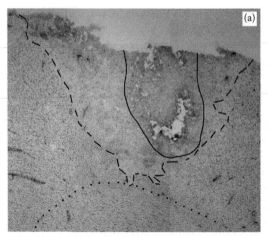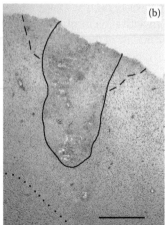

FIGURE 14.2 Sections of a control rat brain needle lesion treated with Pluronic F-127 gel alone (a) and a lesion treated with Pluronic gel containing antisense oligodeoxynucleotides specific to Cx43 (b). The sections were Nissl stained and antibody labeled for the neuronal marker Neuronal-N. The images are of the midpoint of lesions serially sectioned in a coronal plane. The lesioning needle tract in each is marked with a solid line and the extent of lesion spread, demarcated by reduced Neuronal-N antibody labeling, with a dashed line. The upper edge of the corpus callosum is marked with a dotted line. The control lesion is larger with irregular spreading edges, and has spread downwards toward the corpus callosum within 24 h of lesioning. The antisense oligodeoxynucleotide-treated lesion is compact with only slight spreading near the top of the lesion. Viable neurons are seen right to the edge of the needle tract and there is virtually no downward spread, even after 48 h as in this example. Scale bar = 1 mm.

reactive oxygen species, and ammonia (Norenberg et al. 2004). As with the brain, the expression of Cx43 changes in response to spinal cord injury. Levels of Cx43 mRNA and protein increase within hours and reach three times the normal level at 4 weeks postinjury, while no change in expression of connexin32 or connexin36 was reported (Lee et al. 2005; Theriault et al. 1997). The increased Cx43 expression is seen in activated microglia (Eugenin et al. 2001), but it is also seen with astrogliosis and in blood vessel endothelium (Cronin et al. 2008). As for other CNS injury, it has been suggested that gap junction-mediated transfer of Ca^{2+}, nitric oxide, and other death signals destroys local cell neighbors and that hemichannels release ATP, glutamate, and other signals to contribute to lesion spread (Decrock et al. 2009). Theriault et al. (1997), for example, reported that Cx43 immunoreactivity is highest in areas associated with neuronal loss while Cx43 labeling is normal in areas where neurons are preserved.

Cx43-specific AsODN has been used to regulate Cx43 function in *in vivo* spinal cord injury hemisection and compression wound models (Cronin et al. 2008). In this study, Cx43 AsODN was first applied after hemisection of the rat spinal cord. Treated wounds showed less swelling and hemorrhagic inflammation at 24 h than corresponding controls, and markedly reduced tissue disruption was seen in tissue transections. Increased Cx43 expression was seen in astrocytes and small vessels around spinal cord compression injury sites within 6 h after injury and there was a marked upregulation in GFAP expression. The increase in both Cx43 expression and astrocytosis was attenuated by the Cx43 AsODN treatment. One very clear functional effect of this reduced inflammation was demonstrated by a reduction in blood vessel leakiness to FITC-conjugated bovine serum albumin. In control wounds (sense oligodeoxynucleotides or Pluronic gel only), extensive extravasation occurred up to 4 mm rostral and caudal to the injury site; AsODN treatment reduced both the range and extent of vessel leak. Similarly, neutrophil invasion (measured by myeloperoxidase staining) was reduced fivefold compared to control and the same reduction was observed for activated (macrophage phenotype compared within resting stellate form) microglia, the resident phagocytic cells (demonstrated by OX42 staining).

The attenuation of the post-spinal cord injury inflammatory state leads to a better functional outcome, which was demonstrated using the Basso, Beattie, Bresnahan (BBB) locomotor rating scale (Basso, Beattie, and Bresnahan 1995). This scale ranges from 0 (total lower body paralysis) to 21 (normal behavior). Twenty-four hours after a compression wound, control animals had a median BBB score of 5, but Cx43-specific AsODN-treated animals had a significantly higher locomotor rating of 8.75 ($p < 0.05$) (Cronin et al. 2008). This indicated that lesion spread over the first 24 h had been significantly reduced. Over the following 14 days, control animals showed recovery to a median BBB score of 15 after which no further recovery occurred through until day 28 when the experiments were terminated. In contrast, AsODN-treated animal continued to improve over the full 28-day period, maintaining 2–4 BBB points over the controls at all time points, with a widening gap from days 14–21. The difference for the five test points combined showed statistical significance of $p < 0.004$.

More recently, the mimetic peptides that showed reduced swelling, reduced inflammation, and improved neuronal survival in an *ex vivo* model (described in Section 14.3.3) have shown similar benefits in a somewhat more severe impact injury model. It appears, therefore, that gap junction channel regulation after spinal cord injury could be a valuable target for therapeutic intervention. This applies equally to any attempt at spinal cord repair where intervention itself has the potential to cause an inflammatory response, lesion spread, and cyst or scar formation.

14.4.4 Cerebral Malaria

Sepsis is the leading cause of mortality in critically ill patients, with septic shock and sequential organ failure correlating with poor outcome (Matsuda and Hattori 2007). As described in Chapter 1 and 7, the interaction of leukocytes with endothelial cells is an early event in acute and chronic inflammation, and wound defense and repair (for review, see McIntyre et al. 2003). The presence of leukocytes correlates with gap junction Cx43 expression in both the endothelium and the inflammatory cells and, in general, Cx43 changes are associated with macrophage infiltration and endothelial cell activation (see Oviedo-Orta and Evans 2002). However, there is growing evidence for the impact of vascular dysfunction with sepsis, with an emphasis on endothelial failure (discussed in Section 14.2.3). The onset of endothelial dysfunction, plasma leak, and hemorrhaging is also symptomatic of cerebral malaria (Combes et al. 2005; van der Heyde et al. 2006; Martini et al. 2007). Of patients infected with the genus *Plasmodium* parasites, the cause of malaria, over 80% will exhibit profound thrombocytopenia. Patients presenting with impaired consciousness and obtundation (cerebral malaria) often exhibit retinal hemorrhaging, which is an indicator of poor prognosis. Autopsy reveals petechial hemorrhaging in the brain. In experimental cerebral malaria models, vascular leak and petechial hemorrhaging reflect the impairments observed in humans (Combes et al. 2005). In experimental models too, malarial thrombocytopenia can reach life-threatening levels and is believed to be due to antibodies targeting platelets for destruction by the reticuloendothelial system. However, antibodies account for at most 15% of platelet destruction as *Plasmodium berghei*-infected B cell-deficient mice exhibit profound thrombocytopenia (83%) as do C57BL/6 controls (98%) (Gramaglia et al. 2005). No significant increase in antibodies bound to intact platelets is observed during infection. The same authors have shown that impairment of functional capillary density, rather than tissue hypoxia, is the major lethal event in a hamster severe malaria model (Martini et al. 2007). Thus, a unified hypothesis has been proposed for the genesis of cerebral malaria with sequestration, inflammation, and hemostasis leading to microcirculatory dysfunction (van der Heyde et al. 2006).

Cx43 is upregulated in the vascular bed of skin wounds within 6 h of injury (Qiu et al. 2003; Coutinho et al. 2003), and results in vascular leakiness as demonstrated following retinal ischemia (Figure 14.1) or after spinal cord injury (Cronin et al. 2008). A common pathway to vascular disruption resulting from any infection, therefore, may result from (1) endothelial cell gap junction hemichannel (connexon) opening resulting in cytoplasmic edema, cell swelling, and rupture and (2) gap junction-mediated bystander effect signaling (Lin et al. 1998) from damaged endothelial cells or petechial foci to healthy neighboring cells, leading to further endothelial cell loss and damage spread.

To test this, C57BL/6 mice were infected with *Plasmodium berghei*, by injecting 1×10^6 parasitized erythrocytes when mice were between 6 and 10 weeks of age. The extent of parasitemia was confirmed by counting the ratio of infected erythrocytes in Giemsa-stained thin blood smears. At day 5–6, mice generally succumb to cerebral malaria and are scored for clinical signs of severe malaria based upon their gripping ability (score 0–5 where 5 is normal) and righting ability combined with overall appearance (breathing, ruffled fur) (score of 0–5 where 5 is normal). The two scores are summed to give a total clinical score. The objective was to reduce or delay vascular leak and organ failure resulting from infection-induced vessel dysfunction using Cx43 mimetic peptides delivered by intraperitoneal injection to achieve systemic delivery to sites of vascular leak. The treatment arm was injected with 200 μL of saline containing Cx43-specific mimetic peptide (peptide 5, O'Carroll et al. 2008) such that the final circulating concentration was 50 μM (assuming a 2 mL blood volume and total take up of the peptide from the peritoneum). A control group was injected with 200 μL of saline. Peptide (or control saline) was injected at the beginning of day 5 and again 12–16 h later. Four control and four treated animals had a third injection at 36 h. Animals were assessed 24, 36, or 48 h later and their weight and clinical scores recorded. There were nine control and nine treated animals in total.

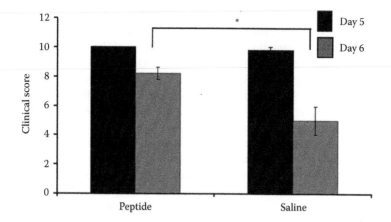

FIGURE 14.3 Graph showing clinical scores for mouse cerebral malaria based upon combined scores of gripping ability, righting ability, and overall appearance. The maximum (normal) score is 10. Animals had an intraperitoneal injection of saline or connexin mimetic peptide at the beginning of day 5 and again 12–16 h later. By the end of day 5 (black bars) control (saline injected) animals are showing a decline, which is severe by the end of day 6 (gray bars). Connexin peptide-treated animals show an attenuated decline.

In this trial experiment, clinical scores indicated a significant benefit using the mimetic peptides. The change in clinical scores was significantly reduced in peptide-treated animals (Figure 14.3; $p = 0.03$) indicating that treated animals were not succumbing to the clinical symptoms as rapidly as controls, and indicating that the window for antimalarial treatment can potentially be extended by at least 12 h, and possibly 24–48 h. The need for an adjunct therapy, which could delay patient decline and organ failure while antimalarial therapy kills circulating blood-stage *Plasmodium*, is well recognized; about 20% of children succumb to cerebral malaria despite rapid administration of chemotherapy (discussed in van der Heyde et al. 2006). While the connexin peptides were targeted at vascular dysfunction as a result of the infection, the mechanism remains to be verified. However, vascular and endothelial dysfunction, fluid leak and edema resulting from malarial infection is common to infection and infectious diseases in general, and the use of connexin-specific regulators to ameliorate the symptoms induced, whether viral, bacterial, or parasitic, may offer new therapeutic opportunities.

14.4.5 STROKE AND ISCHEMIA

Stroke is the most frequent cause of persistent neurological disability in modern Western society. Brain inflammation is an important pathophysiological mechanism involved in stroke and impacts profoundly on the severity and extent of cell loss, as well as injury progression. In a rodent model of stroke, the gap junction inhibitor octanol was reported to reduce infarct size (Rawanduzy et al. 1997) while the treatment of rat brain slices with antisense oligonucleotides, designed against a number of different connexins, was said to reduce neuronal cell death 48 h after an episode of hypoxia and glucose deprivation (Frantseva et al. 2002b). Cx43 expression increases in reactive, proliferating astrocytes following ischemic injury, and can result in glial scar formation. The retina forms part of the CNS and an upregulation of Cx43 transcripts has been shown using a DNA microarray and real-time PCR analysis following mechanical injury to the zebra fish retina (Cameron et al. 2005). In an *in vitro* model of retinal ischemia induced by the chemical agent cobalt chloride, a novel gap junction inhibitor based on primaquine (PQ1) prevented activation of caspase-3 in R28 neuro-retinal cells in culture (Das et al. 2008). The authors postulated that dying retinal cells pass gap junction-permeant apoptotic signals to adjacent cells, resulting in a spread of damage (Das et al. 2008). Strong experimental evidence to support the presence of gap junction-mediated bystander killing in the retina comes from an elegant study by Cusato et al. (2003). Targeted cell death was induced via intracellular cytochrome *c* injection of neurons in the inner retina of retinal whole mounts (Cusato et al. 2003). The injected and surrounding cells were then scrape loaded and examined for apoptotic morphology or caspase-3 cleavage (Cusato et al. 2003). Cytochrome *c* injection induced bystander killing in neighboring cells, which was significantly reduced by the gap junction inhibitors octanol and carbenoxolone (Cusato et al. 2003). In a recent *in vivo* retinal ischemia—reperfusion study, Cx43 upregulation was shown to cause endothelial cell loss and vascular leak, resulting in localised inflammation (astrocytosis) and downstream retinal ganglion cell death (Danesh-Meyer et al. 2012). A single intraperitoneal injection of hemichannel blocking connexin mimetic peptides at the time of reperfusion reduced vascular leak to 14% that of controls, and downstream retinal ganglion cell loss assessed 7 and 21 days later was only one third of that in controls.

14.5 HUMAN CONDITIONS

14.5.1 Chronic Inflammatory Disorders: Neurodegenerative Diseases

In addition to connexin modulation for CNS trauma, there is strong evidence that inflammation exacerbates, maintains, and accelerates a number of chronic diseases. Regardless of genetic predisposition or external stimulus, these neurodegenerative, cardiovascular, and metabolic disorders, once triggered, share common features, including sustained inflammatory cell activation and vascular disruption leading to widespread circulation of inflammatory cytokines, making them effectively persistent "wounds." There is also increasing evidence that the presence of one disease can cause activation of another apparently unrelated disease, leading to multiple disorders via activation of an immune response that "fast forwards" disease progression (Green and Nicholson 2008). As described above, in a number of CNS injury models, including the brain, optic nerve, and spinal cord, astrocytosis is observed, along with blood vessel disruption, neutrophil invasion, and activation of the microglia macrophage phenotype. Treatment with Cx43-specific AsODN or peptidomimetics reduces these inflammatory responses.

In excised human hippocampus from patients with medial temporal lobe epilepsy, and Alzheimer's and Parkinson's disease brain tissue, astrocytosis and Cx43 upregulation correlate with the disease progression. In mesial temporal lobe epilepsy, characterized by reactive astrocytic proliferation in the hippocampus, a highly significant increase in astrocytic Cx43 protein levels is seen (Fonseca et al. 2002). In separate studies, increased expression of Cx43 has been reported at the site of amyloid plaques in Alzheimer's disease associated with reactive astrocytes (Nagy et al. 1996), and increased Cx43 expression has been described in conjunction with astrocytosis in the caudate nucleus of Huntington diseased human brain (Vis et al. 1998). Parkinson's disease is a common neurodegenerative disease caused by the loss of dopaminergic neurons in the substantia nigra-striatum. In a 1-methyl-4-phenyl-1,2,3,6-tetrahydropyridine (MPTP) model of Parkinson's disease, an increase in Cx43 expression was observed in the striatum (Rufer et al. 1996).

Neurodegenerative diseases have two features in common: the presence of misfolded proteins and neuroinflammatory markers (Minghetti 2005). It is possible that misfolded proteins are unable to be degraded and removed and are being recognized as foreign, inducing an inflammatory response. In the brains of Alzheimer's patients, activated microglia have been found in areas where senile plaques, containing the misfolded amyloid protein, are present (McGeer and McGeer 2001). This study also showed the presence of proteins involved in the complement system, cytokines and free radicals, all of which are present in high levels in inflammatory areas (McGeer and McGeer 2001). Multiple studies have also been carried out to determine the cause of ongoing cell loss in Parkinson's disease, and it is now clear that inflammation plays a major part. Postmortem studies show significant loss of pigmented dopaminergic cells in the substantia nigra pars compacta with an associated gliosis indicative of neuroinflammation (Langston et al. 1999). Another study has demonstrated that in the substantia nigra pars compacta of patients with Parkinson's disease, major histocompatibility complex (MHC) proteins, particularly class II, are upregulated and that it is the microglia that are positive for these proteins (McGeer et al. 1988). Another sign of inflammation is the high level of cytokines in the cerebrospinal fluid of Parkinson's patients, including the proinflammatory markers TNF-α, IL-1β, and IL-6 (Hunot and Hirsch 2003). Increased cytokine levels have also been found in the substantia nigra pars compacta and striatum but not in cortical regions, suggesting that the increased expression is occurring only in the affected nigrostriatal pathway (Hunot and Hirsch 2003).

The presence of these cytokines, in addition to the reported T-cell infiltration, supports the idea of a compromised blood–brain barrier in neurodegenerative diseases and chronic inflammation. Cytokines, including TNF-α, IL-1β, and IFN-γ, are known to induce expression of Cx43. TNF-α has been shown to induce increased functional gap junction coupling in various cells, including monocytes, macrophages, and endothelial cells (Eugenin et al. 2003). Unpublished microarray data of our own shows that the expression of a number of connexins is altered in Alzhemier's, Parkinson's, and Huntington's disease human brain tissue from brain areas secondarily affected in the disease, with Cx43 (predominantly on astrocytes) and connexin37 (endothelial cells) being increased consistently across all three diseases. It is noteworthy that this change in connexin expression is accompanied by a highly significant increase in the expression of the inflammatory markers CD11A, CD11B, IL-1β, and IL-6 in all areas.

14.5.2 Chronic Inflammatory Disorders: Diseases of the Eye

Glaucoma occurs at all ages but increases with age, affecting 10% of adults over 70 years old. Alterations in Cx43 levels have been described in several *in vitro* (Danesh-Meyer et al. 2008) and *in vivo* (Johnson et al. 2000; Johnson et al. 2007) models of optic nerve disease. Using real-time quantitative PCR, higher levels of Cx43 gene expression have also been demonstrated in astrocytes from glaucomatous human optic nerve heads compared to age-matched controls (Hernandez et al. 2008). Astrocytes are the major glial cell type in the nonmyelinated optic nerve head. Previous studies have demonstrated glial activation in experimental glaucoma (Inman and Horner 2007; Wang et al. 2000; Kanamori et al. 2005; Yucel et al. 1999; Hayreh et al. 1999; Lam et al. 2003; Kielczewski et al. 2005) and in human primary open angle glaucoma (Varela and Hernandez 1997; Wang et al. 2002; Neufeld et al. 1997; Neufeld 1999). In the prelaminar region of the optic nerve head, increased GFAP immunoreactivity is observed in glaucomatous eyes compared to normal donors using immunohistochemical techniques (Wang et al. 2002). Intense GFAP labeling has also been described in the lamina cribrosa of human glaucomatous eyes (Varela and Hernandez 1997). Importantly, these astrocytes remained active, even in the end stages of chronic disease (Varela and Hernandez 1997). Another potentially important consequence of Cx43 upregulation in glaucomatous eyes may be associated with its role in the maintenance of perivascular barriers, which are weakened in glaucomatous eyes. Patients with glaucoma exhibit vascular leakage on fluorescein angiography (Arend et al. 2004) and peripapillary chorioretinal atrophy is associated with areas of blood–retinal barrier breakdown (Ahn, Kang, and Park 2004). Furthermore, optic disk hemorrhages may represent areas of perivascular barrier dysfunction at the edge of the optic nerve head in glaucoma (Bengtsson et al. 1981). Potentially, breakdown of the blood–retinal barrier may alter the extracellular milieu and permit the extravasation of circulating peptides, such as the potent vasoconstrictor endothelin-1 (Grieshaber and Flammer 2007). This may explain why patients with increased plasma levels of endothelin-1 are at greater risk of progression (Emre et al. 2005).

Age-related macular degeneration (AMD) is the leading cause of irreversible visual impairment in people over the age of 60 years in developed nations. AMD, predominantly irreversible and in many cases untreatable, affects the macula, an area of the retina that subserves fine and detailed vision (Anderson et al. 2010). The conventional understanding of AMD is that it affects four functionally interrelated layers in the eye: the photoreceptors, the retinal pigment epithelium (RPE), Bruch's membrane, and the choriocapillaris (Shelley et al. 2009). Research has primarily focused on the degeneration of the RPE, which leads to irreversible degeneration of the photoreceptors (Nowak 2006). Along with this degeneration, Drusen deposits can accumulate within the RPE and Bruch's membrane. These deposits can be associated with abnormal blood vessel ingrowth, which in turn can lead to leakage, bleeding, and ultimately scar formation at the macula (de Jong 2006; Finger et al. 1999). Although these changes occurring at the level of the RPE have been considered to be the initiating event in the development of AMD (Campochiaro et al. 1999), other pathological sequelae have been recognized. Many of the proteins identified in Drusen deposits can be characterized into either the process of inflammation or its aftermath (Anderson et al. 2002; Hageman et al. 2001). An increase in the systemic inflammatory marker CRP has been observed in patients with AMD (Seddon et al. 2004), giant multinucleated cells and leukocytes have been identified in the RPE of AMD patients (Dastgheib and Green 1994; Penfold et al. 1986), autoantibodies directed against RPE or Drusen have been detected in the sera of AMD patients (Penfold et al. 1990), and abnormal vasculature occurs on the choroid of AMD donor retinae associated with changes in Cx43 expression (CR Green and WR Good, personal observation). These findings may indicate that RPE degeneration and Drusen development are downstream consequences of AMD, rather than primary causes, with alterations in choridal blood flow and inflammation being contributory factors (Zarbin 2004; Charbel Issa et al. 2010) as it is with retinal ischemia (Danesh-Meyer et al. 2012). There is potential for connexin-specific treatments targeting choroidal inflammation to ameliorate the development of AMD.

14.6 CONCLUSIONS

CNS injury presents multiple challenges, not least of which is secondary damage comprising ischemia, inflammation, edema, oxidative injury, and glutamatergic toxicity. Alterations in connexin expression levels, in particular Cx43, are associated with both trauma and chronic conditions. Connexin upregulation has been implicated in secondary damage spread, greater vulnerability to injury, edema, white blood cell infiltration, glial and endothelial cell activation, and blood–brain barrier permeability. We have described a number of *in vitro*, *ex vivo*, and *in vivo* models, which indicate that transiently reducing connexin expression or blocking gap junction channels may provide therapeutic opportunities. Specific gap junction modulators such as AsODN and connexin peptidomimetics show real promise, in all cases reducing inflammation (a reduction in edema, inflammatory cell activation or migration to sites of injury, and prevention of inflammation-mediated vessel leak), reducing lesion size, and/or providing improved functional outcomes.

ACKNOWLEDGMENTS

The authors acknowledge the assistance of Cameron Johnson and Elizabeth Eady (University of Auckland) for work shown in Figure 14.1, Vicki Milham (University of Auckland) for work shown in Figure 14.2, and Irene Gramaglia (La Jolla Infectious Diseases Institute) for work shown in Figure 14.3.

The authors also acknowledge funding from The Wellcome Trust, The International Spinal Research Trust (UK), The Royal Society of New Zealand Marsden Fund, The Auckland Medical Research Foundation, The CatWalk Trust (New Zealand), the New Zealand Save Sight Society, and the National Institutes of Health (USA).

REFERENCES

Ahn, J. K., J. H. Kang, and K. H. Park. 2004. Correlation between a disc hemorrhage and peripapillary atrophy in glaucoma patients with a unilateral disc hemorrhage. *J Glaucoma* 13 (1):9–14.

Albowitz, B., P. König, and U. Kuhnt. 1997. Spatiotemporal distribution of intracellular calcium transients during epileptiform activity in guinea pig hippocampal slices. *J Neurophysiol* 77:491–501.

Aldskogius, H., and E. N. Kozlova. 1998. Central neuron-glial and glial-glial interactions following axon injury. *Prog Neurobiol* 55 (1):1–26.

Anderson, D. H., R. F. Mullins, G. S. Hageman, and L. V. Johnson. 2002. A role for local inflammation in the formation of drusen in the aging eye. *Am J Ophthalmol* 134 (3):411–31.

Anderson, D. H., M. J. Radeke, N. B. Gallo, E. A. Chapin, P. T. Johnson, C. R. Curletti, L. S. Hancox et al. 2010. The pivotal role of the complement system in aging and age-related macular degeneration: Hypothesis re-visited. *Prog Retin Eye Res* 29 (2):95–112.

Arend, O., N. Plange, W. E. Sponsel, and A. Remky. 2004. Pathogenetic aspects of the glaucomatous optic neuropathy: Fluorescein angiographic findings in patients with primary open angle glaucoma. *Brain Res Bull* 62 (6):517–24.

Bahr, B. A. 1995. Long-term hippocampal slices: A model system for investigating synaptic mechanisms and pathologic processes. *J Neurosci Res* 42 (3):294–305.

Basso, D. M., M. S. Beattie, and J. C. Bresnahan. 1995. A sensitive and reliable locomotor rating scale for open field testing in rats. *J Neurotrauma* 12 (1):1–21.

Becker, D. L., J. S. Lin, and Green C. R. 1999. Pluronic gel as a means of antisense delivery. In *Antisense Techniques in the Central Nervous System*. R. A. Leslie, A. J. Hunter, and H. A. Robertson (eds). New York: Oxford University Press.

Bengtsson, B. O., Catharina Holmin, and C. E. T. Krakau. 1981. Disc haemorrhage and glaucoma. *Acta Ophthalmol* 59 (1):1–14.

Bennett, M. V., and R. S. Zukin. 2004. Electrical coupling and neuronal synchronization in the mammalian brain. *Neuron* 41 (4):495–511.

Berman, R. S., P. E. Martin, W. H. Evans, and T. M. Griffith. 2002. Relative contributions of NO and gap junctional communication to endothelium-dependent relaxations of rabbit resistance arteries vary with vessel size. *Microvasc Res* 63 (1):115–28.

Berthoud, V. M., E. C. Beyer, and K. H. Seul. 2000. Peptide inhibitors of intercellular communication. *Am J Physiol Lung Cell Mol Physiol* 279 (4):L619–22.

Blanc, E. M., and A. J. Bruce-Keller. 1998. Astrocytic gap junctional communication decreases neuronal vulnerability to oxidative stress-induced disruption of Ca^{2+} homeostasis and cell death. *J Neurochem* 70 (3):958–70.

Blomstrand, F., C. Giaume, E. Hansson, and L. Ronnback. 1999. Distinct pharmacological properties of ET-1 and ET-3 on astroglial gap junctions and Ca(2+) signaling. *Am J Physiol* 277 (4 Pt 1):C616–27.

Boitano, S., and W. H. Evans. 2000. Connexin mimetic peptides reversibly inhibit Ca(2+) signaling through gap junctions in airway cells. *Am J Physiol Lung Cell Mol Physiol* 279 (4):L623–30.

Braet, K., L. Cabooter, K. Paemeleire, and L. Leybaert. 2004. Calcium signal communication in the central nervous system. *Biol Cell* 96 (1):79–91.

Braet, K., W. Vandamme, P. E. Martin, W. H. Evans, and L. Leybaert. 2003. Photoliberating inositol-1,4,5-trisphosphate triggers ATP release that is blocked by the connexin mimetic peptide gap 26. *Cell Calcium* 33 (1):37–48.

Bragin, A., M. Penttonen, and G. Buzsáki. 1997. Termination of epileptic afterdischarge in the hippocampus. *J Neurosci* 17:2567–79.

Bruzzone, S., L. Guida, E. Zocchi, L. Franco, and A. de Flora. 2001. Connexin 43 hemi channels mediate Ca^{2+}-regulated transmembrane NAD^{+} fluxes in intact cells. *FASEB J* 15 (1):10–2.

Budd, S. L., and S. A. Lipton. 1998. Calcium tsunamis: Do astrocytes transmit cell death messages via gap junctions during ischemia? *Nat Neurosci* 1 (6):431–2.

Cameron, D. A., K. L. Gentile, F. A. Middleton, and P. Yurco. 2005. Gene expression profiles of intact and regenerating zebrafish retina. *Mol Vis* 11:775–91.

Campochiaro, P. A., P. Soloway, S. J. Ryan, and J. W. Miller. 1999. The pathogenesis of choroidal neovascularization in patients with age-related macular degeneration. *Mol Vis* 5:34.

Cancilla, P. A., J. Bready, J. Berliner, H. Sharifi-Nia, A. W. Toga, E. M. Santori, S. Scully, and J. deVellis. 1992. Expression of mRNA for glial fibrillary acidic protein after experimental cerebral injury. *J Neuropathol Exp Neurol* 51 (5):560–5.

Carlen, P. L., F. Skinner, L. Zhang, C. C. G. Naus, M. Kushnir, and J. L. Perez Velazquez. 2000. The role of gap junctions in seizures. *Brain Res Brain Res Rev* 32 (1):235–41.

Charbel Issa, P., N. Victor Chong, and H. P. Scholl. 2010. The significance of the complement system for the pathogenesis of age-related macular degeneration—Current evidence and translation into clinical application. *Graefes Arch Clin Exp Ophthalmol* 249 (2): 163–74.

Charles, A. C., C. C. G. Naus, D. Zhu, G. M. Kidder, E. R. Dirksen, and M. J. Sanderson. 1992. Intercellular calcium signaling via gap junctions in glioma cells. *J Cell Biol* 118 (1):195–201.

Chaytor, A. T., P. E. M. Martin, W. H. Evans, M. D. Randall, and T. M. Griffith. 1999. The endothelial component of cannabinoid-induced relaxation in rabbit mesenteric artery depends on gap junctional communication. *J Physiol* 520 (2):539–50.

Chaytor, A. T., W. H. Evans, and T. M. Griffith. 1997. Peptides homologous to extracellular loop motifs of connexin 43 reversibly abolish rhythmic contractile activity in rabbit arteries. *J Physiol* 503 (Pt 1):99–110.

Chew, S. S., C. S. Johnson, C. R. Green, and H. V. Danesh-Meyer. 2010. Role of Cx43 in central nervous system injury. *Exp Neurol* 225 (2):250–61.

Combes, V., J. B. De Souza, L. Rénia, N. H. Hunt, and G. E. Grau. 2005. Cerebral malaria: Which parasite? Which model? *Drug Discov Today Disease Models* 2 (2):141–7.

Conhaim, R. L., K. E. Watson, C. A. Spiegel, W. F. Dovi, and B. A. Harms. 2008. Bacteremic sepsis disturbs alveolar perfusion distribution in the lungs of rats. *Crit Care Med* 36 (2):511–7.

Contreras, J. E., H. A. Sanchez, L. P. Veliz, F. F. Bukauskas, M. V. Bennett, and J. C. Saez. 2004. Role of connexin-based gap junction channels and hemichannels in ischemia-induced cell death in nervous tissue. *Brain Res Brain Res Rev* 47 (1–3):290–303.

Cornell-Bell, A. H., and S. M. Finkbeiner. 1991. Ca^{2+} waves in astrocytes. *Cell Calcium* 12 (2–3):185–204.

Cotrina, M. L., J. Kang, J. H. Lin, E. Bueno, T. W. Hansen, L. He, Y. Liu, and M. Nedergaard. 1998. Astrocytic gap junctions remain open during ischemic conditions. *J Neurosci* 18 (7):2520–37.

Coutinho, P., C. Qiu, S. Frank, K. Tamber, and D. Becker. 2003. Dynamic changes in connexin expression correlate with key events in the wound healing process. *Cell Biol Int* 27 (7):525–41.

Coutinho, P., C. Qiu, S. Frank, C. M. Wang, T. Brown, C. R. Green, and D. L. Becker. 2005. Limiting burn extension by transient inhibition of Cx43 expression at the site of injury. *Br J Plast Surg* 58 (5):658–67.

Cronin, M., P. N. Anderson, J. E. Cook, C. R. Green, and D. L. Becker. 2008. Blocking Cx43 expression reduces inflammation and improves functional recovery after spinal cord injury. *Mol Cell Neurosci* 39 (2):152–60.

Cronin, M., P. N. Anderson, C. R. Green, and D. L. Becker. 2006. Antisense delivery and protein knockdown within the intact central nervous system. *Front Biosci* 11:2967–75.

Cusato, K., A. Bosco, R. Rozental, C. A. Guimaraes, B. E. Reese, R. Linden, and D. C. Spray. 2003. Gap junctions mediate bystander cell death in developing retina. *J. Neurosci.* 23 (16):6413–22.

Dahl, G. 2007. Gap junction-mimetic peptides do work, but in unexpected ways. *Cell Commun Adhes* 14 (6):259–64.

Danesh-Meyer, H. V., R. Huang, L. F. B. Nicholson, and C. R. Green. 2008. Cx43 antisense oligodeoxynucleotide treatment down-regulates the inflammatory response in an *in vitro* interphase organotypic culture model of optic nerve ischaemia. *J Clin Neurosci* 15 (11): 1253–63.

Danesh-Meyer, H. V., N. M. Kerr, J. Zhang, E. K. Eady, S. J. O'Carroll, L. F. Nicholson, C. S. Johnson, and C. R. Green. 2012. Connexin43 mimetic peptide reduces vascular leak and retinal ganglion cell death following retinal ischaemia. *Brain* 135(2): 506–20.

Das, S., D. Lin, S. Jena, A. Shi, S. Battina, D. H. Hua, R. Allbaugh, and D. J. Takemoto. 2008. Protection of retinal cells from ischemia by a novel gap junction inhibitor. *Biochem Biophys Res Commun* 373 (4):504–8.

Dastgheib, K., and W. R. Green. 1994. Granulomatous reaction to Bruch's membrane in age-related macular degeneration. *Arch Ophthalmol* 112 (6):813–8.

Decrock, E., E. De Vuyst, M. Vinken, M. Van Moorhem, K. Vranckx, N. Wang, L. Van Laeken et al. 2009. Connexin 43 hemichannels contribute to the propagation of apoptotic cell death in a rat C6 glioma cell model. *Cell Death Differ* 16 (1):151–63.

de Curtis, M., A. Manfridi, and G. Biella. 1998. Activity-dependent pH shifts and periodic recurrence of spontaneous interictal spikes in a model of focal epileptogenesis. *J Neurosci* 18:7543–51.

de Jong, P. T. 2006. Age-related macular degeneration. *N Engl J Med* 355 (14):1474–85.

de Pina-Benabou, M. H., V. Szostak, A. Kyrozis, D. Rempe, D. Uziel, M. Urban-Maldonado, S. Benabou et al. 2005. Blockade of gap junctions *in vivo* provides neuroprotection after perinatal global ischemia. *Stroke* 36 (10):2232–7.

De Vuyst, E., E. Decrock, M. De Bock, H. Yamasaki, C. C. Naus, W. H. Evans, and L. Leybaert. 2007. Connexin hemichannels and gap junction channels are differentially influenced by lipopolysaccharide and basic fibroblast growth factor. *Mol Biol Cell* 18 (1):34–46.

DiStasi, M. R., and K. Ley. 2009. Opening the flood-gates: How neutrophil-endothelial interactions regulate permeability. *Trends Immunol* 30 (11):547–56.

Dora, K. A., P. E. M. Martin, A. T. Chaytor, W. H. Evans, C. J. Garland, and T. M. Griffith. 1999. Role of heterocellular gap junctional communication in endothelium-dependent smooth muscle hyperpolarization: Inhibition by a connexin-mimetic peptide. *Biochem Biophys Res Commun* 254:27–31.

Emre, M., S. Orgul, T. Haufschild, S. G. Shaw, and J. Flammer. 2005. Increased plasma endothelin-1 levels in patients with progressive open angle glaucoma. *Br J Ophthalmol* 89 (1):60–3.

Enkvist, M. O. K., and K. D. McCarthy. 1992. Activation of protein kinase C blocks astroglial gap junction communication and inhibits the spread of calcium waves. *J Neurochem* 59 (2):519–26.

Eugenin, E. A., M. C. Branes, J. W. Berman, and J. C. Saez. 2003. TNF-alpha plus IFN-gamma induce Cx43 expression and formation of gap junctions between human monocytes/macrophages that enhance physiological responses. *J Immunol* 170 (3):1320–8.

Eugenin, E. A., D. Eckardt, M. Theis, K. Willecke, M. V. Bennett, and J. C. Saez. 2001. Microglia at brain stab wounds express connexin 43 and *in vitro* form functional gap junctions after treatment with interferon-gamma and tumor necrosis factor-alpha. *Proc Natl Acad Sci U S A* 98 (7):4190–5.

Evans, W. H., and S. Boitano. 2001. Connexin mimetic peptides: Specific inhibitors of gap-junctional intercellular communication. *Biochem Soc Trans* 29 (Pt 4):606–12.

Evans, W. H., E. De Vuyst, and L. Leybaert. 2006. The gap junction cellular internet: Connexin hemichannels enter the signalling limelight. *Biochem J* 397 (1):1–14.

Evans, W. H., and L. Leybaert. 2007. Mimetic peptides as blockers of connexin channel-facilitated intercellular communication. *Cell Commun Adhes* 14 (6):265–73.

Evans, W. H., and P. E. Martin. 2002. Gap junctions: Structure and function (Review). *Mol Membr Biol* 19 (2):121–36.

Farahani, R., M. H. Pina-Benabou, A. Kyrozis, A. Siddiq, P. C. Barradas, F. C. Chiu, L. A. Cavalcante, J. C. Lai, P. K. Stanton, and R. Rozental. 2005. Alterations in metabolism and gap junction expression may determine the role of astrocytes as "good samaritans" or executioners. *Glia* 50 (4):351–61.

Farkas, E., and P. G. Luiten. 2001. Cerebral microvascular pathology in aging and Alzheimer's disease. *Prog Neurobiol* 64 (6):575–611.

Finger, P. T., A. Berson, T. Ng, and A. Szechter. 1999. Ophthalmic plaque radiotherapy for age-related macular degeneration associated with subretinal neovascularization. *Am J Ophthalmol* 127 (2):170–7.

Finkbeiner, S. 1992. Calcium waves in astrocytes-filling in the gaps. *Neuron* 8 (6):1101–8.

Fonseca, Carissa G., Colin R. Green, and Louise F. B. Nicholson. 2002. Upregulation in astrocytic connexin 43 gap junction levels may exacerbate generalized seizures in mesial temporal lobe epilepsy. *Brain Res* 929 (1):105–16.

Frame, M. K., and A. W. de Feijter. 1997. Propagation of mechanically induced intercellular calcium waves via gap junctions and ATP receptors in rat liver epithelial cells. *Exp Cell Res* 230 (2):197–207.

Frantseva, M. V., L. Kokarovtseva, C. G. Naus, P. L. Carlen, D. MacFabe, and J. L. Perez Velazquez. 2002a. Specific gap junctions enhance the neuronal vulnerability to brain traumatic injury. *J Neurosci* 22 (3):644–53.

Frantseva, M. V., L. Kokarovtseva, and J. L. Perez Velazquez. 2002b. Ischemia-induced brain damage depends on specific gap-junctional coupling. *J Cereb Blood Flow Metab* 22 (4):453–62.

Gaietta, G., T. J. Deerinck, S. R. Adams, J. Bouwer, O. Tour, D. W. Laird, G. E. Sosinsky, R. Y. Tsien, and M. H. Ellisman. 2002b. Multicolor and electron microscopic imaging of connexin trafficking. *Science* 296 (5567):503–7.

Gigout, S., J. Louvel, H. Kawasaki, M. D'Antuono, V. Armand, I. Kurcewicz, A. Olivier et al. 2006. Effects of gap junction blockers on human neocortical synchronization. *Neurobiol Dis* 22 (3):496–508.

Goodenough, D. A., and D. L. Paul. 2003. Beyond the gap: Functions of unpaired connexon channels. *Nat Rev Mol Cell Biol* 4 (4):285–94.

Gosain, A., A. M. Matthies, J. V. Dovi, A. Barbul, R. L. Gamelli, and L. A. DiPietro. 2006. Exogenous pro-angiogenic stimuli cannot prevent physiologic vessel regression. *J Surg Res* 135 (2):218–25.

Gramaglia, I., H. Sahlin, J. P. Nolan, J. A. Frangos, M. Intaglietta, and H. C. van der Heyde. 2005. Cell- rather than antibody-mediated immunity leads to the development of profound thrombocytopenia during experimental *Plasmodium berghei* malaria. *J Immunol* 175 (11):7699–707.

Green, C. R., L. Y. Law, J. S. Lin, and D. L. Becker. 2001. Spatiotemporal depletion of connexins using antisense oligonucleotides. *Meth Mol Biol* 154:175–85.

Green, C. R., and L. F. Nicholson. 2008. Interrupting the inflammatory cycle in chronic diseases—Do gap junctions provide the answer? *Cell Biol Int* 32 (12):1578–83.

Grieshaber, M. C., and J. Flammer. 2007. Does the blood-brain barrier play a role in glaucoma? *Surv Ophthalmol* 52 (6, Supplement 1):S115–21.

Guan, J., D. Pavlovic, N. Dalkie, H. J. Waldvogel, S. J. O'Carroll, C. R. Green, and L. F. Nicholson. 2012. Vascular degeneration in Parkinson's disease. *Brain Pathol*. Aug 16. doi: 10.1111/j.1750-3639.2012.00628.x. [Epub ahead of print].

Hageman, G. S., P. J. Luthert, N. H. Victor Chong, L. V. Johnson, D. H. Anderson, and R. F. Mullins. 2001. An integrated hypothesis that considers drusen as biomarkers of immune-mediated processes at the RPE-Bruch's membrane interface in aging and age-related macular degeneration. *Prog Retin Eye Res* 20 (6):705–32.

Haupt, C., O. W. Witte, and C. Frahm. 2007. Up-regulation of Cx43 in the glial scar following photothrombotic ischemic injury. *Mol Cell Neurosci* 35 (1):89–99.

Hayreh, S. S., J. Pe'er, and M. B. Zimmerman. 1999. Morphologic changes in chronic high-pressure experimental glaucoma in rhesus monkeys. *J Glaucoma* 8 (1):56–71.

Hernandez, M. R., H. Miao, T. Lukas, L. Cerulli, N. N. Osborne, C. Nucci, and B. Giacinto. 2008. Astrocytes in glaucomatous optic neuropathy. *Prog Brain Res* 173: 353–73.

Hossain, M. Z., J. Peeling, G. R. Sutherland, E. L. Hertzberg, and J. I. Nagy. 1994. Ischemia-induced cellular redistribution of the astrocytic gap junctional protein Cx43 in rat brain. *Brain Res* 652 (2):311–22.

Hozumi, I., F. C. Chiu, and W. T. Norton. 1990. Biochemical and immunocytochemical changes in glial fibrillary acidic protein after stab wounds. *Brain Res* 524 (1):64–71.

Hunot, S., and E. C. Hirsch. 2003. Neuroinflammatory processes in Parkinson's disease. *Ann Neurol* 53 (Suppl 3):S49–58; discussion S58–60.

Inman, D. M., and P. J. Horner. 2007. Reactive nonproliferative gliosis predominates in a chronic mouse model of glaucoma. *Glia* 55 (9):942–53.

Jahromi, S. S., K. Wentlandt, S. Piran, and P. L. Carlen. 2002. Anticonvulsant actions of gap junctional blockers in an *in vitro* seizure model. *J Neurophysiol* 88:1893–902.

Johnson, E. C., L. M. H. Deppmeier, S. K. F. Wentzien, I. Hsu, and J. C. Morrison. 2000. Chronology of optic nerve head and retinal responses to elevated intraocular pressure. *Invest Ophthalmol Vis Sci* 41 (2):431–42.

Johnson, E. C., L. Jia, W. O. Cepurna, T. A. Doser, and J. C. Morrison. 2007. Global changes in optic nerve head gene expression after exposure to elevated intraocular pressure in a rat glaucoma model. *Invest Ophthalmol Vis Sci* 48 (7):3161–77.

Kanamori, A., M. Nakamura, Y. Nakanishi, Y. Yamada, and A. Negi. 2005. Long-term glial reactivity in rat retinas ipsilateral and contralateral to experimental glaucoma. *Exp Eye Res* 81 (1):48–56.

Kielczewski, J. L., M. E. Pease, and H. A. Quigley. 2005. The effect of experimental glaucoma and optic nerve transection on amacrine cells in the rat retina. *Invest Ophthalmol Vis Sci* 46 (9):3188–96.

Kielian, T. 2008. Glial connexins and gap junctions in CNS inflammation and disease. *J Neurochem* 106 (3):1000–16.

Kimelberg, H. K. 1992. Astrocytic edema in CNS trauma. *J Neurotrauma* 9 (Suppl 1):S71–81.

Köhling, R., S. J. Gladwell, E. Bracci, M. Vreugdenhil, and J. G. R. Jefferys. 2001. Prolonged epileptiform bursting induced by 0-Mg^{2+} in rat hippocampal slices depends on gap junctional coupling. *Neuroscience* 105 (3):579–87.

Krutovskikh, V. A., C. Piccoli, and H. Yamasaki. 2002. Gap junction intercellular communication propagates cell death in cancerous cells. *Oncogene* 21 (13):1989–99.

Kwak, B. R., and H. J. Jongsma. 1999. Selective inhibition of gap junction channel activity by synthetic peptides. *J Physiol* 516 (Pt 3):679–85.

Laird, D. W. 2006. Life cycle of connexins in health and disease. *Biochem J* 394 (Pt 3):527–43.

Lam, T. T., J. M. K. Kwong, and M. O. M. Tso. 2003. Early flial responses after acute elevated intraocular pressure in rats. *Invest Ophthalmol Vis Sci* 44 (2):638–45.

Langston, J. W., L. S. Forno, J. Tetrud, A. G. Reeves, J. A. Kaplan, and D. Karluk. 1999. Evidence of active nerve cell degeneration in the substantia nigra of humans years after 1-methyl-4-phenyl-1,2,3,6-tetrahydropyridine exposure. *Ann Neurol* 46 (4): 598–605.

Law, L. Y., W. V. Zhang, N. S. Stott, D. L. Becker, and C. R. Green. 2006. *In vitro* optimization of antisense oligodeoxynucleotide design: An example using the connexin gene family. *J Biomol Tech* 17 (4):270–82.

Lee, I. H., L. Eva, K. Ole, W. Johan, and O. Lars. 2005. Glial and neuronal connexin expression patterns in the rat spinal cord during development and following injury. *J Comp Neurol* 489 (1):1–10.

Lee, S. H., W. T. Kim, A. H. Cornell-Bell, and H. Sontheimer. 1994. Astrocytes exhibit regional specificity in gap-junction coupling. *Glia* 11 (4):315–25.

Leithe, E., A. Brech, and E. Rivedal. 2006. Endocytic processing of Cx43 gap junctions: A morphological study. *Biochem J* 393 (Pt 1):59–67.

Leybaert, L., K. Braet, W. Vandamme, L. Cabooter, P. E. Martin, and W. H. Evans. 2003. Connexin channels, connexin mimetic peptides and ATP release. *Cell Commun Adhesion* 10 (4–6):251–7.

Li, J., H. Shen, C. C. G. Naus, L. Zhang, and P. L. Carlen. 2001. Upregulation of gap junction connexin 32 with epileptiform activity in the isolated mouse hippocampus. *Neuroscience* 105 (3):589–98.

Lin, J. H., H. Weigel, M. L. Cotrina, S. Liu, E. Bueno, A. J. Hansen, T. W. Hansen, S. Goldman, and M. Nedergaard. 1998. Gap-junction-mediated propagation and amplification of cell injury. *Nat Neurosci* 1 (6):494–500.

Margineanu, D. G., and H. Klitgaard. 2001. Can gap-junction blockade preferentially inhibit neuronal hypersynchrony vs. excitability? *Neuropharmacology* 41 (3):377–83.

Marmarou, A. 2007. A review of progress in understanding the pathophysiology and treatment of brain edema. *Neurosurg Focus* 22 (5):E1.

Martin, P. E., C. Wall, and T. M. Griffith. 2005. Effects of connexin-mimetic peptides on gap junction functionality and connexin expression in cultured vascular cells. *Br J Pharmacol* 144 (5):617–27.

Martini, J., I. Gramaglia, M. Intaglietta, and H. C. van der Heyde. 2007. Impairment of functional capillary density but not oxygen delivery in the hamster window chamber during severe experimental malaria. *Am J Pathol* 170 (2):505–17.

Matsuda, N., and Y. Hattori. 2007. Vascular biology in sepsis: Pathophysiological and therapeutic significance of vascular dysfunction. *J Smooth Muscle Res* 43 (4):117–37.

McCormick, D. A., and D. Contreras. 2001. On the cellular and network bases of epileptic seizures. *Annu Rev Physiol* 63:815–47.

McGeer, P. L., S. Itagaki, and E. G. McGeer. 1988. Expression of the histocompatibility glycoprotein HLA-DR in neurological disease. *Acta Neuropathol* 76 (6):550–7.

McGeer, P. L., and E. G. McGeer. 2001. Inflammation, autotoxicity and Alzheimer disease. *Neurobiol Aging* 22 (6):799–809.

McIntyre, T. M., S. M. Prescott, A. S. Weyrich, and G. A. Zimmerman. 2003. Cell-cell interactions: Leukocyte-endothelial interactions. *Curr Opin Hematol* 10 (2):150–8.

Minghetti, L. 2005. Role of inflammation in neurodegenerative diseases. *Curr Opin Neurol* 18 (3):315–21.

Mori, R., K. T. Power, C. M. Wang, P. Martin, and D. L. Becker. 2006. Acute downregulation of Cx43 at wound sites leads to a reduced inflammatory response, enhanced keratinocyte proliferation and wound fibroblast migration. *J Cell Sci* 119 (Pt 24):5193–203.

Muller, M., B. H. Gahwiler, L. Rietschin, and S. M. Thompson. 1993. Reversible loss of dendritic spines and altered excitability after chronic epilepsy in hippocampal slice cultures. *Proc Natl Acad Sci U S A* 90: pp. 257–61.

Nagy, J. I., W. Li, E. L. Hertzberg, and C. A. Marotta. 1996. Elevated Cx43 immunoreactivity at sites of amyloid plaques in Alzheimer's disease. *Brain Res* 717 (1–2):173–8.

Nakase, T., S. Fushiki, and C. C. G. Naus. 2003. Astrocytic gap junctions composed of connexin 43 reduce apoptotic neuronal damage in cerebral ischemia. *Stroke* 34 (8): 1987–93.

Nakase, T., G. Sohl, M. Theis, K. Willecke, and C. C. Naus. 2004. Increased apoptosis and inflammation after focal brain ischemia in mice lacking Cx43 in astrocytes. *Am J Pathol* 164 (6):2067–75.

Nakase, T., Y. Yoshida, and K. Nagata. 2006. Enhanced connexin 43 immunoreactivity in penumbral areas in the human brain following ischemia. *Glia* 54 (5):369–75.

Naus, C. C., J. F. Bechberger, S. Caveney, and J. X. Wilson. 1991. Expression of gap junction genes in astrocytes and C6 glioma cells. *Neurosci Lett* 126 (1):33–6.

Neufeld, A. H., M. R. Hernandez, and M. Gonzalez. 1997. Nitric oxide synthase in the human glaucomatous optic nerve head. *Arch Ophthalmol* 115 (4):497–503.

Neufeld, A. H. 1999. Microglia in the pptic nerve head and the region of parapapillary chorio-retinal atrophy in glaucoma. *Arch Ophthalmol* 117 (8):1050–6.

Nilsson, P., L. Hillered, Y. Olsson, M. J. Sheardown, and A. J. Hansen. 1993. Regional changes in interstitial K+ and Ca2+ levels following cortical compression contusion trauma in rats. *J Cereb Blood Flow Metab* 13 (2):183–92.

Nodin, C., M. Nilsson, and F. Blomstrand. 2005. Gap junction blockage limits intercellular spreading of astrocytic apoptosis induced by metabolic depression. *J Neurochem* 94 (4): 1111–23.

Norenberg, M. D., J. Smith, and A. Marcillo. 2004. The pathology of human spinal cord injury: Defining the problems. *J Neurotrauma* 21 (4):429–40.

Nowak, J. Z. 2006. Age-related macular degeneration (AMD): Pathogenesis and therapy. *Pharmacol Rep* 58 (3):353–63.

O'Carroll, S. J., M. Alkadhi, L. F. Nicholson, and C. R. Green. 2008. Connexin 43 mimetic peptides reduce swelling, astrogliosis, and neuronal cell death after spinal cord injury. *Cell Commun Adhes* 15 (1):27–42.

Oguro, K., T. Jover, H. Tanaka, Y. Lin, T. Kojima, N. Oguro, S. Y. Grooms, M. V. Bennett, and R. S. Zukin. 2001. Global ischemia-induced increases in the gap junctional proteins connexin 32 (Cx32) and Cx36 in hippocampus and enhanced vulnerability of Cx32 knock-out mice. *J Neurosci* 21 (19):7534–42.

Orellana, J. A., D. E. Hernandez, P. Ezan, V. Velarde, M. V. Bennett, C. Giaume, and J. C. Saez. 2010. Hypoxia in high glucose followed by reoxygenation in normal glucose reduces the viability of cortical astrocytes through increased permeability of connexin 43 hemichannels. *Glia* 58 (3):329–43.

Oviedo-Orta, E., and W. H. Evans. 2002. Gap junctions and connexins: Potential contributors to the immunological synapse. *J Leukoc Biol* 72 (4):636–42.

Oviedo-Orta, E., and W. H. Evans. 2004. Gap junctions and connexin-mediated communication in the immune system. *Biochim Biophys Acta* 1662 (1–2):102–12.

Ozog, M. A., R. Siushansian, and C. C. G. Naus. 2002. Blocked gap junctional coupling increases glutamate-induced neurotoxicity in neuron-astrocyte co-cultures. *J Neuropathol Exp Neurol* 61 (2):132–41.

Parpura, V., T. A. Basarsky, F. Liu, K. Jeftinija, S. Jeftinija, and P. G. Haydon. 1994. Glutamate-mediated astrocyte-neuron signalling. *Nature* 369:744–7.

Paul, D. L., L. Ebihara, L. J. Takemoto, K. I. Swenson, and D. A. Goodenough. 1991. Connexin46, a novel lens gap junction protein, induces voltage-gated currents in non-junctional plasma membrane of Xenopus oocytes. *J Cell Biol* 115 (4):1077–89.

Penfold, P. L., M. C. Killingsworth, and S. H. Sarks. 1986. Senile macular degeneration. The involvement of giant cells in atrophy of the retinal pigment epithelium. *Invest Ophthalmol Vis Sci* 27 (3):364–71.

Penfold, P. L., J. M. Provis, J. H. Furby, P. A. Gatenby, and F. A. Billson. 1990. Autoantibodies to retinal astrocytes associated with age-related macular degeneration. *Graefes Arch Clin Exp Ophthalmol* 228 (3):270–4.

Perez-Velazquez, J. L., T. A. Valiante, and P. L. Carlen. 1994. Modulation of gap junctional mechanisms during calcium-free induced field burst activity: A possible role for electrotonic coupling in epileptogenesis. *J Neurosci* 14 (7):4308–17.

Putnam, R. W., S. C. Conrad, M. J. Gdovin, J. S. Erlichman, and J. C. Leiter. 2005. Neonatal maturation of the hypercapnic ventilatory response and central neural CO2 chemosensitivity. *Respir Physiol Neurobiol* 149 (1–3):165–79.

Qiu, C., P. Coutinho, S. Frank, S. Franke, L-Y. Law, P. Martin, C. R. Green, and D. L. Becker. 2003. Targeting Cx43 expression accelerates the rate of wound repair. *Curr Biol* 13 (19):1697–703.

Quist, A. P., S. K. Rhee, H. Lin, and R. Lal. 2000. Physiological role of gap-junctional hemichannels: Extracellular calcium-dependent isosmotic volume regulation. *J Cell Biol* 148 (5):1063–74.

Rami, A., T. Volkmann, and J. Winckler. 2001. Effective reduction of neuronal death by inhibiting gap junctional intercellular communication in a rodent model of global transient cerebral ischemia. *Exp Neurol* 170 (2):297–304.

Rawanduzy, A., A. Hansen, T. W. Hansen, and M. Nedergaard. 1997. Effective reduction of infarct volume by gap junction blockade in a rodent model of stroke. *J Neurosurg* 87 (6):916–20.

Rodriguez-Sinovas, A., A. Cabestrero, D. Lopez, I. Torre, M. Morente, A. Abellan, E. Miro, M. Ruiz-Meana, and D. Garcia-Dorado. 2007. The modulatory effects of connexin 43 on cell death/survival beyond cell coupling. *Prog Biophys Mol Biol* 94 (1–2):219–32.

Roelfsema, V., L. Bennet, S. George, D. Wu, J. Guan, M. Veerman, and A. J. Gunn. 2004. Window of opportunity of cerebral hypothermia for postischemic white matter injury in the near-term fetal sheep. *J Cereb Blood Flow Metab* 24 (8):877–86.

Romanov, R. A., O. A. Rogachevskaja, M. F. Bystrova, P. Jiang, R. F. Margolskee, and S. S. Kolesnikov. 2007. Afferent neurotransmission mediated by hemichannels in mammalian taste cells. *EMBO J* 26 (3):657–67.

Ross, F. M., P. Gwyn, D. Spanswick, and S. N. Davies. 2000. Carbenoxolone depresses spontaneous epileptiform activity in the CA1 region of rat hippocampal slices. *Neuroscience* 100 (4):789–96.

Rouach, N., E. Avignone, W. Meme, A. Koulakoff, L. Venance, F. Blomstrand, and C. Giaume. 2002. Gap junctions and connexin expression in the normal and pathological central nervous system. *Biol Cell* 94 (7–8):457–75.

Rufer, M., S. B. Wirth, A. Hofer, R. Dermietzel, A. Pastor, H. Kettenmann, and K. Unsicker. 1996. Regulation of connexin-43, GFAP, and FGF-2 is not accompanied by changes in astroglial coupling in MPTP-lesioned, FGF-2-treated Parkisonian mice. *J Neurosci Res* 46 (5):606–617.

Saez, J. C., J. E. Contreras, F. F. Bukauskas, M. A. Retamal, and M. V. Bennett. 2003. Gap junction hemichannels in astrocytes of the CNS. *Acta Physiol Scand* 179 (1):9–22.

Salameh, A., and S. Dhein. 2005. Pharmacology of gap junctions. New pharmacological targets for treatment of arrhythmia, seizure and cancer? *Biochim Biophys Acta* 1719 (1–2): 36–58.

Samoilova, M., K. Wentlandt, Y. Adamchik, A. A. Velumian, and P. L. Carlen. 2008. Connexin 43 mimetic peptides inhibit spontaneous epileptiform activity in organotypic hippocampal slice cultures. *Exp Neurol* 210 (2):762–75.

Santoro, S. W., and G. F. Joyce. 1997. A general purpose RNA-cleaving DNA enzyme. *Proc Natl Acad Sci U S A* 94 (9):4262–6.

Sarieddine, M. Z., K. E. Scheckenbach, B. Foglia, K. Maass, I. Garcia, B. R. Kwak, and M. Chanson. 2009. Cx43 modulates neutrophil recruitment to the lung. *J Cell Mol Med* 13 (11–12):4560–70.

Sawchuk, M. A., M. Z. Hossain, E. L. Hertzberg, and J. I. Nagy. 1995. *In situ* transblot and immunocytochemical comparisons of astrocytic connexin-43 responses to NMDA and kainic acid in rat brain. *Brain Res* 683 (1):153–7.

Seddon, J. M., G. Gensler, R. C. Milton, M. L. Klein, and N. Rifai. 2004. Association between C-reactive protein and age-related macular degeneration. *JAMA* 291 (6):704–10.

Shelley, E. J., M. C. Madigan, R. Natoli, P. L. Penfold, and J. M. Provis. 2009. Cone degeneration in aging and age-related macular degeneration. *Arch Ophthalmol* 127 (4): 483–92.

Siushansian, R., J. F. Bechberger, D. F. Cechetto, V. C. Hachinski, and C. C. G. Naus. 2001. Cx43 null mutation increases infarct size after stroke. *J Comp Neurol* 440 (4):387–94.

Spray, D. C., R. Rozental, and M. Srinivas. 2002. Prospects for rational development of pharmacological gap junction channel blockers. *Curr Drug Targets* 3 (6):455–64.

Stoppini, L., P. A. Buchs, and D. Muller. 1991. A simple method for organotypic cultures of nervous tissue. *J Neurosci Methods* 37 (2):173–82.

Stout, C. E., J. L. Costantin, C. C. Naus, and A. C. Charles. 2002. Intercellular calcium signaling in astrocytes via ATP release through connexin hemichannels. *J Biol Chem* 277 (12):10482–8.

Suadicani, S. O., C. E. Flores, M. Urban-Maldonado, M. Beelitz, and E. Scemes. 2004. Gap junction channels coordinate the propagation of intercellular Ca2+ signals generated by P2Y receptor activation. *Glia* 48 (3):217–29.

Szente, M., Z. Gajda, K. Said Ali, and E. Hermesz. 2002. Involvement of electrical coupling in the *in vivo* ictal epileptiform activity induced by 4-aminopyridine in the neocortex. *Neuroscience* 115 (4):1067–78.

Tekkok, S. B., and M. P. Goldberg. 2001. Ampa/kainate receptor activation mediates hypoxic oligodendrocyte death and axonal injury in cerebral white matter. *J Neurosci* 21 (12):4237–48.

Theriault, E., U. N. Frankenstein, E. L. Hertzberg, and J. I. Nagy. 1997. Cx43 and astrocytic gap junctions in the rat spinal cord after acute compression injury. *J Comp Neurol* 382 (2):199–214.

Thompson, Roger J., Ning Zhou, and Brian A. MacVicar. 2006. Ischemia opens neuronal gap junction hemichannels. *Science* 312 (5775):924–27.

Towfighi, J., D. Mauger, R. C. Vannucci, and S. J. Vannucci. 1997. Influence of age on the cerebral lesions in an immature rat model of cerebral hypoxia-ischemia: A light microscopic study. *Brain Res Dev Brain Res* 100 (2):149–60.

van der Heyde, H. C., J. Nolan, V. Combes, I. Gramaglia, and G. E. Grau. 2006. A unified hypothesis for the genesis of cerebral malaria: Sequestration, inflammation and hemostasis leading to microcirculatory dysfunction. *Trends Parasitol* 22 (11):503–8.

van der Linden, J. A. M., M. Joëls, H. Karst, A. J. A. Juta, and W. J. Wadman. 1993. Bicuculline increases the intracellular calcium response of CA1 hippocampal neurons to synaptic stimulation. *Neurosci Lett* 155 (2):230–3.

Varela, H. J. M. D., and M. R. D. D. S. Hernandez. 1997. Astrocyte responses in human optic nerve head with primary open-angle glaucoma. *J Glaucoma* 6 (5):303–13.

Velardo, M. J., C. Burger, P. R. Williams, H. V. Baker, M. C. Lopez, T. H. Mareci, T. E. White, N. Muzyczka, and P. J. Reier. 2004. Patterns of gene expression reveal a temporally orchestrated wound healing response in the injured spinal cord. *J Neurosci* 24 (39):8562–76.

Velazquez, J. L. Perez, L. Kokarovtseva, R. Sarbaziha, Z. Jeyapalan, and Y. Leshchenko. 2006. Role of gap junctional coupling in astrocytic networks in the determination of global ischaemia-induced oxidative stress and hippocampal damage. *Eur J Neurosci* 23 (1):1–10.

Vijayan, V. K., Y. L. Lee, and L. F. Eng. 1990. Increase in glial fibrillary acidic protein following neural trauma. *Mol Chem Neuropathol* 13 (1–2):107–18.

Vis, J. C., L. F. Nicholson, R. L. Faull, W. H. Evans, N. J. Severs, and C. R. Green. 1998. Connexin expression in Huntington's diseased human brain. *Cell Biol Int* 22 (11–12):837–47.

Wallraff, A., R. Köhling, U. Heinemann, M. Theis, K. Willecke, and C. Steinhäuser. 2006. The impact of astrocytic gap junctional coupling on potassium buffering in the hippocampus. *J Neurosci* 26:5438–47.

Wang, C. M., J. Lincoln, J. E. Cook, and D. L. Becker. 2007. Abnormal connexin expression underlies delayed wound healing in diabetic skin. *Diabetes* 56 (11):2809–17.

Wang, L., G. A. Cioffi, G. Cull, J. Dong, and B. Fortune. 2002. Immunohistologic evidence for retinal glial cell changes in human glaucoma. *Invest Ophthalmol Vis Sci* 43 (4):1088–94.

Wang, X., S. S-W. Tay, and Y-K. Ng. 2000. An immunohistochemical study of neuronal and glial cell reactions in retinae of rats with experimental glaucoma. *Exp Brain Res* 132 (4):476–84.

Warner, A., D. K. Clements, S. Parikh, W. H. Evans, and R. L. DeHaan. 1995. Specific motifs in the external loops of connexin proteins can determine gap junction formation between chick heart myocytes. *J Physiol* 488 (Pt 3):721–8.

Wilke, S., R. Thomas, N. Allcock, and R. Fern. 2004. Mechanism of acute ischemic injury of oligodendroglia in early myelinating white matter: The importance of astrocyte injury and glutamate release. *J Neuropathol Exp Neurol* 63 (8):872–81.

Ye, Z. C., M. S. Wyeth, S. Baltan-Tekkok, and B. R. Ransom. 2003. Functional hemichannels in astrocytes: A novel mechanism of glutamate release. *J Neurosci* 23 (9):3588–96.

Yoon, J. J., C. R. Green, S. J. O'Carroll, and L. F. Nicholson. 2010a. Dose-dependent protective effect of Cx43 mimetic peptide against neurodegeneration in an *ex vivo* model of epileptiform lesion. *Epilepsy Res* 92 (2–3): 153–62.

Yoon, J. J., L. F. Nicholson, S. X. Feng, J. C. Vis, and C. R. Green. 2010b. A novel method of organotypic brain slice culture using connexin-specific antisense oligodeoxynucleotides to improve neuronal survival. *Brain Res* 1353:194–203.

Yücel, Y. H., M. W. Kalichman, A. P. Mizisin, H. C. Powell, and R. N. Weinreb. 1999. Histomorphometric analysis of optic nerve changes in experimental glaucoma. *J Glaucoma* 8 (1):38–45.

Zarbin, M. A. 2004. Current concepts in the pathogenesis of age-related macular degeneration. *Arch Ophthalmol* 122 (4):598–614.

Zhang, J., S. J. O'Carroll, A. Wu, L. F. Nicholson, and C. R. Green. 2010. A model for *ex vivo* spinal cord segment culture—A tool for analysis of injury repair strategies. *J Neurosci Methods* 192 (1):49–57.

Zipfel, G. J., D. J. Babcock, J. M. Lee, and D. W. Choi. 2000. Neuronal apoptosis after CNS injury: The roles of glutamate and calcium. *J Neurotrauma* 17 (10):857–69.

15 Enhancing Epithelial Tissue Repair and Reducing Inflammation by Targeting Connexins

David L. Becker, Jeremy E. Cook, Peter Cormie,
Colin R. Green, Ariadna Mendoza-Naranjo,
Anthony R. J. Phillips, Antonio E. Serrano, and
Christopher Thrasivoulou

CONTENTS

The epithelial wound-healing process is a complex series of overlapping events in which communication between cells is required to orchestrate their behavior. Injury of any tissue results in a rapid inflammatory response, which in the case of skin lesions, attracts neutrophils as the first line of cell defence to combat bacterial infection and prevent sepsis. Release of proinflammatory signals in and around the wound site attracts macrophages and other phagocytes, which clear up the debris in the wound. In skin wounds it has long been thought that a robust inflammatory response is required to trigger the healing process and to mobilize fibroblasts and keratinocytes to crawl forward and close the wound. In the avascular cornea of the eye, inflammatory cells may migrate from the peripheral limbal vessels or tear film, and stromal keratocytes differentiate into myofibroblasts to repair connective tissue. While much of the research into cell signaling in epithelial wound healing has concentrated on the contribution of growth factors and cytokines to the healing process, we have focused on a novel form of cell signaling in wound healing mediated by gap junctions and hemichannels, which are composed of proteins called "connexins."

A variety of different connexins are expressed in the skin and cornea, with Cx43 being the most ubiquitous. In rodent epidermis, Cx43 is normally strongly expressed in the basal layers while Cx26 and Cx30 are expressed less strongly in the spinous layers, with Cx31.1 in the granular layer. In the rodent corneal epithelium, Cx43 is again the main connexin expressed in the basal layers, with Cx26 also prominent there and in the intermediate layer. As in skin, Cx31.1 is expressed in superficial layers. In both the dermis of skin and the corneal stroma, Cx43 predominates and is found in fibroblasts and corneal keratocytes, in skin and conjunctival blood vessels, and in skin dermal appendages. On wounding, the expression patterns of various connexins changes dynamically during the healing process. Cx43 is decreased in wound-edge keratinocytes in both skin and cornea in the first 24–48 h as they become migratory and crawl forward to close the wound. However, in the skin dermis in and around the wound site, Cx43 expression is rapidly increased in the blood vessels and activated neutrophils as part of the inflammatory response to injury. Likewise, in the corneal stroma, Cx43 is amplified in surviving and proliferating stromal keratocytes. In rodents, application of a Cx43-specific antisense oligodeoxynucleotide (Cx43asODN) to a skin lesion speeds the natural downregulation of Cx43 in keratinocytes and results in more rapid migration and wound closure. At the same time, this treatment prevents the upregulation of Cx43 in skin blood vessels and dampens down all aspects of the inflammatory response throughout the rest of the healing process. Interestingly, in diabetic rats, where healing is impaired, the keratinocyte response to a skin lesion is to increase Cx43 expression rather than to decrease it. Migration of keratinocytes appears to be severely impaired while they are overexpressing Cx43 and migration does not begin until Cx43 downregulation occurs 2–3 days later. Treating diabetic skin lesions with Cx43asODNs can prevent the abnormal upregulation of Cx43 and return healing rates to that of control levels. These findings support the idea of Cx43 as a therapeutic target to promote healing of both normal as well as impaired healing and nonhealing chronic wounds. Thus far, in human clinical trials, the Cx43asODN has proved to be safe and has been shown to promote the healing of chronic venous leg ulcers and nonhealing corneal burns.

15.1 INTRODUCTION

Recent research into improving wound healing in the skin and cornea and much of the associated drug development to improve it, has focused on the effects of the cytokines and chemokines that are released during the inflammatory response. While extracellular signaling through these growth factors is known to have complex and even beneficial effects on healing, no clinically successful drugs have emerged. Unfortunately, chronic wounds such as dermal pressure ulcers, venous leg ulcers, diabetic foot ulcers, or persistent corneal epithelial defects are a growing problem worldwide especially as the numbers of elderly and diabetic patients rise in the populations of the developed world. It is estimated that 1–2% of the Western population develops a chronic skin wound (Gottrup, 2004). These wounds can have significant morbidity attached and can have a profound negative effect on quality of life, none more so than when they result in amputation of lower limbs. The clinical management of such wounds and their associated morbidity is also a burgeoning massive economic burden, with conservative estimates of cost exceeding $25 billion per year (Walmsley, 2002; Sen et al., 2009). Currently there are no totally effective treatments for these chronic wounds, and there remains a significant unmet medical need for new therapeutics. In this chapter, we concentrate on the role of direct cell–cell signaling through gap junction channels in the epithelial wound-healing process in cornea and skin before considering the potential of this biology as a target for therapeutic intervention. In the course of this process, we shall consider the evidence that gap junctions are one of the first forms of intercellular communication to respond to epithelial injury and that they play pivotal roles in the normal acute healing process and in potentially perturbed healing processes.

15.2 CONNEXINS IN THE SKIN

The skin is the most extensive organ in the human body and its integrity is vital to our existence. Gap junctions have been found to play a central role in the normal homeostasis of the skin as well as in the repair process. They allow the passage of ions and small molecules between cells and are of particular importance for the passage of gases and nutrients in avascular tissues such as the epidermis. Their importance in the normal functioning and homeostasis of the epithelium—particularly the skin and ear—is readily demonstrated by the effects of point mutations that result in skin diseases in man such as Vohwinkel syndrome, erythrokeratoderma variabilis (EKV), hystrix-like ichthyosis with deafness (HID) and keratitis–ichthyosis–deafness syndrome (KID) (Common et al., 2002; Richard, 2000, 2005; van Steensel, 2004). The effects of these point mutations have been recently been reviewed by Aasen and Kelsell (2009).

Nearly half of the 21 known gap junction proteins have been reported to be expressed in the normal mammalian skin including connexins, namely, Cx26, 30, 30.3, 31, 31.1, 32, 37, 40, 43, and 45 (Risek et al., 1992; Kamibayashi et al., 1993; Fitzgerald et al., 1994; Salomon et al., 1994; Goliger and Paul, 1994; Lampe et al., 1998; Richard, 2000, 2003, 2005; Di et al., 2001; Aasen and Kelsell, 2009). Most are reported to be expressed at relatively low levels. In contrast, Cx43, the most ubiquitous connexin, is found at relatively high levels in epidermal keratinocytes, dermal fibroblasts, and appendages such as hair follicles, sweat glands, and blood vessels (Coutinho et al., 2003; Richard, 2005). Cx26 and Cx30 predominate in the spinous layers of rodent epidermis and Cx26 is also found in hair follicles. Cx31.1 is found in the cells of the granular layer that are undergoing cell death prior to becoming part of the cornified layer which will eventually be sloughed off (Goliger and Paul, 1994; Coutinho et al., 2003).

15.3   CONNEXIN DYNAMICS AND THE EPIDERMAL RESPONSE TO SKIN WOUNDING

Acute wounding of skin triggers a cascade of overlapping events that together bring about healing, which is essential in order to maintain the integrity of the skin and prevent infection. In response to an incisional or excisional skin lesion there is a rapid and dynamic response in connexin expression. In rodents, Cx43 and Cx31.1 start to become downregulated in wound-edge keratinocytes in the first 6 h after wounding. There is very little of these proteins left within 24–48 h as the cells change their phenotype to a migratory one, protruding lamellipodia in order to crawl forward and close the wound (Goliger and Paul, 1995; Saitoh et al., 1997; Coutinho et al., 2003). In contrast, Cx26 and Cx30, which are normally found at low levels in the spinous layers, are greatly upregulated in rodent wound-edge keratinocytes within the first 24–48 h and this elevated expression is maintained throughout the period of active migration (Coutinho et al., 2003).

We note that Cx26 and Cx30 can form hybrid heterotypic gap junctions with each other but not with Cx43. The coupling properties of Cx26 and Cx30 are also different from those of Cx43 in that they do not transfer the dye Lucifer yellow very well (Marziano et al., 2003). In normal rodent wound healing, there is a strong expression of Cx26 and Cx30 in migrating keratinocytes and presumably gap junctional communication among themselves, but possibly not with the following (proliferating) cells due to the interaction properties with Cx43. It would seem, therefore, that normal healing and leading edge keratinocyte migration does not require a complete absence of gap junction communication, as the downregulation of Cx43 might suggest, but instead perhaps a change in communication characteristics associated with other connexins. Meanwhile, several rows of cells back from the advancing front line of migrating keratinocytes, Cx43 is maintained at fairly high levels in cells that are highly communicative and proliferative (Goliger and Paul, 1995; Lampe et al., 1998; Coutinho et al., 2003). After several days of regeneration, once the two epidermal sides of the wound meet up their union is somehow sensed and Cx26 and Cx30 are downregulated and replaced by Cx43, which is transiently expressed at higher levels than normal while the epidermis hyperthickens. Then Cx43 gradually returns to normal, reverting to the laminar distribution characteristic of intact skin (Coutinho et al., 2003).

Why this dramatic series of changes in connexin expression takes place is not clear. The lack of interaction between Cx43 and Cx26 and Cx30 might allow for the formation of an isolated communication compartment among the migrating cells with distinctive properties from those of the normal communicating cells behind them. Indeed, gap junctional communication compartments have been proposed to be involved in coordinating both wound healing and morphogenic movements in embryonic development (Clark, 1985; Gailit and Clark, 1994; Martin, 1997; Richards et al., 2004). It has also been reported that protein kinase C modulates communication during wound healing by phosphorylation of Cx43 on serine368 in wound-edge basal keratinocytes, thereby decreasing communication as measured by dye transfer (Lampe et al., 2000; Solan et al., 2003; van Zeijl et al., 2007; Richards et al., 2004; Pollok et al., 2010).

Alternatively, the downregulation of Cx43 may relate to its role in the center of a protein "nexus." Here it binds either directly by the PDZ domain on its C-terminus tail or indirectly by alternative locations to an array of other junctional membrane proteins such as ZO-1 and cadherins as well as α- and β-catenin and cytoskeletal components including microtubules and actin (Giepmans et al., 2001; Duffy et al., 2002; Theiss and Meller, 2002; Butkevich et al., 2004; Wei et al., 2004, 2005; Li et al., 2005; Shaw et al., 2007). These binding events may, in turn, directly affect the migratory ability of the affected cells. Taken together, such key regulators of migration could influence both adhesion and cytoskeletal dynamics in ways that alter the healing process. In this context, it has also been proposed that Cx43 acts as a "master gene" that can influence the expression of over 300 other genes in a predictable way (Iacobas et al., 2007; Spray and Iacobas, 2007). In this way, it is possible to conceive of Cx43 acting as a "nexus" not only at the protein interaction level but also at the transcriptional level. This would give it a secondary role that is largely independent of its ability to form gap junctions. It is, therefore, possible that Cx43 downregulation also acts indirectly upon wound healing, influencing many other genes and proteins, including those yet to be determined.

15.4 INFLAMMATION AND THE DERMAL RESPONSE TO SKIN WOUNDING

In rodent models it has also been shown that dermis wound-edge fibroblasts also naturally downregulate Cx43 protein as they become migratory and start to form the granulation tissue (Mori et al., 2006). However, a central part of the early dermal response to injury is inflammatory, and this process is accompanied by a rapid upregulation of Cx43 in blood vessels in and around the wound site. This is seen as early as 4 h after injury, and the upregulation is maintained throughout the period of inflammation in the dermis, just as it has been found to do in the injured central nervous system (Coutinho et al., 2003, Mori et al., 2006; Cronin et al., 2008). In this inflammatory phase, blood vessels become more permeable and the subsequent release of intravascular fluids and protein then changes the extracellular osmotic balance, causing the dermal tissue to swell. A cascade of proinflammatory signals is released from the degranulating platelets, while the enzymatic action of thrombin on fibrinogen generates a fibrin clot that functions to plug the wound and temporarily protect it from the external environment (Baum and Arpey, 2005). Degranulating platelets also release growth factors such as platelet-derived growth factor (PDGF), and this component of the inflammatory response attracts Cx43-expressing activated neutrophils, which start to invade the wound site and surrounding tissues, releasing further proinflammatory signals. These in turn serve to recruit more neutrophils and, later, macrophages (Werner and Grose, 2003; Martin and Leibovich, 2005; Shaw and Martin, 2009). The early inflammatory response is thought crucial in order to dispose of any bacterial infection and prevent sepsis and thereby ensure the organism's survival. This early response is followed by the entry of monocytes that will differentiate into macrophages, clearing up spent neutrophils and debris in the wound site (Singer and Clark, 1999). The macrophages themselves release an array of proinflammatory signals similar to those of neutrophils but with the notable addition of fibroblast growth factor 2 (FGF-2) and transforming growth factor beta 1 (TGF-β1).

Generation of a robust inflammatory response has long been thought to be required to promote the process of tissue repair by stimulating keratinocytes to crawl forward and reepithelialize the wound, and fibroblasts to migrate into the wound-bed, proliferate, and form granulation tissue (Singer and Clark, 1999). However, this long-held concept has been questioned over the last decade (reviewed Martin and Leibovich, 2005). At variance with this concept, animal and human early-embryonic wounds heal without any inflammatory response and lack significant scarring (Whitby and Ferguson, 1991; Adzick and Longaker, 1992; Hopkinson-Woolley et al., 1994). However, a scar does form if a wound is made later in embryonic development, when the embryo is capable of mounting a robust inflammatory response. This suggests that inflammation may be involved in scar formation (review: Redd et al., 2004). Supporting evidence comes from wound-healing studies in the neonatal PU-1 knockout mouse, which lacks macrophages, mast cells, and functional neutrophils. In these mice, wounds heal at a normal rate in the complete absence of a circulatory inflammatory response (Martin et al., 2003). Similarly Smad3 knockout mice, which have a reduced inflammatory response to wounding, manifested as fewer neutrophils, heal with accelerated reepithelialization (Ashcroft et al., 1999). Indeed it is likely that the invading neutrophils at a wound site may destroy more than bacteria: their release of toxic free radicals may also kill healthy resident cells when they invade in large numbers (Martin and Leibovich, 2005).

On the other hand, a number of studies have reported benefits from the addition of proinflammatory growth factors such as FGF, PDGF, and TGFβ-1 to wounds, which have been said to promote healing (review: Mustoe, 2004). Different growth factors have been reported to promote proliferation and migration of keratinocytes and fibroblasts as well as enhancing granulation tissue formation and maturation. For example, application of TGFβ-1 to a wound was reported to increase inflammation while at the same time enhancing granulation tissue formation and maturation (Roberts et al., 1986).

Conversely, slow healing wounds may represent the result of some derangement in wound-healing mediators. Several different and potentially therapeutic growth factors have been taken through to human clinical trials to promote the healing of chronic wounds but with little success thus far (Reviewed: Mustoe, 2004; Schreml et al., 2010). The use of PDGF in the form of Regranex (becaplermin), has perhaps been one of the more successful, being found to modestly increase the rate of closure of diabetic ulcers by 10% in large clinical trials (Mustoe, 2004). It may be that ensuring the level of the inflammatory response is appropriate for a given wound by promoting its positives while reducing its negative effects will be a critical part of trying to improve the outcome of the wound-healing process. In any event, given the amount of work in this field, it is evidently not a simple or obvious process to manipulate or control in the complex background of wound healing.

15.5   EFFECTS OF TRANSIENT CX43 KNOCKDOWN ON SKIN INFLAMMATION AND WOUND HEALING

The observations of the almost ubiquitous presence of Cx43 in tissues, its short half-life, dynamic changes with wounding coupled with its key functional roles in cell-to-cell communication, make its role in wound healing an interesting subject for study.

Cx43 protein expression in keratinocytes and fibroblasts is tightly regulated in normal wound healing, with Cx43 downregulating during the period of migration before returning to normal levels when the wound is closed. Connexin proteins, including Cx43, have a short half-life, typically about 1.5–2 h (Gaietta et al., 2002; Leithe et al., 2006), and are prime targets for rapid downregulation with antisense. Using a series of rodent models, we discovered that wound healing can be significantly improved by the topical application of a Cx43-specific antisense oligodeoxynucleotide (Cx43asODN) gel to the wound (Qiu et al., 2003; Mori et al., 2006; Wang et al., 2007; Nakano et al., 2008). Similar improvements in the healing of tail-skin lesions have been seen in an inducible Cx43-deletion mouse, where Cx43 was genetically ablated (Kretz et al., 2003, 2004). However, Cx43 protein expression in keratinocytes and fibroblasts is tightly regulated in normal wound healing, with Cx43 downregulating during the period of migration before returning to normal levels when the wound is closed. We have developed the use of our Cx43asODNs, to produce a topical and transient Cx43 protein knockdown at the wound site, and we have shown that it results in improved healing in a variety of wound-healing models (Qiu et al., 2003; Mori et al., 2006). The ODN used in the above studies is a 30mer with an unmodified backbone, specific to Cx43 mRNA (Becker et al., 1999; Green et al., 2001; Law et al., 2006). Being unmodified, it is rapidly broken down within cells with a short intracellular half-life of 20–30 min and is naturally degraded even faster (minutes) in the bloodstream (Wagner, 1994) so it only acts where it is topically placed. The fact that it is unmodified also has the distinct advantage that it is not as "sticky" as phosphothioate backbone-modified ODNs and it therefore works at significantly lower concentrations without the potential problems of nonspecific binding to serum proteins and the generation of toxic effects or any inflammatory response (Agrawal and Kandimalla, 1999).

To counteract the short half-life, the asODN is formulated in Pluronic™ gel to provide a sustained topical delivery to the wound site (Becker et al., 1999; Green et al., 2001; Qiu et al., 2003; Mori et al., 2006; Wang et al., 2007). Pluronic gel is a thermoreversible compound that is liquid in the range 0–4°C and can run freely into the wound whereupon it rapidly changes phase within several seconds as it warms to become a gel. The gel then provides a consistent and continuous delivery of the asODNs to the wound site over a period of hours. Its mild surfactant properties are thought to aid asODN penetration into the local surrounding tissue.

Cx43asODNs applied to rodent skin lesions can efficiently downregulate Cx43 protein in the wound-edge keratinocytes and dermal fibroblasts within 2 h, rather than the 24 h that this takes naturally (Figure 15.1). The more rapid downregulation of Cx43 at the wound site has the effect of accelerating the wound-healing process by promoting early cell migration and wound closure (Qiu et al., 2003; Mori et al., 2006). At the same time, the Cx43asODNs can suppress the upregulation of Cx43 that would naturally occur in inflamed blood vessels, reducing their leakiness (Cronin et al., 2008) and the edematous swelling around the wound site (Qiu et al., 2003; Coutinho et al., 2005).

The macroscopic effects of Cx43asODN application to wounds can be seen as soon as 6 h after injury, when treated wounds appear to be less reddened and swollen, to

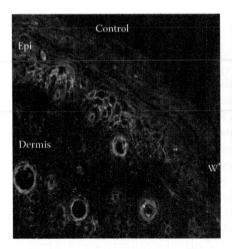 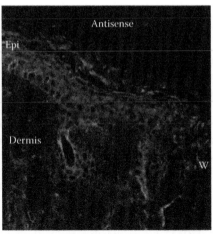

FIGURE 15.1 **(See color insert.)** Confocal microscope images of mouse incision skin wounds (W) 2 hours after injury. In control wounds, staining for Cx43 can be seen as bright green puncta in the epidermis (Epi) and dermis, whereas Cx43 protein levels are significantly reduced in the epidermis and dermis of Cx43asODN treated wounds. (Adapted from Qiu C et al. 2003. *Curr Biol* 13:1697–1703.)

have less exudate and consequently gape less than control wounds (Figure 15.2) (Qiu et al., 2003). These beneficial macroscopic effects are much more evident 24–48 h after injury. Microscopic analysis of the wounds shows that keratinocyte and fibroblast outgrowth is significantly faster in the treated wounds, which reepithelialize almost twice as fast and also form granulation tissue quicker (Mori et al., 2006). The final result of targeting Cx43 expression at the very start of the wound-healing process is a significant increase in the rate of healing and a scar that is smaller, thinner, and flatter, with a smaller underlying granulation tissue area. Within the granulation tissue, also, the collagen bundles of extracellular matrix tend to adopt a more normal appearance with a more interwoven organization, and are less apt to form the parallel pattern of bundles of collagen that characterizes scar tissue. Thus this therapeutic approach, targeting the earliest events in wound healing, appears to jump-start part of the process of repair while damping down inflammation, with beneficial effects on subsequent stages in the healing process.

15.6   CONNEXIN RESPONSES AND EFFECTS OF CX43 KNOCKDOWN ON SKIN THERMAL INJURIES

As will be described in this section, thermal injuries to the skin respond differently from incisional or excisional wounds but in a similar way to central nervous system injuries, showing significant elevations of Cx43 protein that correlate with a spread of damage, over the first 24–48 h, from the initial site of injury outward into the healthy surrounding tissue extending the lesion in both depth and lateral extent similar in ways to the "bystander effect" of spreading cell death observed in CNS injuries (Coutinho et al., 2005; Cronin et al., 2008).

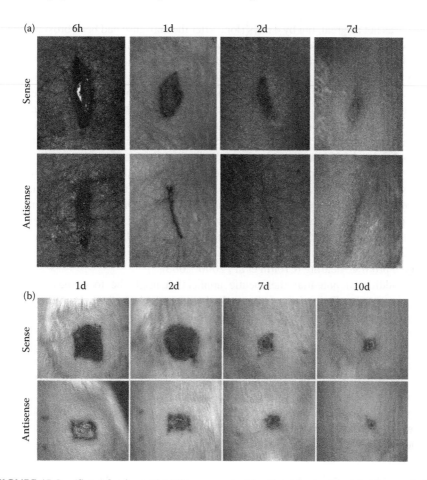

FIGURE 15.2 **(See color insert.)** (a) Time course of healing of paired incisional wounds in neonatal mouse skin treated with sense control or Cx43 antisense. Antisense treated wound can be seen to have a reduced exudate and swelling as early as 6 h, followed by reduced gape, accelerated wound closure and a thinner flatter scar. (b) Time course of healing of paired excisional wounds in adult mouse skin treated with sense control or Cx43 antisense. Antisense treated wounds show reduced swelling and faster wound closure. (Adapted from Qiu C et al. 2003. *Curr Biol* 13:1697–1703.)

Following thermal injury to the skin of neonatal mice resulted in Cx43 levels that decreased in the first 24 hours in the injured epidermis, which blistered, and in the underlying dermis, which was destined to die. However, in the adjacent intact epidermis and the dermis surrounding the injury, Cx43 was greatly increased within 24 h and was maintained at high levels for at least 4 days, during which massive cell death took place within the original site of injury and spread laterally into the previously intact tissue in a "bystander type effect" (Coutinho et al., 2005). Application of a single topical treatment of Cx43asODN to the site of injury had several effects. The size of the blister and the swelling of the surrounding tissue were reduced in the first 6–24 h. At this time, the elevation of Cx43 levels in the tissues surrounding the injury was not as great, but by 4 days (long after the antisense had been broken down) there was no detectable difference between control and treated wounds. However, by 4 days after injury, treated wounds appeared to have reduced cell death at the site of injury and histological analysis showed reepithelialization to be significantly accelerated, with increased cell proliferation in the nascent epidermis. Together this resulted in faster wound closure with smaller flatter scars. Examination of the inflammatory response in terms of neutrophil numbers showed that these were significantly reduced in treated wounds at 24 h after injury but not at 48 h (after the antisense was broken down). Further optimization of dose or regimen frequency may aid healing in burns, which represent a more complex pathological state compared to a simple skin wound.

A similar situation was found in studies of CNS injury to the rat spinal cord. In this setting it has been found that, while 1 µM Cx43asODN is sufficient to reduce Cx43 protein levels in the uninjured intact spinal cord, it was not a high enough concentration (or duration) to prevent the pathological upregulation of Cx43 that takes place after a compression injury, for which a minimum of 15 µM Cx43asODN was found to promote healing (Cronin et al., 2006, 2008).

An additional potential therapeutic application might be to tissues suffering from unwanted radiation injury after radiotherapy, which, in the case of skin, often becomes hypertrophic and hyperplastic. Radiation exposure seems to be a powerful stimulus for Cx43 expression as within a week of the start of daily irradiation of mouse skin with 3 Gy, enhanced levels of Cx43 could be seen in the basal cells and lower spinous layer of the epidermis (Liu et al., 1997). Cx43 levels were found to increase further in the epidermis and dermis by 3 weeks as the skin became hypertrophic. Changing the irradiation regime to a single 10 Gy dose also induced Cx43 in the epidermis but this time within days. It remains to be seen whether targeting Cx43 following irradiation injury is able to reduce the inflammation and damage that the tissues undergo following the insult.

15.7 EFFECTS OF CX43asODNs ON THE CELL BIOLOGY OF SKIN WOUND HEALING

A more detailed analysis of the cell-biological changes that follow active Cx43 downregulation shows that nearly the entire cascade of overlapping wound-healing events is affected in various ways long beyond the immediate action of the Cx43asODN itself. Antisense has a half-life of minutes in the cell and at most the knockdown effect from the drug is expected to last 24–48 h depending on dosing duration. Conversely, it has effects which are of much longer duration. Cell proliferation, for instance, was found to be significantly enhanced in both the nascent epidermis and nascent granulation tissue (Mori et al., 2006). Importantly, the inflammatory response to injury also appears to be dampened down. First, there are significantly fewer neutrophils both at the wound site with about 20% reduction, and in the surrounding tissues with about 80% reduction (Qiu et al., 2003; Mori et al., 2006). Depleting neutrophil numbers with antisera at a wound site has been reported to enhance the reepithelialization process (Dovi et al., 2003), which is also a feature seen in Cx43asODN-treated wounds. As previously noted, invading neutrophils destroy any bacteria in the wound but, by their exuberant actions, can also damage intact tissues in and around the wound site. Reducing the numbers of neutrophils, most significantly in the tissues around the wound therefore, can reduce secondary damage while still preventing infection. In turn, 2–3 days later, there are also a reduced number of macrophages entering the wound to clean up the debris and the spent neutrophils (Qiu et al., 2003; Mori et al., 2006). The neutrophils and macrophages between them release a powerful cocktail of pro-inflammatory cytokines and chemokines into the wound, which serve to amplify the inflammatory response. Cx43 is known to be expressed in macrophages (Beyer and Steinberg, 1991), neutrophils (Zahler et al., 2003) and activated leukocytes (Oviedo-Orta and Evans, 2004) and is involved in the release of cytokines and immunoglobulins. Indeed, it has also been reported that in Cx43asODN-treated wounds, by the second day after wounding, along with reductions in neutrophils and macrophages, there was also a significant reduction of CC chemokine ligand 2 (Ccl2) expression in the wounds (Mori et al., 2006). Likewise, the cytokine tumor necrosis factor alpha (TNF-α) was also reduced at day 7 (Mori et al., 2006). Both Ccl2 and TNF-α) are chemoattractants for neutrophils and macrophages, respectively (Rossi and Zlotnik, 2000). It would appear that in this instance, as observed in the PU-1 knockout mouse and in the fetus and early embryos, that healing not only continues, but is also promoted and improved by a reduced inflammatory response.

Many events at different stages in the wound-healing process have been reported to be promoted by transforming growth factor beta (TGF-β1), such as cell proliferation, migration (Postlethwaite et al., 1987), collagen production (Shah et al., 1994), granulation tissue formation (Desmoulière et al., 1993), contraction (Liu et al., 2001), angiogenesis (Roberts et al., 1986), and reepithelialization (Chesnoy et al., 2003). However, there remain a considerable number of contradictory reports on the effects of applying TGF-β1 directly to wounds depending on the models used (Hebda, 1988; Mustoe et al., 1991). We have found TGF-β1 mRNA and protein levels to be significantly elevated 2 days after Cx43asODN treatment of wounds, this elevation being seen in nascent keratinocytes, newly forming granulation tissue and neutrophils (Mori et al., 2006). Elevated TGF-β1 mRNA expression correlated well with the enhanced migration of keratinocytes and fibroblasts and more rapid formation of granulation tissue after treatment. Not only did the fibroblasts appear to populate the newly forming granulation tissue more rapidly but they also deposited extracellular matrix faster. Significantly enhanced levels of mRNA for collagen I were found on both days 2 and 7 after treatment and enhanced collagen protein was present on day 7, again consistent with known actions of TGF-β1 (Mori et al., 2006).

The formation of granulation tissue appears to have a 2–3 days "head start" in Cx43asODN-treated wounds, and this advantage persists. Myofibroblasts expressing smooth muscle actin appear throughout the granulation tissue, several days ahead of those in control wounds, contracting down the tissue and disappearing, still 2–3 days ahead of those in control wounds (Qiu et al., 2003; Mori et al., 2006). A similar enhancement of angiogenesis of granulation tissue was seen in treated wounds. Very fine blood vessels were found throughout the granulation tissue as early as 7 days after injury in treated wounds, at which time nascent vessels could be detected only at the edge of this tissue in control wounds. Over the next few days, blood vessels in control wounds pervaded the whole of the granulation tissue and the blood vessels in treated wounds thickened up and became very similar in appearance to control wounds over the same period (Mori et al., 2006). While TGF-β1 has previously been reported to promote granulation tissue maturation and angiogenesis (Desmoulière et al., 1993; Liu et al., 2001; Roberts et al., 1986), we did not find enhanced TGF-β1 expression at these later time points so how the enhancement of healing is brought about remains to be determined.

15.8 CONNEXIN EXPRESSION AND MANIPULATION IN DIABETIC SKIN WOUNDS

The onset of diabetes results in significant changes in connexin expression in the skin and other tissues throughout the body. In the skin of rats with streptozotocin-induced diabetes (STZ rats), within 2 weeks of the onset of diabetes, Cx43 and Cx26 proteins were significantly downregulated in the epidermis, whereas Cx43 was upregulated in the dermis, indicating that the diabetic condition does not uniformly repress connexin expression but can also enhance it (Wang et al., 2007). The changes in connexin protein expression were matched by decreased and increased gap junction dye transfer in the epidermis and dermis, respectively (Wang et al., 2007). Of note the altered connexin expression pattern in the skin does not further change 8–12 weeks after the onset of diabetes, by which time peripheral neuropathy has also set in.

Wound healing in diabetic skin is notoriously slow and, in humans often results in the formation of chronic diabetic ulcers that are very hard to heal (Mustoe, 2004; Dinh and Veves, 2005; Falanga, 2005). Interestingly the dynamic response of connexins to skin lesions in diabetic STZ rats is strikingly abnormal. Whereas Cx43 is expressed at significantly subnormal levels in intact STZ rat skin, once this is wounded by an incision, instead of downregulating Cx43 further as would occur in normal skin wounds, conversely the wound edge keratinocytes upregulate Cx43 (Figure 15.3) in the first 24 h after injury (Wang et al., 2007). These wound-edge keratinocytes then form a thickened bulb of nonmigrating cells at the wound margin and do not start migrating until Cx43 downregulates, which in these circumstances generally occurs some 48 h after wounding in diabetic rats and at least 24 h after normal skin downregulation would occur in a similar wound. The diabetic condition does not impact on the upregulation of Cx26 at the wound edge, which occurs with normal timing over the first 24 h but in a broader zone of cells than usual (Wang et al., 2007). When Cx43 eventually becomes downregulated, reepithelialization proceeds, but the healing process is retarded by a couple of days. It is not known why Cx43 is initially abnormally upregulated in the wound edge of the STZ rat. It may relate to abnormally high glucose levels, which are known to be able to increase or decrease connexin expression depending on cell type (Poladia et al., 2005; Hills et al., 2006; Bobbie et al., 2010).

Application of 1 μM Cx43asODN to normal acute rodent wounds is sufficient to promote downregulation of Cx43 at the wound edge and prevent the usual upregulation in blood vessels. As STZ rats greatly increase their Cx43 expression, a single 10 μM application of Cx43asODN has been used to treat these wounds, and has proved to be sufficient to prevent the Cx43 upregulation after wounding (Figure 15.3). This treatment also prevented the formation of a bulb of nonmigratory keratinocytes

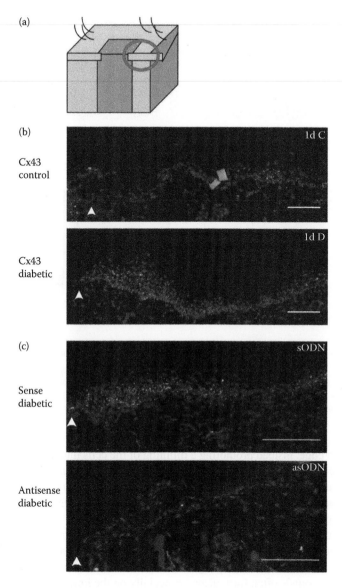

FIGURE 15.3 (**See color insert.**) (a) Cartoon of an excisional wound on rat skin with a red ring marking the epidermal wound edge where the confocal images in (b) and (c) were taken. (b) Epidermal wound edge (arrow head) stained to show nuclei in blue and Cx43 in green, 1 day after wounding. The control wound 1d C shows a thinning epidermal tongue of cells toward the wound with reduced levels of Cx43, while the 1d D diabetic wound shows a thickened bulb of keratinocytes at the wound edge with elevated levels of Cx43 protein. (c) Epidermal diabetic wound edge (arrow head) 1 day after wounding treated with control sense ODNs, showing a thickened bulb of keratinocytes at the wound edge with elevated levels of Cx43 protein compared to the diabetic wound treated with asODNs which shows significantly reduced levels of Cx43 protein and the migratory phenotype of a thinned tongue of keratino- cytes. Scale bar 100 μm. (Adapted from Wang CM et al. 2007. *Diabetes* 56:2809–17.)

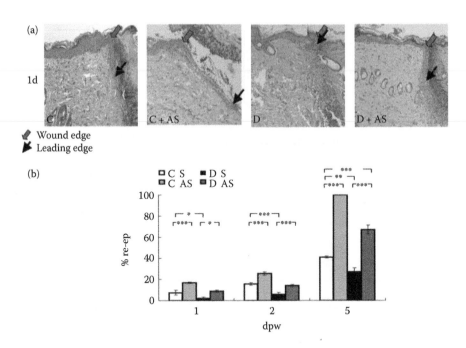

FIGURE 15.4 (**See color insert.**) (a) H&E staining of control (C) and diabetic (D) 1 day wound edges treated with sense control ODNs or antisense ODNs (AS). The epidermal wound edge is marked by a fat blue arrow and the leading edge by a thin black arrow. (b) Graphs showing the relative % reepithelialization at 1, 2, and 5 days after wounding in the different treatments. *$P < 0.05$; **$P < 0.01$; ***$P < 0.001$. (Adapted from Wang CM et al. 2007. *Diabetes* 56:2809–17.)

at the wound edge, instead promoting a thinning of the cells into a migratory tongue like that of controls (Wang et al., 2007). This treatment also promoted the reepithelialization rate of the STZ rats, restoring it to control levels or even above (Figure 15.4). It is unknown whether even higher concentrations of Cx43asODN might further promote healing rates above this level.

Chronic diabetic ulcers are a major and growing problem in the Western world as the number of diabetics and elderly patients in the population increases (Mustoe, 2004; Dinh and Veves, 2005; Falanga, 2005). As well as being very slow to heal, these chronic wounds often form ulcers, especially on the lower limbs where hydrostatic pressure contributes to edema. These ulcers have a considerable impact on the quality of life for patients, reducing their mobility and morbidity while being financially draining for healthcare services. A large proportion of diabetic foot ulcers never heal and the outcome is often amputation and a greatly shortened life expectancy. Although there are various nonbiological treatment options to assist healing in diabetic ulcers such as off-loading foot pressure, these approaches have poor compliance and variable effectiveness unless carefully applied. It is possible that Cx43 may represent a possible therapeutic target. One group has reported expression of Cx43 in wound-edge keratinocytes from biopsies of human diabetic ulcers (Brandner et al., 2004). Human diabetic fibroblasts were reported to have elevated gap junctional communication (Abdullah et al., 1999) and perturbed cell proliferation (Loots et al., 1999). If Cx43 expression in human diabetic wounds is perturbed in a similar way to that in the STZ diabetic rats, then it may be possible to manipulate the healing process in humans as well, and target these debilitating wounds.

15.9 PSORIATIC SKIN LESIONS: A CHRONIC INFLAMMATORY NONHEALING CONDITION

Psoriasis is one of the most prevalent, immune-mediated, adult skin diseases affecting about 2% of the population (Christophers, 2001). Psoriatic plaques are characterized by a thickened epidermis with hyperproliferative keratinocytes and elongated rete pegs (Kruger, 2002). Between the rete pegs are found large numbers of tortuous capillaries, which give the skin a reddened, inflamed appearance. The persistent inflammation around each plaque is marked by leukocytes exiting the capillaries and often migrating to upper strata of the epidermis where they cluster into neutrophil pockets releasing proinflammatory signals. The plaques often crack and these lesions are very slow to heal. The keratinocytes within the plaques appear to have elevated levels of connexins, in particular Cx26 throughout the epidermis and to a lesser extent Cx43 within the stratum spinosum (Labarthe et al., 1998). Upregulation of Cx26 has been previously linked with hyperproliferative skin disorders (Hivnor et al., 2004; Lucke et al., 1999; Wang et al., 2010), but the relationship between connexin expression and the psoriatic condition remains to be determined. It may, in future, be possible to break the pro-inflammatory cycle that persists at the psoriatic plaque, perhaps by targeting leukocyte connexins.

15.10 CONNEXINS IN THE CORNEA

The expression of a number of different connexins has been observed in the rodent and human cornea. The main ones are connexins 26, 30, 31.1, 37, 43, and 50 (Beyer et al., 1989; Matic et al., 1997; Ratkay-Traub et al., 2001; Laux-Fenton et al., 2003; Shurman et al., 2005). In rats and rabbits, Cx43 is found predominantly between epithelial basal cells with immunostaining decreasing in the upper differentiating layers of the epithelium. Lower levels of Cx26 also occur in the basal layer and it is found in the suprabasal wing cells (Laux-Fenton et al., 2003; Matic et al., 1997; Ratkay-Traub et al., 2001). Although Cx50 has also been reported in the rabbit corneal epithelium (Matic et al., 1997) no transcripts were identified in the rat cornea (Laux-Fenton et al., 2003). In the rat Cx30 is found in the peripheral corneal epithelium but disappears toward the center of the cornea. Cx31.1, as in skin, is the most superficial connexin expressed (Laux-Fenton et al., 2003). In the stroma (the corneal equivalent of the dermis) of all species (including rat, rabbit, pig, bovine, and human) Cx43 is the connexin between keratocytes, and in the single cell layer endothelium both Cx43 and Cx26 are seen (Laux-Fenton et al., 2003). In the human corneal epithelium, Cx43 has been reported in most layers except the most superficial, with Cx26 and Cx30 predominantly in the basal and lower suprabasal layers and in wing cells (Shurman et al., 2005). In the human cornea, Cx31.1 is again expressed in the more superficial layers of the epithelium (Chang et al., 2009).

15.11   CONNEXIN DYNAMICS IN THE CORNEA IN
RESPONSE TO WOUNDING

While the skin has been used for many wound-healing studies, there are distinct similarities with the cornea of the eye (Sherwin and Green, 2009). Both have a collagenous stroma. Both then have a stratified squamous cell epithelium, with the main difference being that the healthy corneal epithelium is just five to seven cell layers thick. The corneal stromal keratocytes and skin dermal fibroblasts play equivalent roles, and both undergo phenotypic changes during wound healing. Many of the cytokines involved in wound repair are duplicated, including IL-1, TNF-α, EGF, and PDGF. After wounding keratocyte apoptosis occurs immediately adjacent to the wound, peaking at 4 h after injury and then followed by proliferation of remaining keratocytes around the depletion zone within 12–24 h (Zieske et al., 2001). Chemokines released then attract inflammatory cells from the limbal vessels at the corneal periphery, and from the tear film (Hong et al., 2001) although this response is less marked than blood-derived inflammatory cell invasion into the dermis of the skin. A key event is TGF-β initiation of keratocyte differentiation to myofibroblasts, characterized by smooth muscle actin expression. The keratocytes maintain corneal transparency, and keratocyte to myofibroblast differentiation with altered crystalline expression and disorganized collagen expression is said to account for corneal haze and scarring (Jester et al., 1999). Apart from stromal remodeling, epithelial hyperplasia induced by excessive growth factor release is another source of refractive error after wounding (Wilson et al., 2001).

In sum, connexin changes in response to injury are similar to those seen in the skin. Dynamic changes in connexin expression have been reported in response to wounding of both rat and rabbit corneal epithelium. A rapid downregulation of all the connexins that were studied was observed in the migrating wound edge within 6 h of injury (Matic et al., 1997; Ratkay-Traub et al., 2001; Laux-Fenton, 2006; Rupenthal, 2008; Rupenthal et al., 2011), but relatively high levels of the main connexins were maintained in the following cells behind the leading edge. The reduction in Cx43 expression seen in regenerating migrating corneal epithelial cells is similar to that seen in the wound edge of skin epithelial cells (Goliger and Paul, 1995; Coutinho et al., 2003), if a little more rapid in the dynamics. An overexpression of Cx43 can result after wound closure leading to epithelial hyperplasia (Laux-Fenton, 2006; Rupenthal, 2008; Rupenthal et al., 2011). In the stroma, wounding results in an initial loss of Cx43 at the wound site itself as stromal keratocytes die back, but adjacent to the wound where keratocytes are proliferating and differentiating into the myofibroblast phenotype increased levels of Cx43 occur. As noted above, blood vessels become more permeable in wounded skin with the subsequent release of fluids and protein changing the osmotic balance, causing the dermis to swell. However, connexin hemichannel-mediated cytoplasmic swelling has been reported in the central nervous system tissue (Rodriguez-Sinovas et al., 2007) and this is also a factor in the cornea where edema is pronounced after wounding despite the tissue being avascular. Connexin hemichannel-mediated edema may be more pronounced than appreciated in many tissues.

As in skin, application of Cx43asODNs to wounded corneas significantly improves the rate and quality of healing. In photorefractive keratectomy laser ablated rat corneas, Cx43asODN application increases the rate of Cx43 decrease in cells at the wound edge, and as in skin, increases the rate of epithelial recovery (Laux-Fenton, 2006; Rupenthal et al., 2011). Hyperplasia is also reduced in treated corneas. Control laser injured rat corneas swell to more than double the normal stromal thickness within 24 h, but the edema is blocked with Cx43asODN, as is the proliferation and differentiation of inflammatory cells (myofibroblasts). In five of six patients with nonhealing ocular burns (chemical and thermal) treated on a compassionate use basis Cx43asODN treatment triggered ocular surface recovery. A notable feature was recovery of the sclera vascular bed and limbal reperfusion that preceded epithelial recovery. In the five patients, limbal ischemia and ocular inflammation were significantly improved within 1–2 days of commencing treatment (Ormonde et al., 2012) and connexin modulation with Cx43asODN eventually led to full recovery despite a hopeless prognosis (blindness).

In the corneal endothelium of the rat, Cx43 is strongly expressed in a uniform pattern. However, within 3 h of injury, Cx43 starts to be downregulated as the cells become migratory, only to return to normal when the wound is closed and normal morphology has returned (Nakano et al., 2008). Applying Cx43asODNs or Cx43 siRNA to the anterior chamber prior to wounding accelerated healing of the endothelium and increased cell proliferation; conversely, if Cx43 is overexpressed following infection with adenoviral CMV-Cx43-mRFP1 wound healing was delayed (Nakano et al., 2008). These findings are consistent with those of skin and corneal epithelial wounds indicating that wound closure is enhanced when Cx43 is downregulated but retarded if Cx43 is present or elevated.

Cx31.1, expressed in the more superficial layers of the corneal epithelium (Laux-Fenton et al., 2003), appears unique in that it forms a nonfunctional channel (Harris, 2001), yet appears to play a key role in the regulation of epithelial cell apoptosis which leads to sloughing of epithelial cells into the tear film to maintain corneal homeostasis. In rat and human cornea, Cx31.1-specific antisense oligonucleotides that knockdown Cx31.1 expression decrease epithelial cell apoptosis. In both *ex vivo* rat and human cornea Cx31.1 reduction brings about a corresponding increase in the number of cell layers (22% within 24 h in rat) (Chang et al., 2009). In mouse skin wound healing Cx31.1 expression is reduced after wounding, returning only after 7 days although the wounds were fully closed after 4 days (Coutinho et al., 2003). The dynamics of Cx31.1 after corneal wounding has not been explored.

15.12   EFFECTS OF MIMETIC PEPTIDES ON SKIN WOUND-HEALING MODELS

Hemichannels comprise the six connexin protein units covalently bonded to each other and embedded across the plasma membrane but are not docked to a corresponding hemichannel in another cell. Under certain conditions, for example, ischemic injury, these "half channels" can open to communicate directly with the extracellular milieu (Evans et al., 2006). The control of hemichannel opening is still being determined but the process is now increasingly being invoked in pathological processes, including the inflammatory response of some tissues (Evans et al., 2006; Froger et al., 2010).

While longer term downregulation of hemichannels is possible through the actions of antisense gap junctional communication and hemichannel signaling can be modified more acutely by the application of small peptides. These are designed to match the amino acid sequences from the extracellular loops of connexins, which are responsible for the docking of two connexons to form a gap junction channel (Warner et al., 1995; Boitano and Evans, 2000; Braet et al., 2003).

These peptides act rapidly to block hemichannel opening but with longer incubation times they have been reported to inhibit gap junctional communication (Leybaert et al., 2003). Because the sequences of extracellular loops are widely conserved the mimetic peptides tend not to be specific to connexin subtypes but the originals were derived from sequences from Cx32 and Cx43 (Warner et al., 1995). The most commonly used are known as GAP26 and GAP27. There are several useful reviews on their use and effects of these peptides so we shall not go into details here (Reviewed: Evans and Boitano, 2001; Spray et al., 2002; Evans and Leybaert, 2007; Dahl, 2007; O'Carroll et al., 2008).

Despite considerable research, the precise modes of action of mimetic peptides are not entirely clear. They bind to unapposed hemichannels and act to rapidly block their opening and their ability to release molecules such as glutamate or ATP under pathological states. With longer incubation times they have also been reported to inhibit gap junctional communication, presumably possibly by interfering with the docking of apposed hemichannels to form functional gap junction channels (Evans and Boitano, 2001; Leybaert et al., 2003). However, it has also been suggested that these peptides may diffuse into preexisting gap junction plaques and somehow alter their channel conductance of the channels without disrupting their overall structure (Martin et al., 2005). For the present, therefore, we must assume that applying mimetic peptides to cells or tissues has the potential to influence hemichannel activity, gap junction formation, and/or gap junctional conductance mediated by a wide variety of connexins, although it has been proposed that the peptide GAP27 may be more specific in its actions on Cx43 than GAP26, which may influence a wider variety of connexins (Wright et al., 2009).

The use of mimetic peptides GAP26 and GAP27 have been applied to scratch wound-healing models in confluent monolayer cultures of human keratinocytes and fibroblasts over a 5-day period (Wright et al., 2009). Both GAP26 and GAP27 peptides were reported to reduce gap junctional communication in these cultures after 90 min, although communication returned after 8 h. When the peptide was replaced every 12 h over the 5-day period of a wound-healing assay, either peptide was found to increase the rate of migration of both keratinocytes and fibroblasts. Interestingly, as in a skin lesion, Cx43 was found to downregulate naturally in the leading-edge cells of scratch wound keratinocytes and fibroblasts during the migration process (Wright et al., 2009).

More recently, such studies have been extended to additional human culture models and 3D organotypic porcine models of wounding (Kandyba et al., 2008; Pollok et al., 2011). Comparisons of the effect of GAP27 on the migration rates of infant and adult human keratinocytes and fibroblasts following scratch assays of confluent cultures showed that they were all accelerated in migration by the treatment (Pollok et al., 2011). However, early cultures of adult diabetic keratinocytes or fibroblasts failed to respond to the GAP27 treatment and an enhancement of migration could only be elicited in cells from very late passages (Pollok et al., 2011). In these instances, enhanced migration was found to correlate with enhanced cell proliferation following GAP27 treatment.

Examination of Cx43 levels at wound margins of human cell cultures showed a natural reduction, during migration that was not altered by application of GAP27 peptide, and no differences were found in Cx43 protein levels between diabetic and nondiabetic cultured cells (Pollok et al., 2011). However, treatment with GAP27 did result in a significant increase in phosphorylation of Cx43 at serine368 in all of the different cell types treated. Phosphorylation of Cx43 at serine368 and alteration of conductance has also been reported in wound-edge keratinocytes in cultures of human foreskin (Richards et al., 2004). It was, therefore, suggested that alterations in Cx43 conductance rather than loss of Cx43 protein was required for normal healing (Pollok et al., 2011). These findings are quite intriguing but do not really explain why the keratinocytes and fibroblasts from diabetic patients expressed normal levels of Cx43 protein and responded to GAP27 treatment by phosphorylation of serine 368 but still failed to migrate faster following this treatment. Clearly there is still more work that needs to be done to understand the problems associated with diabetic wound healing.

15.13 CONCLUDING REMARKS

We often think of ourselves, and our bodies as being highly evolved, and optimized for our lifestyle and environment, such that it should normally be hard to improve on any biological process. However, human cultural evolution has vastly outpaced biological evolution, allowing us to occupy environments that are much cleaner and more sterile than those of our ancestors. The evolutionary wound-healing priority of any animal would be to close a wound rapidly, preventing infection and death by sepsis. The quality and visual appearance of the healed wound would not be a top priority and would not feature. It is very likely, therefore, that we have evolved wound-healing responses that are geared in favor of robust inflammatory responses, promoting survival at the expense of perfect regenerative healing. There may, therefore, be scope for us to tinker and improve the rate and quality of healing of acute wounds, including surgical wounds that now represent a significant source of cosmetic scaring. However, a much more important target for optimization is the chronic wound, which is frequently inflamed and often infected, and where healing has stalled. We seem poorly evolved to deal with common age- or disease-related chronic wound conditions. If we are able to find out precisely how healing is perturbed in chronic wounds, and why it stalls, we may be able to promote their repair.

While a multifactorial solution may be required, many indications point toward Cx43 as a potential therapeutic target for improved epithelial healing of human wounds. The technology, of the Cx43asODN in Pluronic gel is currently being developed by CoDaTherapeutics Inc. as Nexagon®. Nexagon has been through phase 1 clinical trials in eye and skin and has recently finished the first phase 2 trial for the treatment of venous leg ulcers where initial results look promising, with one-third of wounds completely healing in just 4 weeks' time with only 3 Nexagon doses administered. An investigator led phase 2 trial for corneal burns is also underway. As we learn more about tissue injury and inflammation in humans it is possible that more targets for therapeutic intervention may become apparent. The anti-inflammatory properties of the Nexagon treatment may also have applications in other inflammatory conditions in the future.

ACKNOWLEDGMENTS

Research in the Becker lab has been supported by the BBSRC, Wellcome Trust, and AMRC (and the Henry Smith Charity).

REFERENCES

Aasen T and Kelsell DP. 2009. Connexins in skin biology. In *Connexins: a Guide*. Eds. A. Harris and D. Locke. Humana Press, pp. 307–321.

Abdullah KM, Luthra G, Bilski JJ, Abdullah SA, Reynolds LP, Redmer DA, Grazul-Bilska AT. 1999. Cell-to-cell communication and expression of gap junctional proteins in human diabetic and nondiabetic skin fibroblasts: Effects of basic fibroblast growth factor. *Endocrine* 10:35–41.

Adzick NS, Longaker MT. 1992. Scarless fetal healing. Therapeutic implications. *Ann Surg* 215:3–7.

Agrawal S and Kandimalla ER. 1999. Medicinal chemistry of antisense oligonucleotides. In *Antisense Techniques in the CNS. A Practical Approach*. Eds. R. Leslie, A.J. Hunter and H.A. Robertson. Oxford University Press, Oxford, New York, pp. 108–136.

Ashcroft GS, Yang X, Glick AB, Weinstein M, Letterio JL, Mizel DE, Anzano M et al. 1999. Mice lacking Smad3 show accelerated wound healing and an impaired local inflammatory response. *Nat Cell Biol* 1:260–6.

Baum CL, Arpey CJ. 2005. Normal cutaneous wound healing: Clinical correlation with cellular and molecular events. *Dermatol Surg* 31:674–86.

Becker DL, Lin JS, Green CR. 1999. Pluronic gel as a means of antisense delivery. In *Antisense Techniques in the CNS. A Practical Approach*. Eds. R. Leslie, A.J. Hunter and H.A. Robertson. Oxford University Press, Oxford, New York, pp. 149–157.

Beyer EC, Kistler J, Paul DL, Goodenough DA. 1989. Antisera directed against connexin43 peptides react with a 43-kD protein localized to gap junctions in myocardium and other tissues. *J Cell Biol* 108:595–605.

Beyer EC, Steinberg TH. 1991. Evidence that the gap junction protein connexin-43 is the ATP-induced pore of mouse macrophages. *J Biol Chem* 266:7971–4.

Boitano S, Evans WH. 2000. Connexin mimetic peptides reversibly inhibit Ca(2+) signaling through gap junctions in airway cells. *Am J Physiol Lung Cell Mol Physiol* 279: L623–30.

Braet K, Vandamme W, Martin PE, Evans WH, Leybaert L. 2003. Photo liberating inositol-1,4,5-trisphosphate triggers ATP release that is blocked by the connexin mimetic peptide gap26. *Cell Calcium* 33:37–48.

Brandner JM, Houdek P, Husing B, Kaiser C, Moll I. 2004. Connexins 26, 30, and 43: Differences among spontaneous, chronic, and accelerated human wound healing. *J Invest Dermatol* 122:1310–1320.

Bobbie MW, Roy S, Trudeau K, Munger SJ, Simon AM. 2010. Reduced connexin 43 expression and its effect on the development of vascular lesions in retinas of diabetic mice. *Invest Ophthalmol Vis Sci* 51:3758–63.

Butkevich E, Hülsmann S, Wenzel D, Shirao T, Duden R, Majoul I. 2004. Drebrin is an ovel connexin-43 binding partner that links gap junctions to the submembrane cytoskeleton. *Curr Biol* 14:650–8.

Chang C-Y, Laux-Fenton WT, Lay L-Y, Becker DL, Sherwin T, Green CR. 2009. Antisense downregulation of connexin31.1 reduces apoptosis and increases thickness of human and rat corneal epithelium. *Cell Biol Int* 33:376–385.

Chesnoy S, Lee PY, Huang L. 2003. Intradermal injection of transforming growth factor-beta 1 gene enhances wound healing in genetically diabetic mice. *Pharm Res* 20:345–50.

Christophers E. 2001. Psoriasis—Epidemiology and clinical spectrum. *Clin Exp Dermatol* 26:314–20.

Clark RA. 1985. Cutaneous tissue repair: Basic biologic considerations. I. *J Am Acad Dermatol* 13:701–25.

Common JE, Becker D, Di WL, Leigh IM, O'Toole EA, Kelsell DP. 2002. Functional studies of human skin disease-and deafness-associated connexin 30 mutations. *Biochem Biophys Res Commun* 298:651–6.

Cronin M, Anderson P, Green CR, Becker DL. 2006. Antisense delivery to the intact central nervous system—Specific knockdown of the gap junction protein, connexin 43, in the rat spinal cord. *Front Biosci* 11, 2967–2975.

Cronin M, Anderson PN, Cook JE, Green CR, Becker DL. 2008. Blocking connexin43 expression reduces inflammation and improves functional recovery after spinal cord injury. *Mol Cell Neurosci* 39:152–60.

Coutinho P, Qiu C, Frank S, Wang CM, Brown T, Green CR, Becker DL. 2005. Limiting burn extension by transient inhibition of Connexin43 expression at the site of injury. *Br J Plast Surg* 58:658–67.

Coutinho P, Qiu C, Frank S, Tamber K, Becker DL. 2003. Dynamic changes in connexin expression correlate with key events in the wound healing process. *Cell Biol Int* 27:525–541.

Dahl G. 2007. Gap junction-mimetic peptides do work, but in unexpected ways. *Cell Commun Adhes* 14:259–64.

Di WL, Rugg EL, Leigh IM, Kelsell DP. 2001. Multiple epidermal connexins are expressed in different keratinocyte subpopulations including connexin 31. *J Invest Dermatol*. 117:958–64.

Desmoulière A, Geinoz A, Gabbiani F, Gabbiani G. 1993. Transforming growth factor-beta1 induce salpha-smooth muscle actin expression in granulation tissue myofibroblasts and inquiescent and growing cultured fibroblasts. *J Cell Biol*. 122:103–11.

Dinh TL, Veves A. 2005. A review of the mechanisms implicated in the pathogenesis of the diabetic foot. *Int J Low Extrem Wounds* 4:154–9.

Dovi JV, He LK, DiPietro LA. 2003. Accelerated wound closure in neutrophil-depleted mice. *J Leukoc Biol* 73:448–55.

Duffy HS, Delmar M, Spray DC. 2002. Formation of the gap junction nexus: Binding partners for connexins. *J Physiol Paris* 96:243–9.

Evans WH, Boitano S. 2001. Connex in mimetic peptides: Specific inhibitors of gap-junctional intercellular communication. *Biochem Soc Trans* 29:606–12.

Evans WH, De Vuyst E, Leybaert L. 2006. The gap junction cellular internet: Connexin hemichannels enter the signalling limelight. *Biochem J* 397:1–14.

Evans WH, Leybaert L. 2007. Mimetic peptides as blockers of connexin channel-facilitated intercellular communication. *Cell Commun Adhes* 14:265–73.

Falanga V. 2005. Wound healing and its impairment in the diabetic foot. *Lancet* 366:1736–1743.

Fitzgerald DJ, Fusenig NE, Boukamp P, Piccoli C, Mesnil M, Yamasaki H. 1994. Expression and function of connexin in normal and transformed human keratinocytes in culture. *Carcinogenesis* 15:1859–65.

Froger N, Orellana JA, Calvo CF, Amigou E, Kozoriz MG, Naus CC, Saez JC, Giaume C. 2010. Inhibition of cytokine-induced connexin43 hemichannel activity in astrocytes is neuroprotective. *Mol Cell Neurosci* 45:37–46.

Gaietta G, Deerinck TJ, Adams SR, Bouwer J, Tour O, Laird DW, Sosinsky GE, Tsien RY, Ellisman MH. 2002. Multicolor and electron microscopic imaging of connexin trafficking. *Science* 296:503–7.

Gailit J, Clark RA. 1994. Wound repair in the context of extra cellular matrix. *Curr Opin Cell Biol.* 6:717–25.

Giepmans BN, Verlaan I, Hengeveld T, Janssen H, Calafat J, Falk MM, Moolenaar WH. 2001. Gap junction protein connexin-43 interacts directly with microtubules. *Curr Biol* 11:1364–8.

Goliger JA, Paul DL. 1994. Expression of gap junction proteins Cx26, Cx31.1, Cx37, and Cx43 in developing and mature rat epidermis. *Dev Dyn* 200:1–13.

Goliger JA, Paul DL. 1995. Wounding alters epidermal connexin expression and gap junction-mediated intercellular communication. *Mol Biol Cell* 6:1491–1501.

Gottrup F. 2004. Oxygen in wound healing and infection. *World J Surg* 28:312–5.

Green CR, Law LY, Lin JS, Becker DL. 2001. Spatiotemporal depletion of connexins using antisense oligonucleotides. Techniques in the study of gap junctions. Connexin methods and protocols 154 175–185. Eds R. Bruzzone and C. Giuame.

Harris AL. 2001. Emerging issues of connexin channels: Biophysics fills the gap. *Quart Rev Biophys* 34:325–472.

Hebda PA. 1988. Stimulatory effects of transforming growth factor-beta and epidermal growth factor on epidermal cell outgrowth from porcine skin explant cultures. *J Invest Dermatol* 91:440–5.

Hills CE, Bland R, Wheelans DC, Bennett J, Ronco PM, Squires PE. 2006. Glucose-evoked alterations in connexin43-mediated cell-to-cell communication in human collecting duct: A possible role in diabetic nephropathy. *Am J Physiol Renal Physiol* 291:F1045–51.

Hivnor C, Williams N, Singh F, VanVoorhees A, Dzubow L, Baldwin D, Seykora J. 2004. Gene expression profiling of porokeratosis demonstrates similarities with psoriasis. *J Cutan Pathol* 31(10): 657–64.

Hopkinson-Woolley J, Hughes D, Gordon S, Martin P. 1994. Macrophage recruitment during limb development and wound healing in the embryonic and foetal mouse. *J Cell Sci.* 107:1159–67.

Hong JW, Liu JJ, Lee JS, Mohan RR, Mohan RR, Woods DJ, He YG, Wilson SE. 2001. Proinflammatory chemokine induction in keratocytes and inflammatory cell infiltration into the cornea. *Invest Ophthalmol Vis Sci* 42(12):2795–803.

Iacobas DA, Iacobas S, Spray DC. 2007. Connexin-dependent transcellular transcriptomic networks in mouse brain. *Prog Biophys Mol Biol.* 94:169–85.

Jester JV, Moller-Pederson WM, Huang J, Sax CM, Kays WT, Cavanagh HD et al. 1999. The cellular basis for corneal transparency: Evidence for "corneal cyrstallins". *J Cell Sci* 112:613–22.

Kamibayashi Y, Oyamada M, Oyamada Y, Mori M. 1993. Expression of gap junction proteins connexin 26 and 43 is modulated during differentiation of keratinocytes in newborn mouse epidermis. *J Invest Dermatol* 101:773–8.

Kandyba EE, Hodgins MB, Martin PE. 2008. A murine living skin equivalent amenable to live-cell imaging: Analysis of the roles of connexins in the epidermis. *J Invest Dermatol* 128:1039–49.

Kretz M, Euwens C, Hombach S, Eckardt D, Teubner B, Traub O, Willecke K, Ott T. 2003. Altered connexin expression and wound healing in the epidermis of connexin-deficient mice. *J Cell Sci* 116:3443–52.

Kretz M, Maass K, Willecke K. 2004. Expression and function of connexins in the epidermis, analyzed with transgenic mouse mutants. *Eur J Cell Biol.* 83:647–54.

Krueger JG. 2002.The immunologic basis for the treatment of psoriasis with new biologic agents. *J Am Acad Dermatol* 46:1–23; quiz 23–26.

Labarthe MP, Bosco D, Saurat JH, Meda P, Salomon D. 1998. Upregulation of connexin 26 between keratinocytes of psoriatic lesions'. *J Invest Dermatol* 111, 72–6.

Lampe PD, Kurata WE, Warn-Cramer BJ, Lau AF. 1998. Formation of a distinct connexin 43 phosphoisoform in mitotic cells is dependent upon p34cdc2kinase. *J Cell Sci* 111:833–41.

Lampe PD, TenBroek EM, Burt JM, Kurata WE, Johnson RG, Lau AF. 2000. Phosphorylation of connexin 43 on serine 368 by protein kinase C regulates gap junctional communication. *J Cell Biol* 149:1503–12.

Laux-Fenton W. 2006. The role of connexins in corneal homeopstasis and repair: Mastering the connections to improve repair. PhD thesis, University of Auckland, New Zealand.

Laux-Fenton WT, Donaldson PJ, Kistler J, Green CR. 2003. Connexin expression patterns in the ratcornea: Molecular evidence for communication compartments. *Cornea* 22:457–64.

Law LY, Zhang W, Stott S, Becker DL, Green CR. 2006. *In vitro* optimisation of antisense oligodeoxynucleotide design: An example using the connexin gene family. *J Biomol Tech* 17:270–82.

Leithe E, Brech A, Rivedal E. 2006. Endocytic processing of connexin 43 gap junctions: A morphological study. *Biochem J.* 393:59–67.

Leybaert L, Braet K, Vandamme W, Cabooter L, Martin PE, Evans WH. 2003. Connexin channels, connexin mimetic peptides and ATP release. *Cell Commun Adhes.* 10:251–7.

Li W, Hertzberg EL, Spray DC. 2005. Regulation of connexin43-protein binding in astrocystein response to chemical ischemia/hypoxia. *J Biol Chem.* 280:7941–8.

Liu K, Kasper M, Bierhaus A, Langer S, Müller M and Trott K-R. 1997. Connexin 43 expression in normal and irradiated mouse skin. *Radiat Res* 147:437–441.

Liu XD, Umino T, Ertl R, Veys T, Skold CM, Takigawa K, Romberger DJ et al. 2001. Persistence of TGF-beta1 induction of increased fibroblast contractility *in vitro*. *Cell Dev Biol Anim* 37:193–201.

Loots MA, Lamme EN, Mekkes JR, Bos JD, Middelkoop E. 1999. Cultured fibroblasts from chronic diabetic wounds on the lower extremity (non-insulin-dependent diabetes mellitus) show disturbed proliferation. *Arch Dermatol Res* 291:93–9.

Lucke T, Choudhry R, Thom R, Selmer IS, Burden AD, Hodgins MB. 1999. Upregulation of connexin 26 is a feature of keratinocyte differentiation in hyperproliferative epidermis, vaginal epithelium, and buccal epithelium. *J Invest Dermatol* 112:354–61.

Martin P. 1997. Wound healing—Aiming for perfect skin regeneration. *Science* 276:75–81.

Martin P, Leibovich SJ. 2005. Inflammatory cells during wound repair: The good, the bad and the ugly. *Trends Cell Biol* 15:599–607.

Martin P, D'Souza D, Martin J, Grose R, Cooper L, Maki R, McKercher SR. 2003. Wound healing in the PU.1 nullmouse—Tissue repair is not dependent on inflammatory cells. *Curr Biol* 13:1122–8.

Martin PE, Wall C, Griffith TM. 2005. Effects of connexin-mimetic peptides on gap junction functionality and connexin expression in cultured vascular cells. *Br J Pharmacol* 144(5):617–27.

Marziano N, Casalotti SO, Portelli AE, Becker DL, Forge A. 2003. Deafness-related mutations in gap junction protein connexin 26 have a dominant negative effect on connexin 30. *Hum Mol Genet* 203:805–12.

Matic M, Petrov IN, Rosenfeld T, Wolosin JM. 1997. Alteration s in connexin expression and cell communication in healing corneal epithelium. *Invest Ophthalmol Vis Sci* 38:600–9.

Mori R, Power KT, Wang CM, Martin P, Becker DL. 2006. Acute downregulation of connexin43 at wound sites leads to a reduced inflammatory response, enhanced keratinocyte proliferation and wound fibroblast migration. *J Cell Sci* 119:5193–5203.

Mustoe TA, Pierce GF, Morishima C, Deuel TF. 1991. Growth factor-induced acceleration of tissue repair through direct and inductive activities in a rabbit dermal ulcer model. *J Clin Invest* 87:694–703.

Mustoe T. 2004. Understanding chronic wounds: A unifying hypothesis on their pathogenesis and implications for therapy. *Am J Surg* 187:65S–70S.

Nakano Y, Oyamada M, Dai P, Nakagami T, Kinoshita S, Takamatsu T. 2008. Connexin43 knock down accelerates wound healing but inhibits mesenchymal transition after corneal endothelial injury in vivo. *Invest Ophthalmol Vis Sci* 49:93–104.

O'Carroll SJ, Alkadhi M, Nicholson LF, Green CR. 2008. Connexin 43 mimetic peptides reduce swelling, astrogliosis, and neuronal cell death after spinal cord injury. *Cell Commun Adhes* 15:27–42.

Ormonde S, Chou C-Y, Goold L, Petsoglou C, Al-Taie R, Sherwin T, McGhee CNJ, Green CR. 2012. Regulation of connexin43 gap junction protein triggers vascular recovery and healing in human ocular persistent epithelial defect wounds. *J Membr Biol* 245(7):381–8.

Oviedo-Orta E, Howard Evans W. 2004. Gap junctions and connexin-mediated communication in the immune system. *Biochim Biophys Acta* 1662:102–12.

Poladia DP, Schanbacher B, Wallace LJ, Bauer JA. 2005. Innervation and connexin isoform expression during diabetes-related bladder dysfunction: Early structural vs. neuronal remodelling. *Acta Diabetol* 42:147–52.

Pollok S, Pfeiffer AC, Lobmann R, Wright CS, Moll I, Martin PE, Brandner JM. 2011. Connexin 43 mimetic peptide Gap27 reveals potential differences in the role of Cx43 in wound repair between diabetic and non-diabetic cells. *J Cell Mol Med* 15(4):861–73.

Postlethwaite AE, Keski-Oja J, Moses HL, Kang AH. 1987. Stimulation of the chemotactic migration of human fibroblasts by transforming growth factor beta. *J Exp Med* 165:251–6.

Qiu C, Coutinho P, Frank S, Franke S, Law LY, Martin P, Green CR, Becker DL. 2003. Targeting connexin43 expression accelerates the rate of wound repair. *Curr Biol* 13:1697–1703.

Ratkay-Traub I, Hopp B, Bor Z, Dux L, Becker DL, Krenacs T. 2001. Regeneration of rabbit cornea following excimer laser photorefractive keratectomy: A study on gap junctions, epithelial junctions and epidermal growth factor receptor expression in correlation with cell proliferation. *Exp Eye Res* 73:291–302.

Redd MJ, Cooper L, Wood W, Stramer B, Martin P. 2004. Wound healing and inflammation: Embryos reveal the way to perfect repair. *Philos Trans R Soc Lond B Biol Sci* 359:777–84.

Richard G. 2000. Connexins: A connection with the skin. *Exp Dermatol* 9:77–96.

Richard G. 2003. Connexin gene pathology. *Clin Exp Dermatol* 28(4):397–409.

Richard G. 2005. Connexin disorders of the skin. *Clin Dermatol* 23:23–32.

Richards TS, Dunn CA, Carter WG, Usui ML, Olerud JE, Lampe PD. 2004. Protein kinase C spatially and temporally regulates gap junctional communication during human wound repair via phosphorylation of connexin 43 on serine 368. *J Cell Biol* 167:555–62.

Risek B, Klier FG, Gilula NB. 1992. Multiple gap junction genes are utilized during rat skin and hair development. *Development* 116:639–51.

Roberts, AB, Sporn, MB, Assoian, RK, Smith JM, Roche JM, Wakefield NS, Heine LM, Liotta UI, Falanga LA, V and Kehrl JH. 1986. Transforming growth factor type beta: Rapid induction of fibrosis and angiogenesis *in vivo* and stimulation of collagen formation in vitro. *Proc Natl Acad Sci* 83:4167–4171.

Rodriguez-Sinovas A, Cabestrero A, Lopez D, Torre I, Morente M, Abellan A, Miro E, Ruiz-Meana E, Garcia-Dorado D. 2007. The modulatory effects of connexin 43 on cell death/survival beyond cell coupling. *Prog Biophys Mol Biol* 94(1–2):219–32.

Rossi D, Zlotnik A. 2000. The biology of chemokines and the irreceptors. *Annu Rev Immunol* 18:217–42.

Rupenthal ID. 2008. Ocular delivery of antisense oligonucleotides using colloidal carriers: Improving the wound repair after corneal surgery. PhD thesis, University of Auckland, New Zealand.

Rupenthal ID, Alany RG, Green CR. 2011. Ion-activated *in situ* gelling systems for antisense delivery to the ocular surface. *Mol Pharm* 8(6):2282–90.

Saitoh M, Oyamada M, Oyamada Y, Kaku T, Mori M. 1997. Changes in the expression of gap junction proteins (connexins) in hamster tongue epithelium during wound healing and carcinogenesis. *Carcinogenesis* 18:1319–28.

Salomon D, Masgrau E, Vischer S, Ullrich S, Dupont E, Sappino P, Saurat JH, Meda P. 1994. Topography of mammalian connexins in human skin. *J Invest Dermatol* 103:240–7.

Schreml S, Szeimies RM, Prantl L, Landthaler M, Babilas P. 2010. Wound healing in the 21st century. *J Am Acad Dermatol* 63:866–81.

Sen CK, Gordillo GM, Roy S, Kirsner R, Lambert L, Hunt TK, Gottrup F, Gurtner GC, Longaker MT. 2009. Human skin wounds: A major and snowballing threat to public health and the economy. *Wound Repair Regen* 17:763–71.

Shah M, Foreman DM, Ferguson MW. 1994. Neutralising antibody to TGF-beta 1,2 reduces cutaneous scarring in adult rodents. *J Cell Sci.* 107:1137–57.

Shaw RM, Fay AJ, Puthenveedu MA, von Zastrow M, Jan YN, Jan LY. 2007. Microtubule plus-end-tracking proteins target gap junctions directly from the cell interior to adherens junctions. *Cell* 128:547–60.

Shaw TJ, Martin P. 2009. Wound repair at a glance. *J Cell Sci* 122:3209–13.

Sherwin T, Green CR. 2009. Stromal wound healing. In *Corneal Surgery- Theory, Technique and Tissue* (4th edition). Eds F.S. Brightbill, P.J. McDonnell, C.N.J. McGhee, A.A. Farjo, and O.N. Serdarevic. Elsevier, pp. 45–56.

Shurman DL, Glazewski L, Gumpert A, Zieske JD, Richard G. 2005. *In vivo* and *in vitro* expression of connexins in the human corneal epithelium. *Invest Ophthalmol Vis Sci* 46:1957–65.

Spray DC, Iacobas DA. 2007. Organizational principles of the connexin-related brain transcriptome. *J Membr Biol* 218:39–47.

Solan JL, Fry MD, TenBroek EM, Lampe PD. 2003. Connexin 43 phosphorylation at S368 is acute during S and G2/Mandin response to protein kinase C activation. *J Cell Sci* 116:2203–11.

Singer AJ, Clark RA. 1999. Cutaneous wound healing. *N Engl J Med* 341:738–46.

Spray DC, Rozental R, Srinivas M. 2002. Prospects for rational development of pharmacological gap junction channel blockers. *Curr Drug Targets* 3:455–64.

Theiss C, Meller K. 2002. Microinjected anti-actin antibodies decrease gap junctional intercellular communication in cultured astrocytes. *Exp Cell Res* 281:197–204.

van Steensel MA. 2004. Gap junction diseases of the skin. *Am J Med Genet C Semin Med Genet* 131C(1):12–9.

van Zeijl L, Ponsioen B, Giepmans BN, Ariaens A, Postma FR, Varnai P, Balla T, Divecha N, Jalink K, Moolenaar WH. 2007. Regulation of connexin43 gap junctional communication by phosphatidylinositol 4,5-bisphosphate. *J Cell Biol* 177:881–91.

Wagner RW. 1994. Gene inhibition using antisense oligodeoxynucleotides. *Nature* 372:333–5.

Walmsley S. 2002. *Advances in Wound Management: Executive Summary.* London: PJB Publications, Ltd.

Wang CM, Lincoln J, Cook JE, Becker DL. 2007. Abnormal connexin expression underlies delayed wound healing in diabetic skin. *Diabetes* 56:2809–17.

Wang X, Ramirez A, Budunova I. 2010. Overexpression of connexin26 in the basal keratinocytes reduces sensitivity to tumor promoter TPA. *Exp Dermatol* 19:633–40.

Warner A, Clements DK, Parikh S, Evans WH, DeHaan RL. 1995. Specific motifs in the external loops of connexin proteins can determine gap junction formation between chick heart myocytes. *J Physiol* 488:721–8.

Wei CJ, Francis R, Xu X, Lo CW. 2005. Connexin43 associated with an N-cadherin-containing multiprotein complex is required for gap junction formation in NIH3T3 cells. *J Biol Chem* 280:19925–36.

Wei CJ, Xu X, Lo CW. 2004. Connexins and cell signaling in development and disease. *Annu Rev Cell Dev Biol* 20:811–38.

Werner S, Grose R. 2003. Regulation of wound healing by growth factors and cytokines. *Physiol Rev.* 83:835–70.

Whitby DJ, Ferguson MW. 1991. Immunohistochemical localization of growth factors in fetal wound healing. *Dev Biol* 147:207–15.

Wilson S, Mohan R, Hong J, Lee J. 2001. The wound healing response after laser *in situ* keratomileusis and photorefractive keratectomy: Elusive control of biological variability and effect on custom laser vision correction. *Archives Ophthalmol* 119:889–96.

Wright CS, van Steensel MA, Hodgins MB, Martin PE. 2009. Connexin mimetic peptides improve cell migration rates of human epidermal keratinocytes and dermal fibroblasts *in vitro*. *Wound Repair Regen* 17:240–9.

Zahler S, Hoffmann A, Gloe T, Pohl U. 2003. Gap-junctional coupling between neutrophils and endothelial cells: A novel modulator of transendothelial migration. *J Leukoc Biol* 73:118–126.

Zieske JD, Guimaraes SR, Hutcheon AE. 2001. Kinetics of keratocyte proliferation in response to epithelial debridement. *Exp Eye Res* 72:33–39.

Index